3Quantum: Algebra Geometry Information

(QQQ Conference 2012)

Journal of Physics: Conference Series Volume 532

Tallinn, Estonia
10-13 July 2012

ISBN: 978-1-63439-112-2
ISSN: 1742-6588

Printed from e-media with permission by:

Curran Associates, Inc.
57 Morehouse Lane
Red Hook, NY 12571

Some format issues inherent in the e-media version may also appear in this print version.

Printed by Curran Associates, Inc. (2014)

For permission requests, please contact the Institute of Physics at the address below.

Institute of Physics
Dirac House, Temple Back
Bristol BS1 6BE UK

Phone: 44 1 17 929 7481
Fax: 44 1 17 920 0979

techtracking@iop.org

Additional copies of this publication are available from:

Curran Associates, Inc.
57 Morehouse Lane
Red Hook, NY 12571 USA
Phone: 845-758-0400
Fax: 845-758-2634
Email: curran@proceedings.com
Web: www.proceedings.com

Preface

Papers

Journal of Physics: Conference Series **532** (2014) 011001 doi:10.1088/1742-6596/532/1/011001

3Quantum: Algebra Geometry Information

E Paal[1], P Kuusk[2] and A Stolin[3]

[1]Tallinn University of Technology, Ehitajate tee 5, 19086 Tallinn, Estonia
[2]Institute of Physics, University of Tartu, Ravila 14c, 50411 Tartu, Estonia
[3]University of Gothenburg, SE-405 30 Gothenburg, Sweden

Abstract. The basic information about the international *QQQ* Conference *3Quantum: Algebra Geometry Information* (Tallinn, July 2012) [*AstrAlgo cWeb* 2012 QQQ II] is presented and the concise review of the *QQQ* Proceedings is given by the proceedings volume editors.

1. Introduction

This proceedings volume presents results obtained by the participants of the 6th ECM satellite *QQQ* Conference *3Quantum: Algebra Geometry Information* [1] held at the *Tallinn University of Technology* [2], Estonia, 10-13 July 2012. The meeting was organized by joint initiative of two networks: ITGP n/w of ESF and AGMP [3] Baltic-Nordic n/w, under the general endorsement and sponsorship of the *European Science Foundation* (ESF) and it was dedicated to memory of *Professor Jean-Louis Loday* (1946-2012).

In the *QQQ* Conference scientific program, the invited thematic session *Gravitation and Cosmology* was dedicated to centenary of *Professor Harald Keres* (1912-2010) who was the initiator of Einstein's gravity studies in Estonia.

The *QQQ* Conference international Organizing Committee was: E. Paal (chair), A. Fialowski, J. Fuchs, M. Schlichenmaier, V. Shumeiko, A. Stolin and P. Zusmanovich.

The *QQQ* Conference was an official satellite meeting of:

- 6th *European Congress of Mathematics* ECM6
  Associated Mini-symposia *25 Years of Quantum Groups: From Definition to Classifications* was organized by A. Stolin with invited speakers: *Y. I. Manin*, *M. Semenev-Tian-Shansky*, *S. Woronowicz* and *E. Zelmanov*
- 17th *International Congress of Mathematical Physics* ICMP12
- 13th *Marcel Grossmann Meeting on General Relativity* MG13

2. Description of *QQQ* Conference

The main purpose of the *QQQ* Conference was to stimulate and promote interactions between three major research areas: (quantum) algebra, (quantum) geometry and gravitation, and (quantum) information theory. Thus, the basic concept

$$geometry \equiv gravitation \quad \text{and} \quad \text{the cubic acronym} \quad QQQ$$

were fixed, that determined the face of the *QQQ* Conference & Proceedings.

In total, 96 scientists from 27 countries attended the *QQQ* Conference. The scientific program of the *QQQ* Conference contained 79 registered talks, including 20 plenary talks, and 59 talks in the following 8 invited thematic sessions:

- *Deformation Theory and Quantization* with 7 speakers
- *Geometry and Topology in Quantum Gravity* with 9 speakers
- *Gravitation and Cosmology* with 12 speakers
  dedicated to centenary of *Professor Harald Keres* (1912-2010)
- *Hopf Algebra, Quantum Algebra* with 11 speakers
- *Jordan Structures in Mathematical Physics* with 4 speakers
- *Modularity and Non-Perturbative Strings* with 5 speakers
- *Nonassociative Algebra, Hom-Algebra, Applications* with 6 speakers
- *Operads, Noncommutative Algebra and Geometry* with 5 speakers

The reader can find more information, including the *QQQ* Conference Abstracts and other related materials, in [1].

3. Review of *QQQ* Proceeding

The research articles, carefully selected for the *QQQ* Proceedings volume cover a wide range of topics representing the main directions of the conference and the topics of the special thematic sessions. All articles submitted for this volume were refereed, and the Editors take this opportunity to thank all referees for their work as well as all contributors for the collaboration.

Among contributors to the volume there are 49 researchers from many different countries. The *QQQ* Proceedings volume includes 30 contributions, including 5 contributions from the *QQQ* Conference plenary speakers – *L. Faddeev, P. Kulish, O. A. Laudal, Y. I. Manin* and *S. Waldmann* – that we review here in the alphabetical order by the first author, by paying more attention on the contributions of the *QQQ* Conference plenary speakers, and apology for the monotonic presentation.

D. V. Artamonov and V. A. Golubeva in the article *Knizhnik-Zamolodchikov type equations for the root system B and Capelli central elements* turn attention that the construction of the famous Knizhnik-Zamolodchikov equations uses the central element of the second order in the universal enveloping algebra for some Lie algebra. But in the universal enveloping algebra there exist central elements of higher orders as well. Thus, it seems desirable to use these elements for the construction of Knizhnik-Zamolodchikov type equations. In the present paper they give a construction of such Knizhnik-Zamolodchikov type equations for the root system B associated with Capelli central elements in the enveloping algebra for the orthogonal algebra. In the article *On quantum deformations of (anti-)de Sitter algebras in (2+1) dimensions*, A. Ballesteros, F. J. Herranz and F. Musso revisit deformations of (anti-)de Sitter (A)dS algebras in (2+1) dimensions, and review several features of these quantum structures. In particular, the classification problem of (2+1) (A)dS Lie bialgebras is presented and the associated noncommutative quantum (A)dS spaces are also analysed. Moreover, the flat limit (or vanishing cosmological constant) of all these structures leading to (2+1) quantum Poincaré algebras and groups is simultaneously given by considering the cosmological constant as an explicit Lie algebra parameter in the (A)dS algebras. By making use of this classification, a three-parameter generalization of the κ-deformation for the (2+1) (A)dS algebras and quantum spacetimes is given. Finally, the same problem is studied in (3+1) dimensions, where a two-parameter generalization of the κ-(A)dS deformation that preserves the space isotropy is found. L. Borsten in the article *3-qubit entanglement: A Jordan algebraic perspective* turns attention that it is by now well known that three qubits can be totally entangled in two physically distinct ways. The author reviews work classifying the physically distinct forms of -qubit entanglement using the elegant framework of Jordan algebras, the Freudenthal-Kantor triple systems (F-KTS) and groups of type E_7. In particular, it is shown that the four Freudenthal-Kantor ranks correspond precisely to the four 3-qubit entanglement classes: (1) Totally separable A-B-C, (2) Biseparable A-BC, B-CA, C-AB, (3) Totally entangled W, (4) Totally entangled GHZ. The rank 4 GHZ class is regarded as maximally entangled in the sense that it has non-vanishing quartic norm, the defining invariant of the Freudenthal-Kantor triple system. While this framework is specific to three qubits, the author shows here how the essential features may be naturally generalised to an arbitrary number of qubits. In the article *Stringlike structures in the real and complex Kerr-Schild geometry*, A. Burinskii studies the four-dimensional Kerr-Schild (KS) geometry that displays remarkable relationships with quantum world and theory of superstrings. In particular, the Kerr-Newman (KN) solution has gyromagnetic ratio $g = 2$, as that of the Dirac electron and represents a consistent background for gravitational and electromagnetic field of the electron. As a consequence of very big spin/mass ratio, black hole horizons disappear, exposing the naked Kerr singular ring. The author considers four-decade history of development of this structure which took finally the form of a point-string-membrane-bubble complex which is reminiscent of the enhancon model of string/M-theory. A complex string obtained in the complex structure of the Kerr geometry gives an extra dimension to the world-sheet of the real Kerr string, forming a membrane by analogue with the string/M-theory unification. By analysis of the orientifold parity of the complex Kerr string, the author obtains that the principal null congruence of the Kerr geometry determined by the Kerr theorem is described by a quartic equation in projective twistor space CP^3, and therefore, it creates the known Calabi-Yau twofold (K3 surface) in twistor space of the 4d KS geometry. The author connects it with N=2 superstring which has (complex) critical dimension two and may be embedded into complex KS geometry.

In the article

***3J*-Symbol for the Modular Double $SL_q(2,\mathbb{R})$ Revisited**

the authors **S. E. Derkachev** and **L. D. Faddeev** first recall that the modular double of quantum group $\mathrm{SL}_q(2,\mathbb{R})$ with $|q| = 1$ has a series of selfadjoint irreducible representations π_s parameterized by $s \in \mathbb{R}_+$. Ponsot and Teschner in [2001 *Comm. Math. Phys.* vol 224 p 613] considered a decomposition of the tensor product $\pi_{s_1} \otimes \pi_{s_2}$ into irreducibles. In the present article, authors give the detailed derivation and some new proofs.

S. Groote and R. Saar in the article *Group theory aspects of chaotic strings* first note that the chaotic strings are a special type of non-hyperbolic coupled map lattices, exhibiting a rich structure of complex dynamical phenomena with a surprising correspondence to physical contents and are generated by the Chebyshev maps $T_2(\phi)$ and $T_3(\phi)$. In this paper the authors connect the Chebyshev maps via the Galois theory to the cyclic groups Z_2 and Z_3 and give some ideas how this fundamental connection might lead to the emergence of the familiar Lie group structure of the particle physics and, in particular, even to the emergence of space-time. The authors present here the most important results associated with these papers. In the article *Lorentz invariance and gauge equivariance*, S. Groote, R. Saar and H. Liivat turn attention that when trying to place Lorentz and gauge transformations on the same foundation, it turns out that the first one generates invariance, the second one equivariance, at least for the abelian case. The authors note that this similarity is not an hypothesis but is supported by and is a consequence of the path integral formalism in quantum field theory. S. Groote, H. Veermäe and C. Beck in the article *Renormalization aspects of chaotic strings* state that chaotic strings are a class of non-hyperbolic coupled map lattices, exhibiting a rich structure of complex dynamical phenomena with a surprising correspondence to physical contents. In this article the authors introduce different types and models for chaotic strings, where 2B-strings with finite length are considered in detail and demonstrate possibilities to extract renormalized quantities, which are expected to describe essential properties of the string. In the article *Aspects of multimetric gravity*, M. Hohmann presents a class of gravity theories containing $N \geq 2$ metric tensors and a corresponding number of standard model copies. In the Newtonian limit gravity is attractive within each standard model copy, but different standard model copies mutually repel each other. The author discusses several aspects of these multimetric gravity theories, including cosmology, structure formation, the post-Newtonian limit and gravitational waves. The most interesting feature the author finds is an accelerating expansion of the universe that naturally becomes small at late times. P. Jastrzębski and A. Tralle in the article *A note on Clifford-Klein forms* consider the difficult problem of finding the Clifford-Klein forms in a class of homogeneous spaces determined by inclusions of real Lie algebras of a special type which they call strongly regular. This class of inclusions is described in terms of their Satake diagrams. For example, the complexifications of such inclusions contain the class of subalgebras generated by automorphisms of finite order. Authors show that the condition of strong regularity implies the restriction on the real rank of subalgebras. This in part explains why the known examples of Clifford-Klein forms are rare. The authors make detailed calculations of some known examples from the point of view of the Satake diagrams. In the article *Parametrizations in scalar-tensor theories of gravity and the limit of general relativity*, L. Järv, P. Kuusk, M. Saal and O. Vilson consider a general scalar-tensor theory of gravity and review briefly different forms it can be presented The authors investigate the conditions under which its field equations and the parametrized post-Newtonian parameters coincide with those of general relativity and demonstrate that these so-called limits of general relativity are independent of the parametrization of the scalar field, although the transformation between scalar fields may be singular at the corresponding value of the scalar field. In particular, the limit of general relativity can equivalently be determined and investigated in the commonly used Jordan and Einstein frames.

In the article

Deformed Richardson-Gaudin Model

the authors **P. Kulish**, **A. Stolin** and **L. H. Johannesson** study the Richardson-Gaudin model that describes strong pairing correlations of fermions confined to a finite chain. The integrability of the Hamiltonian allows the algebraic construction of its eigenstates. The authors show that the quantum group theory provides a possibility to deform the Hamiltonian preserving integrability. More precisely, the authors use the so-called Jordanian r-matrix to deform the Hamiltonian of the Richardson-Gaudin model. In order to preserve its integrability, one needs to insert a special nilpotent term into the auxiliary L-operator which generates integrals of motion of the system. Moreover, the quantum inverse scattering method enables to construct the exact eigenstates of the deformed Hamiltonian. These states have a highly complex entanglement structure which certainly requires further investigation.

In the article

The Structure of $Ph^{\star}$, Generalized de Rham, and Entropy

the author **O. A. Laudal** continues the study of the non-commutative phase space functor, $Ph(A)$, defined for any associative algebra A, and its derived differential co-simplicial algebra, $Ph^{\star}(A)$. The main focus of this research is on its relationship to the classical de Rham complex, to the dynamics of the finite dimensional $Ph^{\infty}(A)$-modules, and to the notion of Entropy. These subjects are treated within the set-up of author's recently published monograph [*Geometry of Time-Spaces* 2011 (World Scientific)] and in the forthcoming monograph [*Cosmos and its Furniture*].

A. Leibak and P. Puusemp in the article *On determinability of some classes of medial quasigroups by their endomorphisms* study the endomorphisms of idempotent medial quasigroups and determinability of some classes of medial quasigroups by their endomorphisms. The authors introduce the endomorphism algebra of idempotent medial quasigroup and prove that if the endomorphism algebras of quasigroups are isomorphic, then the corresponding quasigroups are isomorphic as well. In the article *A short survey of Lie antialgebras*, S. Leidwanger and S. Morier-Genoud follow the presentation at the *QQQ* conference. Authors give a brief survey of the existing theory of Lie antialgebras and suggest open questions. R. K. Loide and P. Suurvarik in the article *Supersymmetry: superfield equations of motion* consider supersymmetry and superfields in connection with the Poincaré superalgebras. The original formalism of the projection operators for deriving the wave equations for ordinary fields and superfields is developed. The superfield equations of motion in the case of massive and massless fields are presented together with an application in linear supergravity. In the article *Drinfeld twisting elements on* Hom-*bialgebras*, A. Makhlouf and B. Torrecillas have the aim to introduce the concept of a twisting element based on a Hom-bialgebra and to use it to provide twists or deformations of Hom-associative algebras. Also, the authors review the module theory in Hom-setting and, in particular, show that a twisting element based on a bialgebra gives rise to a twisting element based on a Hom-bialgebra.

The *QQQ* Conference key speaker **Y. I. Manin** presents in the article

COMPLEXITY vs ENERGY: Theory of Computation and Theoretical Physics

a thorough survey based upon author's plenary lecture [4] at the *QQQ* Conference as well as his recent research results in this field. It is dedicated to a deep analogy between the notions of *complexity* in theoretical computer science and *energy* in physics. The author emphasizes that this analogy is not metaphorical and describes three precise mathematical contexts, suggested recently, in which mathematics related to (un)computability is inspired by and to a degree reproduces formalisms of statistical physics and quantum field theory.

H. Melikyan and P. Zusmanovich in the article *Melikyan algebra is a deformation of a Poisson algebra* prove, using computer, that the restricted Melikyan algebra of dimension 125 is a deformation of a Poisson algebra. In the article *Categorical generalization of spinfoam models*, A. Miković and M. Vojinović discuss problems of the nonrenormalizability of the general relativity and consider the main research directions which have the aim to resolve this problem. The attention of authors focuses on the approach of loop quantum gravity, specifically on the spinfoam models. These models have some issues concerning the semiclassical limit and coupling of matter fields. The recent developments in category theory provide the necessary formalism to introduce a new action for general relativity and perform covariant quantization. Z. Oziewicz in the article *Centre-of-mass for the finite speed of light* first recalls that already in 1632 Galilei was aware of relativity of velocity and that this implies relativity of spaces-of-locations and during centuries the latter was ignored. Author notes that *Professor Harald Keres* (1912-2010) considered the space-of-locations as a congruence of world-lines and there is no universal absolute three-dimensional space-of-locations. Author shows that the centre-of-mass of many-body interacting (bound) system for the case of finite light-speed is a well defined concept within the group-free approach using algebra epimorphisms as splits and considers the Keres space-of-locations as the Grassmann factor-algebra of differential forms. In the article *Note on homological modeling of the electric circuits*, E. Paal and M. Umbleja explain on a simple example how the homological analysis may be used for modeling of the electric circuits. The homological branch, mesh and nodal analyses are presented and the geometrical interpretations are given. E. Paal and J. Virkepu in the article *Operadic quantization as a tool for discrete geometry* use the operadic Lax representations of the harmonic oscillator to construct the quantum counterparts of 3d real Lie algebras in the Bianchi classification. The Jacobi operators of these quantum algebras are studied. The authors show how the energy conservation is related to the Jacobi identity and how the quantization leads to an anomaly – the quantum violation of the Jacobi relations. By using the non-vanishing quantum Jacobi operators, the derivative quantum algebra for a triple of 3d real Lie algebras is constructed. It is proposed that the derivative algebra is the 3d real Heisenberg algebra, from which it follows that in this model only the discrete values of the spatial coordinates are physically allowed. In the article *On quantum groups and Lie bialgebras related to* $sl(n)$, A. Stolin and I. Pop study for given an arbitrary field $\mathbb{F}$ of characteristic 0 the Lie bialgebra structures on $sl(n, \mathbb{F})$, based on the description of the corresponding classical double. It turns out that for any Lie bialgebra structure δ, the classical double $D(sl(n, \mathbb{F}), \delta)$ is isomorphic to $sl(n, \mathbb{F}) \otimes_{\mathbb{F}} A$, where A is either $\mathbb{F}[\varepsilon]$, with $\varepsilon^2 = 0$, or $\mathbb{F} \oplus \mathbb{F}$ or a quadratic field extension of $\mathbb{F}$. In the first case, the classification leads to quasi-Frobenius Lie subalgebras of $sl(n, \mathbb{F})$. In the second and third cases, a Belavin-Drinfeld cohomology can be introduced which enables one to classify the Lie bialgebras on $sl(n, \mathbb{F})$, up to gauge equivalence. The Belavin-Drinfeld untwisted and twisted cohomology sets associated to an r-matrix are computed. E. Randla in the article *PPN parameters for multiscalar-tensor gravity without a potential* proposes a generic multiscalar-tensor gravity action functional in the Jordan conformal frame. After redefining the scalar fields, the parameterised post-Newtonian parameters are calculated, neglecting the potential term. The field equations are shown to be nearly identical with those of scalar-tensor gravity up to the first post-Newtonian approximation, differing only in the definitions of constant terms. In the article *Examples of cosmological rips in the scalar-tensor cosmology*, the authors M. Saal and A. Frantskjavitšius examine whether a cosmological rip scenarios are possible in the scalar-tensor gravity by parametrizing the equation of state of dark energy in an appropriate way. The parameters of the model are constrained by the observational data of type Ia supernovas collected to Union 2.1 compilation.

In the article

Fréchet-Algebraic Deformation Quantizations

the author **S. Waldmann** presents some recent results on the convergence properties of formal star products. Based on a general construction of a Fréchet topology for an algebra with countable vector space basis, the author discusses several examples from deformation quantization: the Wick star product on the flat phase space gives a first example of a Fréchet algebraic framework for the canonical commutation relations. More interesting, the star product on the Poincaré disk can be treated along the same lines, leading to a non-trivial example of a convergent star product on a curved Kähler manifold.

G. P. Wene in the article *Permutations of cubical arrays* notes that the structure constants of an algebra determine a cube called the cubical array associated with the algebra. The permuted indices of the cubical array associated with a finite semifield generate new division algebras. Author does not require that the algebra be finite and asks "Is it possible to choose a basis for the algebra such any permutation of the indices of the structure constants leaves the algebra unchanged?" What are the associated algebras? Author shows that the property "weakly quadratic" is invariant under all permutations of the indices of the corresponding cubical array and presents two algebras for which the cubical array is invariant under all permutations of the indices. In the article *Geometrodynamics and Lorentz symmetry*, D. K. Wise studies dynamics of the gauge theory and general relativity using fields of local observers, thus maintaining local Lorentz symmetry despite a space/time splitting of fields. Author starts with Yang-Mills theory, where the observer fields are defined as normalized future-timelike vector fields, then defines observers without a fixed geometry and finds that these play two related roles in general relativity: splitting fields into spatial and temporal parts, and breaking the gauge symmetry, effectively reducing the spacetime $SO(n, 1)$ connection to an observer-dependent spatial $SO(n)$ connection. In both gauge theory and gravity, the observer field reduces the action to canonical form, without using gauge fixing. In the 4d gravity case, the result is a manifestly the Lorentz covariant counterpart of the Ashtekar-Barbero formulation. A. Zuevsky in the article *Lie-algebraic symmetries of generalized Davey-Stewartson equations* identifies the full Lie-algebraic structure of the generalized Davey-Stewartson system of equations with symmetries of a specific continual Lie algebras. The author shows that these are related to two copies of the Poisson bracket continual Lie algebra.

4. Conclusions

In conclusion, we can state that all main topics and sessions of the *QQQ* Conference are well represented in the *QQQ* Proceedings and we hope that this international interdisciplinary volume will be a source of further inspiration for the international research community working in contemporary physics and mathematics.

Acknowledgments

We are grateful for hospitality to the hosting organization, the *Tallinn University of Technology* [2] which provided the *QQQ* Conference with excellent environment for science.

The *QQQ* Conference & Proceedings were supported by the *European Science Foundation* with ITGP n/w, the *Swedish Research Council*, MP^2 Centre/STINT (grant IG2011-2028), and the *Estonian Research Council* (grants ETF9038 and ETF8837).

We acknowledge *AstrAlgo* at Tallinn for the effective organizational assistance, for the web hosting and the web publishing of the *QQQ* Conference materials [1], and for the technical editing support of the *QQQ* Proceedings as well.

We thank all the participants, the speakers, the co-organizers, the referees and contributors of the *QQQ* Proceedings volume, and the sponsors.

References

[1] Paal E and Zusmanovich P eds 2012 *AstrAlgo cWeb* QQQ II
[2] *Tallinn University of Technology* [www.ttu.ee]
[3] *AstrAlgo cWeb* AGMP
[4] Manin Y I 2012 *AstrAlgo cWeb* QQQ II #872

Journal of Physics: Conference Series **532** (2014) 011002 doi:10.1088/1742-6596/532/1/011002

Contributors of the *QQQ* Proceedings

• **D V Artamonov** Faculty of Economics, Lomonosov Moscow State University 1-46 Leninskie Gory, 119991 Moscow, GSP-1, 1-46, Russia

• **A Ballesteros** Departamento de Física, Universidad de Burgos, 09001 Burgos, Spain

• **C Beck** School of Mathematical Sciences, Queen Mary University of London, End Road, London E1 4NS, U.K.

• **L Borsten** Blackett Laboratory, Imperial College, SW7 2AZ London, U.K.

• **A Burinskii** Theor. Phys. Lab., NSI, Russian Academy of Sciences, B. Tulskaya 52, Moscow 115191, Russia

• **S E Derkachev** St. Petersburg Department of Steklov Mathematical Institute, Russian Academy of Sciences, Fontanka 27, 191023 St. Petersburg, Russia

• **L D Faddeev** St. Petersburg Department of Steklov Mathematical Institute, Russian Academy of Sciences, Fontanka 27, 191023 St. Petersburg, Russia

• **A Frantskjavitšius** Füüsika Instituut, Tartu Ülikool Ravila 14c, 50411 Tartu, Estonia

• **V A Golubeva** Department of Applied Mathematics and Physics, Moscow Aviation Institute, 4 Volokolamskoe Shosse, 125993 Moscow, A-80, GSP-3, Russia

• **S Groote** Institut für Physik der Johannes-Gutenberg-Universität, Staudinger Weg 7, 55099 Mainz, Germany; Füüsika Instituut, Tartu Ülikool, Ravila 14c, 50411 Tartu, Estonia

• **F J Herranz** Departamento de Física, Universidad de Burgos, 09001 Burgos, Spain

• **M Hohmann** Teoreetilise Füüsika Labor, Füüsika Instituut, Tartu Ülikool, Ravila 14c, 50411 Tartu, Estonia

• **P Jastrzębski** Department of Mathematics and Information Technology, University of Warmia and Mazury, Słoneczna str. 54, 10-710 Olsztyn, Poland

• **L Järv** Institute of Physics, University of Tartu, Ravila 14c, 50411 Tartu, Estonia

• **L H Johannesson** University of Gothenburg, Box 100, SE-405 30, Gothenburg, Sweden

• **P Kulish** St. Petersburg Department of Steklov Mathematical Institute, Russian Academy of Sciences, Fontanka 27, 191023 St. Petersburg, Russia

• **P Kuusk** Institute of Physics, University of Tartu, Ravila 14c, 50411 Tartu, Estonia

• **O A Laudal** Matematisk institutt, University of Oslo, Pb. 1053, Blindern, N-0316 Oslo, Norway

• **A Leibak** Matemaatikainstituut, Tallinna Tehnikaülikool, Ehitajate tee 5, 19086 Tallinn, Estonia

• **S Leidwanger** Institut de Mathématique de Jussieu – Paris Rive Gauche UMR 7586, Bâtiment Sophie Germain, case 7012, 75205 Paris Cedex 13, France

• **H Liivat** Füüsika Instituut, Tartu U¨ likool, Ravila 14c, 50411 Tartu, Estonia

• **R K Loide** Department of Physics, Tallinn University of Technology, Ehitajate tee 5, 19086 Tallinn, Estonia

• **A Makhlouf** Université de Haute Alsace, Laboratoire de Mathématique, Informatique et Applications, 4 rue des Fréres Lumiére, F-68093 Mulhouse, France

• **Y I Manin** Max-Planck-Institut für Mathematik, Vivatsgasse 7, 53111 Bonn, Germany

• **H Melikyan** Department of Mathematics and Physics, North Carolina Central University Durham, NC 27707, U.S.A.

• **A Miković** Departamento de Matemática, Universidade Lusófona de Humanidades e Tecnologias, Av. do Campo Grande, 376, 1749-024 Lisboa, Portugal; Grupo de Fisica Matemática da Universidade de Lisboa, Av. Prof. Gama Pinto, 2, 1649-003 Lisboa, Portugal

• **S Morier-Genoud** Institut de Mathématique de Jussieu – Paris Rive Gauche UMR 7586, Université Pierre et Marie Curie Paris 6, 4 place Jussieu, case 247, 75252 Paris Cedex 05, France

• **F Musso** Dipartimento di Matematica e Fisica, Universitá degli Studi di Roma Tre, 00146 Roma, Italy

• **Z Oziewicz** Universidad Nacional Autonoma de México, Facultad de Estudios Superiores,C.P. 54714 Cuautitlan Izcalli, Estado de México

Journal of Physics: Conference Series **532** (2014) 011002 doi:10.1088/1742-6596/532/1/011002

• **E Paal** Tallinn University of Technology, Ehitajate tee 5, 19086 Tallinn, Estonia
• **I Pop** Department of Mathematics, University of Gothenburg, Box 100, SE-405 30, Gothenburg, Sweden
• **P Puusemp** Matemaatikainstituut, Tallinna Tehnikaülikool, Ehitajate tee 5, 19086 Tallinn, Estonia
• **E Randla** Institute of Physics, University of Tartu, Ravila 14c, 50411 Tartu, Estonia
• **M Saal** Füüsika Instituut, Tartu Ülikool, Ravila 14c, 50411 Tartu, Estonia
• **R Saar** Füüsika Instituut, Tartu Ülikool, Ravila 14c, 50411 Tartu, Estonia
• **A Stolin** Department of Mathematics, University of Gothenburg, SE-405 30, Gothenburg, Sweden; Chalmers University of Technology, 412 96 Gothenburg, Sweden
• **P Suurvarik** Department of Physics, Tallinn University of Technology, Ehitajate tee 5, 19086 Tallinn, Estonia
• **B Torrecillas** Department of Algebra and Analysis, University of Almeria, E-04071 Almeria, Spain
• **A Tralle** Department of Mathematics and Information Technology, University of Warmia and Mazury, Słoneczna str. 54, 10-710 Olsztyn, Poland
• **M Umbleja** Tallinn University of Technology, Ehitajate tee 5, 19086 Tallinn, Estonia
• **H Veermäe** Füüsika Instituut, Tartu Ülikool, Ravila 14c, 50411 Tartu, Estonia
• **O Vilson** Institute of Physics, University of Tartu, Ravila 14c, 50411 Tartu, Estonia
• **J Virkepu** Tallinn University of Technology, Ehitajate tee 5, 19086 Tallinn, Estonia
• **M Vojinović** Grupo de Fisica Matemática da Universidade de Lisboa, Av. Prof. Gama Pinto, 2, 1649-003 Lisboa, Portugal
• **S Waldmann** Institute for Mathematics, Julius Maximilian University Würzburg, Emil-Fischer-Straße 31, 97074 Würzburg, Germany
• **G P Wene** Department of Mathematics, University of Texas at San Antonio, 6900 North Loop 1604 West, San Antonio, Texas 78249-0697, U.S.A.
• **D K Wise** Department Mathematik, FAU Erlangen-Nürnberg, Cauerstr. 11, 91058 Erlangen, Germany
• **A Zuevsky** Max-Planck-Institut für Mathematik, Vivatsgasse 7, 53111 Bonn, Germany
• **P Zusmanovich** Institute of Mathematics and Statistics, University of São Paulo, Rua do Mataõ 1010, São Paulo 05508-090, Brazil

Peer review statement

All papers published in this volume of *Journal of Physics: Conference Series* have been peer reviewed through processes administered by the proceedings Editors. Reviews were conducted by expert referees to the professional and scientific standards expected of a proceedings journal published by IOP Publishing.

Knizhnik-Zamolodchikov type equations for the root system B and Capelli central elements

D V Artamonov[1] **and V A Golubeva**[2]

[1]Faculty of Economics, Lomonosov Moscow State Univeristy
1-46 Leninskie Gory, 119991 Moscow, GSP-1, 1-46, Russia

[2]Department of Applied Mathematics and Physics, Moscow Aviation Institute
4 Volokolamskoe Shosse, 125993 Moscow, A-80, GSP-3, Russia

Abstract. The construction of the well-known Knizhnik-Zamolodchikov equations uses the central element of the second order in the universal enveloping algebra for some Lie algebra. But in the universal enveloping algebra there exist central elements of higher orders. It seems desirable to use these elements for the construction of Knizhnik-Zamolodchikov type equations. In the present paper we give a construction of such Knizhnik-Zamolodchikov type equations for the root system B associated with Capelli central elements in the universal enveloping algebra for the orthogonal algebra.

1. Introduction

The Knizhnik-Zamolodchikov equations are a system of differential equations which is satisfied by correlation functions in the WZW theory [1]. Later it turned out that these equation are related with many other areas of mathematics (quantum algebra, isomonodromic deformation).

The Kniznik-Zamolodchikov equations are also interesting as a nontrivial example of an integrable Pfaffian system of Fuchsian type. It must be mentioned that the monodromy representation of this system is known explicitly. Thus we get a solution for the Riemann-Hilbert problem in a very particular case.

The Kniznik-Zamolodchikov equations read

$$dy = \lambda \left(\sum_{i \neq j=1}^{n} \tau_{ij} \frac{d(z_i - z_j)}{z_i - z_j} \right) y$$

where λ is some complex parameter, $y(z_1, \ldots, z_n)$ is a vector function that takes values in a tensor power $V^{\otimes n}$ of a representation space V of a finite-dimensional Lie algebra $\mathfrak{g}$, and τ_{ij} is defined by the formula

$$\tau_{ij} = \sum_s 1 \otimes \ldots \otimes \rho(I_s) \otimes \ldots \otimes \rho(\omega(I_s)) \otimes \ldots \otimes 1 \tag{1.1}$$

Here $\{I_s\}$ is a basis of the Lie algebra $\mathfrak{g}$ and $\{\omega(I_s)\}$ is its dual basis with respect to the Killing form. The elements I_s occur on tensor factors i and j, and $\rho : \mathfrak{g} \to End(V)$ is a representation of the Lie algebra $\mathfrak{g}$. Thus τ_{ij} is the matrix of the linear operator on $V^{\otimes n}$.

The Knizhnik-Zamolodchikov equations have singularities on the planes $z_i = z_j$. It is natural to look for similar systems that have other singular locus.

Several authors constructed different Knizhnik-Zamolodchikov type equations, that have singularities on the reflection hyperplanes of different root system. Thus, in [2] a system of the Knizhnik-Zamolodchikov type equations is constructed that have as the singular set the reflection hyperplanes corresponding to an arbitrary root system. Let us present such a system.

Matsuo considers a root system Δ, the set of positive roots is denoted as Δ^+ and the corresponding Weyl group is denoted as W. The function y takes values in the group algebra $\mathbb{C}[W]$ and the independent variable belong to the dual to the vector space spanned by roots.

Denote a reflection corresponding to the root α as σ_α.

The Matsuo Knizhnik-Zamolodchikov type equations contain arbitrary parameters $\lambda_{|\alpha|}$ depending on the length of the roots α and one additional parameter λ. The system has the following form.

$$\frac{\partial y}{\partial \xi} = \left(\sum_{\alpha \in \Delta^+} \lambda_{|\alpha|}(\alpha, z) \frac{1}{(\alpha, u)} (\sigma_\alpha - 1) \right) y$$

Note that solutions of this system take values in the space $\mathbb{C}[W]$ and no Lie algebra is involved in it's construction explicitly. Nevertheless if one takes in the Matsuo's construction the root system A and some special representation of the Weyl group one can obtain from Matsuo's system an ordinary Knizhnik-Zamolodchikov system corresponding to the algebra $\mathfrak{sl}_n$ and its standard representation. Later systems of Matsuo's type were intensively studied by numerous authors (I. Cherednik [3] and others)

There exist other constructions of the Knizhnik-Zamolodchikov type equations with are closer to the original equations. One of them is the Leibman's construction of the Knizhnik-Zamolodchikov type equations associated with the root system B [4] and Enriquez's cyclotomic Knizhnik-Zamolodchikov type equations [5].

In [6] we have constructed Knizhnik-Zamolodchikov type equations for the root system A, but our construction was based on special central elements in the universal enveloping algebra of the orthogonal algebra, namely the Capelli elements. In the present paper we generalize our construction to the root system B and show that the construction from [6] can be in fact interpreted in some sence as a very special case of the general Leibman's construction.

2. Knizhnik-Zamolodchikov equation associated with the root system B

Let us explain the Leibman's construction of Knizhnik-Zamolodchikov type equations for the root system B. This is a system of type

$$dy = \lambda \left(\sum_{i<j} \frac{d(z_i - z_j)}{z_i - z_j} \tau_{ij} + \sum_{i<j} \frac{d(z_i - z_j)}{z_i - z_j} \mu_{ij} + \sum_i \frac{\nu_i}{z_i} \right) y$$

were τ_{ij} is defined by (1.1). Denote by σ an involution in the considered Lie algebra $\mathfrak{g}$. Then one has

$$\mu_{ij} = \sum_s 1 \otimes \ldots \otimes \rho\left(I_s\right) \otimes \ldots \otimes \rho(\sigma(\omega(I_s))) \otimes \ldots \otimes 1$$

where the elements I_s occur on places i, j and

$$\nu_i = 1/2 \sum_s 1 \otimes \ldots \otimes \rho\left(I_s \sigma(\omega(I_s)) + I_s^2\right) \otimes \ldots \otimes 1 \tag{2.1}$$

Journal of Physics: Conference Series **532** (2014) 012001 doi:10.1088/1742-6596/532/1/012001

where the elements I_s occur on the place i.

Although the Leibman's proof in [4] is done for the case of a simple Lie algebra and is based on calculations in the root base, in [5] there is presented a proof that is valid for an arbitrary finite-dimensional Lie algebra with a fixed central element in $U(\mathfrak{g})$ of the second order.

3. The Lie algebra $\mathcal{T}$

In this section we introduce a Lie algebra $\mathcal{T}$ which plays the crucial role in our construction.

Consider the space of skew-symmetric tensors with $2n$ indices. Let each index of the skew symmetric tensor take values in the set $-n, \dots, n$.

There exists a structure of an associative algebra algebra on this space

$$(e_{a_1}\wedge\dots\wedge e_{a_n}\wedge e_{b_1}\dots\wedge e_{b_n})\cdot(e_{-b_1}\wedge\dots\wedge e_{-b_n}\wedge e_{c_1}\wedge\dots\wedge e_{c_n}) := e_{a_1}\wedge\dots\wedge e_{a_n}\wedge e_{c_1}\wedge\dots\wedge e_{c_n}$$

As a corollary we have a structure of a Lie algebra, denote it as $\mathcal{T}$.

The algebra $\mathcal{T}$ has a representation on the space of skew-symmetric tensors with n indices defined by the formula

$$e_{a_1}\wedge\dots\wedge e_{a_n}\wedge e_{b_1}\wedge\dots\wedge e_{b_n}(e_{-b_1}\wedge\dots\wedge e_{-b_n}) := e_{a_1}\wedge\dots\wedge e_{a_n} \tag{3.1}$$

The algebra has an involution ω defined as follows

$$\omega(e_{a_1}\wedge\dots\wedge e_{a_n}\wedge e_{b_1}\wedge\dots\wedge e_{b_n}) = e_{-a_1}\wedge\dots\wedge e_{-a_n}\wedge e_{-b_1}\wedge\dots\wedge e_{-b_n}$$

In $U(\mathcal{T})$ there is a central element of the second order, namely the element

$$C = \sum_{a_1,\dots,a_n,b_1,\dots,b_n} (e_{a_1}\wedge..\wedge e_{a_n}\wedge e_{b_1}\wedge\dots\wedge e_{b_n})\bullet(e_{-a_1}\wedge..\wedge e_{-a_n}\wedge e_{-b_1}\wedge\dots\wedge e_{-b_n})$$

where $\bullet$ denotes the multiplication in $U(\mathcal{T})$. The proof of this fact is essentially contained in [7].

Using general constructions described in Section (2) one can construct the Knizhnik-Zamolodchikov type equations associated with the root system B based with coefficients in the Lie algebra $\mathcal{T}$.

In the next section we give an interpretation of these equations as the Knizhnik-Zamolodchikov type equations whose construction is based on some certain higher order central elements in $U(\mathfrak{o}_{2n+1})$.

4. Capelli elements and noncommutative Pfaffians

Let us define some certain central elements in the universal enveloping algebra of the orthogonal algebra.

4.1. The split realization of the orthogonal algebra

We use the split realization of the orthogonal algebra. This means that we define the orthogonal algebra as the algebra that preserves the quadratic form with the matrix

$$G = \begin{pmatrix} 0\,0 & \dots & 0\,1 \\ 0\,0 & \dots & 1\,0 \\ \vdots\,\vdots & \ddots & \vdots\,\vdots \\ 0\,1 & \dots & 0\,0 \\ 1\,0 & \dots & 0\,0 \end{pmatrix}$$

The row and columns are indexed by $i, j = -n, -n+1, \dots, n-1, n$. The zero is skipped in the case $N = 2n$ and is included in the case $N = 2n+1$. The algebra $\mathfrak{o}_N$ is generated by matrices

$$F_{ij} = E_{ij} - E_{-j-i}$$

The commutation relations between these generators are the following

$$[F_{ij}, F_{kl}] = \delta_{kj}F_{il} - \delta_{il}F_{kj} + \delta_{i-k}F_{-jl} + \delta_{-lj}F_{k-i}$$

4.2. Noncommutative Pfaffians and Capelli elements

Now describe some special higher order central elements in the universal enveloping for the orthogonal algebra.

Let $\Phi = (\Phi_{ij})$ be a $k \times k$ matrix, where k is even, whose elements belong to some noncommutative ring. The noncommutative Pfaffian is defined by

$$Pf\,\Phi = 2^{-k/2}/(k/2)! \sum_{\sigma \in S_k} (-1)^{\sigma} \Phi_{\sigma(1)\sigma(2)} \dots \Phi_{\sigma(k-1)\sigma(k)}$$

For a subset $I \subset \{-n, \dots, n\}$ define a submatrix $F_I = (F_{ij})_{i,j \in I}$. For this subset put

$$Pf\,F_I := Pf(F_{-i,j})_{-i,j \in I}$$

Definition 4.1. Put

$$C_k = \sum_{I \subset \{1,\dots,N\},\, |I|=k} Pf\,F_I Pf\,F_I$$

The elements C_k are are called the Capelli elements.

Theorem 4.2 ([7])**.** *For odd N the elements C_k are algebraically independent and generate the center, for even N the same is true if one takes instead the highest Capelli element $C_N = (Pf\,F)^2$ the central element $Pf\,F$.*

Below we need two formulas. There proofs can be found in [6].

Lemma 4.3. *We have*

$$Pf\,F_I = (p/2)!\,(q/2)!\Big/(k/2)! \sum_{I=I' \sqcup I'',\, |I'|=p,\, |I''|=q} (-1)^{(I'I'')} Pf\,F_{I'}\, Pf\,F_{I''}$$

where $(-1)^{(I'I'')}$ is a sign of a permutation of the set $I = \{i_1, \dots, i_k\}$ that places first the subset $I' \subset I$ and then the subset $I'' \subset I$. The numbers p, q are even fixed numbers, satisfying $p + q = k = |I|$.

Let Δ be the standard comultiplication in the universal enveloping algebra.

Lemma 4.4. *We have*

$$\Delta Pf\,F_I = \sum_{I' \sqcup I''=I} (-1)^{(I'I'')} Pf\,F_{I'} \otimes Pf\,F_{I''}$$

where $(-1)^{(I'I'')}$ is a sign of a permutation of the set $I = \{i_1, \dots, i_k\}$ that places first the subset $I' \subset I$ and then places the subset $I'' \subset I$.

4.3. The action of Pfaffians on tensors

Let us describe the action of Pfaffians in the tensor representations.

Proposition 4.5. *On the base vectors $e_{-2}, e_{-1}, e_0, e_1, e_2$ of the standard representation of $\mathfrak{o}_5$, the Pfaffians $Pf\,F_I$, where $|I| = 4$, act as zero operators.*

Proof. The proposition is proved by direct calculation using the formulae, where $a \star b = \frac{1}{2}(ab+ba)$:

$$Pf\,F_{\widehat{-2}} = F_{0-1} \star F_{-21} - F_{-1-1} \star F_{-20} + F_{-2-1} \star F_{-10}$$
$$Pf\,F_{\widehat{-1}} = F_{0-2} \star F_{-21} - F_{-1-2} \star F_{-20} + F_{-2-2} \star F_{-10}$$
$$Pf\,F_{\widehat{0}} = F_{1-2} \star F_{-21} - F_{-1-2} \star F_{-2-1} + F_{-2-2} \star F_{-1-1}$$
$$Pf\,F_{\widehat{1}} = F_{1-2} \star F_{-20} - F_{0-2} \star F_{-2-1} + F_{-2-2} \star F_{0-1}$$
$$Pf\,F_{\widehat{2}} = F_{1-2} \star F_{-10} - F_{0-2} \star F_{-1-1} + F_{-1-2} \star F_{0-1}$$

□

Journal of Physics: Conference Series **532** (2014) 012001 doi:10.1088/1742-6596/532/1/012001

Now prove an analogue of the previous statement in an arbitrary dimension.

Proposition 4.6. *On the base vectors $e_{-n}, \dots, e_n$ of the standard representation of $\mathfrak{o}_N$ the Pfaffians $Pf\,F_I$ for $|I| > 2$ act as zero operators.*

Proof. Put $q = 4$, $p = k - 4$ in Lemma (4.3). One has

$$Pf\,F_I e_j = \sum_{I' \sqcup I'' = I, |I'| = k-4, |I''| = 4} (p/2)!\,(q/2)! \big/ (k/2)!\,(-1)^{(I'I'')} Pf\,F_{I'} Pf\,F_{I''} e_j$$

If $j \notin I''$, then obviously $Pf\,F_{I''} e_j = 0$. If $j \in I''$, then using Proposition (4.5) one also obtains $Pf\,F_{I''} e_j = 0$. □

Let us find an action of a Pfaffian of the order k on a tensor product of $< \frac{k}{2}$ vectors, that is on a tensor product $e_{r_2} \otimes e_{r_4} \otimes \dots \otimes e_{r_t}$, where $t < k$.

Proposition 4.7. *We have*

$$Pf\,F_I e_{r_2} \otimes e_{r_4} \dots \otimes e_{r_t} = 0, \quad t < k$$

Proof. The following formulae takes place (see Lemma (4.4)):

$$\Delta Pf\,F_I = \sum_{I' \sqcup I'' = I} (-1)^{(I'I'')} Pf\,F_{I'} \otimes Pf\,F_{I''}$$

By definition one has

$$Pf\,F_I e_{r_2} \otimes e_{r_4} \otimes \dots \otimes e_{r_k} = (\Delta^k Pf\,F_I) e_{r_2} \otimes e_{r_4} \otimes \dots \otimes e_{r_k}$$

Since $t < k$, the comultiplication $\Delta^k Pf\,F_I$ contains only summands in which on some place the Pfaffian stands whose indexing set I satisfies $|I| \geq 4$ (Lemma (4.4)). From Proposition (4.5) it follows that every such a summand acts as a zero operator. □

Find an action of a Pfaffian of the order k on a tensor product of $k/2$ vector, that is on the tensor product $e_{r_2} \otimes e_{r_4} \otimes \dots \otimes e_{r_k}$.

Proposition 4.8. *If $\{r_2, r_4 \dots, r_k\}$ is not contained in I, then*

$$Pf\,F_I e_{r_2} \otimes e_{r_4} \otimes \dots \otimes e_{r_k} = 0$$

Otherwise, take a permutation γ of I, such that

$$(\gamma(i_1), \gamma(i_2), \dots, \gamma(i_k)) = (r_1, r_2, r_3, \dots, r_{k-1}, r_k)$$

Then

$$Pf\,F_I\, e_{r_2} \otimes \dots \otimes e_{r_k} = (-1)^{\gamma} (-1)^{k(k-1)/2} \times$$
$$\times \sum_{\delta \in Aut(r_1, r_3, \dots, r_{k-1})} (-1)^{\delta} e_{-\delta(r_1)} \otimes e_{-\delta(r_3)} \otimes \dots \otimes e_{-\delta(r_{k-3})} \otimes e_{-\delta(r_{k-1})}$$

Proof. By definition one has

$$Pf\,F_I e_{r_2} \otimes e_{r_4} \otimes \ldots \otimes e_{r_k} = (\Delta^k Pf\,F_I) e_{r_2} \otimes e_{r_4} \otimes \ldots \otimes e_{r_k}$$

Applying many times the formulae for comultiplication one obtains

$$\Delta^{\frac{k}{2}} Pf\,F_I = \sum_{I^1 \sqcup \ldots \sqcup I^k} (-1)^{(I^1 \ldots I^k)} Pf\,F_{I^1} \otimes \ldots \otimes Pf\,F_{I^k}$$

Using Proposition (4.7) one gets that, only the summands for which $|I^j| = 2, j = 1, \ldots, k$ are nonzero operators. Hence, the summation over divisions can be written in the following way:

$$\begin{aligned} Pf\,F_I e_{r_2} \otimes e_{r_4} \otimes \ldots \otimes e_{r_k} &= 1/2^{-k/2} \sum_{\sigma \in S_k} (-1)^\sigma F_{-\sigma(i_1)\sigma(i_2)} \otimes \ldots \otimes F_{-\sigma(i_{k-1})\sigma(i_k)} (e_{r_2} \otimes \ldots \otimes e_{r_k}) \\ &= 1/2^{-k/2} \sum_{\sigma \in S_k} (-1)^\sigma F_{-\sigma(i_1)\sigma(i_2)} e_{r_2} \otimes \ldots \otimes F_{-\sigma(i_{k-1})\sigma(i_k)} e_{r_k} \end{aligned}$$

Consider the expression $F_{-\sigma(i_1)\sigma(i_2)} e_{r_2}$. This is $e_{-\sigma(i_1)}$ if $\sigma(i_2) = r_2$, this is $-e_{-\sigma(i_2)}$ if $\sigma(i_1) = r_2$ and zero otherwise. Thus the summand is nonzero only if the permutation σ satisfies the following condition. In each pair $(\sigma(i_{2t-1}), \sigma(i_{2t}))$ either $\sigma(i_{2t-1}) = r_{2t}$ or $\sigma(i_{2t}) = r_{2t}$.

Show that one can consider only the permutations σ such that $\sigma(i_{2t}) = r_{2t}$, that is the permutations of type $(\sigma(i_1), r_2, \sigma(i_2), r_3 \ldots, \sigma(i_{k-1}), r_k)$. But when only summands corresponding to such permutations are considered one must multiply the resulting sum on $2^{k/2}$. It is enough to prove that the permutations

$$\sigma = (\sigma(i_1), \sigma(i_2) = r_2, \sigma(i_3), \ldots, \sigma(r_k)), \quad \sigma' = (\sigma(i_2) = r_2, \sigma(i_1), \sigma(i_3), \ldots, \sigma(r_k))$$

give the same input. Recall that the input for σ is

$$(-1)^\sigma F_{-\sigma(i_1)\sigma(i_2)} e_{r_2} \otimes \ldots \otimes F_{-\sigma(i_{k-1})\sigma(i_k)} e_{r_k}$$

One has from one hand that

$$F_{-\sigma(i_1)\sigma(i_2)} e_{r_2} = e_{-\sigma(i_1)}$$

and from the other hand that

$$F_{-\sigma'(i_1)\sigma'(i_2)} e_{r_2} = -e_{-\sigma'(i_2)} = -e_{-\sigma(i_1)}$$

Also one has $(-1)^\sigma = -(-1)^{\sigma'}$. Thus the inputs corresponding to σ and σ' are the same. Hence one can consider the only the permutations σ of type $(\sigma(i_1), r_2, \sigma(i_2), r_3 \ldots, \sigma(i_{k-1}), r_k)$ but multiplying the resulting sum on $2^{k/2}$.

For the permutation σ of type $(\sigma(i_1), r_2, \sigma(i_2), r_3 \ldots, \sigma(i_{k-1}), r_k)$ using the definition of γ one gets

$$\begin{aligned} (-1)^\sigma F_{-\sigma(i_1)\sigma(i_2)} e_{r_2} \otimes \ldots \otimes F_{-\sigma(i_{k-1})\sigma(i_k)} e_{r_k} &= (-1)^{(\sigma(i_1) r_2, \ldots, \sigma(i_{k-1}) r_k)} e_{-\sigma(i_1)} \otimes e_{-\sigma(i_3)} \otimes \ldots \otimes e_{-\sigma(i_k)} \\ &= (-1)^{k(k-1)/2} (-1)^\gamma (-1)^\delta e_{-\delta(r_1)} \otimes \ldots \otimes e_{-\delta(r_{k-1})} \end{aligned}$$

where δ is a permutation of the set $\{r_1, r_3, \ldots, r_{k-3}, r_k\}$ and the equality

$$(-1)^{k(k-1)/2} (-1)^\delta (-1)^\gamma = (-1)^\sigma$$

was used. Performing summation over all permutations δ, one gets

$$Pf\ F_I e_{r_2} \otimes e_{r_4} \otimes \ldots \otimes e_{r_k} = (-1)^{k(k-1)/2}(-1)^{\gamma} \sum_{\delta \in Aut(r_1,\ldots,r_{k-1})} (-1)^{\delta} e_{-\delta(r_1)} \otimes \ldots \otimes e_{-\delta(r_{k-1})}$$

Finally, from the formula

$$Pf\ F_I e_{r_2} \otimes e_{r_4} \otimes \ldots \otimes e_{r_t} = (\Delta^t Pf\ F_I) e_{r_2} \otimes e_{r_4} \otimes \ldots \otimes e_{r_t}$$

as in the proof of Proposition (4.8), one gets the formulae of the action on an arbitrary tensor $e_{r_2} \otimes \ldots \otimes e_{r_t}$. □

Proposition 4.9. *We have*

$$Pf\ F_I e_{r_2} \otimes e_{r_4} \ldots \otimes e_{r_t} = \sum_{\{j_2,j_4\ldots,j_k\} \subset \{2,4\ldots,t\}} Pf^{j_2,j_4\ldots,j_t} F_I e_{r_2} \otimes e_{r_4} \otimes \ldots \otimes e_{r_t}$$

where $Pf^{j_2,j_4\ldots,j_k} F_I$ *acts on the tensor multiples with numbers* $j_2, j_4 \ldots, j_k$. *It's action is described by Proposition* (4.8)

4.4. Pfaffians and representation of the algebra $\mathcal{T}$

Let us give a relation between the representation of the Lie algebra $\mathcal{T}$ defined by the formula (3.1) through the action of noncommutative Pfaffians.

As a corollary of Proposition (4.9) we get the following proposition

Proposition 4.10. *In the case* $\mathfrak{o}_{2n+1}$, *if*

$$I \subset J, \quad |J| = 2n, \quad v = e_{i_1} \otimes \ldots \otimes e_{i_n} \quad \text{then} \quad Pf\ F_I Pf\ F_J v = C Pf\ F_J v$$

where the constant C *depends only on* n.

Now let $N = 2n+1$ and $|I| = 2n$. Then a subset I which consists of $2n$ elements is of type $\{1,\ldots,N\} \setminus i$. Put $Pf\ F_I := Pf\ F_{\widehat{i}}$. Then $Pf\ F_{Pf\ F_{I''} J} = 0$ for $I'' > 2$ and

$$[Pf\ F_{\widehat{i}}, Pf\ F_{\widehat{j}}] = 1/n \sum_{k \neq i,j} (-1)^{(I'I'')} Pf\ F_{\widehat{ikj}} Pf\ F_{\widehat{k}}$$

where the sign $(-1)^{(I'I'')}$ is defined as follows. For an index s denote as $\overline{s}$ either s for $s < i$ or $s-1$ for $s > i$. Then $(-1)^{(I'I'')}$ equals $(-1)^{n-\overline{j}+(n-1)-\overline{k}} = (-1)^{\overline{j}+\overline{k}-1}$ in the case $j < k$ and $(-1)^{\overline{j}+\overline{k}}$ in the case $j > k$. By denoting this sign as s_{jk}, one has $s_{jk} = -s_{kj}$. Using the theorem (4.10) one obtains that for a vector v from a representation of $\mathfrak{o}_{2n+1}$ with the highest weight $(1,\ldots,1)$ the following holds.

Lemma 4.11. *One has*

$$[Pf\ F_{\widehat{i}}, Pf\ F_{\widehat{j}}] v = C \sum_{k \neq i,j} s_{jk} Pf\ F_{\widehat{k}} v$$

where C *is some constant.*

The following theorem is proved.

Theorem 4.12. *The representation of the algebra* $\mathcal{T}$ *on the space of skew-symmetric tensors with* n *indices, given by the formula* (3.1) *is given also by the formula*

$$e_{a_1} \wedge \ldots \wedge e_{b_n} \mapsto 1/\sqrt{C}\ \ Pf\ F_{\{a_1,\ldots,b_n\}}$$

where the constant is taken from Lemma (4.11) *and the Pfaffian is considered as an operator acting on the space of skew-symmetric tensors with* n *indices.*

5. Knizhnik-Zamolodchikov equations and Capelli elements

Now let us give an interpretation of the Kniznik-Zamolodchikov type equations associated with the root system B constructed for the algebra $\mathfrak{T}$ and its representation (3.1) as a Kniznik-Zamolodchikov type equations constructed for higher order Capelli central elements. This fact is an immediate corollary of Theorem (4.12).

Introduce the elements

$$\tau_{ij} := \sum_{I,|I|=2n} 1\otimes\ldots\otimes\rho(Pf\,F_I)\otimes\ldots\otimes\rho(Pf\,F_{-I})\otimes\ldots\otimes 1 \tag{5.1}$$

where the Pfaffians occur on places i,j, and ρ is a representation of $\mathfrak{o}_{2n+1}$ on the space of skew-symmetric tensors with n indices,

$$\mu_{ij} := \sum_s 1\otimes\ldots\otimes\rho(Pf\,F_I)\otimes\ldots\otimes\rho(Pf\,F_I)\otimes\ldots\otimes 1 \tag{5.2a}$$

$$\nu_i := 1/2\sum_s 1\otimes\ldots..\otimes\rho(Pf\,F_I Pf\,F_{-I} + Pf\,F_I Pf\,F_I)\otimes\ldots\otimes 1 \tag{5.2b}$$

Theorem 5.1. *The action of elements* (1.1)-(2.1) *and* (5.1)-(5.2b) *on the space of skew-symmetric tensors with n indices coincide.*

As a corollary we get

Theorem 5.2. *The elements* (5.1)-(5.2b) *satisfy the commutation relation for the coefficients of the Kniznik-Zamolodchikov type equations associated with the root system B.*

References

[1] Knizhnik V G and Zamolodchikov A B 1984 *Nucl. Phys.* B **247** 83
[2] Matsuo A 1992 *Invent. Math.* **110** 95
[3] Cherednik I 2005 *Double Affine Hecke algebras* (Cambridge Univ. Press)
[4] Leibman A 1994 *Comm. Math. Phys.* **164** 293
[5] Enriquez B 2008 *Selecta Math.* **13** 391
[6] Artamonov D V and Golubeva V A 2012 *Mat. Sb.* **203** 5
[7] Molev A 2007 *Yangians and Classical Lie algebras* Math. Surveys Monog. **143** (Providence, RI: AMS)

On quantum deformations of (anti-)de Sitter algebras in (2+1) dimensions

A Ballesteros[1], F J Herranz[1] and F Musso[2]

[1]Departamento de Física, Universidad de Burgos, 09001 Burgos, Spain
[2]Dipartimento di Matematica e Fisica, Universitá degli Studi di Roma Tre, 00146 Roma, Italy

Abstract. Quantum deformations of (anti-)de Sitter (A)dS algebras in (2+1) dimensions are revisited, and several features of these quantum structures are reviewed. In particular, the classification problem of (2+1) (A)dS Lie bialgebras is presented and the associated noncommutative quantum (A)dS spaces are also analysed. Moreover, the flat limit (or vanishing cosmological constant) of all these structures leading to (2+1) quantum Poincaré algebras and groups is simultaneously given by considering the cosmological constant as an explicit Lie algebra parameter in the (A)dS algebras. By making use of this classification, a three-parameter generalization of the κ-deformation for the (2+1) (A)dS algebras and quantum spacetimes is given. Finally, the same problem is studied in (3+1) dimensions, where a two-parameter generalization of the κ-(A)dS deformation that preserves the space isotropy is found.

1. Introduction

Quantum groups and algebras were introduced in the eighties as Hopf algebra deformations of Lie groups and algebras, respectively, see [1, 2, 3, 4, 5, 6, 7] and references therein. Since then, the construction of quantum deformations of the kinematical groups of spacetimes opened the possibility of introducing consistent proposals of the mathematical descriptions of the 'quantum' spacetime that is commonly assumed to arise in quantum gravity theories when the Planck energy is approached, see [8]. Essentially, this idea has been worked out so far in the literature under two dual approaches:

- *Quantum kinematical algebras*: they provide quantum deformations of the symmetry algebras of spacetimes, see for instance [9, 10, 11], in which the 'quantum' deformation parameter would be a second invariant scale related with the Planck energy (or length) and in which q-deformed Casimir operators generate the kind of modified dispersion relations that are expected to arise in the Planck regime. This viewpoint gave rise to the so-called 'double special relativity' theories [12, 13, 14, 15, 16].
- *Quantum kinematical groups*: they are the Hopf algebra dual of the previous quantum algebras [4, 5, 17] and provide a self-consistent mathematical foundation of the so-called noncommutative spacetimes, in which the noncommutativity of the spacetime coordinates is generated by a non-vanishing quantum deformation parameter that should describe Planck scale effects. As a consequence, uncertainty relations between these noncommuting coordinate operators arise naturally, and the intrinsic Planck scale 'fuzziness', 'discretization' or 'quantization' of the spacetime itself can be mathematically described, see [18, 19, 20].

At this point it is important to stress that for a given Lie group or algebra there exist (many) different Hopf algebra deformations. In fact, quantum algebras are in correspondence with Lie bialgebra structures (and quantum groups with Poisson-Lie structures), and the explicit classification problem for Lie bialgebras (or, equivalently, for Poisson-Lie groups) is by no means a simple problem from a computational viewpoint. Therefore, it seems natural that some criteria should be taken into account in order to select the type of quantum deformation that could be suitable in the abovementioned specific physical settings.

In the particular case of $(2+1)$ quantum gravity, it was (heuristically) stated in [21] that the perturbations of the vacuum state of a Chern-Simons quantum gravity theory with cosmological constant Λ are invariant under transformations that close a certain quantum deformation of the (A)dS algebra. In fact, the low energy regime/zero-curvature limit of this algebra was found to be the known κ-Poincaré quantum algebra [9, 22, 23, 24].

Also, it is well known that Poisson-Lie (PL) structures on the isometry groups of (2+1) spaces with constant curvature seem to play a relevant role as phase spaces in Chern-Simons theory. Moreover, in the Lorentzian case, these PL structures (given by certain classical r-matrices that have to be 'compatible' with the Chern-Simons pairings) are just the classical counterparts of certain quantum deformations of (A)dS and Poincaré groups in (2+1) dimensions (see [25, 26, 27, 28] and references therein). These PL groups are in one-to-one correspondence with certain three-dimensional Lie bialgebras and their associated Drinfel'd doubles, in such a way that an explicit connection between (2+1) Chern-Simons symmetries and Drinfel'd doubles coming from three-dimensional quantum deformations has been recently proposed [29, 30]. In this setting the (2+1) (A)dS and Poincaré r-matrices associated to such Drinfel'd doubles are not the ones defining κ-deformations and, therefore, quantum groups corresponding to these (multiparametric) r-matrices seem to be physically interesting to construct and analyse.

The aim of this paper is to address the classification problem for the quantum deformations of the (2+1) (A)dS and Poincaré algebras by considering the three Lie algebras as particular cases of a one-parametric Lie algebra, the (2+1) AdS_ω algebra, in which $\omega = -\Lambda$. In this way, the $\omega \to 0$ limit will provide the Poincaré Lie bialgebra classification, and the sign of ω will distinguish between the AdS ($\omega > 0$) and dS ($\omega < 0$) cases. Moreover, we will show that the type of approach here presented makes feasible to tackle the same problem in (3+1) dimensions.

The paper is organized as follows. In the next section we will briefly recall the basics of quantum deformations and their classification in terms of their associated Lie bialgebras. In section 3 we will review the known results concerning the κ-deformation of the (2+1) AdS_ω algebra, including the full PL structure that provides the semiclassical (Poisson) counterpart of the noncommutative (2+1) AdS_ω spacetime, where the nonvanishing cosmological constant Λ generates nonlinear commutation relations among the quantum coordinates. In section 4 the most generic Lie bialgebra for the (2+1) AdS_ω algebra is computed, and this result gives a clear idea of the plethora of different quantum deformations that there exist. The possible generalizations of the κ-deformation are identified in section 5 by imposing on the generic deformation that the rotation J and time translation P_0 generators have to be primitive ones, a constraint derived from dimensionality arguments. In this way we find that two more parameters can be added to the κ-deformation, and we explicitly describe the corresponding first-order noncommutative spacetimes, which are generalizations of the κ-Minkowski spacetime. The same approach is used in section 6 in order to study the possible generalizations of κ-deformations in (3+1) dimensions. In this case we find two disjoint families of two-parametric deformations that preserve the primitive nature of the generators P_0 and J_3. Finally, we demonstrate that one of these generalized κ-deformations preserve space isotropy, since under some appropriate automorphism the symmetric role of the three rotation generators can be manifestly shown.

2. Quantum deformations and Lie bialgebras

2.1. Quantum algebras

Quantum algebras are Hopf algebra deformations of universal enveloping algebras. This means that we consider the algebra $U_z(g)$ of formal power series in the deformation parameter z and coefficients in the universal enveloping algebra $U(g)$ of a given Lie algebra g, and we endow it with a Hopf algebra structure by finding an algebra homomorphism called the coproduct map

$$\Delta_U : U_z(g) \longrightarrow U_z(g) \otimes U_z(g)$$

as well as its associated counit ϵ_U and antipode γ_U mappings.

Any quantum universal enveloping algebra can be thought of as a Hopf algebra deformation of $U(g)$ 'in the direction' of a certain Lie bialgebra (g, δ). Such a Lie bialgebra provides the first-order deformation (in z) of the coproduct

$$\Delta_U = \Delta_0 + \delta + o[z^2], \quad \Delta_0(X) = X \otimes 1 + 1 \otimes X$$

where $\delta(X)$, the so-called cocommutator, is linear in z. Therefore, a precise Lie bialgebra structure (g, δ) will be associated to each possible quantum coproduct Δ_U, and the equivalence classes (under automorphisms) of Lie bialgebra structures on a given Lie algebra g will provide the chart of all possible quantum deformations of $U(g)$.

Complementarily, quantum groups (A, Δ_A) are just noncommutative algebras of functions defined as the dual Hopf algebras to the quantum algebras $(U_z(g), \Delta_U)$. More explicitly, if we denote by m_A and m_U the noncommutative products in A and $U_z(g)$, respectively, the duality between the Hopf algebras (A, m_A, Δ_A) and $(U_z(g), m_U, \Delta_U)$ is established through the existence of a canonical pairing $\langle\,\rangle : A \times U_z(g) \to \mathbb{C}$ between them such that

$$\langle m_A(f \otimes g), a\rangle = \langle f \otimes g, \Delta_U(a)\rangle \tag{2.1a}$$
$$\langle \Delta_A(f), a \otimes b\rangle = \langle f, m_U(a \otimes b)\rangle \tag{2.1b}$$

where $a, b \in U_z(g)$, $f, g \in A$, and $\langle f \otimes g, a \otimes b\rangle = \langle f, a\rangle \langle g, b\rangle$. It is important to stress that the duality relation (2.1a) implies that the product in A is defined by the coproduct in $U_z(g)$, and conversely relation (2.1b) implies that the coproduct in A is given by the product in $U_z(g)$. Since the first-order deformation of the coproduct Δ_U is defined by the Lie bialgebra map δ, the first-order noncommutativity for the quantum group A will be given by (the dual of) δ. Therefore, the Lie bialgebra structure associated to a given quantum algebra provides immediately the first-order information about the noncommutative algebra of quantum group coordinates. Moreover, the Lie bialgebra (g, δ) is in one-to-one correspondence with the PL structure whose quantization provides the full quantum group (A, Δ_A). As a consequence, the classification of all possible Lie bialgebra structures for g constitute the first and most relevant step for the analysis and explicit construction of its quantum algebra deformations.

2.2. Lie bialgebras

Let us summarize all the basic facts about Lie bialgebras that will be needed in the sequel. A Lie bialgebra (g, δ) is Lie algebra g with the structure tensor c_{ij}^k,

$$[X_i, X_j] = c_{ij}^k X_k \tag{2.2}$$

that is endowed with a skewsymmetric cocommutator map $\delta : g \to g \otimes g$ fulfilling the following two conditions:

(i) δ is a 1-cocycle, i.e

$$\delta([X, Y]) = [\delta(X),\, Y \otimes 1 + 1 \otimes Y] + [X \otimes 1 + 1 \otimes X,\, \delta(Y)], \quad \forall X, Y \in g$$

(ii) The dual map $\delta^* : g^* \otimes g^* \to g^*$ is a Lie bracket on g^*.
Therefore any cocommutator δ will be of the form

$$\delta(X_i) = f_i^{jk}\, X_j \wedge X_k \tag{2.3}$$

where f_i^{jk} is the structure tensor of the dual Lie algebra g^* defined by

$$[\hat{\xi}^j, \hat{\xi}^k] = f_i^{jk}\, \hat{\xi}^i \tag{2.4}$$

where $\langle \hat{\xi}^j, X_k \rangle = \delta_k^j$. Therefore, a Lie bialgebra is a pair of 'matched' Lie algebras of the same dimension, since the 1-cocycle condition implies the following compatibility equations

$$f_k^{ab} c_{ij}^k = f_i^{ak} c_{kj}^b + f_i^{kb} c_{kj}^a + f_j^{ak} c_{ik}^b + f_j^{kb} c_{ik}^a \tag{2.5}$$

As it could be expected, for some Lie bialgebras the 1-cocycle δ is a coboundary

$$\delta(X) = [X \otimes 1 + 1 \otimes X, r], \quad \forall X \in g \tag{2.6}$$

where r is a skewsymmetric element of $g \otimes g$ (the classical r-matrix)

$$r = r^{ab}\, X_a \wedge X_b$$

that has to be a solution of the modified classical Yang-Baxter equation (mCYBE)

$$[X \otimes 1 \otimes 1 + 1 \otimes X \otimes 1 + 1 \otimes 1 \otimes X, [[r,r]]\,] = 0, \quad \forall X \in g \tag{2.7}$$

in which the Schouten bracket $[[r,r]]$ is defined as

$$[[r,r]] := [r_{12}, r_{13}] + [r_{12}, r_{23}] + [r_{13}, r_{23}] \tag{2.8}$$

where

$$r_{12} = r^{ab} X_a \otimes X_b \otimes 1, \quad r_{13} = r^{ab} X_a \otimes 1 \otimes X_b, \quad r_{23} = r^{ab} 1 \otimes X_a \otimes X_b$$

Recall that $[[r,r]] = 0$ is just the classical Yang-Baxter equation (CYBE).

For semisimple Lie algebras all Lie bialgebra structures are coboundaries, and that is also the case for the Poincaré algebra in (2+1) and (3+1) dimensions. On the contrary, for solvable and nilpotent Lie algebras many of their Lie bialgebra structures are non-coboundaries (see [31, 32] and references therein).

2.3. *Classification of Lie bialgebra structures on g*

With the previous definitions in mind, the algorithm for the explicit computation and classification of all the possible Lie bialgebra structures for a given Lie algebra g (2.2) in the basis X_i can be sketched as follows:

- The most generic (pre)-cocommutator map is given in the form (2.3).
- The 1-cocycle condition (2.5) is imposed onto f_i^{jk}.
- Further, the (quadratic) condition that the dual map $\delta^* : g^* \otimes g^* \to g^*$ is a Lie bracket on g^* is imposed. Note that this constraint is nothing but the Jacobi identity for f_i^{jk}.
- Finally, the solutions for f_i^{jk} can be classified into equivalence classes under automorphisms of the Lie algebra g.

Note that since Lie bialgebra structures for (2+1) (A)dS and Poincaré algebras are known to be always coboundaries, this procedure is completely equivalent to the classification of constant classical skewsymmetric r-matrices on these algebras.

Journal of Physics: Conference Series **532** (2014) 012002 doi:10.1088/1742-6596/532/1/012002

3. The κ-deformation of the AdS$_\omega$ algebra

In this section we exemplify the theory of quantum deformations by recalling the main results concerning the so-called κ-deformation of the (2+1) (anti-)de Sitter and Poincaré algebras. We will use a unified approach to the three Lie algebras by making use of the six-dimensional Lie algebra AdS$_\omega$ which is given in terms of the generators $\{J, P_0, P_i, K_i\}$ ($i = 1, 2$) (rotation, time translation, space translations and boosts) as

$$
\begin{array}{lll}
[J, P_i] = \epsilon_{ij} P_j, & [J, K_i] = \epsilon_{ij} K_j, & [J, P_0] = 0 \\
[P_i, K_j] = -\delta_{ij} P_0, & [P_0, K_i] = -P_i, & [K_1, K_2] = -J \\
[P_0, P_i] = \omega K_i, & [P_1, P_2] = -\omega J &
\end{array}
$$

where $i, j = 1, 2$, $\epsilon_{12} = 1$ and the parameter ω is just the constant sectional curvature of the corresponding classical spacetime (which is related to the cosmological constant through $\omega = -\Lambda$). Therefore, these Lie brackets encompass:

- The AdS Lie algebra, $so(2, 2)$, when $\omega = +1/R^2 > 0$.
- The dS Lie algebra, $so(3, 1)$, when $\omega = -1/R^2 < 0$.
- The Poincaré Lie algebra, $iso(2, 1)$, when $\omega = 0$. Note that this case corresponds to the flat contraction obtained via the limit of the universe radius $R \to \infty$ connecting $so(2, 2) \to iso(2, 1) \leftarrow so(3, 1)$.

The two Casimir invariants of the AdS$_\omega$ algebra are given by

$$
\mathcal{C} = P_0^2 - \mathbf{P}^2 + \omega(J^2 - \mathbf{K}^2), \quad \mathcal{W} = -JP_0 + K_1P_2 - K_2P_1
$$

where $\mathcal{C}$ comes from the Killing-Cartan form and it is related to the energy of the particle, while $\mathcal{W}$ is the Pauli-Lubanski vector.

3.1. The Lie bialgebra of the κ-deformation

As we mentioned before, all Lie bialgebra structures for the (A)dS and Poincaré Lie algebras in (2+1) dimensions are coboundaries. This means that the Lie bialgebra is fully specified by a certain skewsymmetric solution of the mCYBE (or of the CYBE if the Schouten bracket vanishes). In the case of the κ-deformation such a classical r-matrix reads

$$
r = z(K_1 \wedge P_1 + K_2 \wedge P_2) \tag{3.1}
$$

where the quantum deformation parameter z is related to the usual Planck energy scale κ through $z = 1/\kappa$. Once r is fixed, the cocommutator δ providing the first-order deformation of the coproduct is given by the coboundary relation (2.6), and turns out to be

$$
\delta(P_0) = 0,\ \delta(J) = 0,\ \delta(P_i) = z(P_i \wedge P_0 - \omega \epsilon_{ij} K_j \wedge J),\ \delta(K_i) = z(K_i \wedge P_0 + \epsilon_{ij} P_j \wedge J) \tag{3.2}
$$

A 'quantum group local coordinate' $\hat{y}^i$ will be defined as dual to the generator Y_i through the Hopf algebra pairing $\langle \hat{y}^i, Y_j \rangle = \delta^i_j$. In particular, we will denote $\{\hat{\theta}, \hat{x}_\mu, \hat{\xi}_i\}$ the noncommutative coordinates which are dual to the generators $\{J, P_\mu, K_i\}$ ($\mu = 0, 1, 2$), respectively. Now, if we make use of the dual Lie algebra (2.4) induced by the first-order deformation (in z) of the coproduct given by δ (3.2), we get the first-order quantum group relations for the κ-deformation of the AdS$_\omega$ algebra, namely:

$$
\begin{array}{llll}
[\hat{x}_0, \hat{x}_i] = -z\hat{x}_i, & [\hat{\theta}, \hat{x}_i] = z\epsilon_{ij}\hat{\xi}_j, & [\hat{x}_1, \hat{x}_2] = 0, & [\hat{\theta}, \hat{x}_0] = 0 \\
[\hat{\theta}, \hat{\xi}_i] = -z\omega\epsilon_{ij}\hat{x}_j, & [\hat{x}_0, \hat{\xi}_i] = -z\hat{\xi}_i, & [\hat{\xi}_1, \hat{\xi}_2] = 0, & [\hat{x}_i, \hat{\xi}_j] = 0
\end{array}
$$

We realize that in these relations the 'quantum' time and space translation parameters close a non-Abelian subalgebra

$$[\hat{x}_0, \hat{x}_i] = -z\hat{x}_i, \quad [\hat{x}_1, \hat{x}_2] = 0, \quad i = 1, 2 \tag{3.3}$$

that in the case of the quantum Poincaré group is known as the (2+1) κ-Minkowski noncommutative spacetime $\mathbf{M}_z^{2+1}$ [9, 22, 23, 24], since no higher order corrections have to be incorporated when the full quantum Poincaré group is constructed. Note that commutativity is always recovered in the limit $z \to 0$, i.e , when the quantum deformation vanishes. It is also worth stressing that the first-order relations (3.3) do not depend on the curvature ω, so the three first-order (A)dS and Minkowskian noncommutative spacetimes coincide. As we shall see later, higher order corrections depending on ω will appear when the full quantum (A)dS groups are considered.

3.2. The quantum κ-AdS$_\omega$ algebra

The full (all orders in z) quantum universal enveloping algebra corresponding to the κ-deformation of AdS$_\omega$ was constructed for the first time in [10]. Its quantum coproduct reads (hereafter we will use the ρ parameter defined as $\omega = \rho^2$)

$$\begin{aligned}
\Delta(P_0) &= P_0 \otimes 1 + 1 \otimes P_0 \\
\Delta(J) &= J \otimes 1 + 1 \otimes J \\
\Delta(P_i) &= P_i \otimes \mathrm{e}^{\frac{z}{2}P_0} \cosh\left(\frac{z}{2}\rho J\right) + \mathrm{e}^{-\frac{z}{2}P_0} \cosh\left(\frac{z}{2}\rho J\right) \otimes P_i \\
&\quad - \rho\, \epsilon_{ij} K_j \otimes \mathrm{e}^{\frac{z}{2}P_0} \sinh\left(\frac{z}{2}\rho J\right) + \rho\, \mathrm{e}^{-\frac{z}{2}P_0} \sinh\left(\frac{z}{2}\rho J\right) \otimes \epsilon_{ij} K_j \\
\Delta(K_i) &= K_i \otimes \mathrm{e}^{\frac{z}{2}P_0} \cosh\left(\frac{z}{2}\rho J\right) + \mathrm{e}^{-\frac{z}{2}P_0} \cosh\left(\frac{z}{2}\rho J\right) \otimes K_i \\
&\quad + \epsilon_{ij} P_j \otimes \mathrm{e}^{\frac{z}{2}P_0} \frac{\sinh\left(\frac{z}{2}\rho J\right)}{\rho} - \mathrm{e}^{-\frac{z}{2}P_0} \frac{\sinh\left(\frac{z}{2}\rho J\right)}{\rho} \otimes \epsilon_{ij} P_j
\end{aligned}$$

and the compatible set of deformed commutation rules are

$$\begin{aligned}
&[J, P_i] = \epsilon_{ij} P_j, \quad [J, K_i] = \epsilon_{ij} K_j, \quad [J, P_0] = 0 \\
&[P_i, K_j] = -\delta_{ij} \frac{\sinh(zP_0)}{z} \cosh(z\rho J), \quad [P_0, K_i] = -P_i \\
&[K_1, K_2] = -\cosh(zP_0) \frac{\sinh(z\rho J)}{z\rho}, \quad [P_0, P_i] = \omega K_i \\
&[P_1, P_2] = -\omega \cosh(zP_0) \frac{\sinh(z\rho J)}{z\rho}
\end{aligned}$$

and the deformed Casimir operators have the form

$$\begin{aligned}
\mathcal{C}_z &= 4\cos(z\rho) \left\{ \frac{\sinh^2\left(\frac{z}{2}P_0\right)}{z^2} \cosh^2\left(\tfrac{z}{2}\rho J\right) + \frac{\sinh^2\left(\frac{z}{2}\rho J\right)}{z^2} \cosh^2\left(\tfrac{z}{2}P_0\right) \right\} - \frac{\sin(z\rho)}{z\rho} (\mathbf{P}^2 + \omega \mathbf{K}^2) \\
\mathcal{W}_z &= -\cos(z\rho) \frac{\sinh(z\rho J)}{z\rho} \frac{\sinh(zP_0)}{z} + \frac{\sin(z\rho)}{z\rho} (K_1 P_2 - K_2 P_1)
\end{aligned}$$

It is worth mentioning that this is exactly the quantum (A)dS algebra proposed in [21] as the symmetry algebra of the vacuum excitations in (2+1) quantum gravity. By construction, the limit $\omega \to 0$ of this quantum algebra is always well defined and leads to the well known (2+1) κ-Poincaré algebra. Note also that the physical dimensions of the quantum deformation parameter

z are inherited from P_0, since $[z] = [P_0]^{-1}$. If $c = 1$, this means that z can be interpreted as a fundamental length parameter, which in the usual double special relativity theories is considered to be of the order of the Planck length l_p. Algebraically, this link between z and P_0 is directly related to the fact that P_0 and J remain nondeformed (primitive) at the level of the coproduct, which allows the deformation to contain formal power series of these two generators (see [33] for a dimensional analysis of the deformation parameters from the viewpoint of contraction theory).

3.3. The κ-AdS$_\omega$ Poisson-Lie group

The full κ-AdS$_\omega$ quantum group could be obtained by computing the Hopf algebra dual to the quantum κ-AdS$_\omega$ algebra that we have just introduced. This requires a lengthy and cumbersome computation, but many features of this quantum group can be extracted from its semiclassical limit, i.e , the PL structure on the classical AdS$_\omega$ group that is in one-to-one correspondence with the Lie bialgebra that characterizes this quantum deformation.

The construction of the κ-AdS$_\omega$ PL group was fully performed by Ballesteros, Herranz and Bruno in [arXiv:hep-th/0401244], where we refer to the interested reader for details. In particular, under the appropriate parametrization of the group, it is found that the local coordinate functions corresponding to the translation parameters close the following Poisson subalgebra

$$\{x_0, x_1\} = -z\,\frac{\tanh \rho x_1}{\rho \cosh^2 \rho x_2}, \qquad \{x_0, x_2\} = -z\,\frac{\tanh \rho x_2}{\rho}, \qquad \{x_1, x_2\} = 0$$

whose quantization (in the usual 'quantum-mechanical' sense) would give rise to the full noncommutative κ-AdS$_\omega$ spacetime. Since $\{x_1, x_2\} = 0$, no ordering ambiguities appear in the quantization process, and the quantum (2+1) noncommutative AdS$_\omega$ spacetime reads

$$\begin{aligned}
[\hat{x}_0, \hat{x}_1] &= -z\,\frac{\tanh \rho \hat{x}_1}{\rho \cosh^2 \rho \hat{x}_2} = -\, z\hat{x}_1 + \tfrac{1}{3} z\omega \hat{x}_1^3 + z\omega \hat{x}_1 \hat{x}_2^2 + o(\omega^2) \\
[\hat{x}_0, \hat{x}_2] &= -z\,\frac{\tanh \rho \hat{x}_2}{\rho} = -\, z\hat{x}_2 + \tfrac{1}{3} z\omega \hat{x}_2^3 + o(\omega^2) \\
[\hat{x}_1, \hat{x}_2] &= 0
\end{aligned}$$

Therefore, the κ-Minkowski space $\mathbf{M}_z^{2+1}$ is the first-order noncommutative spacetime for all the AdS$_\omega$ quantum groups. However, when the curvature is not zero ($\omega \neq 0$) higher order contributions appear, and the quantum (A)dS spacetimes turn out to be nonlinear noncommutative algebras. Note also that, in any case, see [arXiv:hep-th/0401244] for the full expressions, the quantum spacetime coordinates $\{\hat{x}_0, \hat{x}_1, \hat{x}_2\}$ close a subalgebra.

4. The generic AdS$_\omega$ deformation

The aim of the present section is to show that the κ-deformation is a very particular one-parametric choice amongst all the possible quantum deformations of the AdS$_\omega$ algebra. In the sequel we will present the chart of Lie bialgebra structures of the AdS$_\omega$ algebra, that can be obtained by taking into account that the Lie bialgebra structures of $so(2, 2)$, $so(3, 1)$ and the (2+1) Poincaré algebra are always coboundary structures. Therefore, they come from classical r-matrices, and the most general form for a constant classical r-matrix on AdS$_\omega$ is

$$\begin{aligned}
r =\,& a_1 J \wedge P_1 + a_2 J \wedge K_1 + a_3 P_0 \wedge P_1 + a_4 P_0 \wedge K_1 + a_5 P_1 \wedge K_1 + a_6 P_1 \wedge K_2 \\
&+ \; b_1 J \wedge P_2 + b_2 J \wedge K_2 + b_3 P_0 \wedge P_2 + b_4 P_0 \wedge K_2 + b_5 P_2 \wedge K_2 + b_6 P_2 \wedge K_1 \\
&+ \; c_1 J \wedge P_0 + c_2 K_1 \wedge K_2 + c_3 P_1 \wedge P_2
\end{aligned}$$

where we have initially 15 possible 'deformation parameters', that will have to fulfil the constraints coming form the mCYBE (2.7). We recall that pure non-standard/twist AdS_ω deformations will be obtained when the Schouten bracket $[[r,r]] = 0$ (i.e , if their r-matrices are solutions of the CYBE).

Now, the mCYBE (2.7) leads to the following nonlinear constraints involving the 15 (real) Lie bialgebra parameters and the curvature ω:

$$
\begin{aligned}
&(a_1a_5 - a_4a_6 + a_6b_1 - a_1b_5 + a_4b_6 + b_1b_6 + a_3c_2 - \omega a_3c_3) = 0\\
&(a_2a_3 - a_4b_4 + b_1b_4 + a_5b_6 + b_5b_6 + a_5c_1 - b_5c_1 - \omega a_3b_3) = 0\\
&(-a_2a_5 - a_6b_2 + a_2b_5 - b_2b_6 + a_4c_2 + \omega a_3a_6 - \omega a_3b_6 - \omega a_4c_3) = 0\\
&(-a_1a_5 + a_4a_6 - a_6b_1 - a_5b_4 + a_4c_1 + b_1c_1 + b_2c_3 + \omega a_3c_3) = 0\\
&(a_2a_4 + a_2b_1 - a_1b_2 - a_5c_2 + \omega a_1a_3 - \omega a_5c_3) = \omega(a_1a_3 + a_4b_3 + b_1b_3 - a_3b_4 - a_5c_3 - b_5c_3)\\
&(a_1a_2 + b_1b_2 + a_2b_4 - a_6c_2 - c_1c_2 + \omega a_3b_1 - \omega a_6c_3 + \omega c_1c_3) = 0\\
&(-a_2a_5 - b_2b_6 + b_2c_1 + a_4c_2 - b_1c_2 + \omega a_5b_3 - \omega a_3b_6 - \omega a_3c_1) = 0\\
&(a_1a_6 - a_5b_1 - a_6b_4 + b_1b_5 + a_1b_6 + b_4b_6 + b_3c_2 - \omega b_3c_3) = 0\\
&(a_1a_4 - a_5a_6 + b_2b_3 + a_4b_4 - a_6b_5 - a_5c_1 + b_5c_1 + \omega a_3b_3) = 0\\
&(a_2a_6 - a_5b_2 + b_2b_5 + a_2b_6 - b_4c_2 - \omega a_6b_3 + \omega b_3b_6 + \omega b_4c_3) = 0\\
&(-a_4b_5 + b_1b_5 + a_1b_6 + b_4b_6 + a_1c_1 - b_4c_1 + a_2c_3 - \omega b_3c_3) = 0\\
&(-a_2b_1 + a_1b_2 - b_2b_4 + b_5c_2 - \omega b_1b_3 + \omega b_5c_3) = -\omega(a_1a_3 + a_4b_3 + b_1b_3 - a_3b_4 - a_5c_3 - b_5c_3)\\
&(a_1a_2 - a_4b_2 + b_1b_2 + b_6c_2 - c_1c_2 - \omega a_1b_3 + \omega b_6c_3 + \omega c_1c_3) = 0\\
&(-a_2a_6 - b_2b_5 - a_2c_1 + a_1c_2 + b_4c_2 + \omega a_6b_3 - \omega a_3b_5 - \omega b_3c_1) = 0\\
&(a_2a_4 + b_2b_4 - a_5c_2 - b_5c_2 + \omega a_4b_3 - \omega a_3b_4) = \omega(a_1a_3 + a_4b_3 + b_1b_3 - a_3b_4 - a_5c_3 - b_5c_3)\\
&a_1^2 + b_1^2 - a_3b_2 + a_2b_3 - a_5b_5 + a_6b_6 - a_6c_1 + b_6c_1 - \omega c_3^2\\
&\qquad = (a_4^2 - a_5^2 + a_2b_3 - a_1b_4 - a_6b_6 + a_6c_1 + b_6c_1 - c_2c_3 + \omega a_3^2)\\
&(-a_2^2 - b_2^2 + c_2^2 - \omega a_4b_1 + \omega a_1b_4 + \omega a_5b_5 - \omega a_6b_6 - \omega a_6c_1 + \omega b_6c_1)\\
&\qquad = -\omega(a_4^2 - a_5^2 + a_2b_3 - a_1b_4 - a_6b_6 + a_6c_1 + b_6c_1 - c_2c_3 + \omega a_3^2)\\
&(a_4b_1 - a_3b_2 + b_4^2 - b_5^2 - a_6b_6 - a_6c_1 - b_6c_1 - c_2c_3 + \omega b_3^2)\\
&\qquad = (a_4^2 - a_5^2 + a_2b_3 - a_1b_4 - a_6b_6 + a_6c_1 + b_6c_1 - c_2c_3 + \omega a_3^2) \qquad (4.1)
\end{aligned}
$$

Therefore, the most generic cocommutator is given by (2.6), namely:

$$
\begin{aligned}
\delta(J) &= a_1 J\wedge P_2 + a_2 J\wedge K_2 + a_3 P_0\wedge P_2 + a_4 P_0\wedge K_2 + (b_5 - a_5)(K_1\wedge P_2 - P_1\wedge K_2)\\
&\quad + (a_6 + b_6)(K_1\wedge P_1 + P_2\wedge K_2) + b_1 P_1\wedge J + b_2 K_1\wedge J + b_3 P_1\wedge P_0 + b_4 K_1\wedge P_0\\
\delta(P_0) &= \omega a_1 J\wedge K_1 + a_2 P_1\wedge J + \omega a_3 P_0\wedge K_1 + a_4 P_1\wedge P_0 + (a_6 - b_6)(P_2\wedge P_1 + \omega K_1\wedge K_2)\\
&\quad + \omega b_1 J\wedge K_2 + b_2 P_2\wedge J + \omega b_3 P_0\wedge K_2 + b_4 P_2\wedge P_0 + (\omega c_3 - c_2)(K_1\wedge P_2 + P_1\wedge K_2)\\
\delta(P_1) &= (b_6 + c_1)(P_0\wedge P_2 + \omega K_1\wedge J) + \omega b_5 K_2\wedge J - (a_2 - \omega b_3)(J\wedge P_0 + P_2\wedge K_1) + \omega c_3 J\wedge P_1\\
&\quad + \omega b_4 K_2\wedge K_1 + \omega a_3 P_1\wedge K_1 + c_2 K_2\wedge P_0 + b_2 K_2\wedge P_2 + a_5 P_0\wedge P_1 + a_1 P_1\wedge P_2\\
\delta(P_2) &= \omega a_5 J\wedge K_1 + (a_6 - c_1)(\omega J\wedge K_2 + P_0\wedge P_1) + (b_2 + \omega a_3)(P_0\wedge J + P_1\wedge K_2)\\
&\quad + \omega c_3 J\wedge P_2 + \omega a_4 K_1\wedge K_2 + c_2 P_0\wedge K_1 + a_2 P_1\wedge K_1 + b_5 P_0\wedge P_2 + b_1 P_1\wedge P_2 + \omega b_3 P_2\wedge K_2\\
\delta(K_1) &= c_2 J\wedge K_1 + (a_1 + b_4)(J\wedge P_0 + P_1\wedge K_2) + (a_6 + c_1)(J\wedge P_1 + P_0\wedge K_2) + b_5 J\wedge P_2\\
&\quad + a_2 K_1\wedge K_2 + a_5 P_0\wedge K_1 + c_3 P_0\wedge P_2 + a_4 P_1\wedge K_1 + b_3 P_1\wedge P_2 + b_1 P_2\wedge K_2\\
\delta(K_2) &= c_2 J\wedge K_2 + b_2 K_1\wedge K_2 + a_1 K_1\wedge P_1 + (a_4 - b_1)(P_0\wedge J + P_2\wedge K_1) + b_5 P_0\wedge K_2\\
&\quad + (b_6 - c_1)(P_0\wedge K_1 + P_2\wedge J) + a_5 P_1\wedge J + c_3 P_1\wedge P_0 + b_4 P_2\wedge K_2 + a_3 P_2\wedge P_1
\end{aligned}
$$

where all the parameters appearing in these expressions have to satisfy the constraints (4.1). Obviously, some of the parameters could be shown to be inessential through the appropriate automorphisms, and physical restrictions on the parameters should be also imposed in order to obtain some tractable deformations. Anyhow, these expressions give a clear idea about the size of the zoo of possible quantum deformations of the AdS$_\omega$ algebra. Note also that the dual of δ would provide the most generic first-order AdS$_\omega$ noncommutative spacetime.

5. Beyond κ-deformation in (2+1) dimensions

The generalizations of the κ-deformation that are contained in the previous characterization of all the quantum AdS$_\omega$ deformations can be identified by imposing that, like in (3.2),

$$\delta(J) = 0, \quad \delta(P_0) = 0$$

These conditions imply that the deformation parameters have to fulfil the relations

$$a_i = b_i = 0 \quad (i = 1, 2, 3, 4), \quad a_5 = b_5, \quad a_6 = b_6 = 0, \quad c_2 = \omega c_3$$

Under these constraints the resultant r-matrix only depends on *three* parameters

$$r = b_5(P_1 \wedge K_1 + P_2 \wedge K_2) + c_1 J \wedge P_0 + c_3(\omega K_1 \wedge K_2 + P_1 \wedge P_2) \tag{5.1}$$

where the κ-AdS$_\omega$ Planck length parameter is $b_5 = -z$, see (3.1). These three parameters are free ones, since (5.1) is already a solution of the mCYBE. Moreover, the Schouten bracket for the r-matrix (5.1) reads

$$\begin{aligned}[[r, r]] = &-2b_5c_3(P_1 \wedge P_2 \wedge P_0 + \omega P_0 \wedge K_1 \wedge K_2 + \omega P_1 \wedge K_1 \wedge J + \omega P_2 \wedge K_2 \wedge J) \\ &- (b_5^2 + \omega c_3^2)(P_1 \wedge K_1 \wedge P_0 + P_1 \wedge P_2 \wedge J + \omega K_1 \wedge K_2 \wedge J + P_0 \wedge P_2 \wedge K_2)\end{aligned}$$

and this implies that if $\omega \neq 0$, non-standard (twist) deformations are obtained when $b_5 = c_3 = 0$ (i.e , the term $J \wedge P_0$ generates a Reshetikhin twist). In the Poincaré case ($\omega = 0$), the $P_1 \wedge P_2$ term provides a second twist.

The full cocommutator that corresponds to the three-parametric r-matrix (5.1) is

$$\begin{aligned}
\delta(J) &= \delta(P_0) = 0 \\
\delta(P_1) &= c_1(P_0 \wedge P_2 + \omega K_1 \wedge J) - z(\omega K_2 \wedge J + P_0 \wedge P_1) + \omega c_3(J \wedge P_1 + K_2 \wedge P_0) \\
\delta(P_2) &= -c_1(\omega J \wedge K_2 + P_0 \wedge P_1) - z(\omega J \wedge K_1 + P_0 \wedge P_2) + \omega c_3(J \wedge P_2 + P_0 \wedge K_1) \\
\delta(K_1) &= c_1(J \wedge P_1 + P_0 \wedge K_2) - z(J \wedge P_2 + P_0 \wedge K_1) + c_3(P_0 \wedge P_2 + \omega J \wedge K_1) \\
\delta(K_2) &= -c_1(P_0 \wedge K_1 + P_2 \wedge J) - z(P_0 \wedge K_2 + P_1 \wedge J) + c_3(P_1 \wedge P_0 + \omega J \wedge K_2)
\end{aligned}$$

which leads to the first-order noncommutative spacetime given by

$$[\hat{x}_0, \hat{x}_1] = -z\hat{x}_1 - c_1\hat{x}_2 - c_3\hat{\xi}_2, \quad [\hat{x}_0, \hat{x}_2] = -z\hat{x}_2 + c_1\hat{x}_1 + c_3\hat{\xi}_1, \quad [\hat{x}_1, \hat{x}_2] = 0$$

thus providing a two-parametric generalization of the κ-Minkowski spacetime (3.3).

To the best of our knowledge, the full quantum AdS$_\omega$ algebra generated by the r-matrix (5.1) is not known yet and constitutes and interesting open problem. As a first step, the Poisson counterpart of the corresponding all-orders noncommutative spacetimes could be obtained by computing the PL brackets on the AdS$_\omega$ group through the Sklyanin bracket given by (5.1). Again, the spacetime brackets $\{x_i, x_j\}$ so obtained will be no longer linear when the curvature ω is turned on.

6. The (3+1)-dimensional case

The same Lie bialgebra approach can be applied to study the quantum deformations of the (3+1) AdS$_\omega$ Lie algebras, with generators $\{J_i, P_0, P_i, K_i\}$, $(i = 1, 2, 3)$ and whose commutation rules are

$$[J_i, J_j] = \varepsilon_{ijk} J_k, \qquad [J_i, P_j] = \varepsilon_{ijk} P_k, \qquad [J_i, K_j] = \varepsilon_{ijk} K_k \tag{6.1a}$$
$$[P_i, P_j] = -\omega \varepsilon_{ijk} J_k, \quad [P_i, K_j] = -\delta_{ij} P_0, \quad [K_i, K_j] = -\varepsilon_{ijk} J_k \tag{6.1b}$$
$$[P_0, P_i] = \omega K_i, \qquad [P_0, K_i] = -P_i, \qquad [P_0, J_i] = 0 \tag{6.1c}$$

where $i, j = 1, 2, 3$, $\varepsilon_{123} = 1$ and the ω parameter is again the constant sectional curvature of the associated spacetime such that $\omega = -\Lambda$. Therefore, when $\omega > 0$ we have the (3+1) AdS Lie algebra $so(3, 2)$, when $\omega < 0$ we have the (3+1) dS Lie algebra $so(4, 1)$, and the flat limit $\omega = 0$ gives rise to the (3+1) Poincaré Lie algebra $iso(3, 1)$.

The description of the quantum deformations of the Lie algebra (6.1b) starts by considering the most generic r-matrix, $r = r^{ij} Y_i \wedge Y_j$ that in this case would depend on 45 deformation parameters r^{ij} (we recall that again in the (3+1) case all AdS$_\omega$ Lie bialgebras are coboundaries). In particular, if we are interested in generalizations of the κ-deformation for the AdS$_\omega$ algebra, the minimal constraint that we have to impose is that one deformation parameter has to play the role of a Planck length or, equivalently, that the coproduct of the quantum P_0 generator has to be primitive ($\Delta(P_0) = P_0 \otimes 1 + 1 \otimes P_0$). As we know, this means that we have to impose

$$\delta(P_0) = [P_0 \otimes 1 + 1 \otimes P_0, r^{ij} Y_i \wedge Y_j] = 0$$

and, as a result, this assumption elliminates 30 of the r^{ij} parameters. If, by following the (2+1) κ-AdS$_\omega$ case, we further impose that J_3 also remains primitive under deformation, we get a five-parametric candidate for the r-matrix

$$r = z_1(K_1 \wedge P_1 + K_2 \wedge P_2) + z_2(P_1 \wedge P_2 + \omega K_1 \wedge K_2) + z_3 P_0 \wedge J_3 + z_4 K_3 \wedge P_3 + z_5 J_1 \wedge J_2$$

although now the mCYBE is not automatically fulfilled and leads to the following relations among the quantum deformation parameters:

$$z_1 z_2 = 0, \quad (z_1 - z_4) z_5 - \omega z_2 z_4 = 0, \quad z_2 z_5 + z_1(z_1 - z_4) + \omega z_2^2 = 0$$
$$z_2 z_5 - z_4(z_1 - z_4) = 0, \quad z_5^2 + \omega\,(z_2 z_5 - z_1 z_4) = 0$$

Finally, the solution of these equations leads to two disjoint families of two-parametric AdS$_\omega$ Lie bialgebras, that are generated by the two following classical r-matrices [34]:

$$r_{z_1,z_3} = z_1\left(K_1 \wedge P_1 + K_2 \wedge P_2 + K_3 \wedge P_3 \pm \sqrt{\omega}\, J_1 \wedge J_2\right) + z_3 P_0 \wedge J_3 \tag{6.2a}$$
$$r_{z_2,z_3} = z_2\left(P_1 \wedge P_2 + \omega K_1 \wedge K_2 - \omega\, J_1 \wedge J_2 \pm \sqrt{\omega} P_3 \wedge K_3\right) + z_3 P_0 \wedge J_3 \tag{6.2b}$$

whose Schouten brackets are $(l = 1, 2)$:

$$[[r_{z_l,z_3}, r_{z_l,z_3}]] = A_l\,(\omega\, J_1 \wedge J_2 \wedge J_3 - 1/2\; \varepsilon_{ijk}\,(\omega J_i \wedge K_j \wedge K_k + J_i \wedge P_j \wedge P_k)) + A_l \sum_{i=1}^{3} K_i \wedge P_i \wedge P_0$$

where $A_1 = z_1^2$ and $A_2 = z_2^2 \omega$.

Some comments on these two r-matrices are pertinent:

- The $r_{z_1,0}$ matrix generates the κ-AdS$_\omega$ deformation in (3+1) dimensions, but the explicit form of its all-orders quantization when $\omega \neq 0$ is still an open problem.

- When $\omega = 0$, the r_{z_1,z_3} matrix generates the twisted κ-deformation of the (3+1) Poincaré algebra whose full expressions were given in [35].
- If we compare the (3+1) r-matrix (6.2a) with the (2+1) r-matrix (5.1), we realize that the c_3 and the b_5 terms of the latter are not compatible with the κ-deformation in (3+1) dimensions, and can only be simultaneously considered in the deformation of the second type generated by (6.2b). The full quantization of the latter is also unknown yet.

The first-order noncommutative AdS$_\omega$ spaces associated to these two families of Lie bialgebras are deduced from the dual cocommutators, and are given by the following commutation rules involving the (3+1) quantum coordinates $\{\hat{\theta}, \hat{x}_\mu, \hat{\xi}_i\}$ dual to $\{J, P_\mu, K_i\}$ ($\mu = 0, 1, 2, 3$)

$$
\begin{aligned}
r_{z_1,z_3} &: [\hat{x}_0, \hat{x}_i] = -z_1\,\hat{x}_i + z_3\,\varepsilon_{ij3}\,\hat{x}_j, \quad [\hat{x}_i, \hat{x}_j] = 0 \\
r_{z_2,z_3} &: [\hat{x}_0, \hat{x}_I] = -z_2\,\varepsilon_{IJ3}\,\hat{\xi}_J + z_3\,\varepsilon_{IJ3}\,\hat{x}_J, \quad [\hat{x}_0, \hat{x}_3] = \pm z_2\,\sqrt{\omega}\,\hat{x}_3 \\
&\quad [\hat{x}_I, \hat{x}_3] = z_2\hat{\theta}_I, \quad [\hat{x}_1, \hat{x}_2] = 0, \quad I, J = 1, 2
\end{aligned}
$$

Note that the first deformation gives rise to a generalization of the κ-Minkowski space, that is recovered when $z_3 = 0$. Again, no dependence on the curvature ω appears in this first-order quantum group, although a nonlinearity ruled by the cosmological constant is expected when the full quantum group is constructed. On the contrary, the quantum spacetimes coming from the second deformation have a completely different shape: the deformation linked to z_2 introduces the quantum rotation and boost coordinates into the spacetime commutation rules, that are no longer a subalgebra and, moreover, the cosmological constant is already manifest at the first-order of these quantum group relations.

Finally, we would like to recall that one of the main features of the (3+1) κ-Poincaré algebra is the fact that the rotation subalgebra remains undeformed. Therefore, the isotropy of the 'quantum space' is preserved, since the κ-Minkowski spacetime has the same type of commutation rules for the three quantum space coordinates. This seems to be a quite reasonable physical condition and, apparently, the generalization of the κ-Poincaré deformation provided by r_{z_1,z_3} would break this 'quantum isotropy' property. However, we would like to stress that if we consider the following automorphism of the AdS$_\omega$ algebra

$$
\begin{aligned}
\widetilde{Y}_1 &= -\frac{2}{\sqrt{6}}\,Y_2 + \frac{1}{\sqrt{3}}\,Y_3, & Y_1 &= \frac{1}{\sqrt{2}}\left(\widetilde{Y}_2 - \widetilde{Y}_3\right) \\
\widetilde{Y}_2 &= \frac{1}{\sqrt{2}}\,Y_1 + \frac{1}{\sqrt{6}}\,Y_2 + \frac{1}{\sqrt{3}}\,Y_3, & Y_2 &= \frac{1}{\sqrt{6}}\left(-2\widetilde{Y}_1 + \widetilde{Y}_2 + \widetilde{Y}_3\right) \\
\widetilde{Y}_3 &= -\frac{1}{\sqrt{2}}\,Y_1 + \frac{1}{\sqrt{6}}\,Y_2 + \frac{1}{\sqrt{3}}\,Y_3, & Y_3 &= \frac{1}{\sqrt{3}}\left(\widetilde{Y}_1 + \widetilde{Y}_2 + \widetilde{Y}_3\right) \\
&\text{for } \; Y_i \in \{J_i, P_i, K_i\}, & \widetilde{P}_0 &= P_0
\end{aligned}
$$

where $\tilde{Y}_i$ denote the transformed generators, and if we take $\tilde{z}_3 = z_3/\sqrt{3}$, we get

$$
r_{z_1,\tilde{z}_3} = z_1\left[\tilde{K}_1\wedge\tilde{P}_1 + \tilde{K}_2\wedge\tilde{P}_2 + \tilde{K}_3\wedge\tilde{P}_3 \pm \frac{\sqrt{\omega}}{\sqrt{3}}\left(\tilde{J}_1\wedge\tilde{J}_2 + \tilde{J}_2\wedge\tilde{J}_3 + \tilde{J}_3\wedge\tilde{J}_1\right)\right] + \tilde{z}_3\tilde{P}_0\wedge\left(\tilde{J}_1 + \tilde{J}_2 + \tilde{J}_3\right)
$$

This means that in this new AdS$_\omega$ basis, the 'broken' isotropy of the r_{z_1,z_3} deformation can be restored, since the role of the three rotation generators is exactly the same. This indicates that some new multiparametric deformations of the (A)dS and Poincaré algebras could be considered as reasonable symmetries on a physical basis.

Summarizing, in this work we have shown that the classification of the quantum deformations of the kinematical symmetries of (2+1) and (3+1) relativistic spacetimes can be approached in a very efficient and computationally tractable way by analyzing the first-order deformation given by the corresponding Lie bialgebra structures. We think that this viewpoint could be useful in the near future in order to explore the role that multiparametric quantum groups with non-vanishing cosmological constant could play in a quantum gravity context.

Acknowledgments
This work was partially supported by the Spanish MINECO under grant MTM2010-18556 (with EU-FEDER support).

References

[1] Kulish P P and Reshetikhin N 1981 *Proc. Steklov Math. Inst.* **101** 101
[2] Drinfel'd V G 1987 Quantum Groups *Proc. Int. Congress of Math. (Berkeley 1986)* ed A V Gleason (Providence, RI: AMS) p 798
[3] Jimbo M 1985 *Lett. Math. Phys.* **10** 63
[4] Takhtajan L A 1990 *Introduction to Quantum Group and Integrable Massive Models of Quantum Field Theory* eds Mo-Lin Ge and Bao-Heng Zao (World Sci.) p 69
[5] Reshetikhin N Y, Takhtadzhyan L A and Faddeev L D 1990 *Leningrad Math. J.* **1** 193
[6] Chari V and Pressley A 1994 *A Guide to Quantum Groups* (Cambridge Univ. Press)
[7] Majid S 1995 *Foundations of Quantum Group Theory* (Cambridge Univ. Press)
[8] Garay L J 1995 *Int. J. Mod. Phys.* A **10** 145
[9] Lukierski J, Nowicki A, Ruegg H and Tolstoy V N 1991 *Phys. Lett.* B **264** 331
[10] Ballesteros A, Herranz F J, del Olmo M A and Santander M 1994 *J. Phys.* A **27** 1283
[11] Ballesteros A, Herranz F J, del Olmo M A and Santander M 1995 *J. Phys.* A **28** 941
[12] Amelino-Camelia G 2002 *Int. J. Mod. Phys.* D **11** 35; 2002 **11** 1643
[13] Magueijo J and Smolin L 2002 *Phys. Rev. Lett.* **88** 190403
[14] Kowalski-Glikman J and Nowak S 2002 *Phys. Lett.* B **539** 126; 2003 *Classical Quantum Gravity* **20** 4799
[15] Freidel L, Kowalski-Glikman J and Smolin L 2004 *Phys. Rev.* D **69** 044001
[16] Amelino-Camelia G 2010 *Symmetry* **2** 230
[17] Podleś P and Woronowicz S L 1990 *Comm. Math. Phys.* **130** 381
[18] Maggiore M 1993 *Phys. Lett.* B **319** 83
[19] Doplicher S, Fredenhagen K and Roberts J E 1995 *Comm. Math. Phys.* **172** 187
[20] Chaichian M, Kulish P P, Nishijima K and Tureanu A 2004 *Phys. Lett.* B **604** 98
[21] Amelino-Camelia G, Smolin L and Starodubtsev A 2004 *Classical Quantum Gravity* **21** 3095
[22] Maslanka P 1993 *J. Phys.* A **26** L1251
[23] Majid S and Ruegg H 1994 *Phys. Lett.* B **334** 348
[24] Zakrzewski S 1994 *J. Phys.* A **27** 2075
[25] Alekseev A Y and Malkin A Z 1995 *Comm. Math.Phys.* **169** 99
[26] Fock V V and Rosly A A 1992 *Preprint* ITEP-72-92; 1999 *Amer. Math. Soc. Transl.* **191** 67
[27] Meusburger C and Schroers B J 2008 *J. Math. Phys.* **49** 083510
[28] Meusburger C and Schroers B J 2009 *Nucl. Phys.* B **806** 462
[29] Ballesteros A, Herranz F J and Meusburger C 2010 *Phys. Lett.* B **687** 375
[30] Ballesteros A, Herranz F J and Meusburger C 2013 *Classical Quantum Gravity* **30** 155012
[31] Gomez X 2000 *J. Math. Phys.* **41** 4939
[32] Ballesteros A, Blasco A and Musso F 2012 *J. Phys.* A **45** 175204
[33] Ballesteros A, Gromov N A, Herranz F J, del Olmo M A and Santander M 1995 *J. Math. Phys.* **36** 5916
[34] Ballesteros A, Bruno N R and Herranz F J 2004 *Czech. J. Phys.* **11** 1321
[35] Daszkiewicz M 2008 *Int. J. Mod. Phys.* A **23** 4387

3-qubit entanglement: A Jordan algebraic perspective

L Borsten

Blackett Laboratory, Imperial College, SW7 2AZ London, U.K.

Abstract. It is by now well known that three qubits can be totally entangled in two physically distinct ways. Here we review work classifying the physically distinct forms of 3-qubit entanglement using the elegant framework of Jordan algebras, Freudenthal-Kantor triple systems and groups of type E_7. In particular, it is shown that the four Freudenthal-Kantor ranks correspond precisely to the four 3-qubit entanglement classes: (1) Totally separable *A*-*B*-*C*, (2) Biseparable *A*-*BC*, *B*-*CA*, *C*-*AB*, (3) Totally entangled *W*, (4) Totally entangled GHZ. The rank 4 GHZ class is regarded as maximally entangled in the sense that it has non-vanishing quartic norm, the defining invariant of the Freudenthal-Kantor triple system. While this framework is specific to three qubits, we show here how the essential features may be naturally generalised to an arbitrary number of qubits.

1. Introduction

This contribution relates a talk given at the *QQQ* conference *3Quantum: Algebra Geometry Information* (Tallinn, July 2012) in the sub-theme on Jordan algebras. However, it also incorporated two further sub-themes, quantum information and string theory, and so in some limited sense reflected the interdisciplinary nature of the *QQQ* conference. Given the length constraints, for the sake of clarity we focus here on interplay between Jordan algebras and quantum entanglement. The interested reader may consult [1, 2] and the references therein for the connection to black holes in string theory.

The central idea is that the Freudenthal-Kantor triple system (F-KTS) based on the smallest spin-factor Jordan algebra naturally captures the classification of the physically distinct forms of 3-qubit entanglement as prescribed by the paradigm of *stochastic local operations and classical communication* (SLOCC) [3]. In particular, it is shown that the four *Freudenthal-Kantor ranks* correspond precisely to the four 3-qubit entanglement classes: (1) Totally separable *A*-*B*-*C*, (2) Biseparable *A*-*BC*, *B*-*CA*, *C*-*AB*, (3) Totally entangled *W*, (4) Totally entangled GHZ (Greenberger-Horne-Zeilinger). This agrees perfectly with the SLOCC classification first obtained in [4]. The rank 4 GHZ class is regarded as maximally entangled in the sense that it has non-vanishing quartic norm, the defining invariant of the Freudenthal-Kantor triple system.

Our hope is give an account of this idea which is accessible to both the quantum information and Jordan algebra communities. Accordingly, we begin in section 2 with an elementary introduction to the essential concepts of entanglement and entanglement classification. In particular, a three player non-local game is used to illustrate that three qubits can be totally entangled in two physically distinguishable ways and how the concept of SLOCC is used to classify different forms of entanglement. Then, in section 3 we briskly review the necessary elements of cubic Jordan algebras and the F-KTS before defining the *3-qubit Jordan algebra* and the corresponding *3-qubit F-KTS*. Having laid the foundations we proceed in section 4 to demonstrate how the 3-qubit F-KTS elegantly characterises the 3-qubit entanglement classes via the F-KTS *ranks*. Finally, in section 5, we illustrate how the key features of the 3-qubit F-KTS may be generalised to an arbitrary number of qubits.

2. How entangled is totally entangled?

The importance of entanglement to quantum computing protocols has precipitated a need to characterise, qualitatively and quantatively, the "amount" of entanglement contained in given state. Indeed, what is actually meant by "amount" is somewhat ambiguous and there are many possible measures present in the literature. See, for example, [5, 6]. One might ask, for instance, when can two totally entangled, but ostensibly different, states facilitate the same set of non-local quantum protocols. Two states equivalent is this sense could reasonably be considered to have the same degree of entanglement.

Already for pure 3-qubit states this question is non-trivial. Three qubits can be totally entangled in two physically distinct ways [4]. These two entanglement forms are represented by the states

$$|W\rangle = 1/\sqrt{3}\ (|001\rangle + |010\rangle + |100\rangle), \quad |GHZ\rangle = 1/\sqrt{2}\ (|000\rangle + |111\rangle) \tag{2.1}$$

To those unfamiliar with entanglement classification, what separates these states is perhaps not entirely obvious; they are both totally entangled and permutation symmetric. Nonetheless, they exhibit distinct non-local/information-theoretic properties. One such difference is nicely drawn out by the three-player non-local game introduced in [7]. This is essentially a rephrasing of Mermin's elegant presentation [8] of a "Bell-type theorem without inequalities", the first of which was found by Greenberger, Horne and Zeilinger (GHZ) [9]. The formulation as a non-local game is, however, especially appealing in that it emphasises not only the remarkable non-local properties of these states, but also how they can be used in an information-theoretic sense. More specifically, in the context of cooperative games of incomplete information [10].

A non-local game [10, 7] consists of *players* (Alice, Bob, Charlie...), who act cooperatively in order to win, and a *referee* who coordinates the game. The players may collectively decide on a strategy before the game commences. Once it has begun they may no longer communicate. Whether or not the players win is determined by the referee. To begin the referee randomly selects one question, from a known fixed set $\mathcal{Q}$, to be sent to each player. The players know only their own questions. Each player must then send back a response from the set of answers $\mathcal{A}$. The referee determines whether the players win using the set of sent questions and received answers according to some predetermined rules. These rules are known to the players before the game gets under way so that they may attempt to devise a winning strategy.

For the three-player game [7] the questions sent to Alice, Bob and Charlie, denoted respectively by r, s and t, are taken uniformly at random from the set $\mathcal{Q} = \{0, 1\}$. However, the referee ensures that $rst \in \{000, 110, 101, 011\}$ and the players are aware of this. The answers a, b, c, sent back by Alice, Bob and Charlie, are elements of $\mathcal{A} = \{0, 1\}$. See Figure 1. The players win if $r \vee s \vee t = a \oplus b \oplus c$, where $\vee$ and $\oplus$ respectively denote disjunction and addition mod 2, i.e for question sets $rst = 000, 011, 101$ and 110 the answer set abc must satisfy $a \oplus b \oplus c = 0, 1, 1$ and 1, respectively.

In the quantum version, Alice, Bob and Charlie each possess a qubit, which they may manipulate locally. The 3-qubit state may be entangled and used as a resource to help the players win. However, before examining how this works let us consider how well the players can do classically, i.e. unassisted by entanglement.

A classical strategy amounts to specifying three functions, a, b, c, one for each player, from the question set $\mathcal{Q}$ to the answer set $\mathcal{A}$. The condition that the players win may then be written

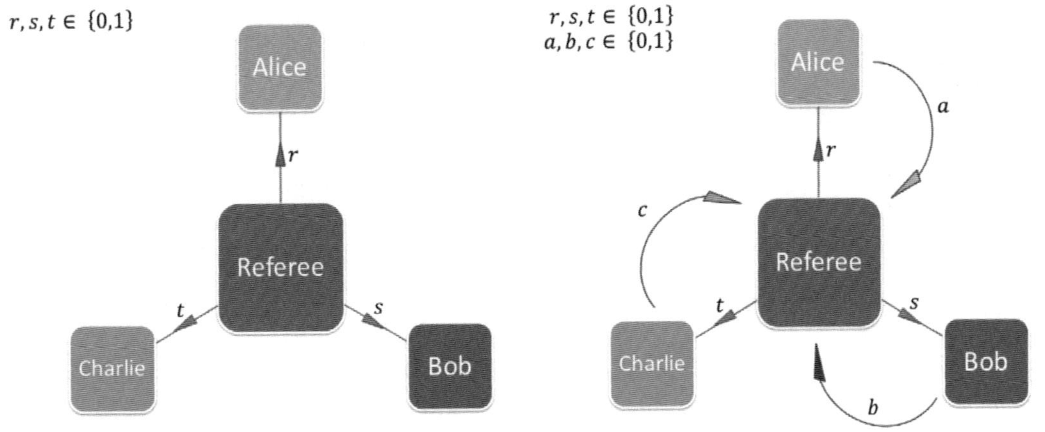

Figure 1. Three player non-local game.

as

$$a(0)\oplus b(0)\oplus c(0)=0,\ a(1)\oplus b(1)\oplus c(0)=1,\ a(1)\oplus b(0)\oplus c(1)=1,\ a(0)\oplus b(1)\oplus c(1)=1 \quad (2.2)$$

This implies that the best one can do is win 75% of the time; the four equations cannot be simultaneously satisfied as can be seen by adding them mod 2, which yields the contradiction $0 = 1$ [7]. On the other hand, the simple strategy that "everyone always answers 1" satisfies three of the four equations so that the 75% upper bound is actually met.

Can this be bettered when equipped with an entangled resource? The answer is a resounding yes: by sharing the GHZ state,

$$|\Psi\rangle = 1/2\ (|000\rangle - |011\rangle - |101\rangle - |110\rangle) \quad (2.3)$$

which is equivalent to (2.1) under a local unitary rotation, they can always win [7].

The winning quantum strategy is remarkably simple. If a player receives the question "0" they measure their qubit in the computational basis $\{|0\rangle, |1\rangle\}$. If a player receives the question "1" they measure their qubit in the Hadamard basis $\{(|0\rangle + |1\rangle)/\sqrt{2}, (|0\rangle - |1\rangle)/\sqrt{2}\}$. Their measurement outcomes are sent back as their answers. By symmetry we need only consider the two cases $rst = 000$ and $rst = 011$. (1) $rst = 000$: All players measure in the computational basis. From (2.3) it is clear that only an odd number of 0's can appear $\Rightarrow a \oplus b \oplus c = 0$. Always win. (2) $rst = 011$: Alice measures in the computational basis, while Bob and Charlie measure in the Hadamard basis. Consulting the locally rotated state,

$$\mathbb{1} \otimes H \otimes H|\Psi\rangle = 1/2\ (|001\rangle + |010\rangle - |100\rangle + |111\rangle)$$

where H is the unitary operator relating the computational basis to the Hadamard basis, it is clear that only an even number of 0's can appear $\Rightarrow a \oplus b \oplus c = 1$. Always win. Hence, using the GHZ entangled resource (2.3), Alice, Bob and Charlie can win 100% of the time, outdoing the best classical strategy by 25%.

One might naively expect to be able to devise a winning strategy when Alice, Bob and Charlie are equipped with a shared W-state, given it is also totally entangled. However, the W-state cannot be used win with certainty even when allowing Alice, Bob and Charlie to each measure in any pair of bases they wish [11]. Hence, despite both being totally entangled, the GHZ-state is "more" entangled than the W-state, in this context at least. Certainly, they constitute physically inequivalent forms of entanglement.

We must therefore conclude that it is not enough to simply say a state is totally entangled -one must also specify in what way it is totally entangled. This is achieved using the paradigm of *local operations and classical communication* (LOCC). See, for example, [5]. Roughly, given a composite quantum system we allow purely *local* quantum operations to be performed on the individual components. These local operations may be supplemented by classical communication: the separated experimenters may communicate via a classical channel, e-mail for example. Any number of LOCC rounds may be performed, which makes the calss of allowed operations difficult to characterise. For an in-depth account of LOCC protocols, see [6]. In this manner arbitrary classical correlations between the constituent subsystems may be generated. However, no quantum correlations may be established -all information exchanged was classical. LOCC protocols cannot increase the amount of entanglement.

This motivates the concept of *Stocastic* LOCC equivalence, introduced in [12, 4]: two states lie in the same SLOCC-equivalence class if and only if they may be transformed into one another with some *non-zero probability* using LOCC operations. The crucial observation is that since LOCC cannot create entanglement any two SLOCC-equivalent states must necessarily possess the same entanglement, irrespective of the particular measure used. It is this property which makes the SLOCC paradigm so suited to the task of classifying entanglement.

We restrict our attention to pure states. For an n-qubit system, two states

$$|\Psi\rangle = a_{A_1\dots A_n}|A_1\dots A_n\rangle, \quad |\Phi\rangle = b_{B_1\dots B_n}|B_1\dots B_n\rangle \tag{2.4}$$

are SLOCC-equivalent if and only if they are related by an element of $\mathrm{SL}_1(2,\mathbb{C})\times\mathrm{SL}_2(2,\mathbb{C})\times\dots\mathrm{SL}_n(2,\mathbb{C})$ [4], which will be referred to as the *SLOCC-equivalence group*. The Hilbert space is partitioned into equivalence classes or orbits under the SLOCC-equivalence group. For the n-qubit system the space of SLOCC-equivalence classes is given by

$$\mathbb{C}^2\otimes\mathbb{C}^2\dots\otimes\mathbb{C}^2/\mathrm{SL}_1(2,\mathbb{C})\times\mathrm{SL}_2(2,\mathbb{C})\times\dots\times\mathrm{SL}_n(2,\mathbb{C}) \tag{2.5}$$

This is the space of physically distinct form of entanglement. Hence, the SLOCC entanglement classification amounts to understanding (2.5).

3. The 3-qubit Jordan algebra and Freudenthal-Kantor triple system

3.1. Cubic Jordan algebras

A Jordan algebra $\mathfrak{J}$ is vector space defined over a ground field $\mathbb{F}$ equipped with a bilinear product satisfying

$$X\circ Y = Y\circ X, \quad X^2\circ(X\circ Y) = X\circ(X^2\circ Y), \quad \forall\, X,Y\in\mathfrak{J} \tag{3.1}$$

The class of *cubic* Jordan algebras are constructed as follows [13]. Let V be a vector space equipped with a cubic norm, which is. a homogeneous map of degree three, $N: V\to\mathbb{F}$ s.t.

$$N(\lambda X) = \lambda^3 N(X), \quad \forall\lambda\in\mathbb{F}, \quad X\in V \tag{3.2}$$

such that

$$N(X,Y,Z) := 1/6\,[N(X+Y+Z)-N(X+Y)-N(X+Z)-N(Y+Z)+N(X)+N(Y)+N(Z)] \tag{3.3}$$

is trilinear. If V further contains a base point $N(c) = 1, c \in V$ one may define the following three maps

$$\begin{aligned} \mathrm{Tr}: V \to \mathbb{F}, &\quad X \mapsto 3N(c,c,X) \\ S: V \times V \to \mathbb{F}, &\quad (X,Y) \mapsto 6N(X,Y,c) \\ \mathrm{Tr}: V \times V \to \mathbb{F}, &\quad (X,Y) \mapsto \mathrm{Tr}(X)\,\mathrm{Tr}(Y) - S(X,Y) \end{aligned} \tag{3.4}$$

A cubic Jordan algebra $\mathfrak{J}$, with multiplicative identity $\mathbb{1} = c$, may be derived from any such vector space if N is *Jordan cubic*. That is: (i) the trace bilinear form (3.4) is non-degenerate (ii) the quadratic adjoint map, $\sharp\colon \mathfrak{J} \to \mathfrak{J}$, uniquely defined by $\mathrm{Tr}(X^\sharp, Y) = 3N(X,X,Y)$, satisfies $(X^\sharp)^\sharp = N(X)X$ for all $X \in \mathfrak{J}$. The Jordan product is then given by

$$X \circ Y = 1/2\left[X \times Y + \mathrm{Tr}(X)Y + \mathrm{Tr}(Y)X - S(X,Y)\mathbb{1}\right] \tag{3.5}$$

where $X \times Y := (X+Y)^\sharp - X^\sharp - Y^\sharp$.

Definition 3.1 (structure group $\mathrm{Str}(\mathfrak{J})$)**.** Invertible $\mathbb{F}$-linear transformations σ preserving the cubic norm up to a fixed scalar factor,

$$\mathrm{Str}(\mathfrak{J}) := \{\sigma \in \mathrm{Iso}_{\mathbb{F}}(\mathfrak{J}) | N(\sigma A) = \lambda N(A), \lambda \in \mathbb{F}\} \tag{3.6}$$

The reduced structure group $\mathrm{Str}_0(\mathfrak{J}) \subset \mathrm{Str}(\mathfrak{J})$ preserves the norm exactly.

The exceptional Jordan algebra of 3×3 Hermitian octonionic matrices, denoted $\mathfrak{J}_3^{\mathbb{O}}$, is perhaps the most important and well known example. The reduced structure group $\mathrm{Str}_0(\mathfrak{J}_3^{\mathbb{O}})$ in this case is given by the 78-dimensional exceptional Lie group $E_{6(-26)}$.

Any cubic Jordan algebra element may be assigned a $\mathrm{Str}(\mathfrak{J})$-invariant Rank [14]. The ranks partition the V.

Definition 3.2 (cubic Jordan algebra rank)**.** A non-zero element $A \in \mathfrak{J}$ has a rank given by

$$\mathrm{Rank}\, A = 1 \Leftrightarrow A^\sharp = 0,\ \mathrm{Rank}\, A = 2 \Leftrightarrow N(A) = 0,\ A^\sharp \neq 0,\ \mathrm{Rank}\, A = 3 \Leftrightarrow N(A) \neq 0 \tag{3.7}$$

3.2. The Freudenthal-Kantor triple system

In 1954, Freudenthal [15] found that the 133-dimensional exceptional Lie group E_7 could be understood in terms of the automorphisms of a construction based on the 56-dimensional E_7-module built from the exceptional Jordan algebra of 3×3 Hermitian octonionic matrices. A key feature of this construction is a triple product. Later, the Freudenthal construction was essentially extended and elaborated, in particular, by Kantor [16], and then also by Yamaguti, Asano, Ono [17, 18] and Kamiya [19], hence the term *Freudenthal-Kantor triple system* (F-KTS) [18] is in use.

Given a cubic Jordan algebra $\mathfrak{J}$ defined over a field $\mathbb{F}$, one is able to construct an F-KTS by defining the vector space $\mathfrak{F}(\mathfrak{J})$, $\mathfrak{F}(\mathfrak{J}) = \mathbb{F} \oplus \mathbb{F} \oplus \mathfrak{J} \oplus \mathfrak{J}$.An arbitrary element $x \in \mathfrak{F}(\mathfrak{J})$ is conventionally written as a "2×2 matrix" $x = \left(\begin{smallmatrix} \alpha & A \\ B & \beta \end{smallmatrix}\right)$, where $\alpha, \beta \in \mathbb{F} A, B \in \mathfrak{J}$,but for notational convenience we will often also write $x = (\alpha, \beta, A, B)$.

The F-KTS comes equipped with a non-degenerate bilinear antisymmetric quadratic form, a quartic form and a trilinear triple product [15, 20]:

(i) Quadratic form $\{x, y\}$: $\mathfrak{F} \times \mathfrak{F} \to \mathbb{F}$

$$\{x, y\} = \alpha\delta - \beta\gamma + \mathrm{Tr}(A, D) - \mathrm{Tr}(B, C) \tag{3.8a}$$

where $x = (\alpha, \beta, A, B)$ and $y = (\gamma, \delta, C, D)$.

Journal of Physics: Conference Series **532** (2014) 012003 doi:10.1088/1742-6596/532/1/012003

(ii) Quartic form $q : \mathfrak{F} \to \mathbb{F}$

$$q(x) = -2[\alpha\beta - \mathrm{Tr}(A, B)]^2 - 8[\alpha N(A) + \beta N(B) - \mathrm{Tr}(A^\sharp, B^\sharp)] \tag{3.8b}$$

(iii) Triple product $T : \mathfrak{F} \times \mathfrak{F} \times \mathfrak{F} \to \mathfrak{F}$ which is uniquely defined by

$$\{T(x, y, w), z\} = q(x, y, w, z) \tag{3.8c}$$

where $q(x, y, w, z)$ is the full linearisation of $q(x)$ such that $q(x, x, x, x) = q(x)$.

Definition 3.3 (the automorphism group $\mathrm{Aut}(\mathbb{F} - KTS)$ [20]). Invertible $\mathbb{F}$-linear transformations preserving the quartic and quadratic forms:

$$\mathrm{Aut}(\mathbb{F} - KTS) := \{\sigma \in \mathrm{Iso}_{\mathbb{F}}(\mathbb{F} - KTS) | \{\sigma x, \sigma y\} = \{x, y\},\ q(\sigma x) = q(x)\} \tag{3.9}$$

Lemma 3.4 (Brown [20]). *The following transformations generate elements of* $\mathrm{Aut}(\mathbb{F} - KTS)$*:*

$$\begin{aligned}
\varphi(C) &: \begin{pmatrix} \alpha & A \\ B & \beta \end{pmatrix} \mapsto \begin{pmatrix} \alpha + \mathrm{Tr}(B, C) + \mathrm{Tr}(A, C^\sharp) + \beta N(C) & A + \beta C \\ B + A \times C + \beta C^\sharp & \beta \end{pmatrix} \\
\psi(D) &: \begin{pmatrix} \alpha & A \\ B & \beta \end{pmatrix} \mapsto \begin{pmatrix} \alpha & A + B \times D + \alpha D^\sharp \\ B + \alpha D & \beta + \mathrm{Tr}(A, D) + \mathrm{Tr}(B, D^\sharp) + \alpha N(D) \end{pmatrix} \\
\widehat{\tau} &: \begin{pmatrix} \alpha & A \\ B & \beta \end{pmatrix} \mapsto \begin{pmatrix} \lambda\alpha & \tau A \\ {}^t\tau^{-1} B & \lambda^{-1}\beta \end{pmatrix}
\end{aligned} \tag{3.10}$$

where $C, D \in \mathfrak{J}$ *and* $\tau \in \mathrm{Str}(\mathfrak{J})$ *s.t.* $N(\tau A) = \lambda N(A)$. *For convenience we also define* $\mathcal{Z} = \phi(-\mathbb{1})\psi(\mathbb{1})\phi(-\mathbb{1})$,

$$\mathcal{Z} : (\alpha, \beta, A, B) \mapsto (-\beta, \alpha, -B, A) \tag{3.11}$$

The archetypal example is given by setting $\mathfrak{J} = \mathfrak{J}_3^{\mathbb{O}}$, in which case elements of $\mathbb{F} - KTS(\mathfrak{J}_3^{\mathbb{O}})$ transform as the **56** of E_7. Decomposing under reduced structure group $\mathrm{Str}_0(\mathfrak{J}_3^{\mathbb{O}}) = E_6$ gives the branching $\mathbf{56} \to \mathbf{1} + \mathbf{1} + \mathbf{27} + \mathbf{27}'$ (neglecting the SO(2) weights), where α, β comprise the singlets and A, B transform as the $\mathbf{27}, \mathbf{27}'$.

The conventional concept of matrix rank may be generalised to Freudenthal-Kantor triple systems in a natural and $\mathrm{Aut}(\mathfrak{F})$-invariant manner.

Definition 3.5 (The F-KTS Rank [21, 22]). The rank of a non-zero element $x \in \mathfrak{F}$ is defined by:

$$\begin{aligned}
\mathrm{Rank}\ x = 1 &\quad \Leftrightarrow \quad \Upsilon_x(y) = 0\ \forall y \\
\mathrm{Rank}\ x = 2 &\quad \Leftrightarrow \quad \exists y \text{ s.t. } \Upsilon_x(y) \neq 0,\ T(x, x, x) = 0 \\
\mathrm{Rank}\ x = 3 &\quad \Leftrightarrow \quad T(x, x, x) \neq 0,\ q(x) = 0 \\
\mathrm{Rank}\ x = 4 &\quad \Leftrightarrow \quad q(x) \neq 0
\end{aligned} \tag{3.12}$$

where $\Upsilon_x(y) := 3T(x, x, y) + x\{x, y\}x$.

3.3. The 3-qubit Freudenthal-Kantor triple system

Definition 3.6 (3-qubit cubic Jordan algebra). We define the 3-qubit cubic Jordan algebra, denoted as $\mathfrak{J}_{ABC}$, as the complex vector space $\mathbb{C} \oplus \mathbb{C} \oplus \mathbb{C}$ with elements $A = (A_1, A_2, A_3)$ and cubic norm

$$N(A) = A_1 A_2 A_3 \tag{3.13}$$

Using the cubic Jordan algebra construction (3.4), one finds

$$\mathrm{Tr}(A, B) = A_1B_1 + A_2B_2 + A_3B_3 \tag{3.14}$$

so that, using $\mathrm{Tr}(A^\sharp, B) = 3N(A, A, B)$, the quadratic adjoint is given by

$$A^\sharp = (A_2A_3, A_1A_3, A_1A_2) \tag{3.15}$$

and therefore

$$(A^\sharp)^\sharp = (A_1A_2A_3A_1, A_1A_2A_3A_2, A_1A_2A_3A_3) = N(A)A \tag{3.16}$$

It is not hard to check $\mathrm{Tr}(A, B)$ is non-degenerate and so N is Jordan cubic. In fact, it is the smallest degree three spin-factor Jordan algebra, see for example [13, 23]. The Jordan product is given by $A \circ B = (A_1B_1, A_2B_2, A_3B_3)$.The structure and reduced structure groups are given by $[\mathrm{SO}(2, \mathbb{C})]^3$ and $[\mathrm{SO}(2, \mathbb{C})]^2$ respectively.

Definition 3.7 (3-qubit Freudenthal-Kantor triple system)**.** We define the 3-qubit Freudenthal-Kantor triple system, denoted $\mathbb{F} - KTS_{ABC}$, as the complex vector space

$$\mathbb{F} - KTS_{ABC} := \mathbb{C} \oplus \mathbb{C} \oplus \mathfrak{J}_{ABC} \oplus \mathfrak{J}_{ABC} \tag{3.17}$$

We identify the eight components of $\mathbb{F} - KTS_{ABC}$ with the eight three qubit basis vectors $|ABC\rangle$ so that, for $|\psi\rangle = a_{ABC}|ABC\rangle$,

$$\begin{pmatrix} \alpha & (A_1, A_2, A_3) \\ (B_1, B_2, B_3) & \beta \end{pmatrix} \leftrightarrow \Psi := \begin{pmatrix} a_{111} & (a_{001}, a_{010}, a_{100}) \\ (a_{110}, a_{101}, a_{011}) & a_{000} \end{pmatrix} \tag{3.18}$$

Crucially, the automorphism group of $\mathbb{F} - KTS_{ABC}$, as defined in (3.9), is $\mathrm{SL}(2, \mathbb{C}) \times \mathrm{SL}(2, \mathbb{C}) \times \mathrm{SL}(2, \mathbb{C}) \rtimes S_3$, where S_3 is the three-qubit permutation group. On including the permutation group the three biseparable entanglement classes, A-BC, B-CA and C-AB, are identified.

4. F-KTS entanglement classification

The ranks of the F-KTS are in fact the entanglement classes: all states of a given rank $r = 1, 2, 3$ are SLOCC-equivalent. Rank four states constitute a $\dim_{\mathbb{C}} = 1$ family of equivalent states parametrised by $q(\Psi)$. More specifically, we have: (Rank 1) Totally separable states A-B-C, (Rank 2) Biseparable states A-BC, (Rank 3) Totally entangled W states, (Rank 4) Totally entangled GHZ states. The rank 4 GHZ class is regarded as maximally entangled in the sense that it has non-vanishing quartic norm. See Table 1. To prove this statement we use the following result, which is an extension of lemma 24 in [22].

Table 1. The entanglement classification of three qubits as according to the F-KTS rank system.

Class	Rank	Representative state	F-KTS rank condition	
			vanishing	non-vanishing
Null	0	0	Ψ	–
A-B-C	1	$\lvert 111\rangle$	$\Upsilon_\Psi(\Phi)$	Ψ
A-BC	2	$\lvert 111\rangle + \lvert 100\rangle$	$T(\Psi, \Psi, \Psi)$	$\Upsilon_\Psi(\Phi)$
W	3	$\lvert 111\rangle + \lvert 001\rangle + \lvert 010\rangle$	$q(\Psi)$	$T(\Psi, \Psi, \Psi)$
GHZ	4	$\lvert 111\rangle + \lvert 001\rangle + \lvert 010\rangle + k\lvert 100\rangle$	–	$q(\Psi)$

Lemma 4.1. *Every state* Ψ *is SLOCC-equivalent to the reduced canonical form:*

$$\Psi_{\text{red}} = (1, 0, A, 0) \longleftrightarrow |111\rangle + a_{001}|001\rangle + a_{010}|010\rangle + a_{100}|100\rangle \tag{4.1}$$

where $\operatorname{Det} a = 4a_{011}a_{101}a_{110}$.

Proof. We start with a generic state (α, β, A, B). We may always assume B is non-zero. (If $B = 0$ and $A \neq 0$ use $\mathcal{Z}$. If $B = 0, A = 0$ then we may assume that α non-zero, using $\mathcal{Z}$ if necessary, and apply $\psi(D)$ to get a non-zero B-componant, as required.) Apply $\phi(C)$ with $C^\sharp = 0$ s.t. and $\alpha \mapsto \alpha + \operatorname{Tr}(C, B) = 1$, which is always possible since $B \neq 0$, the trace form is non-degenerate and $\mathfrak{J}_{ABC}$ is spanned by its rank 1 elements. We are left with a new state $(1, \beta', A', B')$. Now apply $\psi(D)$ with $D = -B'$, so that $(1, \beta', A', B') \mapsto (1, \beta'', A'', 0)$.

Hence we may assume from the outset that our state is in the reduced form $(1, \beta, A, 0)$. We now show that we may also set $\beta = 0$. Since the Jordan ranks (3.7) partition $\mathfrak{J}_{ABC}$, there are four subcases to consider: (i) $A = 0$, (ii) $A \neq 0, A^\sharp = 0$, (iii) $A^\sharp \neq 0, N(A) = 0$ and (iv) $N(A) \neq 0$.

(i) Apply $\phi(C)$ with $C \neq 0, C^\sharp = 0$: $(1, \beta, 0, 0) \mapsto (1, \beta, \beta C, 0)$ so that we are now in case (ii).

(ii) We may assume with out loss of generality, $A = (a, 0, 0)$, $a \neq 0$. Let $C = (0, c, 0)$,

$$\phi(C) : (1, \beta, A, 0) \mapsto (1, \beta, A', A \times C)$$

where $A' = (a, \beta c, 0) \Rightarrow A'^\sharp \neq 0, N(A') = 0$. Apply $\psi(D)$ with $D = -A \times C$,

$$\psi(D) : (1, \beta, A', A \times C) \mapsto (1, \beta, A', 0)$$

so that we find ourselves in case (iii).

(iii) Without loss of generality we may assume $A = (0, a_2, a_3)$, $a_2, a_3 \neq 0$. Let $C = (c, 0, 0)$,

$$\phi(C) : (1, \beta, A, 0) \mapsto (1, \beta, A', B')$$

where $A' = (\beta c, a_2, a_3)$ and $B' = (0, ca_3, ca_2)$. Apply $\psi(D)$ with $D = -B'$,

$$\psi(D) : (1, \beta, A', B') \mapsto (1, \beta - 2ca_2a_3, A'', 0)$$

and let $c = \beta/(2a_2a_3)$ to obtain the canonical reduced form $(1, 0, A'', 0)$.

(iv) The augment of case (iii) does not require $N(A) = 0$ and, hence, also applies to the present case. □

Given Theorem 4.1, the entanglement classifications follows almost trivially. The the rank conditions applied to the reduced canonical form (4.1) dramatically simplify and imply the that each rank (up to permutation) is represented a state corresponding to the classes of Table 1:

$$\begin{array}{llll} \text{Rank } \Psi_{\text{red}} = 1 & \Leftrightarrow \; A = 0 & \Rightarrow & \Psi_{\text{red}} = |111\rangle \\ \text{Rank } \Psi_{\text{red}} = 2 & \Leftrightarrow \; A^\sharp = 0, \; A \neq 0, & \Rightarrow & \Psi_{\text{red}} = |111\rangle + |001\rangle \\ \text{Rank } \Psi_{\text{red}} = 3 & \Leftrightarrow \; N(A) = 0, \; A^\sharp \neq 0 & \Rightarrow & \Psi_{\text{red}} = |111\rangle + |001\rangle + |010\rangle \\ \text{Rank } \Psi_{\text{red}} = 4 & \Leftrightarrow \; N(A) \neq 0 & \Rightarrow & \Psi_{\text{red}} = |111\rangle + |001\rangle + |010\rangle + k|100\rangle \end{array} \tag{4.2}$$

where $k \neq 0$. To reach the final form of the representative states on the far left of (4.2), we have scaled using

$$\hat{\tau} \in [\mathrm{SO}(2, \mathbb{C})]^2 \cong \mathrm{Str}_0(\mathfrak{J}_{ABC}) \quad \subset \mathrm{Aut}(\mathbb{F} - KTS_{ABC})$$

see (3.10). Since the ranks partition $\mathbb{F} - KTS_{ABC}$, this completes the orbit (entanglement class) classification of (2.5), which is summarised in Table 1. As claimed, elements of rank 1, 2 and 3 belong to a single orbit corresponding to totally separable, biseparable and W states, respectively. Rank 4 elements belong to a one complex dimensional family of orbits, parametrised by $q(\Psi) = -8k$, and correspond to GHZ states. Note, the rank classification places GHZ above W; they are not merely inequivalent, but ordered, as reflected by three player non-local game [11].

5. Generalising to an n-qubit F-KTS

The success of the F-KTS classification of 3-qubit entanglement naturally raises the question of generalisation. Are there other composite quantum systems amenable to the F-KTS treatment? What about mixed states? Is there an extension to an arbitrary number of qubits? Indeed, it has already been shown that many of the F-KTS based on cubic Jordan algebras provide the pure state SLOCC entanglement classification of composite quantum systems, including mixtures of bosonic and fermonic qudits [24, 1, 25, 26, 2] and [Levay P and Sarosi G arXiv:1206.5066]. Moreover, in the case of three qubits the F-KTS may be used in the mixed state classification [27].

In the subsequent sections we focus on the final question posed above. While there is no arbitrary n-qubit F-KTS *per se*, we may attempt to identify those aspects of the 3-qubit F-KTS which naturally generalise to an arbitrary number of qubits in the hope that these universal features are illuminating. This is the approach taken here.

5.1. The n-qubit state reorganised

Recall, the Jordan algebra formulation of the F-KTS corresponded to decomposing the representation carried by the F-KTS under the $\text{Str}_0(\mathfrak{J}) \subset \text{Aut}(\mathbb{F} - KTS)$. In the case of three qubits we found the state split into the direct sum of four pieces,

$$\alpha = a_{111}, \quad \beta = a_{000}, \quad A = (a_{100}, a_{010}, a_{001}), \quad B = (a_{011}, a_{101}, a_{110}) \tag{5.1}$$

where α, β are $\text{Str}_0(\mathfrak{J})$ singlets. This leads us to the first important observation: α, β, A, B are the closed subsets under the 3-qubit permutation group S_3. Indeed, if we are only interested in the SLOCC entanglement classification up to permutations, as we are, it is only natural to work with S_n-closed subsets as the basic building blocks. Hence, the 2^n state vector coefficients will be collected into the $n+1$ subsets closed under S_n. It will prove convenient to represent these subsets using $n+1$ totally symmetric tensors with ranks ranging from 0 to n,

$$\mathcal{A}_n := \{A_0, A_{i_1}, A_{i_1 i_2}, \dots, A_{i_1 i_2 \dots i_n}\}, \quad \text{where} \quad i_k = 1, 2, \dots, n \tag{5.2}$$

which are vanishing on any diagonal, i.e. $A_{i_1 i_2 \dots i_n} = 0$ if any two indices are the same. The counting of components goes like p-forms, correctly yielding a total of 2^n independent coefficients. Indeed, we could have just as well defined $\mathcal{A}_n$ as a set of totally antisymmetric tensors of ranks 0 to n, avoiding the need to impose tracelessness. However, the 3-qubit F-KTS structure most naturally transfers over using the symmetric formulation, so we will stay with that convention here.

For three qubits we have (with a slight abuse of notation for A_{ijk})

$$A_0 = a_{000}, \quad A_i = \begin{pmatrix} a_{100} \\ a_{010} \\ a_{001} \end{pmatrix}, \quad A_{ij} = \begin{pmatrix} 0 & a_{110} & a_{101} \\ a_{110} & 0 & a_{011} \\ a_{101} & a_{011} & 0 \end{pmatrix}, \quad A_{ijk} = a_{111} \tag{5.3}$$

Note, numbering the qubits from left to right, the values of the indices on the symmetric tensors determine which indices on its corresponding state vector coefficient take the value 1. For example, $A_1 = a_{100}$, $A_2 = a_{010}$, $A_3 = a_{001}$ and $A_{12} = a_{110}$, $A_{13} = a_{101}$ and so on. I am grateful to Duminda Dahanayake for pointing out this rule.

5.2. The n-qubit algebra

The second feature we might hope to generalise is the set of cubic Jordan algebra maps, $A \times B, \text{Tr}(A, B)$, $N(A)$, see subsection 3.1, which played such a key role in the construction of the various covariants and invariants. Recall, group theoretically these maps correspond to picking out certain irreps appearing in the tensor product of the $\text{Str}_0(\mathfrak{J})$-representation carried by $A, B \in \mathfrak{J}$. For example, in the case $\text{Str}_0(\mathfrak{J}_3^{\mathbb{O}}) = E_{6(-26)}$, with A and B transforming as the $\mathbf{27}$, $A \times B$ is the $\mathbf{27}'$ in $\mathbf{27} \times \mathbf{27} = \mathbf{27}'_s + \mathbf{351}_a + \mathbf{351}'_s$ and $N(A)$ is the singlet in $\mathbf{27} \times \mathbf{27} \times \mathbf{27}$. Each of the cubic Jordan algebra maps may be written using the irreducible $E_{6(-26)}$ invariant tensors, d_{ijk} and d^{ijk}, where a downstairs (upstairs) $i = 1, 2, \dots, 27$ transforms as a $\mathbf{27}$ ($\mathbf{27}'$). For example, $(A^\sharp)^i = \frac{1}{2!} d^{ijk} A_j A_k$, and $N(A) = \frac{1}{3!} d^{ijk} A_i A_j A_k$. For the sake of clarity, we will often drop the combinatorial factors in the following. For three qubits, the invariant tensors were simply

$$d_{ijk} = |\epsilon_{ijk}|, \quad d^{ijk} = |\epsilon^{ijk}| \tag{5.4}$$

which naturally suggests the n-qubit generalisation,

$$d_{i_1 \dots i_n} := |\epsilon_{i_1 \dots i_n}|, \quad d^{i_1 \dots i_n} := |\epsilon^{i_1 \dots i_n}| \tag{5.5}$$

This allows us to dualise a rank p tensor,

$$A^{i_1 i_2 \dots i_{n-p}} := 1/p!\, d^{i_1 i_2 \dots i_{n-p} i_{n-p+1} \dots i_n} A_{i_{n-p+1} \dots i_n} \tag{5.6}$$

For an n-qubit state, rank p pairs $A_{i_1 i_2 \dots i_p}, A^{i_1 i_2 \dots i_p}$ are precisely bit-flip related. For example, for three qubits, $A_i = (a_{100}, a_{010}, a_{001})$ and $A^i = (a_{011}, a_{101}, a_{110})$. This is crucial for building S_n invariants. Equipped with $d_{i_1 \dots i_n}, d^{i_1 \dots i_n}$ the n-qubit space $\mathcal{A}_n$ of symmetric tensors may be endowed with a pseudo-algebraic structure: $\mathcal{A}_n$ is closed so long as we compose the tensors by contracting with $d_{i_1 \dots i_n}$ and $d^{i_1 \dots i_n}$.

5.3. *The n-qubit generalised F-KTS transformations*

The F-KTS transformations (3.10) for three qubits in the our new notation are given by

$$\begin{aligned}
\phi(C) &: \begin{pmatrix} A^{ijk} \\ A^{ij} \\ A^i \\ A^0 \end{pmatrix} \mapsto \begin{pmatrix} A^{ijk} & & & \\ A^{ij} & + C_k A^{ijk} & & \\ A^i & + C_j A^{ij} & + C_j C_k A^{ijk} & \\ A^0 & + C_i A^i & + C_i C_j A^{ij} & + C_i C_j C_k A^{ijk} \end{pmatrix} \\
\psi(D) &: \begin{pmatrix} A_0 \\ A_i \\ A_{ij} \\ A_{ijk} \end{pmatrix} \mapsto \begin{pmatrix} A_0 & + D^i A_i & + D^i D^j A_{ij} & + D^i D^j D^k A_{ijk} \\ A_i & + D^j A_{ij} & + D^j D^k A_{ijk} & \\ A_{ij} & + D^k A_{ijk} & & \\ A_{ijk} & & & \end{pmatrix} \\
\hat{\tau}(\lambda) &: \begin{pmatrix} A_0 \\ A_i \\ A^i \\ A^0 \end{pmatrix} \mapsto \begin{pmatrix} d_{lmn}\lambda^l \lambda^m \lambda^n & A_0 \\ \xi^l d_{lmn} \lambda^m \lambda^n & A_i \\ \xi^l \xi^m d_{lmn} \lambda^n & A^i \\ \xi^l \xi^m \xi^n d_{lmn} & A^0 \end{pmatrix}
\end{aligned} \tag{5.7}$$

where $\lambda \in \mathbb{C} - \{0\}$ and $\xi = \lambda^{-1}$. For $\phi(C), \psi(D)$ we have made a judicious choice of dualisations in order to make the correct n-qubit generalisation manifest. Under $\phi(C)$ a rank p tensor $A^{i_1 i_2 \dots i_p}$ transforms into the sum of all tensors $A^{i_1 i_2 \dots i_q}$ with $q \geq p$ contracted with the necessary powers of C_i to give back rank p. Explicitly,

$$\begin{array}{lll}
A^{i_1 i_2 \dots i_n} & \mapsto & [A^{i_1 i_2 \dots i_n}] \\
A^{i_1 i_2 \dots i_{n-1}} & \mapsto & [A^{i_1 i_2 \dots i_{n-1}} + C_{i_n} A^{i_1 i_2 \dots i_n}] \\
A^{i_1 i_2 \dots i_{n-2}} & \mapsto & [A^{i_1 i_2 \dots i_{n-2}} + C_{i_{n-1}} A^{i_1 i_2 \dots i_{n-1}} + C_{i_{n-1}} C_{i_n} A^{i_1 i_2 \dots i_{n-1} i_n}] \\
\vdots & \vdots & \vdots \\
A^0 & \mapsto & [A^0 + C_i A^i + C_i C_j A^{ij} + \dots + C_{i_1} C_{i_2} \cdots C_{i_n} A^{i_1 i_2 \dots i_n}]
\end{array} \tag{5.8}$$

Similarly, under $\psi(D)$ a rank p tensor $A_{i_1i_2\ldots i_p}$ transforms into the sum of all $A_{i_1i_2\ldots i_q}$ with $q \geq p$, contracted with the necessary powers of D^i to give back rank p,

$$\begin{array}{lll} A_0 & \mapsto & [A_0 + D^iA_i + D^iD^jA_{ij} + \cdots + D^{i_1}D^{i_2}\cdots D^{i_n}A_{i_1i_2\ldots i_n}] \\ A_i & \mapsto & [A_i + D^jA_{ij} + \cdots + D^{i_2}D^{i_3}\cdots D^{i_n}A_{ii_2i_3\ldots i_n}] \\ A_{ij} & \mapsto & [A_{ij} + \cdots + D^{i_3}D^{i_4}\cdots D^{i_n}A_{iji_3i_4\ldots i_n}] \\ \vdots & \vdots & \vdots \\ A_{i_1i_2\ldots i_n} & \mapsto & [A_{i_1i_2\ldots i_n}] \end{array} \tag{5.9}$$

The generalised $\hat{\tau}(\lambda)$ may also be concisely written using this notation,

$$\hat{\tau}(\lambda) : A_{i_1i_2\ldots i_p} \mapsto \xi^{j_1}\xi^{j_2}\cdots\xi^{j_p}d_{j_1j_2\ldots j_n}\lambda^{j_{p+1}}\cdots\lambda^{j_{n-1}}\lambda^{j_n}A_{i_1i_2\ldots i_p} \tag{5.10}$$

Hence, adopting the notational convention $A_{[p]}$ ($A^{[p]}$) for a rank p tensor with downstairs (upstairs) indices, the n-qubit transformations ϕ, ψ and $\hat{\tau}$ may be summarised as follows,

$$\begin{array}{rlll} \phi(C_{[1]}): & A^{[p]} & \mapsto & \sum_{k=p}^{n} C_{[1]}^{(k-p)}A^{[k]} \\ \psi(D^{[1]}): & A_{[p]} & \mapsto & \sum_{k=p}^{n} D^{[1](k-p)}A_{[k]} \\ \hat{\tau}(\lambda^{[1]}): & A_{[p]} & \mapsto & \xi^{[1](p)}d_{[n]}\lambda^{[1](n-p)}A_{[p]} \end{array} \tag{5.11}$$

One useful observation that immediately follows is that one can always assume $A_0 = 1, A^i = 0$ under SLOCC. It is also clear that the Jordan ranks (3.7) naturally generalises to a set of rank conditions on $A_{[1]}, A^{[1]}$.

However, we have yet to develop a systematic method for writing covariants/invariants in this scheme. One example, though, defined for n qubits, is given by

$$(\mathcal{A}_n, \mathcal{B}_n) = \sum_{k=0}^{n}(-1)^k/k!\ A_{[k]}B^{[k]} \tag{5.12}$$

This is symmetric (antisymmetric) for even (odd) n. It is simply the determinant in the 2-qubit case and the antisymmetric bilinear form of the F-KTS in the 3-qubit case. There are four algebraically independent 4-qubit permutation and SLOCC-equivalence group invariants [28] of order two, six, eight and twelve, (5.12) being the order two example.

2-qubit example. The 2-qubit state corresponds to

$$|\psi\rangle \leftrightarrow \Psi = \{A_0, A_i, A_{ij}\}, \quad i, j = 1, \ldots, 2 \tag{5.13}$$

where

$$\left\{A_0 = a_{00}, \quad A_i = \begin{pmatrix} a_{10} \\ a_{01} \end{pmatrix}, \quad A^0 = a_{11}\right\} \tag{5.14}$$

It is easy to verify in this scheme that every state $\mathcal{A}_2$ is SLOCC-equivalent to the reduced canonical form: $\mathcal{A}_2^{\text{red}} = \{1, 0, k\} \longleftrightarrow |00\rangle + k|11\rangle$,where $k = (\mathcal{A}_2, \mathcal{A}_2)$. First we may always assume A_i is non-zero. (If $A_i = 0$ then we may assume that A_0 non-zero using $\mathcal{Z}$ if necessary.) Now apply $\phi(C)$ to get a non-zero A_i. Apply $\psi(D)$ with $d_{ij}D^iD^j = 0$ so that $A_0 \mapsto A_0 + D^iA_i$. Choose D s.t. $A_0 + D^iA_i = 1$. Finally, apply $\phi(C)$ with $C_i = -A_i$.

This gives us the well known 2-qubit SLOCC entanglement classification: there are just two classes, separable and entangled, the latter being a family of orbits parametrised by $(\mathcal{A}_2, \mathcal{A}_2) = k$.

Acknowledgements

Many thanks to D. Dahanayake, M. J. Duff, H. Ebrahim and W. Rubens, with whom this work was done. I would like to extend my gratitude to the organisers of the *QQQ* conference *3Quantum: Algebra Geometry Information* (Tallinn, July 2012) and to Prof. R. Iordanescu. The work of LB is supported by an Imperial College Junior Research Fellowship.

References

[1] Borsten L, Dahanayake D, Duff M J, Ebrahim H and Rubens W 2009 *Phys. Rep.* **471** 113
[2] Borsten L, Duff M J and Levay P 2012 *Classical Quantum Gravity* **29** 224008
[3] Borsten L, Dahanayake D, Duff M J, Rubens W and Ebrahim H 2009 *Phys. Rev.* A **80** 032326
[4] Dür W, Vidal G and Cirac J I 2000 *Phys. Rev.* A **62** 062314
[5] Plenio M B and Virmani S 2007 *Quant. Inf. Comp.* **7** 1
[6] Horodecki R, Horodecki P, Horodecki M and Horodecki K 2009 *Rev. Mod. Phys.* **81** 865
[7] Watrous J 2006 *Quantum Computation Lecture Course* CPSC 519/619 (University of Calgary)
[8] Mermin N D 1990 *Amer. J. Phys.* **58** 731
[9] Greenberger D M, Horne M and Zeilinger A 1989 *Bell's Theorem, Quantum Theory and Conceptions of the Universe* (Kluwer)
[10] Cleve R, Hoyer P, Toner B and Watrous J 2004 *Computational Complexity* Proc. 19th IEEE Annual Conference p 236
[11] Borsten L 2013 *J. Phys.* A **46** 455303
[12] Bennett C H, Popescu S, Rohrlich D, Smolin J A and Thapliyal A V 2000 *Phys. Rev.* A **A3** 012307
[13] McCrimmon K 2004 *A Taste of Jordan Algebras* (Springer-Verlag)
[14] Jacobson N 1961 *J. Reine Angew. Math.* **207** 61
[15] Freudenthal H 1954 *Nederl. Akad. Wetensch. Proc. Ser.* **57** 218
[16] Kantor I L 1973 *Dokl. Akad. Nauk.* **208** 1276 (Russian); *Soviet Math. Dokl.* 1973 **14** 154
[17] Yamaguti K and Asano H 1975 *Proc. Japan Acad. Sci.* **51** 253
[18] Yamaguti K and Ono A 1984 *Bull. Faculty Sch. Ed. Hiroshima Univ.* Part II **7** 43
[19] Kamiya N 1987 *J. Algebra* **110** 108
[20] Brown R B 1969 *J. Reine Angew. Math.* **236** 79
[21] Ferrar C J 1972 *Trans. Amer. Math. Soc.* **174** 313
[22] Krutelevich S 2007 *J. Algebra* **314** 924
[23] Borsten L, Duff M J, Ferrara S, Marrani A and Rubens W 2014 *Comm. Math. Phys.* **325** 17
[24] Borsten L 2008 *Fortschr. Phys.* **56** 842
[25] Lévay P and Vrana P 2008 *Phys. Rev.* A **78** 022329
[26] Vrana P and Lévay P 2009 *J. Phys.* A **42** 285303
[27] Szalay S and Kökényesi Z 2012 *Phys. Rev.* A **86** 032341
[28] Briand E, Luque J G and Thibon J Y 2003 *J. Phys.* A **36** 9915

Stringlike structures in the real and complex Kerr-Schild geometry

A Burinskii

Theor. Phys. Lab., NSI, Russian Academy of Sciences, B. Tulskaya 52, Moscow 115191, Russia

Abstract. Four-dimensional Kerr-Schild (KS) geometry displays remarkable relationships with quantum world and theory of superstrings. In particular, the Kerr-Newman (KN) solution has gyromagnetic ratio $g = 2$, as that of the Dirac electron and represents a consistent background for gravitational and electromagnetic field of the electron. As a consequence of very big spin/mass ratio, black hole horizons disappear, exposing the naked Kerr singular ring. We consider four-decade history of development of this structure which took finally the form of a point-string-membrane-bubble complex which is reminiscent of the *enhancon* model of string/M-theory. A complex string obtained in the complex structure of the Kerr geometry gives an extra dimension to the world-sheet of the real Kerr string, forming a membrane by analogue with the string/M-theory unification. By analysis of the *orientifold* parity of the complex Kerr string, we obtain that the determined by the Kerr theorem principal null congruence of the Kerr geometry is described by a quartic equation in projective twistor space CP^3, and therefore, it creates the known Calabi-Yau twofold (K3 surface) in twistor space of the 4d KS geometry. We connect it with N=2 superstring which has (complex) critical dimension two and may be embedded into complex KS geometry.

1. Introduction

It is now generally accepted that black holes are to be associated with elementary particles. New ideas and methods in the physics of black holes (BH) are based on the complex analyticity and conformal field theory, that unites the physics of black holes with superstring theory and particle physics. In spite of these evident relationships, the path from superstring theory to particle physics represents an unsolved problem, and moreover, it is stated that "... realistic model of elementary particles still appears to be a distant dream...", wrote John Schwarz in [arXiv:1201.0981]. Meanwhile, the four-dimensional Kerr-Newman solution for a charged and rotating BH [1] demonstrates in this relation some interesting surprises. Namely, it exhibits wonderful connections with spinning particles and simultaneously a remarkable parallelism with basic structures of superstring theory.

In 1968 Carter noticed that the KN solution has gyromagnetic ratio $g = 2$, as that of the Dirac electron [2], and therefore, at least the asymptotic electromagnetic (em) and gravitational fields of the electron should correspond to the KN solution with great precision. Angular momentum of the KN solution $J = ma$ is proportional to mass m and to Kerr's parameter a, which is radius of the Kerr singular ring. The em and gravitational fields are concentrated near the Kerr ring forming a type of waveguide, which causes its similarity with a closed string of the dual models. It has been assumed in [3, 4] that the em and gravitational traveling waves can propagate along the Kerr ring in analogue with excitations of the dual string models. Masses of elementary particles are very small with respect to their spin, and the Kerr parameter of rotation $a = J/m$, radius of the Kerr singular ring, turns out to be of order of the Compton wave length of the electron $\sim \hbar/(2mc)$. In the dimensionless units $c = G = \hbar = 1$, the ratio a/m is about 10^{44}, which results in disappearance of the BH horizons, and the Kerr singular ring turns out to be naked Other spinning particles have approximately the same ratio a/m, and we arrive to trivial, but very important fact that *black holes cannot apparently be associated with spinning particles.* The relevant to spinning particles geometry should be an over-rotating KN background which contains a closed string-like singularity.

It looks very bad, since instead of the consistent with quantum theory very weak gravitational field in vicinity of the electron, we reveal a singular core in the form of a closed string, which represents a very strong defect of the background. One more trouble of this result is a two-sheeted structure of the resulting Kerr background, since the naked Kerr singular ring represents a branch line of the Kerr space-time into two sheets. This mysterious structure of the over-rotating KN geometry created the problems in modelling of the source of the Kerr and KN solutions, which could consistently regulate this background. Structure of such a source was specified step by step in many works during more than four decades, in particular in the following papers [5, 6, 4, 3, 7, 8, 9, 10, 11, 12]. As a result, the consistent regular source of the external KN solution, generating the necessary very weak gravitational field in the vicinity of the source, acquired the form of a relativistically rotating string-membrane-bubble complex: a highly oblate bubble bounded by an ellipsoidal membrane and by a circular fundamental string positioned on the edge rim of the bubble. The circular string appears as a result of regularization of the Kerr singular ring, and we argue that traveling waves circulating along this string should create a singular point, D0-brane which exhibits "zitterbewegung" of the Dirac electron. It was also shown [12] that wave excitations of the fundamental Kerr string generate also the axial half-strings which are going to infinity serving as carriers of de Broglie waves.

Along with circular fundamental string [11, 3, 13], an open complex string was obtained in the complex structure of the Kerr-Schild geometry [14, 15]. fnBoth of these strings were rediscovered recently by Adamo and Newman [16] by study of the asymptotic form of the geodesic and shear-free congruences. They comment this discovery writing that "...It would have been a cruel god to have layed down such a pretty scheme and not have it mean something deep." It should also be mentioned that the open Kerr string extends the world-sheet of the closed Kerr string to two-brane source, displaying parallelism with the process of string/M-theory unification [17], and moreover, the resulting source of the KN solution is reminiscent in many details the enhancon model used by the superstring/M-theory unification [18]. Finally, it has been obtained recently [19], that the twistorial structure of the four-dimensional KS geometry, being combined with the orientifold structure of the complex Kerr string, creates a K3 surface (the Calabi-Yau twofold) on the projective twistor space, which forms a holographic skeleton of the KS space, determined by the Kerr theorem. It confirms close relationships of the KS geometry with the basic structures of the superstring theory and hints existence of some underlying theory which may unify the Kerr's gravity with physics of elementary particles and superstring theory. Twistor theory plays apparently principal role in this unification, since the complex N=2 critical superstring [20] has also the inherent twistorial structure [21, 22], which allows it to be consistently embedded in the *complex* four-dimensional KS geometry. We start paper by treatment of the real, complex, twistorial and stringy structures of KS geometry along our previous papers [14, 15, 11, 19, 23].

2. Real structure of the Kerr-Newman (KN) geometry

The KN metric is represented in the Kerr-Schild (KS) form [1],

$$g_{\mu\nu} = \eta_{\mu\nu} + 2he^3_\mu e^3_\nu \tag{2.1}$$

where $\eta_{\mu\nu}$ is auxiliary Minkowski background in Cartesian coordinates $\mathrm{x} = x^\mu = (t, x, y, z)$,

$$h = P^2 \frac{mr - e^2/2}{r^2 + a^2 \cos^2\theta}, \quad P = (1 + Y\bar{Y})/\sqrt{2} \tag{2.2}$$

and $e^3(\mathrm{x})$ is a tangent direction to a *Principal Null Congruence (PNC)*, which is determined by the form[1]

$$e^3_\mu dx^\mu = du + \bar{Y} d\zeta + Y d\bar{\zeta} - Y\bar{Y} dv \tag{2.3}$$

via function $Y(\mathrm{x})$, which is controlled by *the Kerr theorem*, [1, 33, 24].

The twisting lightlike rays of the Kerr PNC are focussing in the equatorial plane $\cos\theta = 0$, at the Kerr singular ring, $r = 0$, approaching it tangentially. As a result, the aligned with Kerr PNC metric and the KN electromagnetic potential,

$$A_\mu = -P^{-2}\mathrm{Re}\frac{e}{(r + ia\cos\theta)} e^3_\mu \tag{2.4}$$

concentrate near the Kerr ring and form a closed lightlike gravitational waveguide [3], playing the role of a closed string which may carry excitations in the form of the lightlike traveling waves [4, 12], which take near the ring the form of pp-waves considered as string excitations in low energy string theory.

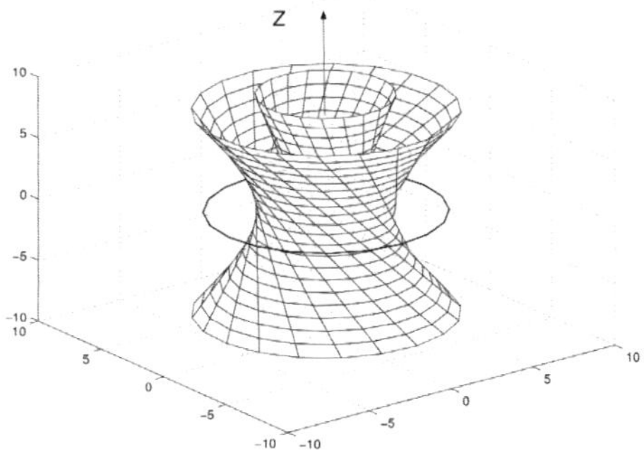

Figure 1. Twistor null lines of the Kerr congruence are focused on the Kerr singular ring which forms a branch line of space in two sheets.

2.1. Regular source of the KN solution

Gyromagnetic ratio of the KN solution is $g = 2$, as that of the Dirac electron [2, 1]. Therefore, the observable parameters of the electron: mass, spin, charge and magnetic moment indicate that the KN gravitational field corresponds to the electron background geometry [12], which created a series of works on the consistent with gravity classical model of the electron [6, 8, 4, 12, 25, 26, 27, 28] and stimulated investigations of models of the source of the Kerr and KN solution.

(1) *The Kerr ring as a closed string.* This model, created first in 1975 [3], by analogue with the Dirac string of magnetic monopole, was supported then by the studies of the fundamental string solutions to low-energy string theory [Dabholkar A *et al* arXiv:hep-th/9511053] In 1992 Sen obtained two important solutions to low-energy string theory: a) solution for fundamental heterotic string [30], and b) analogue of the Kerr solution to low-energy string theory [31]. It has been shown then in [13] that metric and electromagnetic field near

[1] Here $\zeta = (x+iy)/\sqrt{2}$, $\bar{\zeta} = (x-iy)/\sqrt{2}$, $u = (z+t)/\sqrt{2}$, $v = (z-t)/\sqrt{2}$, are the null Cartesian coordinates, r, θ, ϕ are the Kerr oblate spheroidal coordinates, and $Y(\mathrm{x}) = e^{i\phi}\tan\frac{\theta}{2}$ is a projective angular coordinate. The used signature is $(-+++)$.

the Kerr ring in the solution (b) have the lightlike Killing structure similar to that of the fundamental heterotic string solution (a). The only difference consisted in branching of the Kerr singularity related with two sheets of the Kerr geometry.

(2) *String-membrane source.* Alternative line of investigations, related with attempts to remove Kerr's two-sheetedness was started by H. Keres [5], who noticed that the complex radial distance $\tilde{r} = r + ia\cos\theta$, which determines the KN gravitational and electromagnetic potential $\Phi = \mathrm{Re}(m/\tilde{r})$, being expressed in the Cartesian coordinates yields $(\tilde{r})^2 = x^2 + y^2 + (z - ia)^2$, and therefore, "function $\tilde{r}^{-1}$ and its first derivatives are continuous over all space, beside the disk $z = 0$, $x^2 + y^2 \leq a^2$," and "... crossing the disk the derivatives have a jump discontinuity..." indicating that the sources should lie on the disk. Similarly, Israel truncated the Kerr 'negative' Kerr sheet along the disk $r = 0$, and showed in [6] that the resulting jump discontinuity should create a matter distribution over the disk surface. He interpreted the disk-like source of the Kerr geometry as a classical model of electron. Next important step was done by Hamity [7], who showed that Kerr's disk has to be rigidly rotating, reaching the velocity of light on the string-like border of the disk. Finally, López [8] constructed the regular source of the KN solution, excising the singular region $r < r_e = e^2/2m$ (note that r is the Kerr ellipsoidal radial coordinate), and the Kerr-Newman (KN) regular source took the form of a rigidly rotating ellipsoidal membrane, or the disk-like bubble, flat interior of which is matched with the external KN solution along the bubble surface $r = r_e$, being determined unambiguously by the equation $h(r_e) = 0$, where h is given by (2.2).

As a result, the gravitational singularity disappears, and the space around the regularized disk-like source (= Compton region of the dressed electron) turned out to be very close to flat. In the same time, the singular electromagnetic field gets cutoff at $r = r_{reg} = r_e = e^2/2m$ (ellipsoidal radial coordinate) and is regularized It turns the Kerr ring into regular fundamental string positioned at the edge rim of the ellipsoidal bubble. As it was shown in [12], the traveling waves along circular string create the axial singular beams (outgoing pp-wave half-string) which serve as carriers of de Broglie waves.

(3) *Gravitating soliton model.* In [11], the membrane-bubble model was extended to a smooth field model based on the system of chiral superfields and on the suggested in [9] modified form of the KN metric. The regular bubble-source was formed by a domain wall phase transition interpolating between the external KN background and a supersymmetric pseudovacuum state inside the bubble. The model represents a soliton of the oscillon type, since the internal false-vacuum state is built of the oscillating Higgs-like field. The KN vector potential is dragged by the relativistically rotating string and forms a closed Wilson loop, which turns out to be quantized via interaction with the internal Higgs field, resulting in a consistent with KN gravity rotating soliton (for details see [11]).

The KN electromagnetic field is regularized and forms a lightlike fundamental string positioned on the boundary of the disk-like bubble in equatorial plane $\theta = \pi/2$. New important effect is related with circular traveling waves which should propagate along fundamental string distorting the form of the bubble surface. The cutoff r_e is determined by amplitude of the vector-potential, and there appear some nodes where the cutoff r_e approaches zero, and regularization is to be broken. The deformed by waves bubble boundary touches the Kerr singular ring creating singular points, D0-branes which circulate along the bubble in equatorial plane. These singular points form the ends of the fundamental string and can be identified with lightlike quarks or partons. It turns out to be closely related with the fact that the fundamental circular string, being parametrized by the light cone parameter $\sigma = t - n\phi$, cannot indeed be closed, since the end points $\phi = 0$ and $\phi = \pi$ turn out to be disconnected by a time-like gap $x^\mu(t - 0) - x^\mu(t - n\pi)$, which should be closed by the path of a massive mode. These end points are stuck at the bubble, which therefore

Journal of Physics: Conference Series **532** (2014) 012004 doi:10.1088/1742-6596/532/1/012004

acquires the role of a heavy quark which closes this loop It suggests a physical mechanism which can lie beyond "zitterbewegung" of the Dirac electron. On the other hand, it resolves the seeming contradiction between the point-like and extended electron: the naked electron is point-like and represents a circulating singular pole (D0-brane), while the dressed KN electron exhibits an extended structure of the bag model, which is similar to the MIT and SLAC bag models, but is consistent with KN gravity. From geometrical point of view, it represents a *simplicial complex of the Compton size*, consisting of the point, circular string, ellipsoidal membrane and a false-vacuum bubble filled by Higgs field – the complex system of the D0-D1-D2-D3 branes.

(4) *Complex Kerr string.* Along with the circular string structure, the Kerr-Schild geometry contains also a *complex string* [14, 15], which appears in the initiated by Newman complex representation of the Kerr geometry [32]. This string is open and parametrized by angular variable $\theta \in [0, \pi]$. In fact, parameter θ extends the world-sheet of the circular Kerr string (positioned at $\theta = \pi/2$) to the membrane source of the López bubble model [8, 11]. Structure of the regular bubble source of the KN solution represents a copy of the enhancon model used for regularization by the superstring/M-theory unification [18] and represents a counterpart of the transfer from 10-dimensional superstrings to 11d M-theory [17].

2.2. The Kerr theorem and twistorial structure of the KS geometry

The Kerr Theorem [1, 23, 33] determines the shear-free null congruences with tangent direction (2.3) by means of the solution $Y(\mathrm{x})$ of the equation

$$F(T^A) = 0 \tag{2.5}$$

where $F(T^A)$ is an arbitrary holomorphic function in the projective twistor space

$$T^A = \{Y, \quad \lambda^1 = \zeta - Yv, \quad \lambda^2 = u + Y\bar{\zeta}\} \tag{2.6}$$

Using the Cartesian coordinates x^μ, one can rearrange variables and reduce function $F(T^A)$ to the form $F(Y, x^\mu)$, which allows one to get solution of the equation (2.5) in the form $Y(x^\mu)$.

For the Kerr and KN solutions, the function $F(Y, x^\mu)$ turns out to be quadratic in Y,

$$F(Y, x^\mu) = A(x^\mu)Y^2 + B(x^\mu)Y + C(x^\mu) \tag{2.7}$$

and the equation (2.5) represents a *quadric* in the projective twistor space CP^3, with a non-degenerate determinant $\Delta = (B^2 - 4AC)^{1/2}$ which defines the complex radial distance [24, 23]

$$\tilde{r} = -\Delta = -(B^2 - 4AC)^{1/2} \tag{2.8}$$

This case is explicitly solved and yields two solutions

$$Y^{\pm}(x^\mu) = (-B \mp \tilde{r})/2A \tag{2.9}$$

which allows one to restore two PNC by means of (2.3). One can easily obtain from (2.7) and (2.9) that the used in the metric (2.2) and the em potential (2.4) complex radial distance $\tilde{r} = r + ia\cos\theta$ may also be determined from the Kerr generating function by the relation

$$\tilde{r} = -dF/dY \tag{2.10}$$

Therefore, the Kerr singular ring, $\tilde{r} = 0$, is formed as a caustic of the Kerr congruence,

$$dF/dY = 0 \tag{2.11}$$

As a consequence of Vieta's formulas, the quadratic in Y function (2.7) may be expressed via the solutions $Y^{\pm}(x^\mu)$ in the form

$$F(Y, x^\mu) = A\left(Y - Y^+(x^\mu)\right)\left(Y - Y^-(x^\mu)\right) \tag{2.12}$$

It is known [1], that the linear in Y generating functions F correspond to the light-like solutions, while the subsequent analysis of the generating functions of higher degrees in Y showed that they split into products of the blocks of the first and second degrees in Y, and therefore, they should correspond to multi-particle KS solutions [34, 35, 36].

3. Complex Kerr geometry and an open complex string

One can see that the complex radial distance $\tilde{r} = r + ia\cos\theta$ takes in the Cartesian coordinates the form

$$\tilde{r} = \sqrt{x^2 + y^2 + (z + ia)^2} \tag{3.1}$$

and therefore, the scalar component of the vector potential (2.4) may be obtained from the Coulomb solution

$$\phi(\vec{x}) = e/r = e/\sqrt{x^2 + y^2 + z^2} \tag{3.2}$$

by a complex shift $z \to z + ia$, or by the shift of its singular point $\vec{x}_0 = (0, 0, 0)$ in complex region $\vec{x}_0 \to (0, 0, -ia)$. The complex shift was first considered by Appel in 1887 [37], who noticed that the Coulomb solution, being invariant solution to the linear Laplace equation with respect to real shifts of its origin $\vec{x} \to \vec{x} + \vec{a}$, should also be invariant with respect to the complex shift. In spite of triviality of this procedure from complex point of view, it yields very nontrivial consequences in the real section, in particular, the singular point of the Coulomb solution $\vec{x}_0 = (0, 0, 0)$ turns into singular ring

$$x^2 + y^2 + (z + ia)^2 = 0 \tag{3.3}$$

i.e intersection of the sphere

$$x^2 + y^2 + z^2 = a^2 \quad \text{and the plane} \quad z = 0 \tag{3.4}$$

This ring becomes the branch line of the space into two sheets.

Newman showed that the KN electromagnetic field and the linearized KN gravitational field may be obtained from a complex retarded-time construction as the fields generated by a complex source propagating along a complex world line (CWL) [32]. Later on, it has been shown that linearized Newman's representation acquires exact meaning in the Kerr-Schild class of metrics [4, 23, 24], and the KN solution corresponds exactly to the field generated by a *complex source propagating along straight CWL*

$$x_L^\mu(\tau_L) = x_0^\mu(0) + u^\mu \tau_L + ia\left(k_L^\mu - k_R^\mu\right)/2 \tag{3.5}$$

where

$$u^\mu = (1, 0, 0, 0), \quad k_R = (1, 0, 0, -1), \quad k_L = (1, 0, 0, 1), \quad \tau_L = t_L + \sigma_L \tag{3.6}$$

and τ_L is a complex retarded-time parameter. The index L labels it as a Left structure, and we should also add a complex conjugate Right structure

$$x_R^\mu(\tau_R) = x_0^\mu(0) + u^\mu \tau_R - ia(k_L^\mu - k_R^\mu)/2 \tag{3.7}$$

Therefore, from complex point of view the Kerr and Schwarzschild geometries are equivalent and differ only by the relative position of their *real slice*, which for the Kerr solution goes aside of its center. Complex shift turns the Schwarzschild radial directions $\vec{n} = \vec{r}/|r|$ into twisted directions of the Kerr congruence, see Fig. 1.

Principal part of the complex retarded-time construction is a family of the complex light cones $\mathcal{K}$ adjoined to CWL. Taking the Left CWL as a basis for treatment, one can represent the family of the adjoined complex light cones in spinor form

$$\mathcal{K}_L = \left\{ x : x = x^i_L(\tau_L) + \psi^A_L \sigma^i_{A\dot{A}} \tilde{\psi}^{\dot{A}}_R \right\} \tag{KL}$$

The cones are split into two families of null planes: "left" (ψ_L = const.; $\tilde{\psi}_R$ – var.) and "right" ($\tilde{\psi}_R$ = const.; ψ_L – var.). These are the only two-dimensional planes which are wholly contained in the complex null cone. The rays of the principal null congruence of the Kerr geometry are the tracks of these complex null planes (right or left) on the real slice of the Minkowski background. The real null direction e^3, given by (2.3), is a projective version of the spinor form $\psi^A_L \sigma^i_{A\dot{A}} \tilde{\psi}^{\dot{A}}_R$, expressed via the first projective twistor coordinate $Y = \psi^1_L/\psi^0_L$. The real null direction e^3 is completed by two complex conjugate null forms

$$e^1 = d\zeta - Y dv, \quad e^2 = d\bar{\zeta} - \bar{Y} dv \tag{3.8}$$

One sees that the second projective twistor coordinate

$$\lambda^1 = \zeta - Yv = x^\mu e^1_\mu \tag{3.9}$$

represents projection of the space-time point x^μ on the null direction e^1, while the third projective twistor coordinate

$$\lambda^2 = u + Y\bar{\zeta} = x^\mu (e^3_\mu - \bar{Y} e^1_\mu) \tag{3.10}$$

represents a linear combination of the projections on the null directions e^3 and e^1. The determined by the Kerr theorem function $Y(x)$ allows one to restore at each point x the remaining twistor coordinates λ^1 and λ^2, and to fix the incident Left null plane spanned by the null directions e^3 and e^1. As a consequence, the solution of the Kerr theorem $Y(x)$ foliates the Minkowski space-time into the Left null planes, which due to specific structure of the KS metric (2.1) turn out to be null with respect to the curved KS space-time too, and therefore, the Kerr theorem performs a specific linearization of the curved KS space-times, which justifies validity of the twistor version of the Fourier transform [38].

Similarly, the Right null planes are to be spanned by the null directions e^3 and e^2. The twistor null rays of the Kerr congruence are formed as intersections of the complex conjugate Left and Right null planes.

3.1. Complex open string

It was obtained in [14, 15, 21] that the complex world line $x^\mu_0(\tau)$, parametrized by complex time $\tau = t + i\sigma$, represents really a two-dimensional surface which takes an intermediate position between world-line of the particle and the string world-sheet. The corresponding "hyperbolic string" equation [21], $\partial_\tau \partial_{\bar{\tau}} x_0(t, \sigma) = 0$, has the general solution

$$x_0(t, \sigma) = x_L(\tau) + x_R(\bar{\tau}) \tag{3.11}$$

as sum of the analytic and anti-analytic modes $x_L(\tau)$, $x_R(\bar{\tau})$, which are not necessarily complex conjugate. The complex light cones $\mathcal{K}$ adjoined to each point of the world-sheet $x^\mu_0(\tau)$ are split into the Left and Right null planes, forming the linear N=2 structure of the complexified KS geometry. The real KS structure appears as a result of *projection of this structure on the real slice, which is non-linear operation,* resulting in twisted structure of the real KS space-time.

For each real point x^μ, the parameters τ and $\bar{\tau}$ are to be determined by a complex retarded-time construction. Complex source of the KN solution corresponds to two *straight* complex conjugate world-lines (3.5), (3.7). Contrary to the real case, the complex retarded-advanced times $\tau^{\mp} = t \mp \tilde{r}$ may be determined by two different (Left or Right) complex null planes, which are generators of the complex light cone. It yields four different roots for the Left and Right complex structures [24, 23]

$$\tau_L^{\mp} = t \mp (r_L + ia\cos\theta_L), \quad \tau_R^{\mp} = t \mp (r_R + ia\cos\theta_R) \tag{3.12a}$$

The real slice condition determines the relation $\sigma = a\cos\theta$, which connects parameter σ with angular directions of the Kerr congruence $\theta \in [0, \pi]$ and puts the restriction $\sigma \in [-a, a]$ indicating that *the complex string is open*, and its endpoints $\sigma = \pm a$ may be associated with the Chan-Paton charges of a quark-antiquark pair. In the real slice, the complex endpoints of the string are mapped to the north and south twistor null lines, $\theta = 0, \pi$, see Figs. 2 and 3.

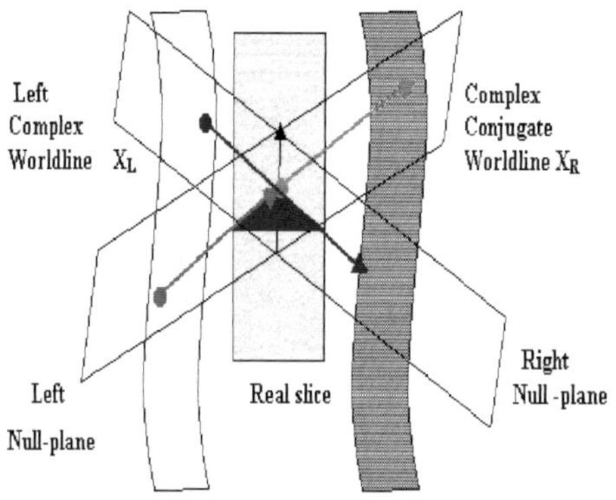

Figure 2. The complex conjugate Left and Right null planes generate the Left and Right retarded and advanced roots.

3.2. Orientifold projection

Boundary conditions of the complex open string cannot consistently be set for its real and complex part simultaneously [14, 15]. Solution of this obstacle requires an *orientifold* structure [17, 20] of the worldsheet of the open complex string. Orientifold turns the open string in a closed but folded one. The world-sheet parity transformation $\Omega : \sigma \to -\sigma$ reverses orientation of the world sheet, and covers it second time in mirror direction. Simultaneously, the Left and Right modes are exchanged [2] The projection Ω is combined with space reflection $R : r \to -r$, resulting in $R\Omega : \tilde{r} \to -\tilde{r}$, which relates the retarded and advanced folds $R\Omega : \tau^+ \to \tau^-$ preserving analyticity of the world-sheet. The string modes $x_L(\tau)$, $x_R(\bar{\tau})$ are extended on the second half-cycle by the well-known extrapolation [17, 20]

$$x_L(\tau^+) = x_R(\tau^-), \quad x_R(\tau^+) = x_L(\tau^-) \tag{3.13}$$

[2] Two oriented copies of the interval $\Sigma = [-a, a]$, $\Sigma^+ = [-a, a]$ and $\Sigma^- = [-a, a]$, are joined, forming a circle $S^1 = \Sigma^+ \bigcup \Sigma^-$, parametrized by θ, and map $\theta \to \sigma = a\cos\theta$ covers the world-sheet twice.

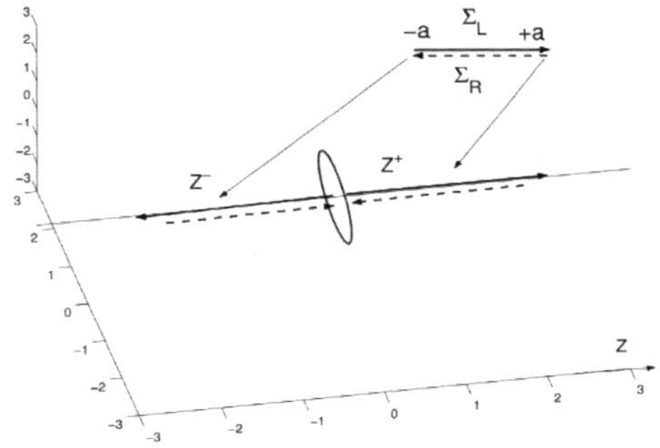

Figure 3. Ends of the open complex string, associated with quantum numbers of quark-antiquark pair, are mapped onto the real half-infinite z^+, z^- axial strings. Dotted lines indicate orientifold projection.

which forms the folded string, in which the retarded and advanced modes are exchanged every half-cycle.

The presented in Fig. 2 diagram shows a crossing symmetry of the four roots $\tau^\pm$ and $\bar\tau^\pm$, for the complex retarded time, which allows one to replace the Right complex conjugate retarded-time structure $x_R(\tau^-)$ by the antipodal Left advanced-time structure $x_L(\tau^+)$ and to work only in terms of the Left complex structures, omitting the index 'L'. For more details on this antipodal relation see [39, 19].

4. Calabi-Yau twofold from the Kerr theorem

4.1. Problem of the non-stationarity

The algebraically special exact KS solutions of the Einstein-Maxwell equations obtained in the remarkable paper [1] have two essential restrictions. First one is requirement of stationarity. In particular, the Kerr and KN solutions are stationary and their nonstationary generalizations are unknown. In the KS formalism [1] stationarity is characterized by a constant Killing direction K^μ, which corresponds to invariance of the metric $g_{\mu\nu}$ with respect to action of the operator $\hat K = K^\mu \partial_\mu$, i.e $\hat K g^{\mu\nu} = 0$. Using Cartesian coordinates, one can express K^μ via coordinates of the straight CWL $x_0^\mu(\tau)$, complex source of the Kerr and KN solutions, and obtain

$$K^\mu = \partial_\tau x_0^\mu(\tau) \tag{4.1}$$

Stationarity of the metric implies stationarity of the Kerr congruence $\hat K e^3 = 0$, which imposes the corresponding restriction on the coefficients A, B, C of the generating function of the Kerr theorem and the corresponding solutions

$$\hat K Y = \hat K \bar Y = 0 \tag{4.2}$$

All these functions turn out to be functions of the coordinates $x_L^\mu(\tau^-)$ and 4-velocity of the CWL,

$$u^\mu(\tau^-) = \dot x_L^\mu(\tau^-) \equiv \partial_\tau x_L^\mu(\tau)|_{\tau^-} \tag{4.3}$$

and finally they should depend on the complex retarded time parameter τ^-. [3] Although the determined by the Kerr theorem roots $Y^\pm(x, \tau^-)$ depend formally on the retarded time τ^-,

[3] Details of these relations are not essential for our treatment here and can be found in [23, 24].

for the straight CWL of the Kerr and KN solutions this dependence drops out from the final expressions, reducing to the dependence of the functions F and Y only on 4-velocity $u^\mu(\tau^-)$, or on the related parameters $K^\mu(\tau)$ [23].

The second restriction is related with idealization of the situation, absence of the external and internal sources.

The non-stationary (accelerating) KS solutions are unknown. However, the known Kinnersley solutions for accelerating non-rotating sources demonstrate that acceleration is always accompanied by radiation. The obtained in [40, 41] exact non-stationary em solutions on the stationary KS background show also that any em excitation of the Kerr geometry creates specific radiation in the form of singular beams (pp-waves) propagating along the null lines of the Kerr congruence. These beams are incoming on the negative sheet of the Kerr geometry, and propagate towards the disk $r = 0$, then passing analytically through the disk, the beam-pulses become outgoing and propagate towards future infinity. Some of the exact non-stationary em solutions do not create additional radiation and do not cause the recoil. However, these exact stationary solutions correspond to an idealized situation with a congruence, which transfers analytically from the negative to positive sheets of KS space-time and corresponds to an isolated stationary system without internal and external sources.

When we consider the regularized KN solution, the idealized analytic KN background is to be replaced by the background with a membrane source separating the bulk of the positive Kerr sheet, $r \geq r_e$, from the bulk of the negative sheet, $r \leq r_e$. The membrane source breaks analyticity of the Kerr solutions, and results in the scattering of the pp-wave radiation creating recoil and non-stationarity of the KS solutions. In the moment of the collision of the beam pulse with the membrane-source which closes the gate from negative to positive sheet at $r = r_e$, there should appear a kink in the parameters of CWL. Parameters of the Kerr ingoing congruence A_{in}, B_{in}, C_{in} should differ from parameters of the outgoing one, $A_{out}, B_{out}, C_{out}$. The in- and out-sources of the Kerr geometry turn out to be decorrelated and should be considered as independent sources, generating different congruences. In the simplest case, we can set $r_e = 0$, which corresponds to Keres-Israel-Hamity disk-like membrane source which blocks the gate from the negative to positive sheet, the ingoing (real or virtual) photon will be scattered by the membrane and produce a recoil, causing a jump, $K^\mu_{in} \neq K^\mu_{out}$, described by a typical Feynman graph.

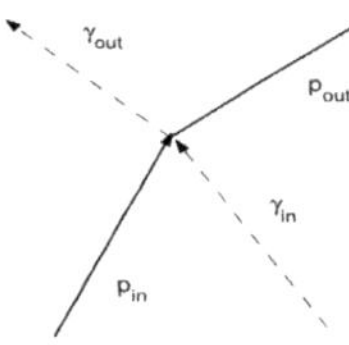

Figure 4. Recoil of the CWL caused by an electromagnetic beam or by a photon propagating from the past infinity through the Kerr ring to the future infinity.

In the modern physics, treatment of the classical problem of radiation reaction has reemerged in the connection with gravitational radiation and problems of quantum gravity. In the seminal work by DeWitt and Breme [42], the so-called 'bi-tensor' fields and Hadamard's 'elementary solution' were considered as a useful instrument for study of this problem in analogue with the two-point Green functions and the Feynman propagator. One can note that orientifold represents a stringy analogue of the 'bi-tensor' fields.

Due to kink of the trajectory at the moment τ_0, the relationship between the retarded and advanced roots is to be be biased (as it is shown in the diagram of Fig. 2), and the retarded coefficients of the Kerr generating function $A(\tau^-)$, $B(\tau^-)$, $C(\tau^-)$ should not match in general with the coefficients $A(\tau^+)$, $B(\tau^+)$, $C(\tau^+)$, corresponding to advanced time τ^+. As a result, the retarded and advanced world lines, $x_L^\mu(\tau^-)$ and $x_L^\mu(\tau^+)$, generating two different functions, F^+ and F^-, and correspondingly, two different twistorial manifolds have to be determined by a two-particle version of the Kerr theorem.

In accordance with the treatment of the multi-particle Kerr-Schild solutions [34, 35, 36], generating function of the two-particle system has to be determined by the function $F^{(2)}(T^A) = F^+(T^A)F^-(T^A)$, composed as a product of two functions F^+ and F^- corresponding to the retarded, $x_L(\tau^-)$, and advanced, $x_L(\tau^+)$, complex sources. The function

$$F^{(2)}(T^A, x_L^+, x_L^-) = F^+(T^A, x_L^+)F^-(T^A, x_L^-) \tag{4.4}$$

represents a bi-scalar, or two-point function associated with the world-sheet parity, $\tau_L^- \leftrightarrow \tau_L^+$.

Both factors, F^+ and F^- are quadratic in T^A, and each of the partial equations $F^+ = 0$ (or $F^- = 0$) generates a *quadric* in the projective twistor space CP^3 corresponding to the usual two-sheeted structure of the Kerr geometry with two different real Killing directions consistent with the Debney-Kerr-Schild formalism.

The 'product' manifold, determined by the equation

$$F^{(2)}(T^A, x_L^+, x_L^-) = 0 \tag{4.5}$$

corresponds to a fourfold described as *a quartic* in the projective twistor space CP^3, which is the Calabi-Yau (complex) twofold, or the well-known K3 surface used in diverse models of the string compactification and also by generalization of superstring theory to M-theory [17, 20]. We obtain here that dynamical generalization of the Kerr geometry requires splitting of the complex source of the Kerr geometry into independent retarded and advanced components described by the orientifold parity of the world-sheet, and application of the Kerr theorem creates the K3 surface in the projective twistor space CP^3.

4.2. Matching of the retarded and advanced fields

One sees that the fields at a fixed *real* point x^μ near the KN source may be represented as a sum of the retarded field generated by the past complex light cone (root τ^-) and the advanced field generated by the future complex light cone (root τ^+). For the stationary case, these points lie on the same straight CWL and generate two congruences of the same 3d form, which differ by their space-orientation. The generated by τ^- congruence is *outgoing* on the physical (positive) sheet, $r > 0$, while the congruence generated by τ^+ is formed by the *ingoing* rays, and therefore, its physical sheet is 'negative', and corresponds to $r < 0$ for the root τ^+. So far as the momenta of the complex Kerr source u_{in}^μ and u_{out}^μ are to be real, there may be set the time-ordering $\tau^- \prec \tau^+$, which shows that the retarded physical sheet precedes the sheet of advanced fields.

The four sheets of these congruences may be marked by the four solutions $Y^\pm(x, \tau^+)$ and $Y^\pm(x, \tau^-)$. For the stationary KN solution, the Kerr generating function $F(Y, x(\tau))$ is independent of τ, and as a result $F^+(Y) = F^-(Y)$, and the four congruences coincide pairwise,

$$Y^\pm(x, \tau^+) = Y^\mp(x, \tau^-) \tag{4.6}$$

forming a *global* orientifold parity for the stationary Kerr geometry.

The orientifold condition (3.13) matches the lines of Kerr congruences at the world-sheet of the complex Kerr string. On the real slice of the Kerr geometry, it induces matching of the Kerr congruence at the membrane-surface of the regularized KN solution which separates the external KN solution from the internal false-vacuum state. As we discussed in Sec. 2, in the simplest Keres-Israel-Hamity model of the KN source, this membrane represents the disk $r = 0$, which separates the 'positive' and 'negative' sheets of the Kerr geometry. The closed lightlike

Kerr string lies at the boundary of the disk at $r = 0$ and $\cos\theta = 0$, and this string is extended to membrane by the extra world-sheet parameter $\theta \in [0, \pi]$. One sees here close parallelism with the superstring/M-theory correspondence, where the heterotic string transfers to a membrane growing by the eleventh dimension [17].

The source of the *gravitating soliton* model represents a false-vacuum bubble, boundary of which represents a domain wall membrane providing smooth phase transition from external KN solution to a flat regular internal space. The membrane-boundary of the bubble has ellipsoidal form determined by the surface $r = r_{reg}$, and therefore, it fixes the boundary of analyticity of the KN geometry, or the boundaries of the 'physical sheets' for the advanced and retarded fields

$$\tau_L^-|_{r=r_{reg}} = t - r_{reg} - ia\cos\theta \equiv t - (r_{reg} + ia\cos\theta) = (\tau_R^-)|^*_{r=r_{reg}} = \tau_L^+|_{r=r_{reg}} \tag{4.7}$$

It means that analyticity of the Kerr congruence is to be broken on the boundary $r = r_{reg}$, where the retarded 'outgoing' Kerr congruence should be replaced by the 'ingoing' advanced congruence. As a result the both retarded and advanced fields should be taken into account for the domain wall boundary conditions at $r = r_{reg}$.

However, the antipodal relation $Y^+ = -1/\bar{Y}^-$ between the *retarded* KN solution on the positive physical sheet $Y^+(x, \tau^-)$ and the *advanced* conjugate solution $\bar{Y}^-$ on the negative sheet (see [39, 19]) allows us to consider these solutions as analytically related, displaying close analogue with formation of the Feynman propagator from a unique analytic multi-sheeted solution.

Using this analogue, one can define the physical space-time of a spinning particle by analytical connection of two of the four physical sheets of the K3 surface. For example, unification of the positive sheet of the retarded F^--solution $Y^+(x, \tau^-)$ with negative sheet of the advanced source (corresponding to F^+) may be considered as a physical space-time of some spinning particle, while the 'physical' spacetime for the corresponding *anti-particle* could be identified with alternative choice of two physical sheets: positive sheet solution $Y^-(x, \tau^+)$ of the advanced generating function F^+, extended analytically on the negative sheet of the retarded generator F^-.

5. Outlook

One sees that the Kerr-Schild geometry displays striking parallelism with basic structures of superstring theory. In particular, one of the striking results is the presence of inherent Calabi-Yau twofold in the complex twistorial structure of the Kerr geometry. In the recent paper [12] we argued that it is not accidental, because gravity is a fundamental part of the superstring theory, and the Kerr-Schild gravity, being based on the twistor theory, Kerr theorem and the wonderful multi-sheeted complex Kerr geometry [34, 35, 36], displays the most deep inherent structures of space-time which may lie beyond quantum physics.

In many respects the Kerr-Schild gravity represents a version of Witten's twistor-string theory [38], which is also four-dimensional, based on twistors and related with experimental particle physics. It has been observed recently [39, 19] that the complex Kerr string has much in common with the N=2 critical superstring [21, 20, 43], which is also related with twistors. The N=2 string is complex and has the complex critical dimension two, which corresponds to four real dimensions, indicating that it may lead to four-dimensional theory of superstrings. However, signature of the N=2 string may only be (2,2) or (4,0), which caused obstacles for embedding of this string in the space-times with Minkowskian signature. Up to our knowledge, the trouble was not resolved, and the initially enormous interest to N=2 string seems to be dampened Meanwhile, embedding of the N=2 string in the complexified Kerr geometry is very natural and hints that stringlike structures of the real and complex Kerr geometry may be related with underlying dynamics of the N=2 superstring [39, 19].

Along with wonderful parallelism with the standard superstring theory, the stringy system of the four-dimensional KN geometry displays very essential peculiarities. First of all note that the discussed regular source of the KN solution provides consistent matching with quantum theory, since it leads to a flat space-time in the Compton region and predicts the Compton size of the dressed electron and some other of its remarkable features [11]. Among them is the discussed above closed Kerr string positioned on the boundary of the Compton region. This string is lightlike and similar to Discrete Light Cone Quantization (DLCQ) circle of M(atrix) theory [Susskind L arXiv:hep-th/9704080]. It has been noted in [44] that such a string cannot be closed indeed This circular string should have the end points (D0-branes) which may be in the same space position, but have to be separated by a time-interval. It is supported by the discussed in Sec. 2 singular pole which appears as a consequence of the wave excitations of the gravitating soliton model. Therefore, along with the stringy D1-brane of the KN source, there appears the adjoined lightlike D0-brane (which is indeed a pair of D0-brane-antibrane system), and the extended source takes the form of a complex of the D0-D1-D2-D3 branes. The singular lightlike D0-component describes zitterbewegung of the point-like electron, while the complex of D1-D2-D3-branes determines its extended dressed shape [12].

The considered stringy structures of the real and complex Kerr geometry set a parallelism between the 4d Kerr geometry and superstring theory, indicating the potential role of this alliance in the particle physics. On the other hand, these relationships show that complexification of the Kerr geometry may serve as an alternative to traditional compactification of higher dimensions in superstring theory.

Acknowledgments

Author thanks T. Nieuwenhuizen, D. Gal'tsov, K. Stepaniants and L. Järv for useful discussions and comments, and also P. Kuusk for kind invitation to the *QQQ* conference *3Quantum: Algebra Geometry Information* (Tallinn 10-13 July 2012) and the financial support. This work was supported by RFBR grant 13-01-00602.

References

[1] Debney G C, Kerr R P and Schild A 1969 *J. Math. Phys.* **10** 1842
[2] Carter B 1968 Global structure of the *Phys. Rev.* **174** 1559
[3] Ivanenko D D and Burinskii A Ya 1975 *Izv. Vuz. Fiz.* **5** 135
[4] Burinskii A Ya 1974 *Sov. Phys. JETP* **39** 193
[5] Keres H 1967 *Zh. Exp Teor. Fiz.* **52** 768 (Russian); *Sov. Phys. JETP* **25** 504 (English)
[6] Israel W 1970 *Phys. Rev.* D **2** 641
[7] Hamity V 1976 *Phys. Lett.* A **56** 77
[8] López C A 1984 *Phys. Rev.* D **30** 313
[9] Gürses M and Gürsey F 1975 *J. Math. Phys.* **16** 2385
[10] Burinskii A, Elizalde E, Hildebrandt S R and Magli G 2002 *Phys. Rev.* D **65**, 064039
[11] Burinskii A 2010 *J. Phys. A: Math. Theor.* **43** 392001
[12] Burinskii A 2012 *J. Phys.: Conf. Series* **361** 012032
[13] Burinskii A 1995 *Phys. Rev.* D **52** 5826
[14] Burinskii A Ya 1994 String-like structures in complex Kerr geometry *Relativity Today* eds R P Kerr and Z Perjés (Budapest: Akadémiai Kiadó) p 149
[15] Burinskii A 1994 *Phys. Lett.* A **185** 441
[16] Adamo T M and Newman E T 2011 *Phys. Rev.* D **83** 044023
[17] Backer K, Backer M and Schwarz J H 2007 *String Theory and M-Theory: A Modern Introduction* (Cambridge Univ. Press)
[18] Dyson L M, Järv L and Johnson C V 2002 *JHEP* **0205** 019
[19] Burinskii A 2013 *Adv. High Energy Phys.* **2013** id 509749
[20] Green M B, Schwarz J H and Witten E 1987 *Superstring Theory* Vol. 1 (Cambridge Univ. Press)
[21] Ooguri H and Vafa C 1991 *Nucl. Phys.* B **361** 469; **367** 83
[22] Berret J, Gibbons G W, Perry M J, Pope C N and Ruback P 1994 *Int. J. Mod. Phys* A **9** 1457
[23] Burinskii A 2003 *Phys. Rev.* D **67** 124024

[24] Burinskii A 2012 *J. Phys.: Conf. Ser.* **343** 012019
[25] Burinskii A 2008 *Grav. Cosmol.* **14** 109
[26] Burinskii A 2004 *Grav. Cosmol.* **10** 50
[27] Burinskii A 2007 Kerr geometry beyond the quantum theory *Beyond the Quantum*, eds Th M Nieuwenhuizen *et al* (World Sci.) p 319
[28] Nieuwenhuizen Th M 2007 *Beyond the Quantum* eds Th M Nieuwenhuizen et al. (World Sci.) p 332
[29] Horowitz G and Steif A 1990 *Phys. Rev. Lett.* **64** 260
[30] Sen A 1992 *Nucl. Phys.* B **388** 457
[31] Sen A 1992 *Phys. Rev. Lett.* **69** 1006
[32] Newman E T 1973 *J. Math. Phys.* **14** 102
[33] Penrose R 1967 *J. Math. Phys.* **8** 345
[34] Burinskii A 2005 *Grav. Cosmol.* **11** 301
[35] Burinskii A 2006 *Grav. Cosmol.* **12** 119
[36] Burinskii A 2007 *Int. J. Geom. Meth. Mod. Phys.* **4** 437
[37] Whittaker E T and Watson G N 1969 *A Course of Modern Analysis* (Cambrige Univ. Press) excercise p 400
[38] Witten E 2004 *Comm. Math. Phys.* **252** 189
[39] Burinskii A 2013 *Theor. Mat. Phys.* **177** 1492
[40] Burinskii A 2009 *Gen. Relativity Gravitation* **41** 2281
[41] Burinskii A 2010 *J. Phys.: Conf. Ser.* **222** 012044
[42] De Witt B S and Breme R W 1960 *Ann. Phys.* **9** 220
[43] D'Adda A and Lizzi F 1987 *Phys. Lett.* B **191** 85
[44] Burinskii A 2003 *Phys. Rev.* D **68** 105004

Journal of Physics: Conference Series **532** (2014) 012005 doi:10.1088/1742-6596/532/1/012005

$3j$-symbol for the modular double $SL_q(2,\mathbb{R})$ revisited

S E Derkachev and L D Faddeev

St. Petersburg Department of Steklov Mathematical Institute, Russian Academy of Sciences
Fontanka 27, 191023 St. Petersburg, Russia

Abstract. Modular double of quantum group $\mathrm{SL}_q(2,\mathbb{R})$ with $|q| = 1$ has a series of selfadjoint irreducible representations π_s parameterized by $s \in \mathbb{R}_+$. Ponsot and Teschner in [*Comm. Math. Phys.* **224** (2001) 613] considered a decomposition of the tensor product $\pi_{s_1} \otimes \pi_{s_2}$ into irreducibles. In our paper we give more detailed derivation and some new proofs.

1. Introduction

Conformal Field Theory is one of the main sources of quantum groups. The very first example of deformed algebra G_q of functions on SL(2) was given by the monodromy matrix for the quantized Lax operator of the Liouville model [1]. The variable τ, entering the deformation parameter $q = e^{i\pi\tau}$ played the role of the coupling constant. The duality $\tau \to -1/\tau$ observed in [2, 3] was formalized in [4] in the notion of modular double.

The irreducible representations of the modular double of $\mathrm{SL}_q(2,\mathbb{R})$, introduced in [4], were investigated in [5, 6]. In particular in a remarkable paper [5] the problem of the decomposition of the tensor product is solved.

There is an intriguing connection of the representations of the modular double of $\mathrm{SL}_q(2,\mathbb{R})$ and primary fields of the Liouville model. Both the representation π_s of the modular double and vertex operators $V_\alpha(x) = \exp(\alpha\phi(x))$ are labeled by the same number $\alpha = 1/2 + is$, $s > 0$. Apparently there should be a correspondence between the operator expansion of $V_{\alpha_1}(x_1)V_{\alpha_2}(x_2)$ and the decomposition of $\pi_{s_1} \otimes \pi_{s_2}$ into irreducibles. Some indications on such connection can be found in [7]. However the work in this direction is still to be done. Having this in mind we decided to rederive the results of the paper of Ponsot and Teschner [5] and supply more details of derivations and proofs. Our paper is a complete exposition of the talks of the second author (L.D.F) in the early summer of 2012. The first author (S.E.D) joined the company in the late summer and his contribution let to important improvement of the full exposition.

2. Modular double of $\mathrm{SL}_q(2,\mathbb{R})$

The algebra has six generators, combined in two mutually commuting triplets E, F, K and $\tilde{E}, \tilde{F}, \tilde{K}$. The usual relations [8, 9] for E, F, K read

$$KE = q^2 EK, \quad KF = q^{-2} FK, \quad EF - FE = \frac{K - K^{-1}}{q - q^{-1}}$$

with $q = \mathrm{e}^{i\pi\tau}$ are supplemented by similar relations for $\tilde{E}, \tilde{F}, \tilde{K}$ with $\tilde{q} = \mathrm{e}^{i\pi/\tau}$. Generators E, F, K and $\tilde{E}, \tilde{F}, \tilde{K}$ commute. The coproduct is given by

$$\Delta(E) = E \otimes K + I \otimes E, \quad \Delta(F) = F \otimes I + K^{-1} \otimes F, \quad \Delta(K) = K \otimes K \tag{2.1}$$

and similarly for $\tilde{E}, \tilde{F}$ and $\tilde{K}$.

We shall use the irreducible representations which are equivalent to the used in [5, 6]. Let u and v realize the Weyl relations $uv = q^2vu$ and act in $L_2(\mathbb{R})$ by the explicit formulae

$$uf(x) = \exp\left(-\tfrac{i\pi x}{\omega}\right) f(x), \quad vf(x) = f(x + 2\omega')$$

Here ω and ω' are half-periods which substitute periods 1 and τ

$$\tau = \omega'/\omega, \quad \omega\omega' = -1/4, \quad \omega'' = \omega + \omega'$$

We shall consider the case $\tau > 0$, so that ω and ω' are pure imaginary with positive imaginary part. Operators u and v are unbounded and can be defined on the dense domain $\mathcal{D}$ consisting of the entire functions $f(x)$, rapidly vanishing at infinity along the lines $\operatorname{Im} x = const$. For instance we can take $f(x)$ in the form

$$f(x) = \mathrm{e}^{-\alpha x^2} \mathrm{e}^{\beta x} P(x)$$

for positive α, arbitrary complex β and polynomial $P(x)$. The operators u and v are nonnegative and essentially selfadjoint.

The representation π_s is given by formulae

$$E = \frac{i}{q - q^{-1}} e(s), \quad F = \frac{i}{q - q^{-1}} f(s)$$

where

$$e(s) = u^{-1}\left(qv + Z\right) = \left(q^{-1}v + Z\right) u^{-1}, \quad f(s) = u\left(1 + qZ^{-1}v^{-1}\right) = \left(1 + q^{-1}Z^{-1}v^{-1}\right) u$$

and

$$Z = \exp\left(-i\pi s/\omega\right), \quad \forall s \in \mathbb{R}, \quad K = v.$$

The representation for the second triple is given by similar formulae in terms of $\tilde{u}, \tilde{v}$ and $\tilde{Z}$

$$\tilde{u} = u^{1/\tau}, \quad \tilde{v} = v^{1/\tau}, \quad \tilde{Z} = Z^{1/\tau}$$

so that

$$\tilde{u}f(x) = \exp\left(-i\pi x/\omega'\right) f(x), \quad \tilde{v}f(x) = f(x + 2\omega)$$

and $\tilde{Z} = \exp\left(-i\pi s/\omega'\right)$. Thus $\tilde{u}, \tilde{v}$ and $\tilde{Z}$ are obtained from u, v and Z by interchange $\omega \rightleftarrows \omega'$.

Operators $\tilde{u}, \tilde{v}$ have the same domain and are nonnegative and essentially selfadjoint.

Let us note that there is the second regime for τ which could be called real form for $\mathrm{SL}_q(2)$. It is the case $|\tau| = 1$ or $\omega' = -\bar{\omega}$. In this case the involution interchanges pairs u, v and $\tilde{u}, \tilde{v}$ by $u^* = \tilde{u}$, $v^* = \tilde{v}$. This regime has many interesting features. In particular it corresponds to the value of central charge of Liouville model in the interval between 1 and 25. However in this paper we shall consider only regime $\tau > 0$.

3. Modular quantum dilogarithm

We shall use this term for the function

$$\gamma(x) = \exp\left\{-\frac{1}{4}\int_{-\infty}^{+\infty}\frac{\mathrm{d}t}{t}\frac{e^{itx}}{\sin(\omega t)\sin(\omega' t)}\right\}$$

where the contour goes above the singularity at $t = 0$. This function has long history, different names, normalizations and applications. The normalization used in this paper is used in [10, 11].

The term modular quantum dilogarithm is used for the function

$$\Phi(u) = \gamma(x), \quad u = \exp\left(-i\pi x/\omega\right)$$

The adjective "modular" is due to the symmetry of $\Phi(u)$ after exchange $\omega \rightleftarrows \omega'$ and term "dilogarithm" is due to the asymptotic for $\tau \to 0$

$$\Phi(u) \to \exp\frac{1}{2\pi i\tau}\mathrm{Li}_2(-u)$$

where

$$\mathrm{Li}_2(u) = \sum_{n=1}^{\infty} u^n/n^2$$

containing the Euler dilogarithm. Finally "quantum" is a conventional term for q-deformation.

The main property is the functional equation

$$\frac{\Phi\left(qu\right)}{\Phi\left(q^{-1}u\right)} = \frac{1}{1+u} \tag{3.1}$$

and similar one for the shift with $\tilde{q}$. In terms of function $\gamma(x)$ these equations look as follows

$$\gamma(x+\omega')/\gamma(x-\omega') = 1+e^{-\frac{i\pi}{\omega}x}, \quad \gamma(x+\omega)/\gamma(x-\omega) = 1+e^{-\frac{i\pi}{\omega'}x}$$

We shall need also the reflection formula

$$\gamma(x)\gamma(-x) = e^{i\beta}e^{i\pi x^2}, \quad \beta = \frac{\pi}{12}\left(\frac{\omega'}{\omega}+\frac{\omega}{\omega'}\right) \tag{3.2}$$

and formula for the complex conjugation

$$\overline{\gamma(x)} = 1/\gamma(\overline{x}) \tag{3.3}$$

Asymptotic behaviour is $\gamma(x) \to 1$ for $\mathrm{Re}(x) \to +\infty$ and reflection formula (3.2) can be used to get asymptotic for $\mathrm{Re}(x) \to -\infty$.

The function $\gamma(x)$ has a pole at the point $x = -\omega''$ and zero at the point $x = +\omega''$. The first terms of the series expansions are [10, 11, 14]

$$\gamma(-\omega''+z) = -\frac{1}{2\pi i c}\frac{1}{z}+\ldots, \quad \gamma(\omega''+z) = \frac{2\pi i}{c}z+\ldots \tag{3.4}$$

Comparison of (3.2) and (3.4) gives

$$c^2 = e^{-i\beta}e^{-i\pi\omega''^2} = ie^{2i\beta}, \quad c = e^{i\beta+\frac{i\pi}{4}}$$

There are main integral identities [10, 11]

$$\int_{\mathbb{R}}\mathrm{d}te^{-2\pi itz}\frac{1}{\gamma(\omega''-i0-t)} = c\gamma(z-\omega'') \tag{3.5a}$$

$$\int_{\mathbb{R}}\mathrm{d}te^{-2\pi itz}\frac{\gamma(x-t)}{\gamma(\omega''-i0-t)} = c\frac{\gamma(x)\gamma(z-\omega'')}{\gamma(x+z)} \tag{3.5b}$$

$$\int_{\mathbb{R}}\mathrm{d}te^{-2\pi itz}\frac{\gamma(x-t)\gamma(y-t)}{\gamma(\omega''-i0-t)\gamma(x+y+z+\omega''-i0-t)} = c\frac{\gamma(x)\gamma(y)\gamma(z-\omega'')}{\gamma(x+z)\gamma(y+z)} \tag{3.5c}$$

Journal of Physics: Conference Series **532** (2014) 012005 doi:10.1088/1742-6596/532/1/012005

3.1. *Intertwining operator*

Let us use the functional equation to show equivalence of the representations π_s and π_{-s}. We should find an operator $A(s)$ such that

$$e(s)A(s) = A(s)e(-s), \quad f(s)A(s) = A(s)f(-s), \quad vA(s) = A(s)v \tag{3.6}$$

The last equation indicates that $A(s)$ is an operator of convolution with kernel of the form

$$A(x, y, s) = A(x - y, s) = \int_{-\infty}^{+\infty} \mathrm{d}t e^{2\pi i t(x-y)} \hat{A}(t)$$

It is advisable to make Fourier transform. Let F be operator

$$[Ff](x) = \int_{-\infty}^{+\infty} \mathrm{d}y e^{-2\pi i x y} f(y) = \hat{f}(x)$$

We have $uF = Fv$ and $vF = Fu^{-1}$ and after conjugation by F we get from (3.1)

$$v^{-1}\left(qu^{-1} + Z\right)\hat{A}(u) = \hat{A}(u)v^{-1}\left(qu^{-1} + Z^{-1}\right)$$
$$v\left(1 + qZ^{-1}u\right)\hat{A}(u) = \hat{A}(u)v\left(1 + qZu\right)$$

We move v^{-1} and v to the right and cancel. After that we obtain the equations

$$\left(q^{-1}u^{-1} + Z\right)\hat{A}(q^2u) = \hat{A}(u)\left(q^{-1}u^{-1} + Z^{-1}\right)$$
$$\left(1 + q^{-1}Z^{-1}u\right)\hat{A}(q^{-2}u) = \hat{A}(u)\left(1 + q^{-1}Zu\right)$$

which are equivalent and differs by the change $u \to q^2u$ so that really we have only one equation

$$\left(1 + qZu\right)\hat{A}(q^2u) = \hat{A}(u)\left(1 + qZ^{-1}u\right)$$

It is evident that solution of this equation is

$$\hat{A}(u) = \frac{\Phi(Zu)}{\Phi(Z^{-1}u)}$$

and finally

$$A(s) = F^{-1}\frac{\Phi(Zu)}{\Phi(Z^{-1}u)}F \tag{3.7}$$

The unitarity of the operator $A(s)$ is evident due to (3.3).

4. The main equation for the 3j-symbol

In this section we shall solve the system of equations for the function $S(x_1, x_2, x_3|s_1, s_2, s_3)$

$$e_{12}(s_1, s_2)S = e_3'(s_3)S \tag{4.1a}$$
$$f_{12}(s_1, s_2)S = f_3'(s_3)S \tag{4.1b}$$
$$K_{12}S = K_3'S \tag{4.1c}$$

where the operators e_{12}, f_{12} and K_{12} act on variables x_1 and x_2 by

$$\Delta(E) = E_{12} = E_1K_2 + E_2, \quad \Delta(F) = F_{12} = F_1 + K_1^{-1}F_2, \quad \Delta(K) = K_{12} = K_1K_2$$

and e'_3, f'_3, K'_3 acts on variable x_3. These operators can be obtained from the transposition $u' = u$, $v' = v^{-1}$ and are given by $K'_3 = v_3^{-1}$,

$$e'_3 = \left(Z_3 + qv_3^{-1}\right) u_3^{-1} = u_3^{-1}\left(Z_3 + q^{-1}v_3^{-1}\right), \quad f'_3 = \left(1 + qZ_3^{-1}v_3\right) u_3 = u_3\left(1 + q^{-1}Z_3^{-1}v_3\right)$$

It is already clear that function S realizes the decomposition of the representation $\pi_{s_1} \otimes \pi_{s_2}$ into irreducibles. More will be said in the end of the section.

Following the previous section we shall often use variables u_i instead of x_i. Equation (4.1c),

$$v_1 v_2 v_3 S = S \tag{4.2}$$

allows to exclude v_2 from the equation (4.1a) and v_2^{-1} from the equation (4.1b) to get the system of equations for $v_1^{-1}S/S$ and $v_3 S/S$ as follows:

$$\begin{aligned}&\left(u_2 + qZ_1^{-1}u_1\right) v_1^{-1}S + \left(qZ_2^{-1}u_2 - q^{-1}Z_3^{-1}u_3\right) v_3 S = \left(u_3 - u_1\right) S\\ &\left(Z_1 u_1^{-1} + qu_2^{-1}\right) v_1^{-1}S + \left(Z_2 u_2^{-1} - q^2 Z_3 u_3^{-1}\right) v_3 S = qu_1^{-1}u_3^{-1}\left(u_1 - u_3\right) S\end{aligned}$$

or after diagonalization

$$\frac{v_1^{-1}S}{S} = qZ_1 \frac{u_1 - u_3}{Z_1 u_3 + qZ_2 Z_3 u_1} \frac{u_2 - Z_2 Z_3 u_1}{qu_1 + Z_1 u_2}, \quad \frac{v_3 S}{S} = qZ_2 Z_3 \frac{u_1 - u_3}{Z_1 u_3 + qZ_2 Z_3 u_1} \frac{qu_2 + Z_1 u_3}{Z_2 u_3 - qZ_3 u_2}$$

More explicitly these equations read

$$\begin{aligned}S(q^2 u_1, u_2, u_3) &= \frac{1}{Z_2 Z_3} \frac{1 - \frac{u_3}{u_1}}{1 + \frac{Z_1}{qZ_2 Z_3}\frac{u_3}{u_1}} \frac{1 - Z_2 Z_3 \frac{u_1}{u_2}}{1 + \frac{q}{Z_1}\frac{u_1}{u_2}} S(u_1, u_2, u_3)\\ S(u_1, u_2, q^{-2}u_3) &= \frac{Z_1}{Z_2} \frac{1 - \frac{u_3}{u_1}}{1 + \frac{Z_1}{qZ_2 Z_3}\frac{u_3}{u_1}} \frac{1 + \frac{q}{Z_1}\frac{u_2}{u_3}}{1 - \frac{qZ_3}{Z_2}\frac{u_2}{u_3}} S(u_1, u_2, u_3)\end{aligned}$$

and can be supplemented by

$$S\left(u_1, q^{-2}u_2, u_3\right) = \frac{1}{Z_1 Z_3} \frac{1 - Z_2 Z_3 \frac{u_1}{u_2}}{1 + \frac{q}{Z_1}\frac{u_1}{u_2}} \frac{1 - \frac{Z_3}{qZ_2}\frac{u_2}{u_3}}{1 + \frac{1}{qZ_1}\frac{u_2}{u_3}} S(u_1, u_2, u_3)$$

which is a corollary of (4.2). Let us look for solution in the form

$$S = \frac{\Phi\left(\alpha_1 \frac{u_2}{u_3}\right) \Phi\left(\alpha_2 \frac{u_3}{u_1}\right) \Phi\left(\alpha_3 \frac{u_1}{u_2}\right)}{\Phi\left(\beta_1 \frac{u_2}{u_3}\right) \Phi\left(\beta_2 \frac{u_3}{u_1}\right) \Phi\left(\beta_3 \frac{u_1}{u_2}\right)} S_1 \tag{4.3}$$

and due to (3.1) the choice

$$\alpha_1 = -qZ_3 Z_2^{-1}, \quad \alpha_2 = -q, \quad \alpha_3 = Z_1^{-1}, \quad \beta_1 = Z_1^{-1}, \quad \beta_2 = Z_1 Z_2^{-1} Z_3^{-1}, \quad \beta_3 = -Z_2 Z_3 q^{-1}$$

reduces the equations to

$$\begin{aligned}&S_1(q^2 u_1, u_2, u_3) = Z_2^{-1} Z_3^{-1} S_1(u_1, u_2, u_3)\\ &S_1(u_1, u_2, q^{-2}u_3) = Z_1 Z_2^{-1} S_1(u_1, u_2, u_3)\\ &S_1\left(u_1, q^{-2}u_2, u_3\right) = Z_1^{-1} Z_3^{-1} S_1(u_1, u_2, u_3)\end{aligned}$$

The dual equations, corresponding to the interchange $\omega \rightleftarrows \omega'$, have the same solution (4.3) and after that the Ansatz

$$S_1(x_1, x_2, x_3) = \exp -2\pi i\,(s_1 x_{23} + s_2 x_{31} + s_3 x_{21})\, S_0(x_1, x_2, x_3)$$

where $x_{ik} = x_i - x_k$, reduce the freedom to double periodic function S_0, which has to be constant.

Thus the solution of the equations (4.1) is given by

$$S(x_1, x_2, x_3) = S_0 \exp -2\pi i\,(s_1 x_{23} + s_2 x_{31} + s_3 x_{21}) \times \\ \times \frac{\gamma(x_{12} - s_1)}{\gamma(x_{12} + s_2 + s_3 + \omega'')} \frac{\gamma(x_{23} + s_3 - s_2 - \omega'')}{\gamma(x_{23} - s_1)} \frac{\gamma(x_{31} - \omega'')}{\gamma(x_{31} + s_1 - s_2 - s_3)} \tag{4.4}$$

The appearance of ω'' is due to sign factors in (4.3),

$$-qu = \exp\left(-\frac{i\pi(x-\omega'')}{\omega}\right)$$

and the singularities here have to be understood as $\omega'' \to \omega'' - i0$, which will be explained in the course of the proof of completeness. The expression for S, equivalent to (4.4), was given in [5] without derivation.

Now we can interpret the result in more details. The solution exists for any triplet of real s_1, s_2, s_3 and is unique up to normalization constant. This means that the representation with "spin" s_3 enter the tensor product $\pi_{s_1} \otimes \pi_{s_2}$ once for any s_3. This can be formalized by the relation

$$\pi_{s_1} \otimes \pi_{s_2} = \int_0^{+\infty} ds_3 \rho(s_3) \pi_{s_3} \tag{4.5}$$

We can consider $S\,(x_1 x_2 x_3)$ as the kernel of the integral operator S defined by

$$\pi_{s_3} \xrightarrow{\mathrm{S}} \pi_{s_1} \otimes \pi_{s_2}, \quad f(x_3) \mapsto [\mathrm{S}f]\,(x_1, x_2) = \int_{\mathbb{R}} dx_3 S\,(x_1 x_2 x_3)\, f(x_3)$$

and equations (4.1) in operator form are

$$e_{12}\mathrm{S} = \mathrm{S}e_3, \quad f_{12}\mathrm{S} = \mathrm{S}f_3, \quad K_{12}\mathrm{S} = \mathrm{S}K_3 \tag{4.6}$$

The complex conjugate function $\overline{S\,(x_1 x_2 x_3)}$ has interpretation as the kernel of the projection operator

$$\pi_{s_1} \otimes \pi_{s_2} \xrightarrow{\mathrm{P}} \pi_{s_3}, \quad f(x_1, x_2) \xrightarrow{\mathrm{P}} [\mathrm{P}f]\,(x) = \int_{\mathbb{R}^2} dx_1 dx_2 \overline{S\,(x_1 x_2 x)} f(x_1, x_2)$$

The measure $\rho(s)$ in (4.5) should be found from normalization condition for the kernel $S(x_1, x_2, x_3)$, which will be obtained in the last section.

5. Undressing of the Casimir

To get the normalization for S it is useful to interpret it as an eigenfunction of the Casimir operator

$$C_{12} = f_{12} e_{12} - q K_{12} - q^{-1} K_{12}^{-1}$$

It is clear from (4.6), that as a function of x_1 and x_2, S satisfies the equation

$$C_{12}S = \left(Z_3 + Z_3^{-1}\right) S$$

where s_3 and x_3 play the role of parameters labeling eigenvalue and multiplicity.

Explicitly C_{12} can be written as

$$C_{12} = Z_2\frac{u_1}{u_2} + Z_2^{-1}\frac{u_2}{u_1} + \left(Z_1 + \frac{1}{q}\frac{u_2}{u_1}\right)\left(1 + \frac{q}{Z_1}\frac{u_1}{u_2}\right)v_2 + \left(Z_2 + \frac{Z_1}{qZ_2}\frac{u_2}{u_1}\right)\left(1 + \frac{q}{Z_1}\frac{u_1}{u_2}\right)v_1^{-1}$$
$$+ \frac{Z_1}{q^2}\frac{u_2}{u_1}\left(1 + \frac{q}{Z_1}\frac{u_1}{u_2}\right)\left(1 + \frac{q^3}{Z_1}\frac{u_1}{u_2}\right)v_1^{-1}v_2$$

or

$$C_{12} = Z_2\frac{u_1}{u_2} + Z_2^{-1}\frac{u_2}{u_1} + \left(Z_1 + \frac{1}{q}\frac{u_2}{u_1}\right)V_2 + \left(Z_2 + \frac{Z_1}{qZ_2}\frac{u_2}{u_1}\right)V_1^{-1} + \frac{Z_1}{q^2}\frac{u_2}{u_1}V_1^{-1}V_2$$

offer the substitution

$$V_1 = v_1\frac{1}{1 + \frac{q}{Z_1}\frac{u_1}{u_2}}, \quad V_2 = \left(1 + \frac{q}{Z_1}\frac{u_1}{u_2}\right)v_2$$

We shall introduce a series of adjoint transformations of C_{12} to reduce it to more simple form. These transformations we shall call "undressing".

The first step is to use operator R_1 to cancel factors in front of v_1 and v_2

$$R_1^{-1}V_1R_1 = v_1, \quad R_1^{-1}V_2R_1 = v_2$$

The solution is a multiplication operator by function

$$R_1 = \Phi\left(\frac{1}{Z_1}\frac{u_1}{u_2}\right)$$

The operator K_{12} is invariant under transformation by R_1, i.e $R_1^{-1}K_{12}R_1 = K_{12}$ and C_{12} transforms into

$$C'_{12} = R_1^{-1}C_{12}R_1 = Z_2\frac{u_1}{u_2} + Z_1v_2 + Z_2v_1^{-1} + Z_2^{-1}u_1^{-1}\left(1 + q^{-1}Z_1v_1^{-1}\right)U_2$$

where

$$U_2 = u_2\left(1 + q^{-1}Z_2v_2\right)$$

Now we find R_2 transforming U_2 to u_2, i.e $R_2^{-1}U_2R_2 = u_2$. It is clear that R_2 is similar to R_1 after interchange u_2 and v_2, which is given by the Fourier transformation which respect variable x_2, so that $R_2 = F_2^{-1}\tilde{R}_2F_2$ where $\tilde{R}_2$ is a multiplication by $\Phi(Z_2u_2)$. The operator K_{12} is invariant under transformation R_2 and operator C'_{12} acquires the form

$$C''_{12} = R_2^{-1}C'_{12}R_2 = Z_2\frac{u_1}{u_2} + \frac{1}{Z_2}\frac{u_2}{u_1} + Z_1V'_2 + \frac{Z_1}{qZ_2}\frac{u_2}{u_1}V_1'^{-1}$$

where

$$V'_1 = v_1\frac{1}{1 + \frac{qZ_2^2}{Z_1}\frac{u_1}{u_2}}, \quad V'_2 = \left(1 + \frac{qZ_2^2}{Z_1}\frac{u_1}{u_2}\right)v_2$$

Now we transform V_1' and V_2' to v_1 and v_2 by multiplication operator

$$R_3 = \Phi\left(\frac{Z_2^2}{Z_1}\frac{u_1}{u_2}\right)$$

which leaves K_{12} invariant and transform C''_{12} into

$$\tilde{C}_{12} = R_3^{-1}C''_{12}R_3 = Z_2\frac{u_1}{u_2} + \frac{1}{Z_2}\frac{u_2}{u_1} + Z_1 v_2 + \frac{Z_1}{qZ_2}\frac{u_2}{u_1}v_1^{-1}$$

Altogether the operator $\mathrm{A} = R_1F_2^{-1}\tilde{R}_2F_2R_3$ gives $\tilde{C}_{12}$ from C_{12} and leaves K_{12} invariant,

$$\tilde{C}_{12} = \mathrm{A}^{-1}C_{12}\mathrm{A}, \quad \mathrm{A}^{-1}K_{12}\mathrm{A} = K_{12}$$

In more explicit form A^{-1} acts on the function of two variables $f(x_1, x_2)$ as follows:

$$\left[\mathrm{A}^{-1}f\right](x_1, x_2) = \frac{1}{c\gamma(x_{12} - s_1 + 2s_2)}\int_{\mathbb{R}} \mathrm{d}te^{2\pi it(s_2-\omega'')}\frac{\gamma(-t-\omega''+i0)}{\gamma(x_1 - t - s_1)}f(x_1, t) \tag{5.1}$$

Operator $\tilde{C}_{12}$ is much more simple then C_{12} and the problem of simultaneous diagonalization of $\tilde{C}_{12}$ and K_{12} allows separation of variables. Consider equations for the corresponding eigenfunctions

$$K_{12}\Psi_p(x_1, x_2) = v_1v_2\Psi_p(x_1, x_2) = e^{\frac{i\pi p}{\omega}}\Psi_p(x_1, x_2)$$

$$\left(Z_2\frac{u_1}{u_2} + \frac{1}{Z_2}\frac{u_2}{u_1} + Z_1v_2 + \frac{Z_1}{qZ_2}\frac{u_2}{u_1}v_1^{-1}\right)\Psi_p(x_1, x_2) = \left(Z_3 + Z_3^{-1}\right)\Psi_p(x_1, x_2)$$

where we parameterize the eigenvalues by p and s_3. The first equation allows to exclude v_1^{-1} from the second to get

$$\left[Z_2\frac{u_1}{u_2} + \frac{1}{Z_2}\frac{u_2}{u_1} + Z_1\left(1 + \frac{e^{-\frac{i\pi p}{\omega}}}{qZ_2}\frac{u_2}{u_1}\right)v_2\right]\Psi_p(x_1, x_2) = \left(Z_3 + Z_3^{-1}\right)\Psi_p(x_1, x_2)$$

The general solution of the first equation is given by

$$\Psi_p(x_1, x_2) = e^{-2\pi ipx_1}\Psi_p(x_{21})$$

where $x_{21} = x_2 - x_1$ and after substitution

$$\Psi_p(x_{21}) = \frac{e^{-2\pi is_1x_{21}}}{\gamma(x_{21} + p - s_2)}\Psi(x_{21})$$

which eliminates the factor in front of v_2, we get

$$\left(Z_2\frac{u_1}{u_2} + \frac{1}{Z_2}\frac{u_2}{u_1} + v_2\right)\Psi(x_{21}) = \left(Z_3 + Z_3^{-1}\right)\Psi(x_{21})$$

Introducing the new operators

$$u = \frac{1}{Z_2}\frac{u_2}{u_1}, \quad v = v_2$$

we rewrite the remaining equation in the form

$$\left(v+u+u^{-1}\right)\Psi=\left(Z_3+Z_3^{-1}\right)\Psi$$

The operator in the LHS is well known in CFT. It appears as a trace of monodromy of Lax operator in the Liouville model [1]. In quantum Teichmüller thery it got the name of the length operator for geodesics. R. Kashaev has shown [12, 13] that this operator has continues spectrum in the interval $[2,\infty]$ with eigenvalues parameterized in the form $Z+Z^{-1}$ with

$$Z=\exp\left(-\tfrac{i\pi s}{\omega}\right),\quad s\geq 0$$

and eigenfunctions are given by

$$\phi(x,s)=e^{-i\pi(x-\omega'')^2}\gamma(x+s-\omega''+i0)\gamma(x-s-\omega''+i0) \tag{5.2}$$

The latter are even functions of s so that these can be considered for any $s\in\mathbb{R}$. Kashaev proved [12] the orthogonality and completeness for $\phi(x,s)$ in the form

$$\int_{\mathbb{R}} dx\overline{\phi(x,s)}\phi(x,s')=\rho^{-1}(s)\left[\delta(s-s')+\delta(s+s')\right] \tag{5.3}$$

$$\int_0^{+\infty} ds\rho(s)\overline{\phi(x,s)}\phi(y,s)=\delta(x-y) \tag{5.4}$$

with $\rho(s)$ given by

$$\rho(s)=M(s)M(-s)=-4\sin\frac{\pi s}{\omega}\sin\frac{\pi s}{\omega'}$$

where $M(s)$, which can be considered as analogue of the Jost function from scattering theory or Harish-Chandra-Gindikin-Karpelevich function from the theory of representations of $\mathrm{SL}(2,\mathbb{R})$, can be taken as

$$M(s)=ce^{-2i\pi s^2-2i\pi s\omega''}\gamma(2s+\omega'')$$

One can say that the operator

$$[Uf](s)=\int_{-\infty}^{+\infty} dxM(s)\phi(x,s)f(x)=F(s)$$

acts from $L_2(\mathbb{R})$ into the subspace of $L_2(\mathbb{R})$, defined by condition $F(s)=S(s)F(-s)$ where the reflection coefficient $S(s)$ is given by $S(s)=M(s)/M(-s)$. Incidentally, the same reflection coefficient appears in the discussion of the zero modes in the Liouville model in [15].

It is evident that integral operator P with the kernel

$$P(s,s')=\frac{1}{2}\left[\delta(s-s')+S(s)\delta(s+s')\right]$$

defines a projection and U maps $L_2(\mathbb{R})$ into subspace $PL_2(\mathbb{R})$. However the natural completeness

$$\int_{-\infty}^{+\infty} dsds'P(s,s')\overline{U(x,s)}U(y,s')=\delta(x-y)$$

reduces to (5.4) due to the fact, that $\phi(x,s)$ is an even function of s and property $\overline{M(s)}=M(-s)$. The inversion $s\to -s$ is evidently connected to the Weyl reflection. The proof of Kashaev results is given in Appendix.

After all we obtain the following expression for the eigenfunction of the undressed Casimir operator

$$\Psi_p(x_1,x_2)=e^{-2\pi ipx_1}\frac{e^{-2\pi is_1x_{21}}}{\gamma(x_{21}+p-s_2)}\phi(x_{21}-s_2,s_3) \tag{5.5}$$

and now we can use them to formulate the orthogonality and completeness for the kernel $S(x_1,x_2,x_3)$.

Journal of Physics: Conference Series **532** (2014) 012005 doi:10.1088/1742-6596/532/1/012005

6. Undressing of the eigenfunctions

First of all we have to find out the connection between $\Psi_p(x_1, x_2)$ and undressed eigenfunction $\mathrm{A}^{-1}S(x_1, x_2, x_3)$. The explicit expression for the undressed eigenfunction reads

$$\mathrm{A}^{-1}S(x_1,x_2,x_3) = S_0 c^{-1} e^{-2\pi i(s_1x_{23}+s_2x_{31}+s_3x_{21})}\frac{1}{\gamma(x_{12}-s_1+2s_2)}\frac{\gamma(x_{31}-\omega'')}{\gamma(x_{31}+s_1-s_2-s_3)}\times$$
$$\times\int_{\mathbb{R}}\mathrm{d}t e^{2\pi i t(s_2-s_1-s_3-\omega'')}\frac{\gamma(-t-\omega''+i0)\gamma(t-x_3+s_3-s_2-\omega''+i0)}{\gamma(x_1-t+s_2+s_3+\omega''+i0)\gamma(t-x_3-s_1)}$$

where undressing operator A^{-1} is given by (5.1). The t-integral is reduced to (3.5c) and can be calculated in explicit form so that we obtain

$$\mathrm{A}^{-1}S(x_1,x_2,x_3) = S_0 e^{2\pi i\omega''^2}\frac{e^{-i\pi(s_2-s_1+s_3)^2}}{\gamma(s_2-s_1-s_3)}\frac{e^{2\pi i(s_1+s_3)x_{12}}}{\gamma(x_{12}+s_2+s_3+\omega'')}\frac{e^{2\pi i(\omega''-s_3)x_{13}}\gamma(x_{23}+s_3-s_2-\omega'')}{\gamma(x_{13}+\omega'')}$$

Next step is the calculation of Fourier transformation with respect variable x_3 using (3.5b):

$$\begin{aligned}
S_p(x_1,x_2) &= \int_{\mathbb{R}}\mathrm{d}x_3 e^{-2\pi i p x_3}\mathrm{A}^{-1}S(x_1,x_2,x_3)\\
&= S_0 e^{2\pi i\omega''^2}\frac{e^{-i\pi(s_2-s_1+s_3)^2}}{\gamma(s_2-s_1-s_3)}\frac{e^{2\pi i(s_1+s_3)x_{12}}}{\gamma(x_{12}+s_2+s_3+\omega'')}\times\\
&\quad\times e^{-2\pi i p x_1}\int_{\mathbb{R}}\mathrm{d}t e^{-2\pi i t(p-s_3+\omega'')}\frac{\gamma(x_{21}+s_3-s_2-\omega''+i0-t)}{\gamma(\omega''-i0-t)}\\
&= S_0 e^{2\pi i\omega''^2} c\frac{e^{-i\pi(s_2-s_1+s_3)^2}\gamma(p-s_3)}{\gamma(s_2-s_1-s_3)}\frac{e^{-2\pi i p x_1}}{\gamma(x_{21}-s_2+p)}\frac{e^{2\pi i(s_1+s_3)x_{12}}\gamma(x_{21}+s_3-s_2-\omega'')}{\gamma(x_{12}+s_2+s_3+\omega'')}
\end{aligned}$$

This expression coincides with (5.5),

$$S_p(x_1,x_2) = \mathrm{Z}(s_1,s_2|s_3,p)\Psi_p(x_1,x_2)$$

up to overall normalization $\mathrm{Z}(s_1,s_2|s_3,p)$,

$$\mathrm{Z}(s_1,s_2|s_3,p) = S_0 i e^{2\pi i\omega''^2}\frac{e^{-i\pi s_3^2-2\pi i s_3(s_2+\omega'')-i\pi(s_2-s_1+s_3)^2}\gamma(p-s_3)}{\gamma(s_2-s_1-s_3)}$$

The special choice of the initial normalization constant S_0 given by

$$S_0 = -ie^{-2\pi i\omega''^2}e^{i\pi s_3^2+2\pi i s_3(s_2+\omega'')+i\pi(s_2-s_1+s_3)^2}\gamma(s_2-s_1-s_3) = e^{2\pi i s_3\omega''}e^{i\phi} \tag{6.1}$$

where ϕ is real phase, leads to simplification

$$\mathrm{Z}(s_1,s_2|s_3,p) = \gamma(p-s_3)$$

so that we obtain properly normalized eigenfunctions

$$S_p(x_1,x_2) = e^{2\pi i s_3 p}\gamma(p-s_3)\frac{e^{-2\pi i p x_1}}{\gamma(x_{21}-s_2+p)}e^{-2\pi i s_1 x_{21}}\phi_{s_3}(x_{21}-s_2) \tag{6.2}$$

Now we shall prove the orthogonality and completeness of the functions $S(x_1, x_2, x_3)$ in the momentum representation

$$S(x_1, x_2, p) = \int_{\mathbb{R}} \mathrm{d}x_3 e^{-2\pi i p x_3} S(x_1, x_2, x_3)$$

It will be sufficient to prove the orthogonality and completeness for the undressed eigenfunctions $S_p(x_1, x_2)$ because the dressed eigenfunction $S(x_1, x_2, p) = \mathrm{A} S_p(x_1, x_2)$ is obtained from the $S_p(x_1, x_2)$ after action of the dressing unitary operator A. Due to unitarity this dressing operator effectively cancels out from considered relations.

Let us begin from orthogonality. There the dressing operator A cancels out on the first step

$$\begin{aligned}
\int_{\mathbb{R}^2} \mathrm{d}x_1 \mathrm{d}x_2 \overline{S\,(x_1 x_2 q)} S\,(x_1 x_2 p) &= \int_{\mathbb{R}^2} \mathrm{d}x_1 \mathrm{d}x_2 \overline{S_q\,(x_1 x_2)} S_p\,(x_1 x_2) \\
&= \frac{\gamma(p-s)}{\gamma(q-s')} \int_{\mathbb{R}^2} \mathrm{d}x_1 \mathrm{d}x_2 e^{-2\pi i (p-q) x_1} \frac{\gamma(x_{21} - s_2 + q)}{\gamma(x_{21} - s_2 + p)} \overline{\phi(x_{21} - s_2, s')} \phi(x_{21} - s_2, s) \\
&= \delta(p-q) \frac{\gamma(p-s)}{\gamma(p-s')} \int_{\mathbb{R}} \mathrm{d}x \overline{\phi(x, s')} \phi(x, s) \\
&= \rho^{-1}(s) \delta(p-q) \left[\delta(s-s') + \delta(s+s') \frac{\gamma(p-s)}{\gamma(p+s)} \right]
\end{aligned}$$

Note that the appearance of the second term containing $\delta\,(s+s')$ and kernel of the intertwining operator (3.7) is the direct consequence of the equivalence of representations π_s and π_{-s}.

The completeness for the undressed eigenfunction can be proven as follows

$$\begin{aligned}
\int_0^{+\infty} \mathrm{d}s \rho(s) \int_{\mathbb{R}} \mathrm{d}p \overline{S_p\,(x'_1 x'_2)} S_p\,(x_1 x_2) &= \int_{\mathbb{R}} \mathrm{d}p e^{-2\pi i p (x_1 - x'_1)} \frac{\gamma(x'_{21} - s_2 + p)}{\gamma(x_{21} - s_2 + p)} \times \\
&\quad \times \int_0^{+\infty} \mathrm{d}s \rho(s) \overline{\phi(x'_{21} - s_2, s)} \phi(x_{21} - s_2, s) \\
&= \int_{\mathbb{R}} \mathrm{d}p e^{-2\pi i p (x_1 - x'_1)} \frac{\gamma(x'_{21} - s_2 + p)}{\gamma(x_{21} - s_2 + p)} \delta(x'_{21} - x_{21}) \\
&= \delta(x'_1 - x_1) \delta(x'_2 - x_2)
\end{aligned}$$

Due to unitarity of the dressing operator the same relation holds for the dressed eigenfunctions,

$$\int_0^{+\infty} \mathrm{d}s \rho(s) \int_{\mathbb{R}} \mathrm{d}p \overline{S\,(x'_1 x'_2 p)} S\,(x_1 x_2 p) = \delta(x'_1 - x_1) \delta(x'_2 - x_2)$$

Appendix A. Orthogonality and completeness of Kashaev eigenfunctions

We shall need the generalization of identity (3.5b) in the form

$$\int_{\mathbb{R}} \mathrm{d}t e^{-2\pi i t s} \frac{\gamma(t+a)}{\gamma(t+b)} = c e^{2\pi i s (b - \omega'')} \frac{\gamma(a - b + \omega'') \gamma(-s - \omega'')}{\gamma(a - b - s + \omega'')} \tag{A.1}$$

$$= c^{-1} e^{2\pi i s (a + \omega'')} \frac{\gamma(b - a + s - \omega'')}{\gamma(b - a - \omega'') \gamma(s + \omega'')} \tag{A.2}$$

where the integral converges under conditions

$$\mathrm{Im}(s) < 0, \quad \mathrm{Im}(a - b - s) < 0 \tag{A.3}$$

The inverse formula is

$$\frac{\gamma(t+a)}{\gamma(t+b)} = \frac{1}{c\gamma(b-a-\omega'')} \int_{\mathbb{R}} \mathrm{d}s e^{2\pi i s(t+a+\omega'')} \frac{\gamma(s+b-a-\omega'')}{\gamma(s+\omega'')} \tag{A.4}$$

and the contour goes below the singularity at $s = 0$.

Appendix A.1. Orthogonality

We take eigenfunctions in the form

$$\phi(x,\lambda) = e^{-i\pi(x-\omega'')^2}\gamma(x+\lambda-\omega'')\gamma(x-\lambda-\omega'')$$

so that

$$\overline{\phi(x,\lambda)}\phi(x,\mu) = e^{4\pi i x\omega''}\frac{\gamma(x+\mu-\omega'')\gamma(x-\mu-\omega'')}{\gamma(x+\lambda+\omega'')\gamma(x-\lambda+\omega'')}$$

and the singularities here have to be understood as $\omega'' \to \omega'' - i0$ We have to calculate

$$I(\lambda,\mu) = \int_{\mathbb{R}} \mathrm{d}x\overline{\phi(x,\lambda)}\phi(x,\mu)$$

This function is even in λ and μ and it is sufficient to calculate it in one quadrant, say $\lambda > 0$ and $\mu < 0$. First we transform the ratio of two γ-functions using (A.4),

$$\frac{\gamma(x+\mu-\omega'')}{\gamma(x+\lambda+\omega'')} = \frac{1}{c\gamma(\lambda-\mu+\omega'')}\int_{\mathbb{R}} \mathrm{d}s e^{2\pi i s(x+\mu)}\frac{\gamma(s+\lambda-\mu+\omega'')}{\gamma(s+\omega'')}$$

and calculate the x-integral using (A.2),

$$\int_{\mathbb{R}} \mathrm{d}x e^{2\pi i x(s+2\omega'')}\frac{\gamma(x-\mu-\omega'')}{\gamma(x-\lambda+\omega'')} = c^{-1}e^{2\pi i\mu(s+2\omega'')}\frac{\gamma(\mu-\lambda-s-\omega'')}{\gamma(\mu-\lambda+\omega'')\gamma(-s-\omega'')}$$

and after these two steps arrive to the following expression for $I(\lambda,\mu)$:

$$I(\lambda,\mu) = \frac{1}{c^2\gamma(\lambda-\mu+\omega'')\gamma(\mu-\lambda+\omega'')}\int_{\mathbb{R}} \mathrm{d}s e^{4\pi i s\mu}\frac{\gamma(s+\lambda-\mu+\omega'')\gamma(\mu-s-\lambda-\omega'')}{\gamma(s+\omega'')\gamma(-s-\omega'')}$$

The ratio of γ-functions is reduced to the simple exponent

$$\frac{\gamma(s+\lambda-\mu+\omega'')\gamma(\mu-s-\lambda-\omega'')}{\gamma(s+\omega'')\gamma(-s-\omega'')} = e^{i\pi(s+\lambda-\mu+\omega'')^2 - i\pi(s+\omega'')^2}$$

due to reflection relation (3.2) so that we obtain for $\lambda > 0$ and $\mu < 0$

$$I(\lambda,\mu) = \frac{e^{i\pi(\lambda-\mu)^2+2i\pi\omega''(\lambda-\mu)}}{c^2\gamma(\lambda-\mu+\omega'')\gamma(\mu-\lambda+\omega'')}\int \mathrm{d}s e^{2i\pi s(\lambda+\mu)}$$

$$= \frac{e^{4i\pi\lambda^2+4i\pi\lambda\omega''}}{c^2\gamma(2\lambda+\omega'')\gamma(-2\lambda+\omega'')}\delta(\lambda+\mu)$$

The full answer is restored by the symmetry

$$\int_{\mathbb{R}} \mathrm{d}x\overline{\phi(x,\lambda)}\phi(x,\mu) = \frac{1}{M(\lambda)M(-\lambda)}\left[\delta(\lambda-\mu)+\delta(\lambda+\mu)\right]$$

where

$$M(\lambda) = ce^{-2i\pi\lambda^2-2i\pi\lambda\omega''}\gamma(2\lambda+\omega'')$$

It is exactly the formula (5.3) with

$$\rho(\lambda) = M(\lambda)M(-\lambda) = e^{-4\pi i\lambda\omega''}\frac{\gamma(2\lambda+\omega'')}{\gamma(2\lambda-\omega'')} = \left(e^{\frac{i\pi\lambda}{\omega}} - e^{-\frac{i\pi\lambda}{\omega}}\right)\left(e^{\frac{i\pi\lambda}{\omega'}} - e^{-\frac{i\pi\lambda}{\omega'}}\right)$$

Appendix A.2. Completeness
Now we have to calculate

$$I(x,y) = \int_0^{+\infty} \mathrm{d}\lambda \rho(\lambda)\overline{\phi(x,\lambda)}\phi(y,\lambda) = \int_{-\infty}^{+\infty} \mathrm{d}\lambda \sigma(\lambda)\overline{\phi(x,\lambda)}\phi(y,\lambda)$$

where we used the symmetry $\lambda \to -\lambda$,

$$\rho(\lambda) = \sigma(\lambda) + \sigma(-\lambda), \quad \sigma(\lambda) = e^{-4\pi i\lambda\omega''} - e^{-4\pi i\lambda(\omega''-2\omega')}$$

We introduce regularization and obtain the following expression

$$I(x,y) = e^{i\pi(x^2-y^2)-2i\pi(x-y)\omega''} \int_{\mathbb{R}} \mathrm{d}\lambda\sigma(\lambda)e^{2\pi\lambda\delta} \frac{\gamma(y+\lambda-\omega''+i\epsilon)\gamma(y-\lambda-\omega''+i\epsilon)}{\gamma(x+\lambda+\omega''-i\epsilon)\gamma(x-\lambda+\omega''-i\epsilon)}$$

where $\epsilon > 0, \delta > 0$ and $\delta > 2\epsilon$. First we transform the ratio of γ-functions using (A.4),

$$\frac{\gamma(y+\lambda-\omega''+i\epsilon)}{\gamma(x+\lambda+\omega''-i\epsilon)} = \frac{1}{c\gamma(x-y+\omega''-2i\epsilon)} \int_{\mathbb{R}} \mathrm{d}s e^{2\pi is(y+\lambda+i\epsilon)} \frac{\gamma(s+x-y+\omega''-2i\epsilon)}{\gamma(s+\omega'')}$$

where the contour goes below the singularity at $s=0$. Let us consider the λ-integral with the first contribution in $\sigma(\lambda)$ and for convenience make the change of the variable $\lambda \to -\lambda$,

$$I_1(s) = \int \mathrm{d}\lambda e^{-2\pi i\lambda(s-2\omega''+i\delta)} \frac{\gamma(y+\lambda-\omega''+i\epsilon)}{\gamma(x+\lambda+\omega''-i\epsilon)}$$

The second condition in (A.3) is fulfilled due to relation $\delta > 2\epsilon$ and using (A.2) we obtain

$$I_1(s) = c^{-1}e^{2\pi i(y+i\epsilon)(s-2\omega''+i\delta)} \frac{\gamma(x-y+s-\omega''+i(\delta-2\epsilon))}{\gamma(x-y+\omega''-2i\epsilon)\gamma(s-\omega''+i\delta)}$$

The same calculation with the second contribution in $\sigma(\lambda)$ gives

$$\begin{aligned} I_2(s) &= \int \mathrm{d}\lambda e^{-2\pi i\lambda(s-2\omega+2\omega'+i\delta)} \frac{\gamma(y+\lambda-\omega''+i\epsilon)}{\gamma(x+\lambda+\omega''-i\epsilon)} \\ &= c^{-1}e^{2\pi i(y+i\epsilon)(s-2\omega+2\omega'+i\delta)} \frac{\gamma(x-y+s+\omega''-2\omega+2\omega'+i(\delta-2\epsilon))}{\gamma(x-y+\omega''-2i\epsilon)\gamma(s+\omega''-2\omega+2\omega'+i\delta)} \end{aligned}$$

The change of variables $s \to s-2\omega'$ in s-integral containing $I_2(s)$ transforms ratio of s-dependent γ-functions to the form

$$\frac{\gamma(x-y+s+\omega-\omega'-2i\epsilon)}{\gamma(s+\omega-\omega')} \frac{\gamma(x-y+s-\omega+\omega'+i(\delta-2\epsilon))}{\gamma(s-\omega+\omega'+i\delta)}$$

where we restored s-dependent γ-functions from the first stage. In the corresponding s-integral containing $I_1(s)$ there is the following ratio of s-dependent γ-functions

$$\frac{\gamma(x-y+s+\omega''-2i\epsilon)}{\gamma(s+\omega'')} \frac{\gamma(x-y+s-\omega''+i(\delta-2\epsilon))}{\gamma(s-\omega''+i\delta)}$$

The formula

$$\gamma(z+\omega-\omega')\gamma(z-\omega+\omega') = \gamma(z+\omega'')\gamma(z-\omega'')$$

allows to transform one expression to another for $\delta = 0$. It means that in situation when it is possible to skip δ-regularization, we obtain the integral over closed contour. Let us deform the contour in integral with $I_1(s)$: the contour which goes above the singularity at $s = 0$ and a small closed contour around $s = 0$ leading to additional contribution

$$2\pi i \mathrm{Res}_{s=0} = \frac{1}{c}\frac{\gamma(x-y+\omega''-2i\epsilon)\gamma(x-y-\omega''-2i\epsilon+i\delta)}{\gamma(-\omega''+i\delta)}$$

In the remaining two integrals it is possible to put $\delta = 0$ and therefore to reduce it to the integral over closed contour without any singularity inside. As a result only the term $2\pi i\mathrm{Res}_{s=0}$ leads to nonzero contribution and restoring all needed factors we obtain

$$\frac{c\gamma(x-y-\omega''-2i\epsilon+i\delta)}{\gamma(x-y+\omega''-2i\epsilon)\gamma(-\omega''+i\delta)} \to \frac{1}{2\pi i}\frac{i\delta}{(x-y-2i\epsilon)(x-y+i\delta-2i\epsilon)} \to \delta(x-y)$$

so that

$$\int_0^{+\infty} \mathrm{d}\lambda \rho(\lambda)\overline{\phi(x,\lambda)}\phi(y,\lambda) = \delta(x-y)$$

Acknowledgments
One of the authors (L.D.F) is grateful to Professor J. Teschner for discussions. In particular, his unpublished notes "The Clebsch-Gordan maps for the modular double" received in the fall of 2012 contain procedure similar to our undressing as a part of Nidaiev and Teschner [arXiv:1302.3454]. Our work is partially supported by RFBR grants 11-01-00570-a, 11-01-12037-ofi-m and the RAS program "Mathematical problems of nonlinear dynamics".

References

[1] Faddeev L D and Takhtajan L A 1986 *Lectur Notes in Phys.* **246** 166
[2] Gervais J-L 1991 *Comm. Math. Phys.* **138** 301
[3] Zamolodchikov Alexander B and Zamolodchikov Alexei B 1996 *Nucl. Phys.* B **477** 577
[4] Faddeev L D 2000 Conf. Moshé Flato 1999 vol I; *Math. Phys. Stud.* **21** (Kluwer) p 149
[5] Ponsot B and Teschner J 2001 *Comm. Math. Phys.* **224** 613
[6] Bytsko A G and Teschner J 2003 *Comm. Math. Phys.* **240** 171
[7] Teschner J 2001 *Classical Quantum Gravity* **18** R153
[8] Jimbo M 1985 *Lett. Math. Phys.* **10** 63
[9] Drinfeld V G 1988 *J. Sov. Math.* **41** 898; 1986 *Zap. Nauchn. Semin.* **155** 18
[10] Faddeev L D, Kashaev R M and Volkov Yu A 2001 *Comm. Math. Phys.* **219** 199
[11] Volkov A Yu 2005 *Comm. Math. Phys.* **258** 257
[12] Kashaev R M 2001 On the spectrum of Dehn twists in quantum Teichmuller theory *Physics and Combinatorics* Nagoya 2000 (World Sci.) p 63
[13] Kashaev R M 2001 The quantum dilogarithm and Dehn twists in quantum Teichmuller theory *Integrable Structures of Exactly Solvable Two-Dimensional Models of Quantum Field Theory* (Kiev, 2000) *NATO Sci. Ser. II Math. Phys. Chem.* **35** (Kluwer) p 211
[14] Kashaev R M 2008 Discrete Liouville equation and Teichmuller theory *Handbook in Teichmuller Theory* IRMA Lect. Math. Theor. Phys. Vol III
[15] Faddeev L D and Volkov A Yu 2008 *J. Phys.* A **41** 194008

Journal of Physics: Conference Series **532** (2014) 012006 doi:10.1088/1742-6596/532/1/012006

Group theory aspects of chaotic strings

S Groote[1,2] and R Saar[2]
[1]Institut für Physik, Johannes-Gutenberg-Universität, Staudinger Weg 7, 55099 Mainz, Germany
[2]Füüsika Instituut, Tartu Ülikool, Ravila 4c, 50411 Tartu, Estonia

Abstract. Chaotic strings are a special type of non-hyperbolic coupled map lattices, exhibiting a rich structure of complex dynamical phenomena with a surprising correspondence to physical contents. Chaotic strings are generated by the Chebyshev maps $T_2(\phi)$ and $T_3(\phi)$. In this paper we connect the Chebyshev maps via the Galois theory to the cyclic groups Z_2 and Z_3 and give some ideas how this fundamental connection might lead to the emergence of the familiar Lie group structure of particle physics and, finally, even to the emergence of space-time. The Z_3-graded cubic and ternary algebras presented here have been introduced by R. Kerner in 1991 and then developed and elaborated in collaboration with many researches. We present here the most important results associated with these papers.

1. Introduction

As far as we understand by now, the relation of chaotic strings with elementary particle physics is given by the energy dependence of the couplings (usually called running) caused by the renormalization group of the theories of the Standard Model. Chaotic strings have a single coupling which, if being identified with the couplings of the Standard Model, unveils a surprising fact: The minima of the self energy of the chaotic string are related to the masses of the elementary particles [1, 2]. The details of this relation, however, are still quite unknown. Even though first progress has been made in understanding the perturbation theory for chaotic strings [3, 4] including also nonperturbative effects [5], the origin of the pattern for the self energy of the chaotic string appearing also for strings of length 3 [6] is still unclear.

In this paper we undertake the attempt to shed light on the relation by employing group theory. As in case of the theories of the Standard Model of elementary particles, group theory helps to organize particle states. However, in looking at the group theory aspects of chaotic strings it turns out that group theory appears prior to quantum mechanics and, in addition, space and time naturally emerge in the course of this. The approach we are following here is motivated more by inspiration than by manifest calculation and proof.

2. Symmetry group of chaotic strings

Chaotic strings are mathematical objects defined by the iterative map. For simplicity, we look only at one special case of chaotic strings given by

$$\phi_i^{n+1} = (1-a)T_N\left(\phi_i^n\right) + \frac{a}{2}\left(\phi_{i+1}^n + \phi_{i-1}^n\right)$$

As can be seen, besides a Laplace coupling between nearest neighbours chaotic strings are based on the Chebyshev polynomials $T_N(x)$, and it turns out that only the degrees $N = 2$ and $N = 3$ are relevant. Interesting for the relation searched for are characteristic points (maxima, minima and zeros) not only in the self energy as mentioned in the introduction but also in the map itself. According to the equation $T_N(\phi) = \cos(N \arccos\phi)$, characteristic points are the real values of $\phi^N = \pm 1$ and $\phi^N = i$, as it is displayed in Figure 1 as symbolic "wheels". To be more precise, the Galois group of the Chebyshev polynomial $T_N(x)$ is S_N, the symmetric group with N elements. The minimal (nontrivial) normal subgroup of S_N, finally, is the cyclic group $Z_N = Z/(NZ)$ which are something like turns of the weel by an angle of 120°.

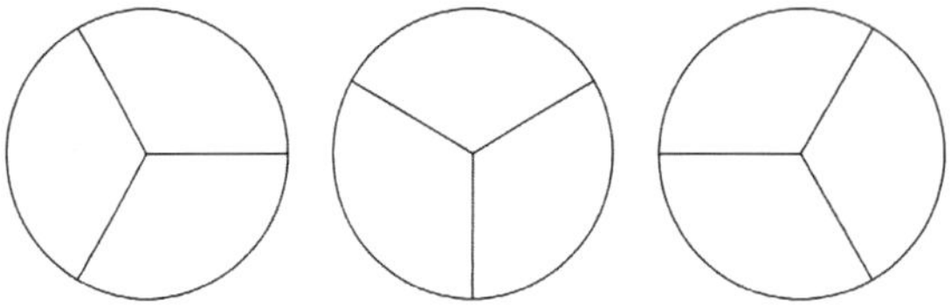

Figure 1. "Wheels" for maxima (left), zeros (middle) and minima (right)

During his short live, Évariste Galois (1811–1832) has enriched the mathematics of his time with interesting considerations about polynomials. One of these was the observation that a polynomial equation is solvable by roots if a chain of normal subgroups exists. This for instance exclude a polynomial equation of degree $N \geq 5$ from being solvable. Modern Galois theory is formulated in terms of (number) field extensions. In case of the Chebyshev polynomials, these field extensions would be $Q(\sqrt{2})$ and $Q(\sqrt{3})$.

3. Graded Z_N algebras

For this paper we stick with the Galois normal subgroups Z_2 and Z_3. For these groups one can define a graded algebra. Cubic and ternary Z_3-graded algebras have been introduced by Richard Kerner in 1991 [7] and developed and elaborated by R. Kerner and his co-authors V. Abramov, B. Le Roy, L. Vainerman and A. Borowiec in a series of publications dating from 1992 till 2012 [8, 9, 10, 11, 12]. If for instance θ is generator of the cyclic group Z_N, the graduation means that $\theta^N = 0$. A physical example of a graded Z_2 algebra is the creation operator θ^A of a quantum state $|A\rangle$, for one has

$$\theta^A|0\rangle = |A\rangle \quad \text{and} \quad (\theta^A)^2|0\rangle = \theta^A|A\rangle = 0$$

where the second equation expressing Pauli's exclusion principle for fermions. If there is more than one generator, we can speak of a graded algebra. In our example, with $M = 2$ generators there will be $M = 2$ different fermionic (spin) states, and the creation operators obey a Grassmann algebra[1]

$$\theta^A\theta^B + \theta^B\theta^A = 0$$

In general, for M states and M creation operators, in mathematical terms one can speak of a graded Z_2 algebra with M generators.

Surprisingly, there is a generalization of the graded Z_2 algebra well known from quantum mechanics to a graded Z_3 algebra which makes sense physically. To start from mathematics, the graduation of generators θ^A of Z_3 is established by distinguishing between elements θ^A of grade 1, $\theta^A\theta^B$ of grade 2, and $\theta^A\theta^B\theta^C$ of grade 3, the last one identified with grade 0. For this graded Z_3 algebra $\mathcal{A}$ a generalized Grassmann algebra holds,

$$\theta^A\theta^B\theta^C = j\theta^B\theta^C\theta^A = j^2\theta^C\theta^A\theta^B$$

[1] including also annihilation operators, one ends up with a Heisenberg algebra.

where $j = e^{2\pi i/3}$ so that $1 + j + j^2 = 0$ [8, 9, 10]. In addition, one can define a graded conjugate Z_3 algebra $\bar{\mathcal{A}}$ with generators $\bar{\theta}^{\bar{A}}$ of grade -1 (which, according to the graduation, is equivalent to grade 2) with the property

$$\theta^A \bar{\theta}^{\bar{B}} = -j \bar{\theta}^{\bar{B}} \theta^A$$

(note the minus sign on the right hand side). The graded (conjugate) Z_3 algebras can be used as model for (anti)quarks with three colours. The generators θ^A of grade 1 in this model can be seen as quarks, the generators $\bar{\theta}^{\bar{A}}$ of grade -1 as antiquarks, forming colour neutral mesonic and baryonic states with grade 0. In the following we will analyse these mesonic and baryonic forms for $M = 2$ and $M = 3$ generators [Kerner R]
[].

3.1. Graded Z_3 algebra with $M = 2$ generators

Using simple combinatorics we can determine how many independent elements are contained in the zero-graded subalgebras $\mathcal{A}_0$, $\bar{\mathcal{A}}_0$ and $(\mathcal{A} \times \bar{\mathcal{A}})_0$. There are

- $(M^3 - M)/3$ independent elements in $\mathcal{A}_0$ and $\bar{\mathcal{A}}_0$ (baryonic states),
- M^2 independent elements in $(\mathcal{A} \times \bar{\mathcal{A}})_0$ (mesonic states).

In case of $M = 2$ (two flavours for instance) one has two 3-forms (baryonic states p, n) and four 2-forms (mesonic states $\pi^\pm$, π^0, ρ). As seen for the baryonic states, the generalized Grassmann algebra implies that there are no three quarks in the same state. This restriction (no *uuu* and *ddd* states, only *uud* (proton) and *udd* (neutron)) can be named accordingly as generalized Pauli principle [11, 12]. Mathematically, these two possibilities are given by two independent choices for the 3-form

$$\psi^\alpha = \psi^\alpha_{ABC} \theta^A \theta^B \theta^C$$

Namely $\psi^1_{121} = 1$ (from which follows $\psi^1_{211} = j$ and $\psi^1_{112} = j^2$) and $\psi^2_{212} = 1$ ($\psi^2_{122} = j$, $\psi^2_{221} = j^2$). Claiming covariance under the transformation $\theta^{A'} = U^{A'}_A \theta^A$ of the generators, the corresponding 3-forms are related by $\psi' = L\psi$ where

$$L^{\alpha'}_\alpha \psi^\alpha_{ABC} = U^{A'}_A U^{B'}_B U^{C'}_C \psi'^{\alpha'}_{A'B'C'}$$

The matrix

$$L = \begin{pmatrix} L^{1'}_1 & L^{1'}_2 \\ L^{2'}_1 & L^{2'}_2 \end{pmatrix} = \begin{pmatrix} U^{1'}_1 & -U^{1'}_2 \\ -U^{2'}_1 & U^{2'}_2 \end{pmatrix} \det U$$

obeys $\det L = (\det U)^3 = 1$. It turns out to be isomorphic to the covering $SL(2, C)$ of the Lorentz group. L is unitary but U is unitary only up to j and j^2. This means that ψ transform as a free spinor (baryon) but the generator transform as bounded spinors (quarks). Looking at the mesonic states, there are four mixed 2-forms

$$\eta^\mu = \eta^\mu_{A\bar{B}} \theta^A \bar{\theta}^{\bar{B}}$$

Due to the reality condition $\eta^\mu_{A\bar{B}} \theta^A \bar{\theta}^{\bar{B}} = \bar{\eta}^\mu_{\bar{B}A} \bar{\theta}^{\bar{B}} \theta^A$ one obtains $\eta^\mu_{A\bar{B}} = -j^2 \bar{\eta}^\mu_{\bar{B}A}$. One possible representation (but not the only one) is via the Pauli spin matrices,

$$\eta^\mu_{A\bar{B}} = \frac{i\sigma^\mu_{A\bar{B}}}{\sqrt{2}}, \quad \bar{\eta}^\mu_{\bar{B}A} = -j \frac{i\bar{\sigma}^\mu_{\bar{B}A}}{\sqrt{2}}$$

where σ^0 is the 2×2 unit matrix. Using the "spinorial metric"

$$\epsilon^{12} = 1 = -\epsilon^{21}, \quad \epsilon^{\bar{1}\bar{2}} = 1 = -\epsilon^{\bar{2}\bar{1}}$$

the given representation makes it easy to see that

$$\eta^{\mu\nu} = \eta^{\mu}_{A\bar{B}}\eta^{\nu A\bar{B}} = \mathrm{diag}\,(+1; -1, -1, -1)$$

Surprising enough, we end up with the metric of flat space-time. One might speculate at this point that the forms η^μ themselves are something like tetrad fields, for instance of the Poincaré gauge theory of gravitation (see e.g. Ref. [13]). Correspondingly, under the transformation of θ^A and $\bar{\theta}^{\bar{B}}$ one has to claim covariance of the form, $\eta' = \Lambda\eta$ with

$$\Lambda^{\mu'}_{\mu}\eta^{\mu}_{A\bar{B}} = U^{A'}_{A}\bar{U}^{\bar{B}'}_{\bar{B}}\eta'^{\mu'}_{A'\bar{B}'}$$

This claim defines the 4×4 Lorentz group with elements Λ for the Lorentz covariant 4-vector η.

3.2. Graded Z_3 algebra for $M = 3$ generators

Given three generators Q^a, one has eight 3-forms (baryonic states) and nine mixed 2-forms (mesonic states). Analysing the transformation, it turns out that the 3-forms transform under the adjoint representation of $SU(3)$, the mixed 2-forms transform under the $3 \otimes \bar{3}$ representation of $SU(3)$, and the generators themselves transform under the fundamental representation of $SU(3)$, for a more detailed consideration we refer to [Kerner R arXiv:0901.3961].

4. Conclusions

We have seen that the graded Z_3 algebra with $M = 2$ generators transform under the Lorentz group while the algebra with $M = 3$ generators transform under the Lie group $SU(3)$. Looking at the whole stream of arguments, one can state that the group properties of the outer and inner degrees of freedom emerge from a graded algebra originating from the Chebyshev polynomials $T_3(\phi)$ and associated with (confined) quark states.

4.1. First the hen, then the egg – or vice versa*?*

Concerning the Lorentz transformations, we obtain a new point of view for the relation between the two representations [11, 12]. The traditional view (we call it "first the hen, then the egg") is that the Lorentz transformations $x \mapsto \Lambda x$ (explicitly $x^\mu \mapsto x'^\mu = \Lambda^\mu_\nu x^\nu$) and $j \to \Lambda j$ with vector currents

$$j^\mu = \langle\psi|\gamma|\psi\rangle \quad \mapsto \quad j'^\mu = \langle\psi'|\gamma^\mu|\psi'\rangle = \Lambda^\mu_\nu j^\nu$$

lead to the Lorentz transformation $\psi(x) \mapsto \psi'(x') = \psi'(\Lambda x) = L(\Lambda)\psi(x)$. The new point of view ("first the egg, then the hen") states that the transformation of spinors

$$\psi(x) \to \psi'(x') = \psi'(\Lambda(L)x) = L\psi(x)$$

is the primary transformation and the space-time transformation is the secondary one. This makes sense from the mathematical point of view because the covering group $SL(2, C)$ of the Lorentz group as the larger group releases information to the particle–antiparticle symmetry in covering the Lorentz group Λ of space-time.

4.2. Outlook

The ideas presented here are linked with a lot of more directions to think which are not included in this paper:

- Group theory seems to be the link between chaotic strings and quantum mechanics. This link should be developed as an instrument to specify properties related to the minima of the self energy of the chaotic strings.
- The considerations are done only for the Chebyshev polynomials which are the "motors" for the chaotic strings. While iterated Chebyshev polynomials are related to chaotic strings at coupling $a = 0$, these iterated polynomials are deformed for non-vanishing coupling $a \neq 0$, resulting in a modification of the Galois group. It is intuitively clear that from this we might obtain tetrad fields for a non-flat space-time and interactions between particles.
- As a link again between group theory and space-time emergence, quantum mechanics turns out to be "space-timeless". This point of view would support a quantum gravity in which the geometry of space-time (via general relativity or possible generalizations) is founded on quantum mechanics – and not vice versa (see, for instance, Ref. [14]).

In any case, new ideas are always welcome to our "think tank".

Acknowledgments

We are grateful to Richard Kerner who provided us with the newest ideas from his side. This work is supported by the Estonian Institutional Research Support No. IUT2-27, and by the Estonian Science Foundation under grant No. 8769.

References

[1] Beck C 2002 *Physica* D **171** 62
[2] Beck C 2002 *Spatio-Temporal Chaos and Vacuum Fluctuations of Quantized Fields* (World Sci.)
[3] Groote S and Beck C 2006 *Phys. Rev.* E **74** 046216
[4] Groote S and Beck C 2007 *Dyn. Sys.* **22** 219
[5] Groote S, Veermäe H and Beck C 2013 *Chaos Sol. Fract.* **53** 18
[6] Groote S and Veermäe H 2009 *Chaos Sol. Fract.* **41** 2354
[7] Kerner R 1991 *C. R. Acad. Sci. Paris* **312** série II 191
[8] Kerner R 1992 *J. Math. Phys.* **33** 403
[9] Abramov V, Kerner R and Le Roy B 1997 *J. Math. Phys.* **38** 1650
[10] Kerner R 1997 *Classical Quantum Gravity* **14** A203
[11] Kerner R 2012 *J. Phys.: Conf. Ser.* **343** 012056
[12] Kerner R 2012 *AIP Conf. Proc.* **1483** 144
[13] Hehl F W, von der Heyde P, Kerlick G D and Nester J M 1976 *Rev. Mod. Phys.* **48** 393
[14] Wüthrich C 2012 *Die allgemeine Relativitätstheorie als Ausgangspunkt einer Theorie der Quantengravitation* (Philosophie der Physik) Hrsg M Esfeld (Berlin: Suhrkamp) p 306

Lorentz invariance and gauge equivariance

S Groote[1,2], R Saar[2] and H Liivat[2]

[1]Institut für Physik, Johannes-Gutenberg-Universität, Staudinger Weg 7, 55099 Mainz, Germany
[2]Füüsika Instituut, Tartu Ülikool, Ravila4c, 50411 Tartu, Estonia

Abstract. Trying to place Lorentz and gauge transformations on the same foundation, it turns out that the first one generates invariance, the second one equivariance, at least for the abelian case. This similarity is not a hypothesis but is supported by and a consequence of the path integral formalism in quantum field theory.

1. Vector field: the background

In the framework of local gauge invariance, the fields are used to compensate the gauge dependence of the particle field. In QED, for instance, the local phase transformation $\psi \to e^{iq\lambda}\psi$ of the particle field (for instance, a Dirac spinor) is accompagnied by the gauge transformation $A \to A + \partial\lambda$ of the electromagnetic vector potential A (for short, we will call this potential the electromagnetic field in what follows). From this point of view, the vector field A is a background field, and it is legitimate to ask whether

- there is a relation between these two *a priori* transformations via a mechanism called "dynamical interaction principle", and whether
- we have a functional dependence of ψ on A similar to the dependence of ψ on x.

In the following we will show how the Poincaré group guides us to find such a "dynamical interaction". The Poincaré group is a semidirect product of the translation group $\mathcal{T}_{1,3}$ and the proper orthochronous Lorentz group (for short: Lorentz group) $\mathcal{L}$. In this way it combines Lorentz transformations Λ and space-time translations a on the linear space $E_{1,3}$ with metric $\eta_{\mu\nu} = \mathrm{diag}\,(+1; -1, -1, -1)$ by the prescription

$$(a, \Lambda) : x^\mu \to \Lambda^\mu{}_\nu x^\nu + a^\mu, \quad \Lambda^\mu{}_\rho \Lambda^\nu{}_\sigma \eta_{\mu\nu} = \eta_{\rho\sigma}$$

As we have the Lorentz and Poincaré group defined on space-time, we have to ask how these groups act on spinors. The answer is given by the commutative diagram

$$\begin{array}{rccc} \psi: & x & \longrightarrow & \psi(x) \\ \tau(\Lambda)\downarrow & \downarrow\Lambda & & \downarrow T(\Lambda) \\ \tau(\Lambda)\psi: & \Lambda x & \longrightarrow & T(\Lambda)\psi(x) \end{array}$$

A commutative diagram of this type is a scheme to express covariance. For the spinor ψ, the commutativity of the diagram means that

$$T(\Lambda)\psi(x) = (\tau(\Lambda)\psi)(\Lambda x) = \psi^\Lambda(\Lambda x)$$

Journal of Physics: Conference Series **532** (2014) 012007 doi:10.1088/1742-6596/532/1/012007

which can be read off from the right lower corner of the diagram. For the parametrization

$$\Lambda(\omega) = \exp\left(-\frac{1}{2}\omega_{\mu\nu}e^{\mu\nu}\right), \quad (e_{\mu\nu})_{\rho\sigma} := \eta_{\mu\sigma}\eta_{\nu\rho} - \eta_{\mu\rho}\eta_{\nu\sigma}$$

one obtains

$$T(\Lambda(\omega)) = \exp\left(-\frac{i}{2}\omega_{\mu\nu}s^{\mu\nu}\right)$$

While $T(\Lambda)$ transforms spinor and argument, the map $\tau(\Lambda)$ in the commutative diagram is given by

$$\tau(\Lambda(\omega)) = \exp\left(-\frac{i}{2}\omega_{\mu\nu}M^{\mu\nu}\right), \quad M_{\mu\nu} := \ell_{\mu\nu} + s_{\mu\nu}, \quad \ell_{\mu\nu} := i(x_\mu\partial_\nu - x_\nu\partial_\mu)$$

Together with the transformations $\exp(ia_\mu P^\mu)$ under the translation group one obtains

$$\tau(a,\Lambda) : \psi(x) \to (\tau(a,\Lambda)\psi)(x) = T(\Lambda)\psi\left(\Lambda^{-1}(x-a)\right)$$

This prescription is known as the Lorentz-Poincaré connection. The Poincaré algebra is constituted by $M^{\mu\nu}$ and P^μ.

2. Spinor field: three possibilities

Using the abbreviation $A^\lambda = A + \partial\lambda$, for the most general transformation $\psi(A) \to \psi^\lambda(A^\lambda)$ there are three different types of "gauge variance", namely

(i) $\psi^\lambda(A^\lambda) = G(\lambda)\psi(A)$ (gauge covariance)
(ii) $\psi^\lambda(A) = G(\lambda)\psi(A)$ (gauge invariance)
(iii) $\psi(A^\lambda) = G(\lambda)\psi(A)$ (gauge equivariance)

The first one (gauge covariance) imposes no constraints. Therefore, we consider the other ones.

2.1. Invariance: the traditional view

The second possibility (gauge invariance) effectively means that there is no dependence of the particle field on the gauge field at all. This is the traditional approach of quantum field theory where the dependence on A is not written explicitly. This approach allows for two independent fields ψ and A which can be quantized easily. While the dependence on A can formally be skipped, we have two *a priori* transformations

$$A \to A^\lambda = A + \partial\lambda, \quad \psi \to \psi^\lambda = e^{iq\lambda}\psi$$

which are actually related (via λ). Our proposal, therefore, is to sacrifice the independence of ψ and A in order to establish a relation between the transformations.

2.2. Equivariance: the dynamical view

Equivariance is known also as form invariance and is given by the commutative diagram

$$\begin{array}{rlcl} \psi: & A & \longrightarrow & \psi(A) \\ \mathrm{id}\downarrow & \downarrow\lambda & & \downarrow G(\lambda) \\ \psi: & A^\lambda = A + \partial\lambda & \longrightarrow & G(\lambda)\psi(A) \end{array}$$

which means that $\psi(A^\lambda) = G(\lambda)\psi(A)$. Note the Identity id on the left hand side of the diagram. Equivariance (or form invariance) is related to the form invariance of linear operators. An example for a linear operator is the covariant derivative $\mathcal{D}(A) = \partial - iqA$. Form invariance means $\mathcal{D}^\lambda = \mathcal{D}$ or, equivalently, the commutativity of

$$\begin{array}{rccc} \mathcal{D}\psi: & A & \longrightarrow & (\mathcal{D}\psi)(A) \\ \mathrm{id}\downarrow & \downarrow\lambda & & \downarrow G(\lambda) \\ \mathcal{D}\psi: & A^\lambda & \longrightarrow & G(\lambda)(\mathcal{D}\psi)(A) \end{array}$$

which means $\mathcal{D}(A^\lambda)\psi(A^\lambda) = G(\lambda)\mathcal{D}(A)\psi(A)$. Using the previous commutative diagram, we obtain

$$\mathcal{D}(A^\lambda)G(\lambda) = G(\lambda)\mathcal{D}(A)$$

on ψ space, i.e as weak relation.

3. Linear operators: the reference

"It is ... logical to assume that the fundamental equations of the elementary particles must be first-order equations of the form $\mathcal{D}\psi = 0$ and that all properties of the particles must be derivable from these without the use of any further subsidiary conditions." Bhabhas [1] statement is in contrast to statements about higher spin theories. However, in earlier works we have shown that these subsidiary conditions are indeed only consequences of the fundamental equations of motion [2, 3]. Instead, we build the "dynamical interaction" solely on the grounds of the Poincaré group [4, 5].

3.1. The Poincaré algebra

Relativistic field theories are based on the invariance under the Poincaré group $\mathcal{P}_{1,3}$. Therefore, linear operators serve as projectors onto $\mathcal{P}_{1,3}$. The Poincaré group $\mathcal{P}_{1,3}$ is based on the Poincaré algebra $p_{1,3}$. Accordingly, linear operators are generators of $p_{1,3}$. If we generalize the Poincaré algebra to a "dynamical" Poincaré algebra $p^d_{1,3}(A)$ depending on the field A, this "dynamical" algebra should obey the same relation as the linear operators,

$$p^d_{1,3}(A^\lambda)G(\lambda) = G(\lambda)p^d_{1,3}(A)$$

The introduction of a field dependent algebra in this way guarantees

- algebraic consistency (no additional constraints),
- causality (no Velo-Zwanziger problem), and
- the same magnetic moment $g = 2$ for arbitrary spin.

3.2. A nonsingular transformation

The simplest way to build a field-dependent algebra is to introduce the external field by a nonsingular transformation [6, 7]

$$p_{1,3} \to p^d_{1,3}(A) = \mathcal{V}(A)p_{1,3}\mathcal{V}^{-1}(A)$$

The condition of equivariance leads to

$$\mathcal{V}(A^\lambda)p_{1,3}\mathcal{V}^{-1}(A^\lambda) = G(\lambda)\mathcal{V}(A)p_{1,3}\mathcal{V}^{-1}(A)G^{-1}(\lambda)$$

which is solved on $p_{1,3}$ by

$$\mathcal{V}(A^\lambda) = G(\lambda)\mathcal{V}(A)$$

4. Lorentz type: the foundation

Assuming that the transformation operator $\mathcal{V}(a)$ obeys the Lorentz-type equation

$$\mathcal{V}(A)s^{\mu\nu}\mathcal{V}^{-1}(A) = V^{\mu}{}_{\rho}(A)V^{\nu}{}_{\sigma}(A)s^{\rho\sigma}$$

for the generators $s^{\mu\nu} \in p_{1,3}$ of the Poincaré group $\mathcal{P}_{1,3}$ with

$$V^{\mu}{}_{\rho}(A)V^{\nu}{}_{\sigma}(A)\eta_{\mu\nu} = \eta_{\rho\sigma}$$

the nonsingular transformation can be determined and, therefore, the Dirac equation can be solved at least in the case of a plane wave electromagnetic field $A = A(\xi)$, $\xi = k \cdot x$ with wave vector k ($k^2 = 0$), $G_{\mu\nu} = k_\mu A_\nu - A_\mu k_\nu$, and P the momentum vector of the particle [8]. One obtains ($k \cdot P \neq 0$ for massive particles)

$$V_{\mu\nu}(A) = \eta_{\mu\nu} - \frac{q}{k \cdot P}G_{\mu\nu} - \frac{q^2}{2(k \cdot P)^2}A^2 k_\mu k_\nu$$

which is an element of the little group of the propagator vector K. The solution of the Dirac equation reads $\psi(x, A) = \mathcal{V}(A)\psi(x)$ with $\mathcal{V}(A) = \mathcal{V}_0(A)\mathcal{V}_s(A)$, where

$$\mathcal{V}_0(A) \;=\; \exp\left\{i\int\frac{d\xi}{2k \cdot P}\left(2q(A \cdot P) - q^2A^2\right)\right\}, \quad \mathcal{V}_s(A) = \exp\left\{\frac{iq}{2k \cdot P}G_{\mu\nu}s^{\mu\nu}\right\} \quad (4.1)$$

5. Possible generalizations

Using the explicit solution in Eq. (4.1), one can see that $\mathcal{V}(A + \partial\lambda) = e^{iq\lambda}\mathcal{V}(a)$, so that $G(\lambda) = e^{iq\lambda}$ is a *consequence* of the gauge transformation. But there are obvious problems:

- By now, only the case of plane waves is treated.
- The dependence of $\mathcal{V}_0$ on A is nonlinear.

In addition, we would like to generalize the approach to more general field theories, even to nonabelian gauge theories. At the end of this paper we show first steps for possible generalizations by using either the Green's function formalism or the path integral formalism.

5.1. Green's function formalism

In the Green's function formalism the Dirac equation

$$(i\gamma^\mu\partial_\mu - m)\psi(x, A) = -q\gamma^\mu A_\mu(x)\psi(x, A)$$

can be solved by

$$\psi(x, A) = \psi(x, 0) + iq\int d^4x' S_F(x - x')\gamma^\mu A_\mu(x')\psi(x', A)$$

where $S_F(x-x')$ is the Feynman propagator of the fermion field. Iterating this integral equation we formally obtain a time ordered exponential function,

$$\psi(x, A) = \mathcal{T}\left\{\exp\left(iq\int d^4x' S_F(x - x')\gamma^\mu A_\mu(x')\right)\right\}\psi(x', 0)$$

where the first factor is something like $\mathcal{V}(A)$. It turns out that

$$\psi(x, A + \partial\lambda) = \exp\left(iq\lambda(x)\right)\psi(x, A)$$

5.2. Path integral formalism

In the path integral formalism the particle propagation is described by

$$\psi(x_b; A) = \int K(x_b, x_a; A)\psi(x_a; A)d^4x_a$$

The transition amplitude between the space-time points x_a and x_b reads

$$K(x_b, x_a; A) = \int_{x_a}^{x_b} \mathcal{D}x \, \exp[iS(x_b, x_a; A)]$$

where the action is given by

$$S(x_b, x_a; A) = -\int_a^b (mc^2 d\tau + qA^\mu dx_\mu)$$

Under a gauge transformation one obtains

$$S(x_b, x_a; A + \partial\lambda) = S(x_b, x_a; A) - q\left[\lambda(x_b) - \lambda(x_a)\right]$$

resulting in

$$K(x_b, x_a; A + \partial\lambda) = e^{-iq\lambda(x_b)} K(x_b, x_a; A) e^{iq\lambda(x_a)}$$

and, finally, in the commutativity of the diagram

$$\begin{array}{ccc} \psi(x_a, A) & \longrightarrow & \psi(x_b, A) \\ \downarrow \lambda & & \lambda \downarrow \\ e^{iq\lambda(x_a)}\psi(x_a, A) & \longrightarrow & e^{iq\lambda(x_b)}\psi(x_b, A) \end{array}$$

Therefore, we again end up with

$$\psi(x, A + \partial\lambda) = e^{iq\lambda(x)}\psi(x, A)$$

6. Conclusions

To conclude, we have shown the similarity between gauge and Lorentz transformations, and we have used this similarity to construct a "dynamical" interaction. In this context the gauge field appears as another kind of "coordinate". The approach presented here for electromagnetic plane waves is in principle generalizable to arbitrary electromagnetic fields and non-abelian gauge fields. At the end we give the word to Hermann Weyl [9] who recapitulates our efforts in the most beautiful way: *The action integral of general relativity is invariant not only under space-time Lorentz transformations but also under the gauge transformation, if this is incorporated consistently.*

Acknowledgments

This work is supported by the Estonian Institutional Research Support No. IUT2-27, and by the Estonian Science Foundation under grant No. 8769.

References

[1] Bhabha H J 1945 *Rev. Mod. Phys.* **17** 200
[2] Saar R, Ots I, Tammelo R and Loide R K 1999 *J. Phys.* A **32** 2499
[3] Ots I, Saar R, Loide R K and Liivat H 2001 *Europhys. Lett.* **56** 367
[4] Saar R, Groote S, Liivat H and Ots I 2011 *Adv. Math. Phys.* **2011** 652126
[5] Saar R, Groote S, Liivat H and Ots I 2011 *Phys. Rev.* D **84** 065022
[6] Chakrabarti A 1968 *Nuovo Cim.* A **56** 604
[7] Beers B and Nickle H H 1972 *J. Math. Phys.* **13** 1592
[8] Wolkow D M 1935 *Z. Phys.* **94** 250
[9] Weyl H 1919 *Annalen der Physik* **364** 101

Renormalization aspects of chaotic strings

S Groote[1,2], H Veermäe[2] and C Beck[3]

[1]Institut für Physik, Johannes-Gutenberg-Universität, Staudinger Weg 7, 55099 Mainz, Germany
[2]Füüsika Instituut, Tartu Ülikool, Ravila 14c, 50411 Tartu, Estonia

[3]School of Mathematical Sciences, Queen Mary University of London
Mile End Road, London E1 4NS, U.K.

Abstract. Chaotic strings are a class of non-hyperbolic coupled map lattices, exhibiting a rich structure of complex dynamical phenomena with a surprising correspondence to physical contents. In this paper we introduce different types and models for chaotic strings, where 2B-strings with finite length are considered in detail. We demonstrate possibilities to extract renormalized quantities, which are expected to describe essential properties of the string.

1. Introduction

Coupled map lattices (CMLs) were introduced in seminal papers by Kaneko and Kapral [1, 2] as a relatively simple model of spatiotemporal chaos. They exhibit a rich structure of complex dynamical phenomena [3, 4, 5, 6, 7] and their applications range from hydrodynamical flows, turbulence, chemical reactions, biological systems to quantum field theories, see e.g reviews in [3, 4]. Of particular interest are CMLs where the local map exhibits chaotic behaviour. For weakly coupled hyperbolic local maps it can be proved [8, 9, 10, 11] that the dynamics is ergodic and there exists a smooth invariant density of the entire CML. However, the case of nonhyperbolic local maps, e.g one-dimensional maps for which the slope is equal to zero at some point, is more complicated from a mathematical point of view and fewer analytical results are known [12, 13, 14, 15, 16, 17, 18].

Chaotic strings are one-dimensional CMLs of diffusively coupled Chebychev maps introduced by Christian Beck [18]. These CMLs are non-hyperbolic, since the Chebychev maps have an extremum with vanishing slope. They can be used to model vacuum fluctuations [4, 19, 20, 21] by replacing the noise in stochastically quantized field theories by a deterministic but chaotic dynamics [22, 23, 24]. A possible physical embedding is to associate the vacuum energy of chaotic strings with dark energy [25]. Surprisingly, these models could also improve our understanding about the observed mass spectrum of the standard model (see Ref. [4] for more details).

This paper, however, will not deal with the physical aspects but will merely concentrate on the mathematical properties of chaotic strings. The main object of study is an interesting selfsimilar dependence of the invariant density on the coupling constant a of the CML. For weakly coupled N-th order Chebyshev maps [18, 24, 25, 26, 27] it was shown that certain scaling properties with respect to the coupling a can be calculated perturbatively [19, 20]. However, the fine structure of the coupling dependence for important characteristics of the chaotic string as for instance the self energy could not be explained by perturbative methods. It has been illustrated numerically that the fine structure does occur only for CMLs of dimension 1, and not on lattices of higher dimension [28]. In Ref. [29] chaotic strings with different boundary conditions were studied

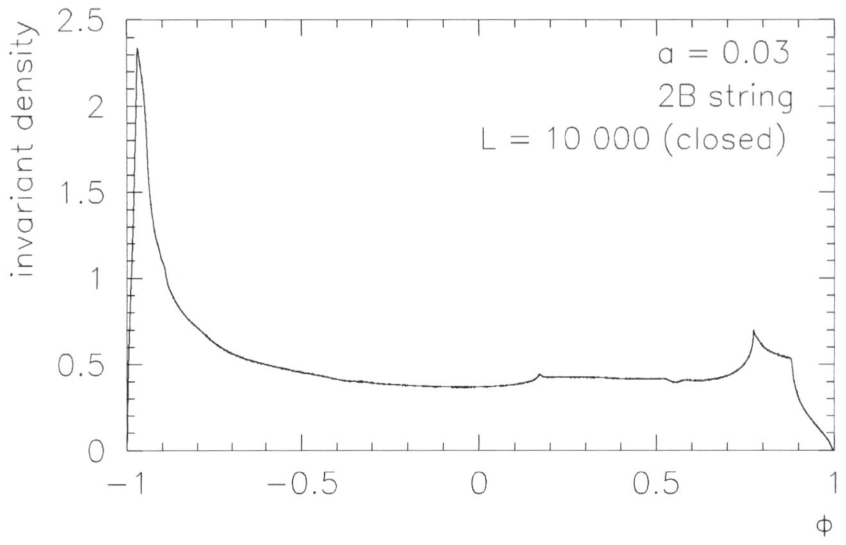

Figure 1. Invariant density for the closed $2B$-string with $L = 10\,000$ at $a = 0.03$

numerically. It was shown that some short strings (with lengths between 3 and 5 lattice sites) show the same fine structure as long strings with lengths of the order of 10 000 sites, while for other short strings the fine structure had a totally different shape.

2. Overview

The dynamics of chaotic strings is given by the recurrence relation

$$\phi_{n+1}^i = (1-a)T_N(\phi_n^i) + s\frac{a}{2}(T_M(\phi_n^{i+1}) + T_M(\phi_n^{i-1})) \tag{2.1}$$

where $T_N(x) = \cos(N \arccos(x))$ is the N-th order Chebychev polynomial of the first kind, $a \in [0,1]$ is a coupling constant, $s = \pm 1$ distinguishes between diffuse and anti-diffuse coupling, and n resp. i are the temporal and spatial indices. The strings are assumed to have infinite spatial extent. The field ϕ_n^i takes values in the interval $[-1,1]$. Chaotic strings are divided up into type A strings with $M = N$ and type B strings with $M = 1$, denoted by NA and NB, respectively. The most important cases are $N = 2, 3$ corresponding to

$$T_2(x) = 2x^2 - 1, \quad T_3(x) = 4x^3 - 3x$$

Due to the ergodicity of the map (2.1), local averages can be given by

$$\langle f(\phi) \rangle_a = \lim_{k\to\infty} \frac{1}{k} \sum_{n=1}^{k} f(\phi_i^n) = \int_{-1}^{1} \mathrm{d}\phi \rho_a(\phi) f(\phi)$$

with $\rho_a(\phi)$ being the invariant density of the field variable ϕ. As an example, in Fig. 1 we show the invariant density for the $2B$-string for a closed chaotic string of length $L = 10\,000$. The density is independent of i, since the map (2.1) is invariant under shifts in i. The interesting objects to study are the coupling dependences of averages $\langle V \rangle_a$ of formal potentials defined by $T_N(x) = -\frac{\partial}{\partial x} V_N(x)$. For $N = 2, 3$ one has

$$V_2(\phi) = -\frac{2}{3}\phi^3 + \phi, \quad V_3(\phi) = -\phi^4 + \frac{3}{2}\phi^2$$

Journal of Physics: Conference Series **532** (2014) 012008 doi:10.1088/1742-6596/532/1/012008

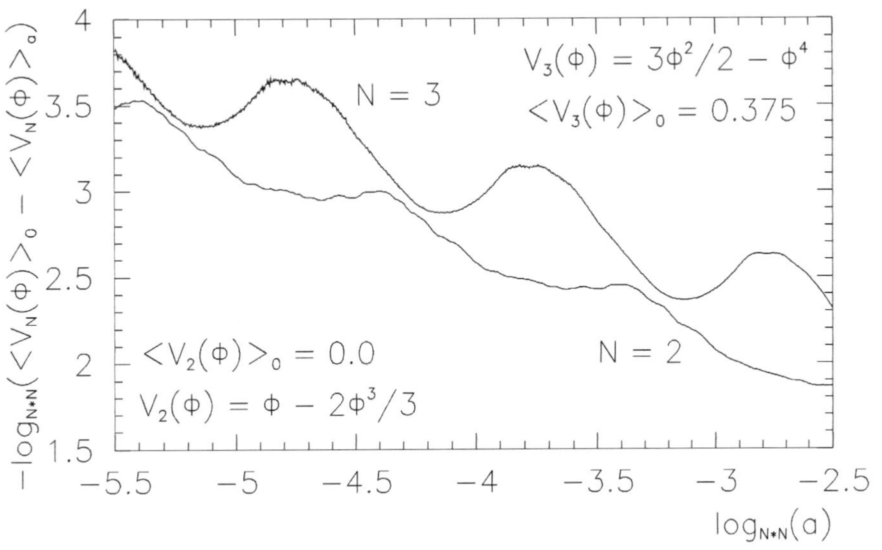

Figure 2. Double logarithmic plot of the self energies of the $2B$- and $3B$-strings in dependence on the coupling a

The quantity $\langle V_N \rangle_a$ can be interpreted as (formal) self energy of the field ϕ. The dependence of the expectation values $\langle V_2 \rangle_a$ and $\langle V_3 \rangle_a$ for the $2B$- and $3B$-strings on a are shown in Fig. 2 in a double logaritmic plot.

In the case of non-coupled maps ($a = 0$) the recurrence relation (2.1) reduces effectively to the Chebyshev map $\phi^{n+1} = T_N(\phi^n)$ for which the dynamics is well understood. In this case the process is ergodic and ϕ behaves as a pseudorandom variable [3, 4] with the invariant density

$$\rho_0(\phi) = \frac{1}{\pi\sqrt{1-\phi^2}} \tag{2.2}$$

which is independent of N.

2.1. Scaling behaviour

In general, the invariant density depends on the coupling in a highly non-trivial way. However, it was shown analytically using perturbative methods, that self energies exhibit a scaling property for weak couplings [19, 20]. To be more precise, the self energies behave in the scaling region $0 \leq a \leq 0.039$ as

$$\langle V \rangle_a - \langle V \rangle_0 = f_N(\ln(a))\sqrt{a}$$

where f_N is periodic function with the period $2\ln(N)$. The function f_N describes the fine structure of the self energies, as shown in Fig. 3 for the $2B$-string in the scaling interval $a \in [1.5, 6.0] \times 10^{-4}$.

2.2. Boundary conditions

The chaotic strings are taken to have infinite spatial extent, but this case is usually complicated to study in practice. The number of possible sites i is called the length L of the string. It is expected that for smaller couplings the spatial correlations should be weaker and, therefore, the

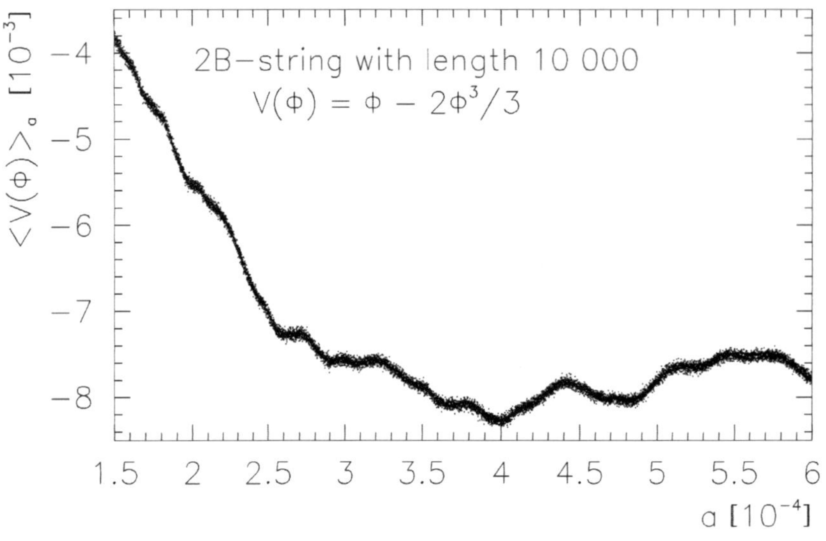

Figure 3. Self energy of the $2B$-string in the scaling interval $a \in [1.5, 6.0] \times 10^{-4}$

invariant density should be less sensitive to the length of the string. The effect of imposing different boundary conditions was studied in Ref. [29] where it was proved numerically that invariant density is relatively insensitive to the length of the string if $L \geq 3$. There are many different possible boundary conditions:

- *Cyclic boundary condition* (closed strings) for which $\phi_1^n = \phi_{L+1}^n$. This boundary condition preserves translational invariance. However, low-L models can have stable fixed points which strongly change the shape of the invariant density. The fixed points disappear when $L \to \infty$. The fine structure of the self energies can be reproduced for $L \geq 3$.
- *Chaotic boundary condition* (open strings) for which the endpoints of the string are uncoupled from their neighbours and evolve as $\phi^{n+1} = T_N(\phi^n)$. This boundary condition mimics the behaviour of the infinite string with negligible long range spatial correlations. The case $L \geq 3$ works well in the scaling region as it reproduces the fine structure.[1] Since the endpoints evolve independently, this boundary condition also avoids the appearance of fixed points.
- *Fixed boundary conditions* (fixed strings) for which the endpoints of the string are given fixed values. These boundary conditions are probably the simplest, but numerical studies show they are able to reproduce the fine structure only for $L \geq 5$.
- *Random boundary conditions* (random strings) for which the endpoints of the string are independent random variables distributed by Eq. (2.2). These boundary conditions approximate the behaviour of $\phi^{n+1} = T_N(\phi^n)$, i.e the chaotic boundary conditions, with the important difference that they also neglect temporal correlations of the endpoints. This model does not reproduce the fine structure for $L = 3$.

These boundary conditions could also be used to study the effects of different correlations of the invariant density. In Fig. 4 we show the self energy of the $2A$-string for two different boundary conditions and two different lengths L.

[1] Note that due to a numerical error, in Ref. [29] is was erroneously stated that the open string of length 3 does not reproduce the fine structure. This will be corrected in an erratum to Ref. [29].

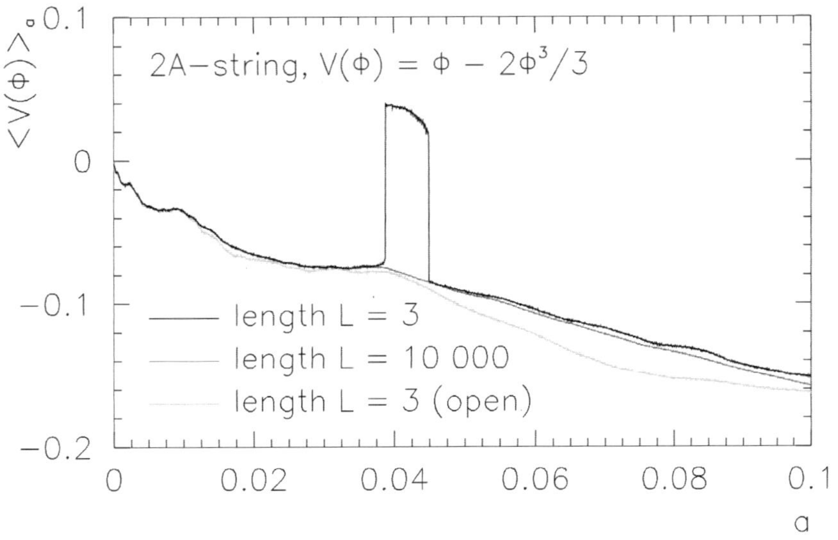

Figure 4. Self energy of the $2A$-string for a closed string of length $L = 3$ (black line), $L = 10\,000$ (red line), and for an open string of length $L = 3$ (green line)

2.3. Perturbative method

The perturbative method developed in Refs. [19, 20] is based on open strings of length $L = 3$ given by

$$\phi^{n+1} = (1-a)T_2(\phi^n) + \frac{a}{2}(\phi_+^n + \phi_-^n), \quad \phi_\pm^{n+1} = T_2(\phi_\pm^n) \tag{2.3}$$

Note that this map does not distinguish between strings of type A and B. The idea is to analytically understand the origin of the scaling property. By choosing the initial distribution to be the uncoupled density (2.2), after a few p iterations the scaling of the invariant density $\rho_a^{(p)}(\phi)$ (like the one in Fig. 5) close to the boundaries $\phi = \pm 1$ can be obtained. As a consequence this method explains the scaling behaviour of the self energy. However, the perturbative method is unable to reproduce the fine structure.

3. Renormalization

In looking at the gradual changes of the invariant density while decreasing the coupling a, we became aware of a possible reason for the fine structure of the self energy of the $2B$-string. Decreasing the coupling, more and more peaks appear from the left hand border of the interval (i.e at $\phi = -1$) and move to the right, while the distance between existing peaks decreases. As a result a selfsimilar structure develops [19, 20]. However, a closer look reveals an opposite movement. Before the peaks are generated at the left hand side, they arrive as wide peaks from the right. When such a newly-formed peak (in Fig. 5 found close to $\phi = -0.15$) passes existing peaks, it modulates the distribution and could thereby cause a minimum in the self energy.

3.1. Partitioning the interval $[-1, 1]$

In extending the perturbative method based on Eq. (2.3) onto the whole interval $[-1, 1]$ we could construct the self energy as a three-fold integral of a polynomial of very high degree. The integration is taken over the central field ϕ and the fields $\phi_\pm$ of the next neighbours, while the

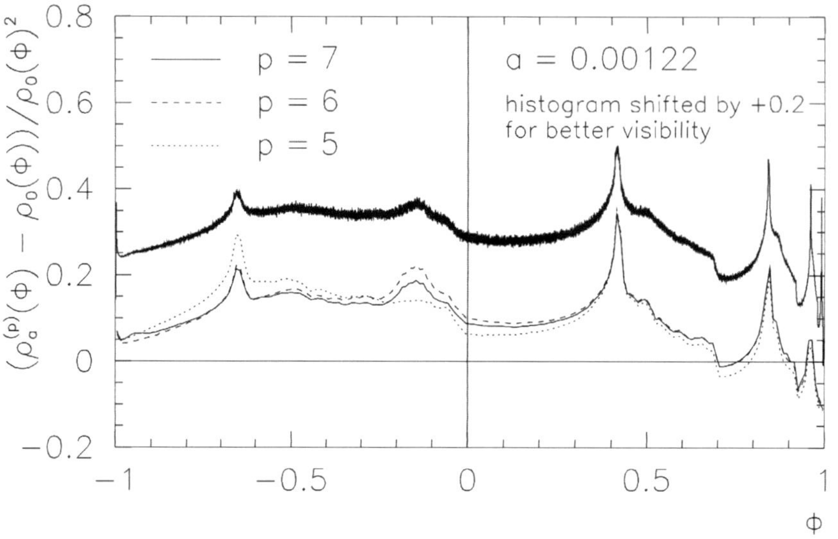

Figure 5. Comparison of the rescaled $\rho_a^{(p)}(\phi)$ $(p = 5, 6, 7)$ with the numerically obtained distribution. In this plot the numerical distribution is shifted by $+0.2$ for better visibility.

integrand is generated by the $(p+1)$-fold iteration of Eq. (2.3). For $a = 0$, the integrand is just the $(p+1)$-fold iteration of the Chebyshev map $T_2^{p+1}(\phi) = T_{2^{p+1}}(\phi)$, displayed on the left hand side of Fig. 6 for $p = 5$. For $a \neq 0$, there are slight deviations from this shape. As shown on the right hand side of Fig. 6 for $a = 0.00122$, some of the maxima (if not all) are damped, i.e reach lower values than for the uncoupled integrand. It turns out that the heights of the maxima are directly related to the positions of the peaks of the invariant density. If the coupling increases, the maximum for instance at $\phi = 0$ will disappear, and the two neighbouring minima will merge and cause the created minima to raise. This corresponds to a peak moving to the left. The same will happen later on to the maxima at $\pm\sin(\pi/4) = \pm 1/\sqrt{2}$ if the coupling is increased further. But what is the role of disappearing maxima? The positions of the maxima constitute a partition of the interval $[-1, 1]$. If one of the maxima disappears, this partition will be changed.

3.2. The model function

In the following we were able to construct two characterizing functions, the draft function $\tilde{d}^{(k)}(t_0; \phi_+, \phi_-)$ and the blunt function $b_{p,k}(\phi_+, \phi_-)$, that measure the gradual displacement resp. damping of the maxima. For this, the maxima are divided into classes with common behaviour. For $a = 0$ the maxima in class k are located at $\sin(\pi t_0/2)$, where $t_0 = \pm(2n+1)/2^k$, $n = 0, 1 \ldots, 2^{k-1} - 1)$ [30]. Using the draft and blunt functions it's possible to construct a model function

$$f(\Delta t; t_0) = \int_{-1}^{1} \rho_0(\phi_+) d\phi_+ \rho_0(\phi_-) d\phi_- \times$$
$$\times \cos\left(2^p \sqrt{\pi^2 \left(\Delta t - a\tilde{d}^{(k)}(t_0; \phi_+, \phi_-) \right)^2 + 4^{1-k} a r_2(-T_{2^k}(\phi_+), -T_{2^k}(\phi_-))} \right)$$

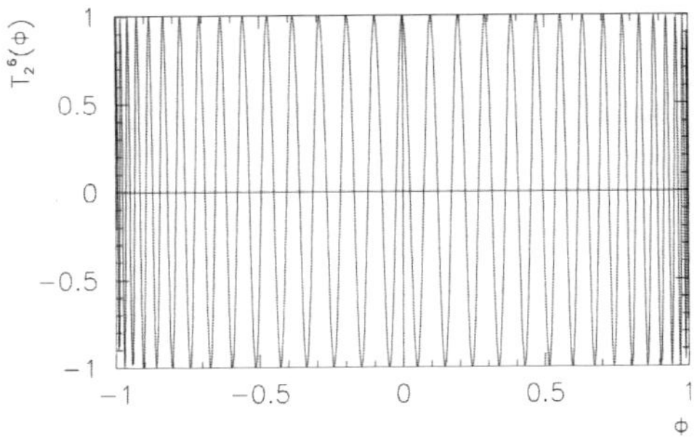

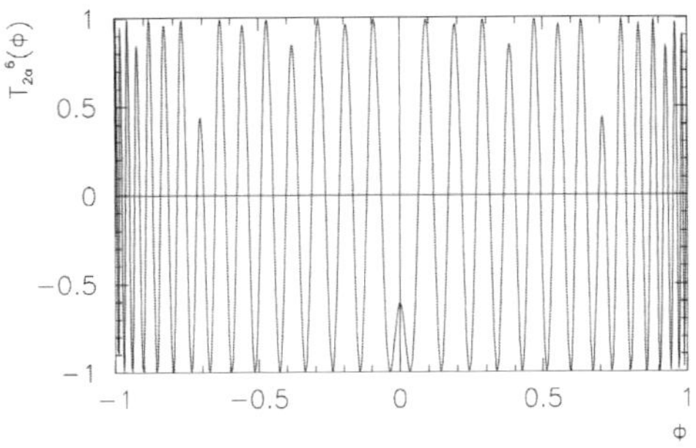

Figure 6. Integrand with $p = 5$ for $a = 0$ (left hand side) and $a = 0.00122$ (right hand side) at $\phi_+ = \phi_- = 1$ as a function of $\phi \in [-1, 1]$

that depends on the coupling a. This is a a quite precise approximation for the integrand $f(t_0 + \Delta t)$ in the neighbourhood of t_0. The draft is given by

$$\tilde{d}^{(k)}(t_0; \phi_+, \phi_-) = \frac{1}{2}\left(\tilde{d}^{(k)}(t_0; \phi_+) + \tilde{d}^{(k)}(t_0; \phi_-)\right)$$

with

$$\tilde{d}^{(k)}(t_0; \phi_\pm) = \tilde{d}^{(k)}(t_0) + \sum_{l=0}^{k-1} T_{2^l}(-\phi_\pm)\tilde{d}_l^{(k)}(t_0)$$

and

$$\tilde{d}^{(k)}(t_0) = 2^{-k} d_{-1}^{(k)}(t_0) - \sum_{l=0}^{k-1}\left(2^{1-k+l} + \cos(2^l \pi t_0)\right)\tilde{d}_l^{(k)}(t_0), \quad \tilde{d}_l^{(k)}(t_0) = \frac{1}{2^l \pi \sin(2^l \pi t_0)}$$

and the blunt is characterized by

$$r_2(\phi_+, \phi_-) := \frac{1}{2}\left(r_2(\phi_+) + r_2(\phi_-)\right), \quad r_2(\phi) := \frac{1}{2}\sum_{p=0}^{\infty}\frac{1 - T_{2^p}(\phi)}{4^p}$$

3.3. The renormalization procedure

The invariant density of the chaotic string is obtained in the limit $p \to \infty$. In order to perform this limit in a scale invariant way, certain parameters have to be replaced by reduced parameters which stay unaltered under this limit. These parameters are the reduced order $\hat{p} = p - 2k$, the reduced coupling $\hat{a} = 4^{p-k}a$, and the reduced deviation $\Delta\hat{t}$. In terms of these parameters the model function can be rewritten as

$$\hat{f}^{(\hat{p})}(\Delta\hat{t}; t_0) := f(2^{-p}\Delta\hat{t}; t_0) = \int \rho_0(\phi_+)d\rho_+\rho_0(\phi_-)d\phi_- \times$$

$$\times \cos\left(\sqrt{\pi^2\left(\Delta\hat{t} - 2^{\hat{p}}\hat{a}\tilde{d}^{(k)}(t_0; \phi_+, \phi_-)\right)^2 + 2\hat{a}r_2^\infty\left(-T_{2^k}(\phi_+), -T_{2^k}(\phi_-)\right)}\right)$$

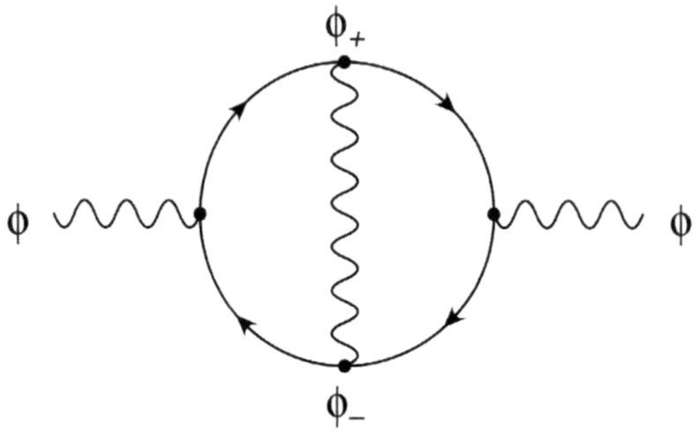

Figure 7. A fish-type diagram containing bosons (twiggles) and fermions (straight lines)

Finally, in performing the limit $p \to \infty$ while keeping $\hat{p}$ constant, one can separate the high-frequency blunt from the low-frequency draft, leading to a (yet unexplained) effective blunt of 8/3 and a model function

$$\tilde{f}^{(\hat{p})}(\Delta\hat{t};t_0) = \int \rho_0(\phi_+)d\phi_+\rho_0(\phi_-)d\phi_- \cos\left(\sqrt{\pi^2\left(\Delta\hat{t}-2^{-\hat{p}}\hat{a}\tilde{d}^{(k)}(t_0;\phi_+,\phi_-)\right)^2+\frac{8}{3}\hat{a}}\right)$$

which in turn can be expressed via Fourier transformation as

$$\begin{aligned}\tilde{f}^{(\hat{p})}(\Delta\hat{t};t_0) &= \int \rho_0(\phi_+)d\phi_+\rho_0(\phi_-)d\phi_- \Bigg[\cos\left(\pi\left(\Delta\hat{t}-2^{-\hat{p}}\hat{a}\tilde{d}^{(k)}(t_0;\phi_+,\phi_-)\right)\right) \\ &- \int_{-1}^{+1}\frac{4\hat{a}\,d\omega}{3\sqrt{8\hat{a}(1-\omega^2)/3}}J_1\left(\sqrt{8\hat{a}(1-\omega^2)/3}\right)\cos\left(\pi\omega\left(\Delta\hat{t}-2^{-\hat{p}}\hat{a}\tilde{d}^{(k)}(t_0;\phi_+,\phi_-)\right)\right)\Bigg]\end{aligned}$$

An interesting observation is the appearance of the Bessel function $J_\lambda(x)$ of degree $\lambda = 1$. A Bessel function of the same degree appears as integration measure for $D = 4$ space-time dimensions (see e.g Ref. [31]). This observation can provide us with some clues about the physical embedding of the chaotic string dynamics into generalized versions of quantum field theory [4]. Could it be possible to consider $\lambda = 1$ as a reference to four-dimensional space-time without the explicit emergence of space-time itself? If we could interpret the integrand of the Bessel function as a correlator in (four-dimensional) configuration space, the integration over ϕ_+ and ϕ_- should mean the insertion of two vertices into the diagram which are connected by a propagator, i.e a virtual particle transition. Still, one has to be careful, since the analogy is far from being exact. However, the fish-type diagram in Fig. 7 is a compactified version of the Feynman web that was used in Ref. [4] to give a physical meaning to the chaotic string dynamics.

4. Conclusions and Outlook

We gave guidelines for a detailed analysis of the dynamics of chaotic strings, taking as an example the $2B$-string. It turned out that an open string of length $L = 3$ proves useful in simplifying numerical and analytical calculations for small couplings, i.e in the scaling region. We have introduced a model function which allows to analyse the scaling behaviour in the stochastic limit $q \to \infty$ using a renormalization procedure. Renormalized quantities are the reduced degree $\hat{p}$, the reduced coupling $\hat{a}$ and the reduced deviation $\Delta\hat{t}$. Results obtained by us earlier via a perturbative approach can be reproduced for larger values of $\hat{p}$ while nonperturbative effects are seen for smaller (even negative) values of $\hat{p}$. These nonperturbative effects are expected to explain the fine structure of the self energy of the $2B$-string.

Acknowledgments
This work is supported by the Estonian Institutional Research Support No. IUT2-27, and by the Estonian Science Foundation under grant No. 8769.

References

[1] Kaneko K 1984 *Progr. Theor. Phys.* **72** 480
[2] Kapral R 1985 *Phys. Rev.* A **31** 3868
[3] Kaneko K 1993 *Theory and Applications of Coupled Map Lattices* (John Wiley & Sons)
[4] Beck C 2002 *Spatio-Temporal Chaos and Vacuum Fluctuations of Quantized Fields* (World Sci.)
[5] Bunimovich A and Sinai Y G 1988 *Nonlinearity* **1** 491
[6] Amritkar R E and Gade P M 1993 *Phys. Rev. Lett.* **70** 3408
[7] Carretero-Gonzáles R, Arrowsmith D K and Vivaldi F 1997 *Physica* D **103** 381
[8] Bunimovich L A 1997 *Physica* D **103** 1
[9] Baladi V and Rugh H H 2001 *Comm. Math. Phys.* **220** 561
[10] Bricmont J and Kupiainen A 1996 *Comm. Math. Phys.* **178** 703
[11] Järvenpää E and Järvenpää M 2001 *Comm. Math. Phys.* **220** 1
[12] Daido H 1984 *Progr. Theor. Phys. Suppl.* **79** 75
[13] Lemaître A and Chaté H 1998 *Phys. Rev. Lett.* **80** 5528
[14] Lemaître A and Chaté H 1999 *Phys. Rev. Lett.* **82** 1140
[15] Yang W, Ding E J and Ding M 1996 *Phys. Rev. Lett.* **76** 1808
[16] Torcini A, Livi R, Politi A and Ruffo S 1997 *Phys. Rev. Lett.* **78** 1391
[17] Mackey M C and Milton J 1995 *Physica* D **80** 1
[18] Beck C 2002 *Physica* D **171** 72
[19] Groote S and Beck C 2006 *Phys. Rev.* E **74** 046216
[20] Groote S and Beck C 2007 *Dyn. Syst.* **22** 219
[21] Beck C 2012 *Mod. Phys. Lett.* B **26** 1250060
[22] Parisi G and Wu Y S 1981 *Sci. Sin.* **24** 483
[23] Damgaard P H and Hüffel H 1988 *Stochastic Quantization* (World Sci.)
[24] Beck C 1995 *Nonlinearity* **8** 423
[25] Beck C 2004 *Phys. Rev.* D **69** 123515
[26] Dettmann C P 2002 *Physica* D **172** 88
[27] Dettmann C P and Lippolis D 2005 *Chaos Sol. Fract.* **23** (2005) 43
[28] Maher M and Beck C 2008 *Chaos Sol. Fract.* **37** 9
[29] Groote S and Veermäe H 2009 *Chaos Sol. Fract.* **41** 2354
[30] Groote S, Veermäe H and Beck C 2013 *Chaos Sol. Fract.* **53** 18
[31] Groote S, Körner J G and Pivovarov A A 2007 *Annals Phys.* **322** 2374

Aspects of multimetric gravity

M Hohmann

Teoreetilise Füüsika Labor, Füüsika Instituut, Tartu Ülikool, Ravila 4c, 50411 Tartu, Estonia

Abstract. We present a class of gravity theories containing $N \geq 2$ metric tensors and a corresponding number of standard model copies. In the Newtonian limit gravity is attractive within each standard model copy, but different standard model copies mutually repel each other. We discuss several aspects of these multimetric gravity theories, including cosmology, structure formation, the post-Newtonian limit and gravitational waves. The most interesting feature we find is an accelerating expansion of the universe that naturally becomes small at late times.

1. Motivation

Among the most important open problems of modern physics are the observed accelerating expansion of the universe [1, 2] and the formation of large scale structures [3]. Their most widely accepted explanation is given by the ΛCDM model, which claims that, in addition to visible matter, the universe must contain dark matter and dark energy in order to match cosmological observations. Measurements of the cosmic microwave background by Planck indicate that their amounts are given by 68.3% dark energy, 26.8% dark matter and only 4.9% visible matter [Ade P A R *et al* arXiv:1303.5076 Planck Collaboration]. However, despite its success in modeling cosmological observations, the ΛCDM model is purely empirical and does not provide explanations for the constituents of dark matter and dark energy.

This shortcoming of the standard model of cosmology invites for various supplemental theories from particle physics and alternative theories from gravitational physics. The model we discuss here belongs to the second class and is based on the idea to introduce a second copy of the standard model, which is governed by a second metric. Both copies are assumed to interact only gravitationally, mediated by an interaction of their respective metrics, so that they mutually appear dark. The gravitational interaction is further assumed to be attractive within each matter sector, but repulsive of equal strength between the two standard model copies. If one finally assumes the universe to be constituted by equal amounts of both matter types, the mutual repulsion may act as the driving force for the observed accelerating expansion, and mimic the presence of dark matter effects by pushing light and visible matter towards visible galaxies. However, it turns out that repulsive gravity cannot be achieved in the bimetric case [4], but can indeed account for an accelerating cosmology for $N \geq 3$ standard model copies and metrics [5]. In the following we will discuss various aspects of these multimetric gravity theories [5, 6, 7, 8].

We will give a concise definition of the class of theories we consider here in section 2, where we also provide a concrete family of multimetric gravity theories. We then enter a detailed discussion of their properties in the remaining sections. In section 3 we derive their post-Newtonian limit and obtain constraints from high-precision solar system experiments. For this constrained class of theories we discuss cosmology in section 4, the formation of large scale structures in section 5 and gravitational waves in section 6. We end with a conclusion in section 7.

Journal of Physics: Conference Series **532** (2014) 012009 doi:10.1088/1742-6596/532/1/012009

2. Multimetric gravity

We begin our discussion with a brief description of the class of multimetric gravity theories we consider. These theories can be characterized by a simple set of assumptions. In this section we first list these assumptions, before we present a generic example theory. Various aspects of this example theory will then be discussed in the remaining sections.

In the following we will assume:

(i) The field content is given by $N \geq 2$ copies $\varphi^1, \ldots, \varphi^N$ of standard model matter and a corresponding number of metric tensors $g^1, \ldots, g^N$.

(ii) The dynamics are governed by a diffeomorphism invariant action of the type

$$S = S_G[g^1, \ldots, g^N] + \sum_{I=1}^{N} S_M[g^I, \varphi^I] \tag{2.1}$$

where S_M denotes the standard model action.

(iii) The field equations are obtained by variation with respect to the metrics $g^1_{ab}, \ldots, g^N_{ab}$, and so are a set of symmetric two-tensor equations of the form $K^I_{ab} = 8\pi G_N T^I_{ab}$.

(iv) The geometry tensor K^I_{ab} contains at most second derivatives of the metric, which can be achieved by a suitable choice of the gravitational action (2.1).

(v) The field equations are symmetric with respect to arbitrary permutations of the sectors (g^I, φ^I), which can be understood as a generalized Copernican principle.

(vi) The vacuum solution is given by a set of flat metrics $g^I_{ab} = \eta_{ab}$. (Poincaré symmetry for all metrics simultaneously implies $g^I_{ab} = \lambda^I \eta_{ab}$ for constants λ^I that, invoking the Copernican principle for the vacuum, should be equal and can be set to $\lambda^I = 1$.)

Assumption *(ii)* implies that each standard model copy φ^I couples only to its corresponding metric tensor g^I_{ab}. This ensures that the dynamics and causality of each standard model copy are governed by a single metric. It further ensures that the interaction between the different standard model copies is mediated only through gravity, so that they appear mutually dark. Variation of the gravitational action S_G with respect to the metrics g^I_{ab} then yields the geometry tensors K^I_{ab}, while variation of the matter action S_M yields the usual energy-momentum tensors T^I_{ab}, as stated in assumption *(iii)*. Assumption *(iv)* is a technical requirement, which we use here in order to restrict the possible terms that may appear in the linearized theory and the post-Newtonian limit. In assumption *(v)* we employ the Copernican principle in the sense that, in addition to the matter sector symmetry imposed by choosing N identical standard model actions, there is no distinction between the different standard model copies by gravitational effects. We finally exclude cosmological constants by assumption *(vi)*.

For the calculations we present here we use a concrete multimetric gravity theory satisfying the assumptions listed above. The theory we consider here is defined by the action

$$\begin{aligned}
S_G = \frac{1}{16\pi}\int d^4x\,\sqrt{g_0}\Bigg[&\sum_{I=1}^{N}\Big(c_1 R^I + g^{I\,ij}\left(c_3\tilde{S}^I{}_i\tilde{S}^I{}_j + c_5\tilde{S}^I{}_k\tilde{S}^{I\,k}{}_{ij} + c_7\tilde{S}^{I\,k}{}_{il}\tilde{S}^{I\,l}{}_{jk}\right)\\
&+ g^{I\,ij}g^{I\,kl}g^I{}_{mn}\left(c_9\tilde{S}^{I\,m}{}_{ik}\tilde{S}^{I\,n}{}_{jl} + c_{11}\tilde{S}^{I\,m}{}_{ij}\tilde{S}^{I\,n}{}_{kl}\right)\Big)\\
&+ \sum_{I,J=1}^{N}\Big(c_2 g^{I\,ij}R^J{}_{ij} + g^{I\,ij}\left(c_4 S^{IJ}{}_i S^{IJ}{}_j + c_6 S^{IJ}{}_k S^{IJ\,k}{}_{ij} + c_8 S^{IJ\,k}{}_{il}S^{IJ\,l}{}_{jk}\right)\\
&+ g^{I\,ij}g^{I\,kl}g^I{}_{mn}\left(c_{10}S^{IJ\,m}{}_{ik}S^{IJ\,n}{}_{jl} + c_{12}S^{IJ\,m}{}_{ij}S^{IJ\,n}{}_{kl}\right)\Big)\Bigg]
\end{aligned} \tag{2.2}$$

where $c_1, \dots, c_{12}$ are constant parameters and we used the connection difference tensors

$$S^{IJ\,i}{}_{jk} = \Gamma^{I\,i}{}_{jk} - \Gamma^{J\,i}{}_{jk} \quad S^{IJ}{}_{j} = S^{IJ\,k}{}_{jk} \quad \tilde{S}^{J\,i}{}_{jk} = \frac{1}{N}\sum_{I=1}^{N} S^{IJ\,i}{}_{jk} \quad \tilde{S}^{J}{}_{j} = \tilde{S}^{J\,k}{}_{jk} \tag{2.3}$$

and the mixed density

$$g_0 = \prod_{I=1}^{N} \left(g^I\right)^{\frac{1}{N}} \tag{2.4}$$

Here we have chosen units in which the Newtonian gravitational constant takes the value $G_N = 1$. This action is both general and simple, in the sense that it contains a large number of different terms, each of which has a simple form. In the following sections we will discuss several aspects of this particular multimetric gravity model.

3. Post-Newtonian consistency

In this section we discuss the experimental consistency of our multimetric gravity theory. For this purpose we make use of the parameterized post-Newtonian (PPN) formalism [9, 10, 11, 12]. In its standard form the PPN formalism applies to gravity theories in which a single metric tensor determines the motion of test bodies, and characterizes each gravity theory by a set of ten parameters. The values of these parameters have been measured with high precision in the solar system. Here we present an extension of the PPN formalism to multimetric gravity theories [6, 8] and use the measured values of the PPN parameters to derive constraints on the concrete theory displayed in the preceding section.

The starting point of our derivation is a perturbative expansion of the metrics g^I_{ab} in orders of the velocity $\vec{v}$ of the source matter in a given frame of reference. Using assumption *(vi)* this is a weak field approximation around the flat vacuum metric η_{ab} in Cartesian coordinates $(x^a) = (t, \vec{x})$,

$$g^I_{ab} = \eta_{ab} + h^I_{ab} = \eta_{ab} + h^{I(1)}_{ab} + h^{I(2)}_{ab} + h^{I(3)}_{ab} + h^{I(4)}_{ab} \tag{3.1}$$

where each term $h^{I(n)}_{ab}$ is of order $|\vec{v}|^n \equiv \mathcal{O}(n)$. However, not all of these terms are necessary in order to describe the motion of test bodies, while others vanish due to Newtonian energy conservation or time reversal symmetry. Here we consider only the relevant, non-vanishing components. Using the Laplacian $\triangle = \partial_\alpha \partial^\alpha$ we can express them in the form

$$h^{I(2)}_{00} = -\sum_{J=1}^{N} \alpha^{IJ} \triangle \chi^J \tag{3.2a}$$

$$h^{I(2)}_{\alpha\beta} = \sum_{J=1}^{N} \left(2\theta^{IJ}\chi^J_{,\alpha\beta} - (\gamma^{IJ} + \theta^{IJ})\triangle\chi^J \delta_{\alpha\beta}\right) \tag{3.2b}$$

$$h^{I(3)}_{0\alpha} = \sum_{J=1}^{N} \left(\sigma^{IJ}_{+} W^{J+}_{\alpha} + \sigma^{IJ}_{-} W^{J-}_{\alpha}\right) \tag{3.2c}$$

$$h^{I(4)}_{00} = \sum_{J=1}^{N} \left(\phi^{IJ}_{p}\Phi^J_p + \phi^{IJ}_{\Pi}\Phi^J_{\Pi} + \sum_{A=1}^{2} \omega^{IJ}_{A}\Omega^J_A\right) + \sum_{J,K=1}^{N} \sum_{A=1}^{7} \psi^{IJK}_{A} \Psi^{JK}_{A} \tag{3.2d}$$

The so-called PPN potentials $\chi^I, W^{I\pm}, \Phi^I_p, \Phi^I_\Pi, \Omega^I_1, \Omega^I_2, \Psi^{IJ}_1, \dots, \Psi^{IJ}_7$ are Poisson-like integrals over the source matter distribution, which is assumed to be a perfect fluid with rest energy density $\rho^I \sim \mathcal{O}(2)$, velocity $v^I_\alpha \sim \mathcal{O}(1)$, internal energy density $\rho^I\Pi^I \sim \mathcal{O}(4)$ and pressure $p^I \sim \mathcal{O}(4)$ for each matter type $I = 1, \dots, N$. The PPN parameters $\alpha^{IJ}, \gamma^{IJ}, \theta^{IJ}, \sigma^{IJ}_\pm, \phi^{IJ}_p, \phi^{IJ}_\Pi, \omega^{IJ}_1, \omega^{IJ}_2, \psi^{IJK}_1, \dots, \psi^{IJK}_7$ are constants which are characteristic for each multimetric gravity theory. They can be determined from a perturbative solution of the gravitational field equations

$$K^I_{ab} = 8\pi T^I_{ab} \tag{3.3}$$

starting from the lowest velocity order $\mathcal{O}(2)$ and up to the highest velocity order $\mathcal{O}(4)$.

The most important asset of the standard PPN formalism for a single metric is the possibility to measure the values of the PPN parameters at high precision using solar system experiments. A comparison between the multimetric PPN formalism displayed here and the standard PPN formalism shows [8] that a multimetric gravity theory is consistent with solar system observations at the post-Newtonian level if its PPN parameters satisfy the relations

$$\begin{aligned} &\alpha^{11} = \gamma^{11} = 1, \quad \sigma^{11}_+ = \psi^{111}_3 = \psi^{111}_6 = -2, \quad \sigma^{11}_- = -\frac{3}{2}, \quad \phi^{11}_\Pi = 2, \quad \phi^{11}_p = 6 \\ &\omega^{11}_1 = 4, \quad \theta^{11} = \omega^{11}_2 = \psi^{111}_1 = \psi^{111}_2 = \psi^{111}_4 = \psi^{111}_5 = \psi^{111}_7 = 0 \end{aligned} \tag{3.4}$$

where we have identified the standard model copy φ^1 with the visible standard model, which is the only matter source and the only type of test masses in the solar system. Hence, these are the only PPN parameters which are accessible to solar system experiments. A thorough analysis shows [8] that among these sixteen numbers are two gauge degrees of freedom and one rescaling freedom. This leaves us with thirteen physical parameters, including the ten known parameters from the standard PPN formalism.

We finally apply our version of the PPN formalism to the multimetric gravity theory defined by the action (2.2). From the requirement that its PPN parameters satisfy the consistency conditions (3.4) and that in the Newtonian limit the gravitational interaction between the different standard model copies becomes repulsive of equal strength we derive the restrictions

$$c_1 = c_4 - \frac{1}{2}c_6 + \frac{1}{2}c_8 - \frac{3}{2}c_{10} - \frac{3N+2}{8(N-2)} \tag{3.5a}$$

$$c_2 = -c_4 + \frac{1}{2}c_6 - \frac{1}{2}c_8 + \frac{3}{2}c_{10} + \frac{3}{8} \tag{3.5b}$$

$$c_3 = -\frac{7}{3}c_4 + \frac{1}{6}c_6 - \frac{1}{6}c_8 - \frac{3}{2}c_{10} - \frac{N+30}{24(N-2)} \tag{3.5c}$$

$$c_5 = -\frac{5}{3}c_4 - \frac{7}{6}c_6 - \frac{5}{6}c_8 - \frac{3}{2}c_{10} - \frac{5N+6}{24(N-2)} \tag{3.5d}$$

$$c_7 = 2c_4 - c_6 - c_8 + 3c_{10} + \frac{N+6}{4(N-2)} \tag{3.5e}$$

$$c_9 = \frac{4}{3}c_4 - \frac{2}{3}c_6 + \frac{2}{3}c_8 - 2c_{10} - \frac{N-3}{3(N-2)} \tag{3.5f}$$

on the constant coefficients $c_1, \dots, c_{12}$. This further yields us the PPN parameters

$$\begin{aligned} &\alpha^{IJ} = 2\delta^{IJ} - 1 && \gamma^{IJ} = 2\delta^{IJ} - 1 && \theta^{IJ} = 0 \\ &\sigma^{IJ}_+ = 2 - 4\delta^{IJ} && \sigma^{IJ}_- = \frac{5}{2} - 4\delta^{IJ} && \omega^{IJ}_1 = 10\delta^{IJ} - 6 \\ &\omega^{IJ}_2 = 2 - 2\delta^{IJ} && \phi^{IJ}_p = 16\delta^{IJ} - 10 && \phi^{IJ}_\Pi = 4\delta^{IJ} - 2 \end{aligned} \tag{3.6}$$

where δ^{IJ} denotes the Kronecker symbol, and lengthy expressions for $\psi^{IJK}_1, \dots, \psi^{IJK}_7$; see [8] for a complete list. In the following sections we will restrict ourselves to a discussion of this experimentally consistent repulsive gravity model.

Journal of Physics: Conference Series **532** (2014) 012009 doi:10.1088/1742-6596/532/1/012009

4. Cosmology and accelerating expansion

We will now discuss the cosmological properties of multimetric gravity [5]. For this purpose we will start from the action (2.2), where we restrict the constant input parameters using the experimental consistency conditions (3.5). Using standard assumptions on cosmological symmetries we derive the cosmological equations of motion for a universe filled with general perfect fluid matter. We provide explicit solutions of these equations for both radiation and dust matter.

The standard assumption on homogeneous and isotropic cosmologies comprises the existence of six Killing vector fields responsible for spatial translations and rotations. The requirement that these fields are symmetry generators for all metrics g^I simultaneously restricts their form to be of Robertson–Walker type,

$$g^I = -(n^I)^2(t)dt \otimes dt + (a^I)^2(t)\gamma_{\alpha\beta}dx^\alpha \otimes dx^\beta \tag{4.1}$$

with lapse functions $n^I(t)$, scale factors $a^I(t)$, and a common purely spatial metric $\gamma_{\alpha\beta}$ of constant curvature $k \in \{-1, 0, 1\}$ and Riemann tensor $R(\gamma)_{\alpha\beta\gamma\delta} = 2k\gamma_{\alpha[\gamma}\gamma_{\delta]\beta}$. The matter content consistent with the cosmological symmetries is given by a set of N homogeneous fluids with density $\rho^I(t)$ and pressure $p^I(t)$. Their energy-momentum tensors can be written as

$$T^{Iab} = (\rho^I + p^I)u^{Ia}u^{Ib} + p^I g^{Iab} \tag{4.2}$$

with velocities normalized by the relevant metrics from their sector so that $g^I_{ab}u^{Ia}u^{Ib} = -1$. These tensors can be decomposed into the components $T^I_{00} = \rho^I(n^I)^2$ and $T^I_{\alpha\beta} = p^I(a^I)^2\gamma_{\alpha\beta}$.

As another assumption on cosmological symmetries we employ the Copernican principle in the sense that we assume equal amounts of matter to be present in all matter sectors. This means we can omit the sector index I from all matter densities and pressure functions. From the symmetry assumption *(v)* it then follows that also the metrics should be equal, hence also the lapse functions and scale factors. We may now rescale the cosmological time so that $n(t) \equiv 1$. It follows that the connection difference tensors in the action (2.2) vanish and the gravitational field equations take the simple form

$$(c_1 + c_2)\left(R_{ab} - \frac{1}{2}Rg_{ab}\right) = 8\pi T_{ab} \tag{4.3}$$

Using the conditions (3.5) on the parameters $c_1, \ldots, c_{12}$ we find the simple expression

$$c_1 + c_2 = 1/(2 - N) \tag{4.4}$$

which becomes singular for $N = 2$ and negative for $N \geq 3$. Inserting the Robertson–Walker form (4.1) of the metric we then find the cosmological equations of motion

$$8\pi\rho = \frac{3}{2-N}\left(\frac{\dot{a}^2}{a^2} + \frac{k}{a^2}\right) \tag{4.5a}$$

$$8\pi p = \frac{1}{N-2}\left(2\frac{\ddot{a}}{a} + \frac{\dot{a}^2}{a^2} + \frac{k}{a^2}\right) \tag{4.5b}$$

Positive matter densities require $k = -1$ and $|\dot{a}| < 1$. We further derive the acceleration equation

$$\frac{\ddot{a}}{a} = \frac{4\pi(N-2)}{3}(\rho + 3p). \tag{4.6}$$

We thus find that the acceleration becomes positive for $N \geq 3$ and standard model matter with $\rho + 3p > 0$.

Journal of Physics: Conference Series **532** (2014) 012009 doi:10.1088/1742-6596/532/1/012009

In order to solve the cosmological equations of motion we further introduce the equation of state $p = w\rho$ with constant barotropic parameter w. The physical reasoning for this choice is the assumption that the matter content of the universe can be approximated by radiation, or relativistic particles, with $w = 1/3$ in the very early universe, and by dust, or non-relativistic particles, with $w = 0$ in the late universe. Using the conformal time η defined by $dt = a\,d\eta$ we then find the solutions

$$a = a_{\text{min}} \left[\cosh\left(\frac{3w+1}{2}(\eta - \eta_{\text{bounce}})\right)\right]^{\frac{2}{3w+1}} \tag{4.7a}$$

$$\rho = \rho_{\text{max}} \left[\cosh\left(\frac{3w+1}{2}(\eta - \eta_{\text{bounce}})\right)\right]^{-\frac{6w+6}{3w+1}} \tag{4.7b}$$

Here $a_{\text{min}} > 0$ and η_{bounce} are integration constants and ρ_{max} is given by

$$\rho_{\text{max}} = \frac{3}{8\pi(N-2)a_{\text{min}}^2} \tag{4.8}$$

From these solutions we find that our simple cosmological model features a big bounce, i.e, a positive minimal value of the scale factor at finite time, as a consequence of the negative effective gravitational constant in equation (4.3). In order to study its late time behavior we must transform the solutions from conformal time η to cosmological time t. By integration we obtain

$$t = -\frac{a_{\text{min}}}{4^{3w+1} e^{\eta - \eta_{\text{bounce}}}}\, {}_2F_1\left(\frac{-1}{3w+1}; \frac{-2}{3w+1}; \frac{3w}{3w+1}; -e^{(3w+1)(\eta - \eta_{\text{bounce}})}\right) \tag{4.9}$$

in terms of the hypergeometric function ${}_2F_1$. Now the big bounce at $\eta = \eta_{\text{bounce}}$ corresponds to $t = 0$. We then find that for late times the acceleration becomes small, $\ddot{a} \to 0$, and the expansion rate approaches the asymptotic value $\dot{a} \to 1$. This also becomes evident when we plot our solutions. Figure 1 shows the evolution of the scale factor $a(t)$. Figure 2 shows the matter density $\rho(t)$. Finally, figure 3 shows the Hubble parameter $H = \dot{a}/a$. In all plots the solution for radiation is displayed by a dashed line, while a solid line shows the dust matter solution.

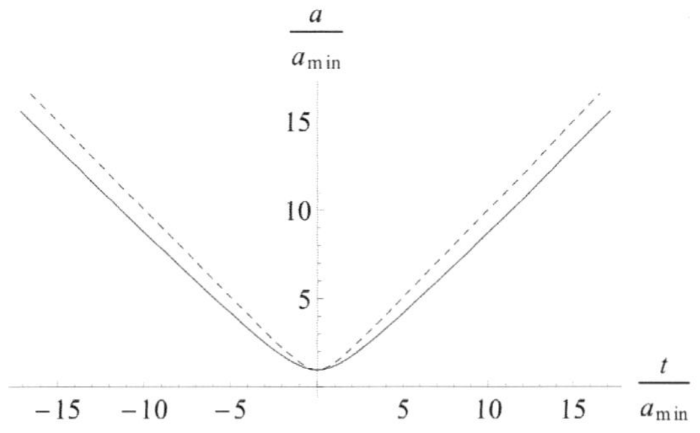

Figure 1. Scale factor a of the radiation-filled universe (dashed line) and the dust-filled universe (solid line) plotted over cosmological time t.

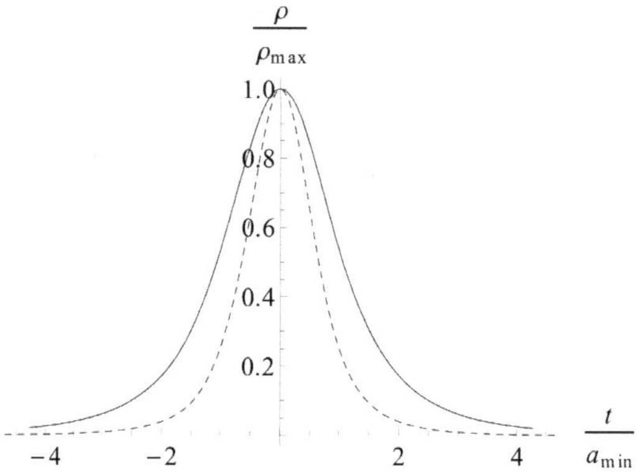

Figure 2. Matter density ρ of the radiation-filled universe (dashed line) and the dust-filled universe (solid line) plotted over cosmological time t.

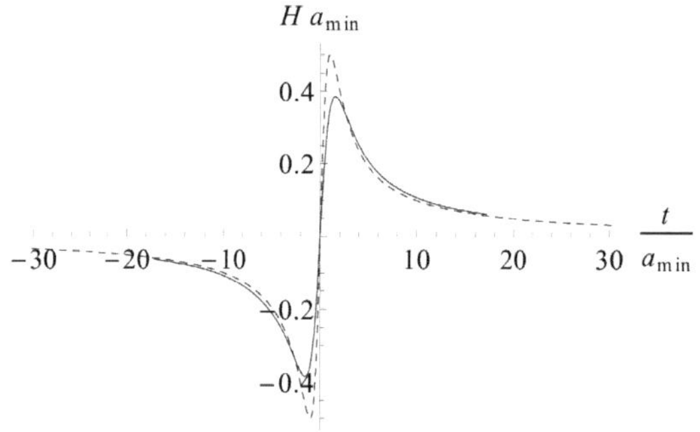

Figure 3. Hubble parameter $H = \dot{a}/a$ of the radiation-filled universe (dashed line) and the dust-filled universe (solid line) plotted over cosmological time t.

5. Formation of large scale structures

In this section we take the cosmological solution shown in the previous section as a basic ingredient for the evolution of large scale structures in the universe. This allows us to construct a simple model of structure formation in multimetric gravity. Our model will be based on a set of well-motivated assumptions, which we display in the first part of this section. We then discuss the resulting equations of motion which govern the dynamics of structure formation. We finally present the result of a simple computer simulation.

We start from the following set of assumptions:

(i) The metrics g^I_{ab} can be written as a small perturbation $g^I_{ab} = g^0_{ab} + h^I_{ab}$ around the cosmological solution

$$g^0 = -dt \otimes dt + a^2(t)\gamma_{\alpha\beta}dx^\alpha \otimes dx^\beta \tag{5.1}$$

(ii) The relevant scale for structure formation is small compared to the curvature radius of the universe, so that we can restrict ourselves to a cube $0 \leq x^\alpha \leq \ell$ of edge length ℓ, for

which we impose periodic boundary conditions, and approximate the spatial part $\gamma_{\alpha\beta}$ of the metric (5.1) by a flat metric $\delta_{\alpha\beta}$ within this cube.

(iii) The matter content of the universe is constituted by an equal number of n point masses of equal mass M per unit volume $(a\ell)^3$ for each of the N standard model copies.

(iv) The mean distance $a\ell/\sqrt[3]{Nn}$ between two point masses is large compared to their Schwarzschild radius $2G_N M$, so that we can treat their gravitational interaction in a weak field approximation.

(v) The spatial velocities $v^\alpha_{Ii} = a\dot{x}^\alpha_{Ii} \ll 1$ of the point masses measured in co-moving coordinates are small.

(vi) The initial state is given by a uniform matter distribution which is at rest with respect to our chosen coordinates, $\dot{x}^\alpha_{Ii}(t=0) = 0$.

It follows from these assumptions that the matter content of our model corresponds to dust with vanishing pressure $p = 0$ and a density

$$\rho = Mn/(a\ell)^3 \tag{5.2}$$

which enters the cosmological equations of motion (4.5). These govern the global dynamics of our model. The local dynamics is governed by the components h^I_{00} of the metric perturbation in the Newtonian limit, which we write in the form

$$h^I_{00} = 2\sum_{J=1}^{N} \alpha^{IJ} U^J \tag{5.3}$$

where the PPN parameter $\alpha^{IJ} = 2\delta^{IJ} - 1$ corresponds to the repulsive Newtonian limit. We further define the Newtonian potentials

$$U^I(t,\vec{x}) = \frac{M}{a}\sum_{i=1}^{n} \frac{1}{d(\vec{x},\vec{x}_{Ii}(t))} \tag{5.4}$$

using the distance function

$$d(\vec{x},\vec{x'}) = \min_{\vec{n}\in\mathbb{Z}^3} |\vec{x} - \vec{x}' + \ell\vec{n}| \tag{5.5}$$

which respects the periodic boundary conditions we imposed. Finally, the dynamics of the point masses is governed by the geodesic equation, which takes the form

$$\ddot{x}^\alpha_{Ii} = \frac{\partial_\alpha h^I_{00}}{2a^2} - 2\frac{\dot{a}}{a}\dot{x}^\alpha_{Ii} \tag{5.6}$$

in the Newtonian limit. These equations provide the full dynamics of our simple model of structure formation. They can be solved using the techniques of n-body simulations. The result of a simple implementation using only direct n-body interactions visualized using Splotch [http://www.mpa-garching.mpg.de/~kdolag/Splotch] is shown in figure 4. In this calculation $n = 2^{16} = 65536$ point masses for each of $N = 16$ matter types have been simulated. The simulation has been performed with the aid of GPU computing.

We can see in this simulation that visible matter forms clusters, which are connected by filament-like structures, in agreement with observations. We further find large, "empty" voids, in the sense that visible matter is absent. In our model it turns out that these are not truly empty, but contain clusters of dark galaxies formed by the additional standard model copies. This result is of particular interest in combination with the calculated value (3.6) of the PPN parameter γ^{IJ}, from which follows that these clusters should have a negative gravitational lensing effect on visible light.

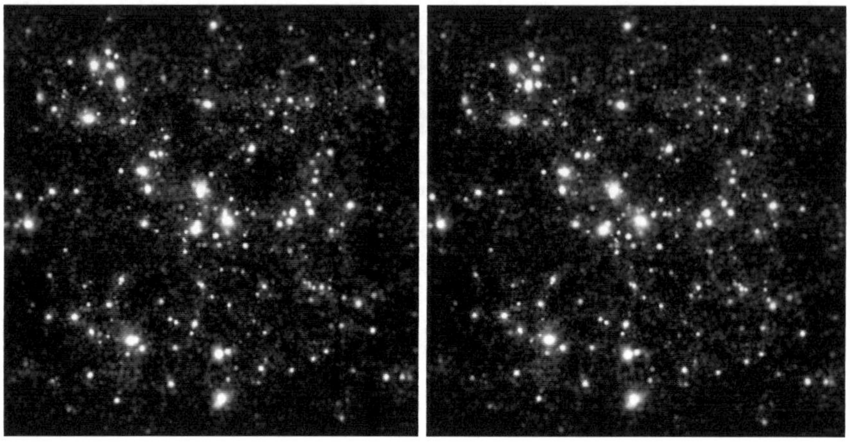

Figure 4. Stereoscopic image from a simulation of structure formation with $N = 16$ matter types and $n = 65536$ point masses for each matter type. Only visible matter is shown. For three-dimensional vision, view the two images from a short distance, so that each eye sees a different image, until they appear to merge. Then slowly increase the viewing distance to achieve focus.

6. Gravitational waves

We finally discuss the propagation of gravitational waves in the context of multimetric gravity [7]. The starting point of our discussion will be the most general linearized field equations of multimetric gravity. We will consider planar wave solution and discuss their propagation velocity. We will then make use of the Newman-Penrose formalism and derive the possible polarizations.

The starting point for our derivation is given by the most general linearized vacuum field equations compatible with our assumptions on multimetric gravity theories. Using the perturbation ansatz (3.1) they take the form

$$0 = K^I_{ab} = \sum_{J=1}^{N} \left(P^{IJ} \partial^p \partial_{(a} h^J_{b)p} + Q^{IJ} \Box h^J_{ab} + R^{IJ} \partial_a \partial_b h^J + M^{IJ} \partial^p \partial^q h^J_{pq} \eta_{ab} + N^{IJ} \Box h^J \eta_{ab} \right) \quad (6.1)$$

where indices are raised with the flat metric η and $\Box = \eta^{ab}\partial_a\partial_b$. The coefficients $P^{IJ}, Q^{IJ}, R^{IJ}, M^{IJ}, N^{IJ}$ are constant parameters which are determined by the concrete gravity theory under consideration. Due to the the symmetry assumption *(v)* they take the form

$$O^{IJ} = O^- + (O^+ - O^-)\delta^{IJ} \quad (6.2)$$

with diagonal entries O^+ and off-diagonal entries O^- for $O = P, Q, R, M, N$. From these expressions it further follows that the linearized vacuum field equations (6.1) decouple,

$$0 = \mathfrak{K}^1_{ab} = P_1 \partial^p \partial_{(a} \mathfrak{h}^1_{b)p} + Q_1 \Box \mathfrak{h}^1_{ab} + R_1 \partial_a \partial_b \mathfrak{h}^1 + M_1 \partial^p \partial^q \mathfrak{h}^1_{pq} \eta_{ab} + N_1 \Box \mathfrak{h}^1 \eta_{ab} \quad (6.3a)$$

$$0 = \mathfrak{K}^i_{ab} = P_0 \partial^p \partial_{(a} \mathfrak{h}^i_{b)p} + Q_0 \Box \mathfrak{h}^i_{ab} + R_0 \partial_a \partial_b \mathfrak{h}^i + M_0 \partial^p \partial^q \mathfrak{h}^i_{pq} \eta_{ab} + N_0 \Box \mathfrak{h}^i \eta_{ab} \quad (6.3b)$$

where $O_0 = O^+ - O^-$ and $O_1 = O^+ + (N-1)O^-$ are the eigenvalues of the corresponding parameter matrix O^{IJ}. The diffeomorphism invariance assumption *(ii)* restricts their values to

$$P_1 = -2Q_1 = -2R_1 = -2M_1 = 2N_1 \quad (6.4)$$

The quantities $\mathfrak{h}^1_{ab}$ and $\mathfrak{h}^i_{ab}$ for $i = 2, \ldots, N$ are linear combinations of the metric perturbations h^I_{ab}. Using the gauge invariant linear perturbation formalism they can be decomposed into physical degrees of freedom and pure gauge quantities. From the planar wave ansatz

$$\mathfrak{h}^I_{ab}(x) = \hat{\mathfrak{h}}^I_{ab} e^{ik_a x^a} \tag{6.5}$$

with constant amplitude $\hat{\mathfrak{h}}^I_{ab}$ and wave covector k_a for a single Fourier mode one can then show [7] that solutions for the physical degrees of freedom exist only if k_a is null, $k_a k^a = 0$. Hence, gravitational waves must propagate at the speed of light.

We now turn our attention to the possible polarizations of gravitational waves. For this purpose we employ the Newman-Penrose formalism [13] and make use of a double null basis $(l^a, n^a, m^a, \bar{m}^a)$ of the tangent space, which can be expressed as

$$l^a = (1,0,0,1), \quad n^a = \frac{1}{2}(1,0,0,-1), \quad m^a = \frac{1}{\sqrt{2}}(0,1,i,0), \quad \bar{m}^a = \frac{1}{\sqrt{2}}(0,1,-i,0) \tag{6.6}$$

in the (x^0, x^α) basis. We consider a plane wave propagating in the positive x^3 direction,

$$\mathfrak{h}_{ab} = \hat{\mathfrak{h}}_{ab} e^{i\omega(x^0 - x^3)} = \hat{\mathfrak{h}}_{ab} e^{i\omega u} \tag{6.7}$$

where $\mathfrak{h}_{ab}$ denotes one of the metric components $\mathfrak{h}^I_{ab}$ in the decoupled field equations (6.3) and we introduced the retarded time $u = x^0 - x^3$. The effect of this wave on test masses depends only on its Riemann tensor $\mathfrak{R}$, which is determined completely by the six electric components

$$\begin{aligned}
&\mathfrak{R}_{nlnl} = -\frac{1}{2}\ddot{\mathfrak{h}}_{ll} \quad \mathfrak{R}_{nln\bar{m}} = \overline{\mathfrak{R}_{nlnm}} = -\frac{1}{2}\ddot{\mathfrak{h}}_{l\bar{m}} = -\frac{1}{2}\overline{\ddot{\mathfrak{h}}_{lm}} \\
&\mathfrak{R}_{n\bar{m}n\bar{m}} = \overline{\mathfrak{R}_{nmnm}} = -\frac{1}{2}\ddot{\mathfrak{h}}_{\bar{m}\bar{m}} = -\frac{1}{2}\overline{\ddot{\mathfrak{h}}_{mm}} \quad \mathfrak{R}_{nmn\bar{m}} = -\frac{1}{2}\ddot{\mathfrak{h}}_{m\bar{m}}
\end{aligned} \tag{6.8}$$

where dots denote derivatives with respect to u. Their presence is restricted by the gravitational field equations. In the Newman-Penrose basis (6.6) we find that the five component equations

$$0 = \mathfrak{K}_{ll} = \mathfrak{K}_{mm} = \mathfrak{K}_{\bar{m}\bar{m}} = \mathfrak{K}_{lm} = \mathfrak{K}_{l\bar{m}} \tag{6.9}$$

are satisfied identically, while the remaining five component equations take the form

$$0 = \mathfrak{K}_{nn} = 2R\ddot{\mathfrak{h}}_{m\bar{m}} - (P + 2R)\ddot{\mathfrak{h}}_{ln} \tag{6.10a}$$

$$0 = \mathfrak{K}_{ln} = -\frac{1}{2}(P + 2M)\ddot{\mathfrak{h}}_{ll} \tag{6.10b}$$

$$0 = \mathfrak{K}_{nm} = -\frac{1}{2}P\ddot{\mathfrak{h}}_{lm} \tag{6.10c}$$

$$0 = \mathfrak{K}_{n\bar{m}} = -\frac{1}{2}P\ddot{\mathfrak{h}}_{l\bar{m}} \tag{6.10d}$$

$$0 = \mathfrak{K}_{m\bar{m}} = M\ddot{\mathfrak{h}}_{ll} \tag{6.10e}$$

where the curvature tensor $\mathfrak{K}_{ab}$ and the eigenvalues P, R, M of the parameter matrices are taken from the same field equation (6.3) as the metric perturbation $\mathfrak{h}_{ab}$ in the plane wave ansatz (6.7).

We now follow the scheme developed in [14, 15] to classify multimetric gravity theories by the allowed polarizations (6.8) of gravitational waves, labeled by representations of the little group E(2). For $P \neq 0$ and $P + 2R = 0$ we find that only the two tensor modes $\mathfrak{h}_{mm}, \mathfrak{h}_{\bar{m}\bar{m}}$ are allowed by the component equations (6.10), the E(2) class is thus N_2. For $P \neq 0$ and $P + 2R \neq 0$ we find an additional scalar mode $\mathfrak{h}_{m\bar{m}}$, corresponding to the E(2) class N_3. In the case $P = 0$ and

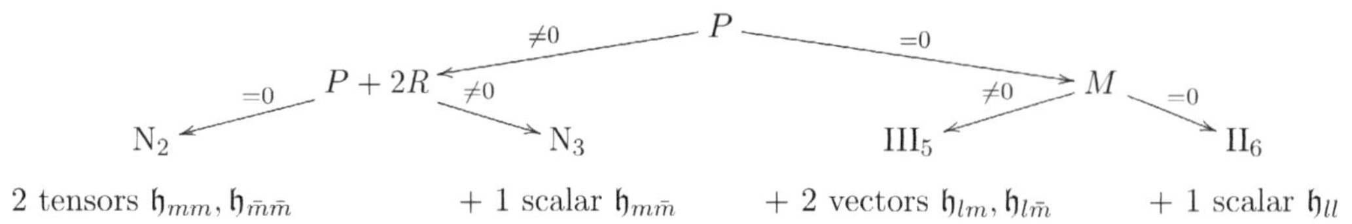

Figure 5. Classification of the different E(2) classes depending on the eigenvalues of the coefficient matrices in the linearized field equations.

$M \neq 0$ we additionally find two vector modes $\mathfrak{h}_{lm}, \mathfrak{h}_{l\bar{m}}$, which yields the E(2) class III_5. Finally, for $P = M = 0$ another scalar mode $\mathfrak{h}_{ll}$ is allowed and we obtain the E(2) class II_6. This is summarized in the diagram in figure 5.

We finally apply the general formalism discussed above to the multimetric gravity theory defined by the action (2.2) together with the experimental consistency constraints (3.5). Here the eigenvalues of the parameter matrices $P^{IJ}, Q^{IJ}, R^{IJ}, M^{IJ}, N^{IJ}$ take the values

$$\begin{aligned} P_1 &= -2Q_1 = -2R_1 = -2M_1 = 2N_1 = \frac{1}{2-N}, \quad Q_0 = -\frac{1}{4} \\ P_0 &= -2R_0 = -2M_0 = \frac{6-N}{6(2-N)} + 4y, \quad N_0 = \frac{N}{12(N-2)} + y \end{aligned} \tag{6.11}$$

where y is given by

$$y = \frac{2}{3}c_4 - \frac{1}{3}c_6 + \frac{1}{3}c_8 + \frac{1}{2}c_{11} + c_{12} \tag{6.12}$$

From this we immediately read off that for the metric perturbation $\mathfrak{h}^1_{ab}$, which is governed by the eigenvalues P_1, Q_1, R_1, M_1, N_1, the E(2) class is always N_2, i.e, the only allowed polarizations are the two tensor modes $\mathfrak{h}_{mm}, \mathfrak{h}_{\bar{m}\bar{m}}$. The same applies to the remaining perturbations $\mathfrak{h}^i_{ab}$ governed by the eigenvalues P_0, Q_0, R_0, M_0, N_0, unless y takes the special value

$$y = -\frac{6-N}{24(2-N)} \tag{6.13}$$

in which case all six polarizations of gravitational waves are allowed and the E(2) class is II_6.

7. Conclusion

We have presented a class of gravity theories with $N \geq 2$ metric tensors and a corresponding number of standard model copies and discussed various aspects of multimetric gravity. In particular we considered theories in which gravity is attractive within each matter sector, but repulsive between the different standard model copies in the Newtonian limit.

As the first aspect we have studied the post-Newtonian limit of multimetric gravity. For this purpose we constructed an extension of the parameterized post-Newtonian (PPN) formalism. Our extended formalism characterizes multimetric gravity theories with a set of constant parameters, thirteen of which can be measured by visible matter experiments in the solar system. Our findings have shown that multimetric gravity is fully consistent with current measurements of the ten standard PPN parameters in solar system experiments.

We continued our discussion in the field of cosmology, where we started from the standard assumptions of isotropy and homogeneity. We showed that under these assumptions multimetric gravity can be described by a single common metric, which obeys the Einstein equations with a negative effective gravitational constant. As a consequence we obtained an accelerating expansion of the universe which naturally becomes small at late times and a big bounce.

Using the exact cosmological solution for dust matter as a background geometry, we constructed a simple model for the formation of large scale structures. From a numerical simulation we found that voids, in which the density of visible matter is low, should not be empty, but contain clusters of the additional, dark, standard model copies. From our post-Newtonian analysis it further follows that these would act as negative gravitational lenses.

Finally we discussed the linearized gravitational field equations in vacuum under the aspect of gravitational waves. We presented a wave ansatz, from which follows that gravitational waves propagate at the speed of light. This allowed us to apply the Newman-Penrose formalism and to calculate the allowed polarizations and the E(2) class of multimetric gravity. We found that in general these can be N_2, N_3, III_5 and II_6, and displayed example theories for N_2 and II_6.

Naturally the question arises whether the number N of metrics can be fixed or bounded by observations. An immediate restriction follows from our no-go theorem for bimetric repulsive gravity, which states that $N = 2$ is excluded in the case that different standard model copies repel each other with equal strength [4]. This assumption appears natural, since it does not introduce a second, undetermined gravitational coupling strength, and has also entered our discussions of cosmology an structure formation. However, we do not find other restrictions on N from the presented calculations or from current observational data.

The work we presented here is still fundamental and further studies are required to test multimetric gravity using observations. Of particular interest would be tests in strongly coupled systems, such as double pulsars, for which high-precision timing measurements of radio pulses exist and observations of gravitational waves are expected for the near future. The cosmic microwave background and the distribution of large scale structures could provide another testbed for multimetric gravity theories, using cosmological perturbation analysis and further simulations of structure formation. Finally, our findings motivate new experimental tests of gravity, such as solar system measurements of the three newly introduced PPN parameters, and the search for negative gravitational lenses in the galactic voids.

Acknowledgments

The author gratefully acknowledges full financial support from the Estonian Research Council through the Postdoctoral Research Grant ERMOS115.

References

[1] Riess A G *et al* 1998 [Supernova Search Team Collaboration] *Astron. J.* **116** 1009
[2] Perlmutter S *et al* 1999 [Supernova Cosmology Project Collaboration] *Astrophys. J.* **517** 565
[3] Davis M, Efstathiou G, Frenk C S and White S D M 1985 *Astrophys. J.* **292** 371
[4] Hohmann M and Wohlfarth M N R 2009 *Phys. Rev.* D **80** 104011
[5] Hohmann M and Wohlfarth M N R 2010 *Phys. Rev.* D **81** 104006
[6] Hohmann M and Wohlfarth M N R 2010 *Phys. Rev.* D **82** 084028
[7] Hohmann M 2012 *Phys. Rev.* D **85** 084024
[8] Hohmann M 2013 *Classical Quantum Gravity* **31** 135003
[9] Nordtvedt K 1968 *Phys. Rev.* **169** 1017
[10] Thorne K S and Will C M 1971 *Astrophys. J.* **163** 595; **163** 611; **169** 125
[11] Will C M 1993 *Theory and Experiment in Gravitational Physics* (Cambridge Univ. Press)
[12] Will C M 2014 *Living Rev. Rel.* **17** 4
[13] Newman E and Penrose R 1962 *J. Math. Phys.* **3** 566
[14] Eardley D M, Lee D L, Lightman A P *et al* 1973 *Phys. Rev. Lett.* **30** 884
[15] Eardley D M, Lee D L and Lightman A P 1973 *Phys. Rev.* D **8** 3308

Journal of Physics: Conference Series **532** (2014) 012010 doi:10.1088/1742-6596/532/1/012010

A note on Clifford-Klein forms

P Jastrzębski and A Tralle

Department of Mathematics and Information Technology, University of Warmia and Mazury Słoneczna str. 54, 10-710 Olsztyn, Poland

Abstract. We consider the problem of finding Clifford-Klein forms in a class of homogeneous spaces determined by inclusions of real Lie algebras of a special type which we call strongly regular. This class of inclusions is described in terms of their Satake diagrams. For example, the complexifications of such inclusions contain the class of subalgebras generated by automorphisms of finite order. We show that the condition of strong regularity implies the restriction on the real rank of subalgebras. This in part explains why the known examples of Clifford-Klein forms are rare. We make detailed calculations of some known examples from the point of view of the Satake diagrams.

1. Introduction

We say that a pair of reductive Lie groups (G, H) is a Clifford-Klein form, if there is a discrete subgroup $\Gamma \subset G$ acting properly discontinuosly on G/H such that the double coset $\Gamma \setminus G/H$ is compact. Riemannian locally symmetric spaces yield a class of examples of such pairs. In this case, H is compact. We are interested in the case when H is non-compact. It is important in geometry to solve the problem of finding Clifford-Klein forms. We refer to the survey [5], as well as research papers [4, 6, 7, 8, 10] and references therein for motivation and known results. The importance of Clifford-Klein forms comes from the fact that such homogeneous spaces G/H may carry various important invariant geometric structures which descend onto compactifications $\Gamma\backslash G/H$. For example this kind of compactification was used in [11, 3] to construct new examples of symplectic structures. The problem of determining whether a pair (G, H) is a Clifford-Klein form is difficult. There is the "Calabi-Markus phenomenon" that homogeneous space G/H with G and H of equal real ranks cannot have infinite discrete subgroups $\Gamma \subset G$ acting freely and uniformly on G/H. The existence of some invariant geometric structures on G/H eliminates the possibility of G/H being a Clifford-Klein form. For example, para-hermitian symmetric spaces cannot be Clifford-Klein forms. Lorentz space forms admit compact forms only in the flat case and in case of negative sectional curvature, if the dimension is odd, see [5]. The following method of proof that (G, H) is a Clifford-Klein form was found in [4]. We restrict ourselves to homogeneous spaces G/H of reductive type, which means that both G and H are reductive and their Cartan decompositions can be made compatible, see details in [4]. To prove that G/H is a Clifford-Klein form, one can consider the Lie algebras $\mathfrak{g} \supset \mathfrak{h}$ and look for a second subalgebra $\mathfrak{l}$ (which also corresponds to a closed Lie subgroup $L \subset G$) with the following properties. Both $\mathfrak{h}$ and $\mathfrak{l}$ are compatible with the Cartan decomposition of $\mathfrak{g}$:

$$\mathfrak{g} = \mathfrak{k} \oplus \mathfrak{p}, \quad \mathfrak{h} = \mathfrak{h} \cap \mathfrak{k} \oplus \mathfrak{h} \cap \mathfrak{p}, \quad \mathfrak{l} = \mathfrak{l} \cap \mathfrak{k} \oplus \mathfrak{l} \cap \mathfrak{p}, \quad \dim \mathfrak{p} = \dim \mathfrak{h} \cap \mathfrak{p} + \dim \mathfrak{l} \cap \mathfrak{p}$$

and maximal Cartan subspaces $\mathfrak{a} \subset \mathfrak{p}, \mathfrak{a}' \subset \mathfrak{h} \cap \mathfrak{p}, \mathfrak{a}'' \subset \mathfrak{l} \cap \mathfrak{p}$ can be chosen in a way that

$$\mathfrak{a}' \cap \cup_{w \in W} w \cdot \mathfrak{a}'' = \{0\}$$

for the little Weyl group

$$W = W(\mathfrak{g}, \mathfrak{a}) = N_G(\mathfrak{a})/Z_G(\mathfrak{a})$$

Thus, given a triple (G, H, L) one can check if it yields a Clifford-Klein form by the described method. However there is no way in sight of looking for triples (G, H, L) satisfying these conditions. The purpose of this note is to propose a procedure of detecting Clifford-Klein forms among a certain class of homogeneous spaces. The idea is to restrict ourselves to a special family of embeddings $H \subset G$ and $L \subset G$ (and, on the Lie algebra level, to inclusions $\mathfrak{h} \subset \mathfrak{g}, \mathfrak{l} \subset \mathfrak{g}$), which we call strongly regular. It seems that singling out much more "tame" class of subgroups is useful for understanding the problem of constructing Clifford-Klein forms. Our inclusions are obtained by a certain procedure applied to "generalized" Dynkin diagrams of the complexification $\mathfrak{g}^c$, these are called the diagrams of type $S(A)$ in [2]. In particular, subalgebras generated by automorphisms of finite order $\varphi : \mathfrak{g}^c \to \mathfrak{g}^c$ belong to this class. Introducing this class of subalgebras allows us to describe the problem of finding Clifford-Klein forms in terms of the Satake diagrams. We make all calculations in detail for some known examples of Clifford-Klein forms. These calculations, as well as the main theorem reveal the algebraic nature of the difficulties arising in constructing Clifford-Klein forms, and explain to some extent why they are so rare. The main result of this paper shows that the strong regularity condition together with the condition of Kobayashi put restrictions on the real rank of subgroups, which should be ≤ 2. It is worth mentioning that we don't know any examples of Clifford-Klein forms obtained by the described method [4] which are not strongly regular.

2. Notation and preliminaries

We use basics of the theory of (semisimple) real and complex Lie algebras without further explanations. Our terminology and notation are close to [9]. Let $\mathfrak{g}$ be a real semisimple Lie algebra. Consider a Cartan decomposition $\mathfrak{g} = \mathfrak{k} \oplus \mathfrak{p}$. Fix a maximal abelian subalgebra $\mathfrak{a} \subset \mathfrak{p}$ and consider the real root system $\Sigma \subset \mathfrak{a}^*$ of $\mathfrak{g}$. The choices made yield the decompositions

$$\mathfrak{g} = \mathfrak{g}_0 \oplus \sum_{\lambda \in \Sigma} \mathfrak{g}_\lambda, \quad \mathfrak{g}_0 = \mathfrak{z}_\mathfrak{g}(\mathfrak{a}), \quad \mathfrak{g}_0 = (\mathfrak{g}_0 \cap \mathfrak{k}) \oplus \mathfrak{a}$$

Let Δ be the root system of $\mathfrak{g}^c$. In what follows we make several observations and choices which relate Σ and Δ. Choose a Cartan subalgebra $\mathfrak{t} \subset \mathfrak{g}$ as a maximal abelian subalgebra in $\mathfrak{g}$ containing $\mathfrak{a}$. This yields the decompositions

$$\mathfrak{t} = \mathfrak{t}^+ \oplus \mathfrak{a}, \quad \mathfrak{t}^c = (\mathfrak{t}^+)^c \oplus \mathfrak{a}^c$$

It is known that $\mathfrak{t}^c$ is an algebraic subalgebra consisiting of semisimple elements, and, therefore

$$\mathfrak{t}^c(\mathbb{R}) = (i\mathfrak{t}^+) \oplus \mathfrak{a}$$

The latter enables one to introduce the restriction map $r : \mathfrak{t}^c(\mathbb{R}) \to \mathfrak{a}^*$, and to show that $r(\Delta \cup \{0\}) = \Sigma \cup \{0\}$. Let

$$\Delta_0 = \{\alpha \in \Delta \,|\, r(\alpha) = 0\}, \quad \Delta_1 = \Delta \setminus \Delta_0$$

The decomposition $\Delta = \Delta_0 \cup \Delta_1$ yields the decompositions

$$(\mathfrak{g}_0 \cap \mathfrak{k})^c = \mathfrak{t}^c \oplus \sum_{\alpha \in \Delta_0} \mathfrak{g}^c_\alpha, \quad \mathfrak{g}^c_\lambda = \sum_{r(\alpha)=\lambda} \mathfrak{g}^c_\alpha, \quad \mathfrak{g}^c = \mathfrak{t}^c \oplus \sum_{\alpha \in \Delta_0} \mathfrak{g}^c_\alpha \oplus \sum_{\lambda \in \Sigma} \sum_{r(\alpha)=\lambda} \mathfrak{g}^c_\alpha, \quad \alpha \in \Delta_1, \ \lambda \in \Sigma$$

It is visible from the latter formula that Δ_0 is a root system for the semisimple part of $(\mathfrak{g}_0 \cap \mathfrak{k})^c$. One assigns to any real semisimiple Lie algebra the *Satake diagram* as follows. It is a Dynkin diagram of $\mathfrak{g}^c$ with vertexes of two types, black and white. Black vertexes α_i are those for which $r(\alpha_i) = 0$. White vertexes α_j and α_k are joined by an arrow if $r(\alpha_j) = r(\alpha_k)$. It is known that two semisimple real Lie algebras are isomorphic if and only if so are their Satake diagrams. As usual, we consider the dual root systems Σ and $\Sigma^\vee$ and relate the corresponding vectors by the formula $\mu(\alpha^\vee) = 2(\mu,\alpha)/(\alpha,\alpha) = \langle \mu \,|\, \alpha\rangle$. The Cartan matrix $A = A_\Sigma$ corresponding to Σ is a matrix with elements $a_{ij} = \langle \alpha_i \,|\, \alpha_j\rangle$, where α_i are simple roots of Σ. Note that α_i is a black root in the Satake diagram, if and only if $\alpha_i^\vee$ centralizes $\mathfrak{a}^c = (\langle \Sigma^\vee\rangle)^c$.

3. On Clifford-Klein forms generated by strongly regular triples

In order to introduce the class of subalgebras we are interested in, we need to describe the notion of the diagram of type $S(A)$ in the sense of [2]. Note that our description is very brief and we recommend the reader to consult [2], Chapter X. Let E be a real vector space of dimension n. A subset $\{\alpha_0, ..., \alpha_n\} \subset E$ is called a subset of type $S(A)$ if

(i) $a_{ij} = \frac{2(\alpha_i,\alpha_j)}{(\alpha_i,\alpha_i)} \in -\mathbb{Z}^+$ for $i \neq j$

(ii) $S(A) = \{\alpha_0, ..., \alpha_n\}$ is an indecomposable system of vectors (that is, $S(A)$ cannot be decomposed into a sum of orthogonal non empty subsets),

(iii) $S(A)$ is a linearly dependent system of vectors generating E, in particular

$$\det(a_{ij}) = \det A = 0$$

One associates to each matrix A a diagram $S(A)$ as follows (note that in the sequel we use the same notation for the set $S(A)$, and for the corresponding diagram). We take $n+1$ vertexes, join the i-th and j-th vertexes by $a_{ij} \cdot a_{ji}$ lines, and if $|a_{ij}| < |a_{ji}|$, these lines have an arrow pointing towards the i-th vertex. The corresponding diagram will be called the diagram of type $S(A)$. All such diagrams are classified and one can find this classification in [2] on page 503. Note that we don't reproduce this table, but just mention that it consists of extended Dynkin diagrams (these are of types $A_n^{(1)}, B_n^{(1)}, C_n^{(1)}, D_n^{(1)}, E_6^{(1)}, E_7^{(1)}, E_8^{(1)}, F_4^{(1)}, G_2^{(1)}$), and two other types: $A_{2n}^{(2)}, D_{n+1}^{(2)}, A_{2n-1}^{(2)}, E_6^{(2)}$, and $D_4^{(3)}$.

Let us mention the relation between the diagrams of type $S(A)$ and the Dynkin diagrams of simple complex Lie algebras. The diagrams of types $S(A)$ correspond to those automorphisms $\nu : \mathfrak{g}^c \to \mathfrak{g}^c$ of simple complex Lie algebras $\mathfrak{g}^c$, which are induced by the automorphisms (of order $k = 1,2,3$) of the Dynkin diagrams. Then, the subalgebra $(\mathfrak{g}^c)^\nu$ of fixed points has the Dynkin diagram determined by the subdiagram $S(A)\setminus\{\rho_0\}$ for a suitable choice of a root ρ_0 (we refer to Section 5 of Chapter X in [2], especially Examples 1 and 2 and arguments on page 508). Moreover, simple roots of $S(A)\setminus\{\rho_0\}$ are restrictions of simple roots from the Dynkin diagram of $\mathfrak{g}^c$ onto a Cartan subalgebra of $(\mathfrak{g}^c)^\nu$. Let us adopt the following notation: Π is the set of simple roots for the root system Δ, $\tilde{\Pi}$ will be reserved for the set $\{\alpha_0, ..., \alpha_n\}$ of roots generating the extended Dynkin diagram of rank n, while $\hat{\Pi}$ will denote *any* of the possible generating sets of diagrams $S(A)\setminus\{\rho_0\}$. Note that $\hat{\Pi}$ can be considered as a subset of $\tilde{\Pi}$, but in general not as a subdiagram.

Definition 3.1. We say that an inclusion of real semisimple Lie algebras is *strongly regular*, if

- the inclusion of their complexifications $\mathfrak{h}^c \subset \mathfrak{g}^c$ is described as follows: the Dynkin diagram of $\mathfrak{h}^c$ is obtained by removing certain vertexes from one of the diagrams of $\hat{\Pi} = S(A)\setminus\{\rho_0\}$,
- if $\Sigma = r(\Delta)$ and $r_1(\Delta') = \Sigma'$ are real root systems of $\mathfrak{g}$, and $\mathfrak{h}$ respectively, then $\Pi'_\Sigma \subset r(\tilde{\Pi})$.

Note that semisimple part of any subalgebra $\mathfrak{h}^c = (\mathfrak{g}^c)^\varphi$ of fixed points of an automorphism φ of finite order satisfies the first condition of the definition. This follows from Theorem 5.15 in [2]. Indeed, the Dynkin diagram of such subalgebra is obtained by removing certain vertexes from $\hat{\Pi}$ (according to a certain rule which is not used here).

Remark 3.2. In the proof of the main result we use only the second condition of the definition. However, in order to calculate examples we need to use the Satake diagrams, and, therefore, some conditions which enable us to calculate the Dynkin diagrams of $\mathfrak{h}^c$ and $\mathfrak{l}^c$. This is the reason why we prefer to define our class using both restrictions (on Π', Π'' and $\Pi'_\Sigma, \Pi''_\Sigma$).

There is the notion of regularity for subalgebras of complex Lie algebras (see [9], Chapter 6). Note that maximal regular subalgebras are obtained by removal one or two vertexes from the extended Dynkin diagram (see Theorem 1.2 in [9]). It is known that maximal semisimple subalgebras of maximal rank are of the form $(\mathfrak{g}^c)^\nu$ for an inner involutive automorphism $\nu : \mathfrak{g}^c \to \mathfrak{g}^c$, and therefore, are obtained from diagrams $S(A)$ as well. We don't know if all regular subalgebras can be obtained from diagrams of types $S(A)$. The triple $(\mathfrak{g}, \mathfrak{h}, \mathfrak{l})$ is called strongly regular, if $\mathfrak{h}$ and $\mathfrak{l}$ are strongly regular subalgebras in $\mathfrak{g}$. We say that a Clifford-Klein form is generated by a strongly regular triple (G, H, L), if the corresponding triple $(\mathfrak{g}, \mathfrak{h}, \mathfrak{l})$ is strongly regular, L acts properly on G/H and the double coset space $L \setminus G/H$ is compact.

Recall that such triple always produces a Clifford-Klein form if there is a cocompact torsionfree lattice $\Gamma \subset L$ [4]. Now we are ready to formulate the main result of this work.

Theorem 3.3. *Assume that G is a simple Lie group which is a non-compact real form of a complex classical Lie group. Any strongly regular triple of semisimple Lie groups (G, H, L) generating the Clifford-Klein form has the property that at least one of the subgroups has real rank ≤ 2.*

Proof. Let Π be the system of simple roots of $\mathfrak{g}^c$, Π' and Π'' be the systems of simple roots for $\mathfrak{h}^c$ and $\mathfrak{l}^c$, respectively. Then, for the restriction maps r_1 and r_2 (built for $\mathfrak{h}$ and $\mathfrak{l}$, respectively), $r_1(\Pi') = \Pi'_\Sigma$, $r_2(\Pi'') = \Pi''_\Sigma$. By the assumption of strong regularity

$$\Pi'_\Sigma \cup \Pi''_\Sigma \quad \subset r(\tilde{\Pi})$$

it follows that

$$\mathfrak{a}' = \langle \Pi'_\Sigma \rangle, \quad \mathfrak{a}'' = \langle \Pi''_\Sigma \rangle, \quad \mathfrak{a} = \langle \Pi_\Sigma \rangle = \langle r(\tilde{\Pi}) \rangle$$

The latter implies that the condition $\cup_{w \in W} w \cdot \mathfrak{a}' \cap \mathfrak{a}'' = \{0\}$ is satisfied if and only if $w(\Pi'_\Sigma) \cup \Pi''_\Sigma$ is linearly independent in $\mathfrak{a}$ for any $w \in W$. We are going to show that the latter implies the condition of the Theorem. This will be done by considering each of the classical Lie groups separately. Let $\mathfrak{g}$ be a non-compact real form of a complex simple Lie algebra. The type of Π_Σ is known for all simple Lie algebras, for example, from [9]. Note also that

$$r(\tilde{\Pi}) = \begin{cases} \Pi_\Sigma, & \text{if} \quad r(\alpha_0) = 0 \\ \Pi_\Sigma \cup \{r(\alpha_0)\}, & \text{otherwise} \end{cases}$$

This implies the inclusions

$$\Pi'_\Sigma \subset \Pi_\Sigma \cup \{r(\alpha_0)\}, \quad \Pi''_\Sigma \subset \Pi_\Sigma \cup \{r(\alpha_0)\}$$

because of the strong regularity assumption. Looking at Table 4 in [9] one can notice that only the following types of Σ are possible:

(i) $\mathfrak{g}^c$ of type A_n: Σ may have types A_n, BC_p and C_p, with $p \leq [\frac{n}{2}] - 1$
(ii) $\mathfrak{g}^c$ of type B_n: Σ has type B_p, with $p < n$
(iii) $\mathfrak{g}^c$ of type C_n: Σ has type C_n, or $BC_p, p < n$
(iv) $\mathfrak{g}^c$ of type D_n: Σ has type C_p or $BC_p, p < n$

Let us begin our case by case analysis.

- **type** B_p In this case,

$$\Pi_\Sigma = \{\alpha_1 = \varepsilon_1 - \varepsilon_2, \alpha_2 = \varepsilon_2 - \varepsilon_3, ..., \alpha_{p-1} = \varepsilon_{p-1} - \varepsilon_p, \alpha_p = \varepsilon_p\}$$

Since $W \cong W_\Sigma$ acts on Π_Σ permuting all ε_i and possibly changing signs $\varepsilon_i \to (\pm)\varepsilon_i$, it has two orbits $\mathcal{O}_1$ consisting of roots $\alpha_1, ..., \alpha_{p-1}$, and $\mathcal{O}_2$ consisting of $\alpha_p = \varepsilon_p$. Note that $r(\alpha_0)$ may belong to Π'_Σ, or to Π''_Σ or may not.

(i) if $r(\alpha_0) \notin \Pi'_\Sigma \cup \Pi''_\Sigma$, then, from the definition of r one can see that

$$\Pi'_\Sigma \cup \Pi''_\Sigma \quad \subset \Pi_\Sigma$$

and the only possibility of their linear independence is that one of these sets is $\mathcal{O}_2$, and the other is a subset of $\mathcal{O}_1$. But the latter means that the rank of the subalgebra coresponding to Π'_Σ must be one.

(ii) if $r(\alpha_0) \in \Pi'_\Sigma$, then the following possibilities may occur:

$$r(\alpha_0) \in \mathcal{O}_1, \; r(\alpha_0) \in \mathcal{O}_2, \; r(\alpha_0) \notin \mathcal{O}_1 \cup \mathcal{O}_2$$

In the first case, the linear independence is achieved, if one of the sets Π'_Σ or Π''_Σ has a form $\{\alpha_n\}$, and the other consists of vectors in $\mathcal{O}_2$. This case again, yields the rank one. If $r(\alpha_0) \in \mathcal{O}_2$, necessarily $\Pi'_\Sigma = \{r(\alpha_0), \alpha_n\}$, and the rank of the corresponding real Lie subalgebra is 2. If $r(\alpha_0)$ does not belong to $\mathcal{O}_1 \cup \mathcal{O}_2$, this means that still the rank is 1 or 2, because in any case one of the subspaces $\mathfrak{a}'$ or $\mathfrak{a}''$ does not contain any vector from $\mathcal{O}_1$.

- **type** C_p, BC_p. The argument is similar. We have

$$\Pi_\Sigma = \langle \alpha_1 = \varepsilon_1 - \varepsilon_2, ..., \varepsilon_{p-1} - \varepsilon_p, \alpha_p = 2\varepsilon_p \rangle$$

It follows that still Π_Σ has two orbits $\mathcal{O}_1 = \{\alpha_1, ..., \alpha_{p-1}\}$ and $\mathcal{O}_2 = \{\alpha_p\}$.

- **type** A_n differs, because in this case W acts transitively on on Π_Σ, and the only possible new orbit of W may appear if $r(\alpha_0)$ does not belong to this orbit. Thus, one of the subalgebras must have real rank 1. □

Note that our theorem contains a bit more than the restriction on the real rank. It contains a procedure of looking for possible regular Clifford-Klein forms. The latter goes as follows.

- Given a (say, simple) classical Lie algebra $\mathfrak{g}$, consider its complexification $\mathfrak{g}^c$ and all the diagrams of type $S(A)$ with the set of vertexes $\hat{\Pi} = S(A) \setminus \{\rho_0\}$, and write down possible strongly regular subalgebras,
- looking at the corresponding Satake diagrams of these subalgebras, choose those which satisfy $\Pi'_\Sigma \subset \Pi_\Sigma \cup \{r(\alpha_0)\}$ and such that one of them has real rank ≤ 2,
- check the condition $\dim \mathfrak{p} = \dim \mathfrak{p}' + \dim \mathfrak{p}''$.

4. Examples

Let us analyze examples constructed by T. Kobayashi in [4, 5] from the point of view of strongly regular triples.

Example 4.1. The triple

$$(G = SU(2n,2), H = SU(2n,1), L = Sp(n,1))$$

is a strongly regular Clifford-Klein form.

To prove this, we pass to the Lie algebra level. Thus

$$\mathfrak{g} = su(2n,2), \quad \mathfrak{h} = su(2n,1), \quad \mathfrak{l} = sp(n,1)$$
$$\mathfrak{g}^c = sl(2n+2,\mathbb{C}), \quad \mathfrak{h}^c = sl(2n+1,\mathbb{C}), \quad \mathfrak{l} = sp(n+1,\mathbb{C})$$

which shows that all considered complex Lie algebras are simple, and have types, respectively, A_{2n+1}, A_{2n} and C_n. Consider the embedding $\mathfrak{h}^c \subset \mathfrak{g}^c$ determined by the corresponding inclusion

$$\Pi' = \{\alpha_2, ..., \alpha_{2n+1}\} \quad \subset \Pi = \{\alpha_1, ..., \alpha_{2n+1}\}$$

Looking at the corresponding Satake diagrams for the real forms $su(2n,2)$ and $su(2n,1)$ we determine Σ and Σ'. Let us begin with the Satake diagram on for $su(2n,2)$:

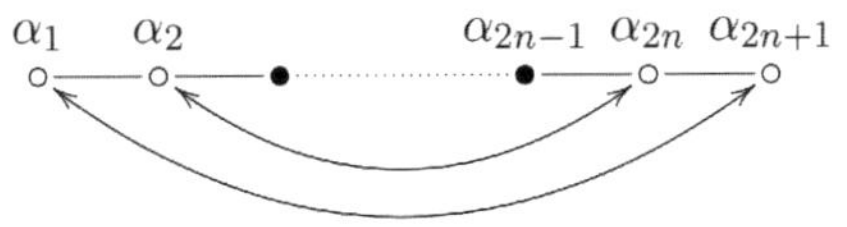

Figure 4.1. Satake diagram for $su(2n,2)$.

We see that $\mathfrak{a} = \langle \Sigma^\vee \rangle$ and

$$\mathfrak{g} = \mathfrak{z}_\mathfrak{g}(\langle \Sigma^\vee \rangle) \oplus \sum_{\lambda \in \Sigma} \mathfrak{g}_\lambda$$

where $\Sigma = \langle \bar{\alpha}_1, \bar{\alpha}_2 \rangle$. Note that the black roots $\alpha_3, ..., \alpha_{2n-1}$ determine the semisimple part of $\mathfrak{z}_\mathfrak{g}(\langle \Sigma^\vee \rangle)$. Consider

$$\Sigma' = \langle \bar{\alpha}_2 \rangle \subset \Sigma = \langle \bar{\alpha}_1, \bar{\alpha}_2 \rangle, \quad \mathfrak{h} = \mathfrak{z}_\mathfrak{g}(\langle (\Sigma')^\vee \rangle) \oplus \sum_{\lambda \in \Sigma'} \mathfrak{g}_\lambda$$

Clearly, $\mathfrak{h} \subset \mathfrak{g}$. In order to determine the Satake diagram of the semisimple part $\mathfrak{h}_s$ of $\mathfrak{h}$, we need to determine the Dynkin diagram for $\mathfrak{h}_s^c$ and vertexes which represent roots α_i, whose duals $\alpha_i^\vee$ centralize $\langle (\Sigma')^\vee \rangle$. Since $\alpha_3^\vee, ..., \alpha_{2n-1}^\vee$ centralize $\langle (\Sigma')^\vee \rangle$, they belong to the Dynkin diagram of $\mathfrak{h}_s^c$, and represent black vertexes of the corresponding Satake diagram. It is known that the Cartan matrix (a_{ij}) of $\mathfrak{g}^c$ has $a_{ii} = 2$, $a_{i,i+1} = -1$ and $a_{ij} = 0, j \neq i$. For example, one can consult [1]. It follows that $\alpha_{2n}(\alpha_2^\vee) = 0$, for $n > 1$. Thus, α_{2n} must belong to the Dynkin diagram of $\mathfrak{h}_s^c$, and must be black. Hence, the Satake diagram for $\mathfrak{h}_s^c$ must contain $\alpha_2, ..., \alpha_{2n}$. Looking at the table of all possible Satake diagrams of simple real Lie algebras, one concludes that there is only one possibility for the Satake diagram which contains the part $\{\alpha_2, ..., \alpha_{2n+1}\}$ with black roots $\alpha_3, ..., \alpha_{2n}$. This is the diagram

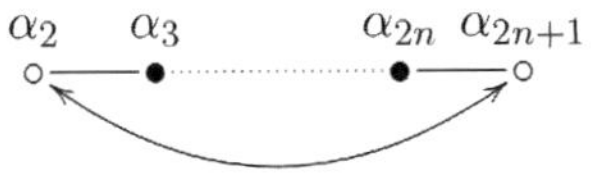

Figure 4.2. Satake diagram for $su(2n,1)$.

The latter is the Satake diagram of $su(2n,1)$, which means that $\mathfrak{h}_s = \mathfrak{h} = su(2n,1)$. We see that $\mathfrak{h}$ is embedded into $\mathfrak{g}$ in a strongly regular way, as required, because $\Pi'_\Sigma \subset r(\Pi) \subset r(\tilde{\Pi})$ (in this case, the diagram of type $S(A)$ is the extended Dynkin diagram with the set of roots $\tilde{\Pi}$).

Consider the case of $\mathfrak{l} = sp(n,1) \subset sl(2n+1,\mathbb{C})$. On the complex Lie algebra level, the Lie algebra of type A_{2n+1} admits an involutive automorphism ν such that $\mathfrak{l}^c = (\mathfrak{g}^c)^\nu$. The root system of $(\mathfrak{g}^c)^\nu$ is obtained from the diagram $S(A)$ of type $A^{(2)}_{2n+1}$ (in the terminology of [2]). This diagram has the form

Figure 4.3. Diagram $S(A)$ of type $A^{(2)}_{2n+1}$.

The Dynkin diagram of $\mathfrak{l}^c$ is obtained from the diagram of type $A^{(2)}_{2n+1}$ by removing vertex α_0. Looking again at the Satake diagrams of $su(2n,2)$ and $sp(n,1)$ one can argue exactly as in the previous case and calculate that $r_2(\alpha_1) = 0, r_2(\alpha_2) = \alpha_2$. Here is the required Satake diagram:

Figure 4.4. Satake diagram for $sp(n,1)$.

Thus, $\Pi''_\Sigma \subset r(\hat{\Pi}) \subset r(\tilde{\Pi})$, where $\hat{\Pi}$ is determined by the diagram of type $A^{(2)}_{2n+1}$. This shows that $\mathfrak{l}$ is a strongly regular subalgebra in $\mathfrak{g}$.

It remains to prove that the corresponding triple yields a Clifford-Klein form. Looking at the Satake diagrams and consulting Table 4 on page 231 in [9] we get the types of the root systems

$$\Sigma = \langle \bar{\alpha}_1, \quad \bar{\alpha}_2 \rangle, \quad \Sigma' = \langle \bar{\alpha}_1 \rangle, \quad \Sigma'' = \langle \bar{\alpha}_2 \rangle$$

where Σ has type BC_2, while Σ' and Σ'' both have rank 1. It follows that

$$\Sigma = \langle \varepsilon_1 - \varepsilon_2, \varepsilon_2 \rangle, \quad \Sigma' = \langle \varepsilon_1 - \varepsilon_2 \rangle, \quad \Sigma'' = \langle \varepsilon_2 \rangle$$

Since the Weyl group W of Σ acts on Σ by permuting indices and changing signs, for any $w \in W$, $w\bar{\alpha}_1, \alpha_2$ always constitute a base for $\mathfrak{a}$, and this is equivalent to the statement we wanted to prove. The condition

$$\dim \mathfrak{p} = \dim \mathfrak{p}' \oplus \dim \mathfrak{p}''$$

is checked directly, in this case one can also use [9], Table 4 on page 229.

Example 4.2. In [5] the author analyzed the problem of existence of Clifford-Klein forms in the class of (non-Riemannian) symmetric spaces and pointed out some of them which are Clifford-Klein. Our method yields similar results (we restrict ourselves to the case when G is a real form of non-compact type of a complex simple Lie group). Here is an example.

Theorem 4.3. *The following triples are strongly regular, satisfy the restrictions of Theorem 3.3 and generate Clifford-Klein forms:*

$$(SU(2,2n), Sp(1,n), U(1,2n))$$
$$(SO(2,2n), U(1,n), SO(1,2n))$$
$$(SO(4,4n), SO(3,4n), Sp(1,n))$$
$$(SO(4,4n), SO(3,4n), Sp(1,n) \times Sp(1))$$
$$(SO(3,4), G_{2(2)}, SO(1,4))$$
$$(SO(8,8), SO(7,8), SO(1,8))$$
$$(SO(4,4), SO(3,4), SO(1,4))$$
$$(SO(4,4), SO(3,4), SO(1,4) \times SO(3))$$

Proof. There are several ways of proof. One of the possibilities is to use the method described in Example 4.1 (the first triple has been already worked out in detail). □

References

[1] Bourbaki N 1968 *Groupes et Algébres de Lie* (Paris: Herman) Chapitre IV
[2] Helgason S 2001 *Differential Geometry, Lie Groups and Symmetric Spaces* (Providence, RI: AMS)
[3] Kędra J, Tralle A and Woike A 2011 *J. Geom. Physics* **61** 462
[4] Kobayashi T 1989 *Math. Ann.* **285** 249
[5] Kobayashi T 2005 *Pure Appl. Math. Quart.* **1** 603
[6] Kobayashi T 1994 *Invent. Math.* **117** 181
[7] Kobayashi T 1992 *Duke Math. J.* **67** 653
[8] Kobayashi T 1996 *J. Lie Theory* **6** 147
[9] Onishchik A L and Vinberg E B 2004 *Lie Groups and Lie Algerbas III* (Springer-Verlag)
[10] Oh H and Witte D 2002 *Geom. Dedicata* **89** 25
[11] Szczepkowska A, Tralle A and Woike A 2011 *Banach Center Publ.* **93** 261

Parametrizations in scalar-tensor theories of gravity and the limit of general relativity

L Järv, P Kuusk, M Saal and O Vilson
Institute of Physics, University of Tartu, Ravila 14c, 50411 Tartu, Estonia

Abstract. We consider a general scalar-tensor theory of gravity and review briefly different forms it can be presented (different conformal frames and scalar field parametrizations). We investigate the conditions under which its field equations and the parametrized post-Newtonian parameters coincide with those of general relativity. We demonstrate that these so-called limits of general relativity are independent of the parametrization of the scalar field, although the transformation between scalar fields may be singular at the corresponding value of the scalar field. In particular, the limit of general relativity can equivalently be determined and investigated in the commonly used Jordan and Einstein frames.

1. Introduction

The still unknown nature of the phenomena of dark matter and dark energy facilitates continued interest in the alternatives to Einstein's general relativity, for a comprehensive review see e.g [1]. A simple and straightforward extension of general relativity (GR) is provided by the Jordan-Brans-Dicke theory [2, 3] and its generalization scalar-tensor gravity (STG) [4, 5], where an additional scalar field participates in the gravitational interaction. As was proposed by Dicke, it is possible via a conformal rescaling of the metric (frame change) and reparametrization of the scalar field to transform the STG action into another representation, equivalent to the original one if the units of measurement are also appropriately rescaled [6]. However, the precise significance of this transformation and the physical and mathematical equivalence of different representations are still a topic for an ongoing debate, for a glimpse of the most recent papers, see e.g [7, 8, 9] and [Stabile A *et al* arXiv:1310.7097]. The present work contributes to the discussion by following our earlier study [10] in asking what happens when the transformation from one parametrization of the scalar field to another is singular.

In the following section we write down the most general STG action involving four free functions, the transformations that leave this action invariant, and the ensuing field equations. In Sec. 3 we recall some of the most used STG frames and parametrizations, and collect some useful relations between them. Then Sec. 4 argues that the limit that reduces the STG field equations into those of general relativity does not depend on the particular frame and parametrization, in particular, it is not affected by a possible singularity in reparametrizing the scalar field. Sec. 5 gives the parametrized post-Newtonian (PPN) parameters for STG in its most general form as well as for the most used special cases introduced before, in order to witness again that the conditions for these parameters to coincide with those of GR are independent of the parametrization. Therefore if in some parametrization there is a certain value of the scalar field that takes STG to its GR limit, there is necessarily a corresponding value (or values) of the reparametrized field, that does the same in another frame and parametrization. This conclusion is illustrated by an example in Sec. 6 and summarized in Sec. 7.

2. General action functional and field equations

The most general action functional for a scalar-tensor theory of gravity including scalar self-interaction only through scalar field but not its derivatives was written down by Flanagan [11],

$$S = \frac{1}{2\kappa^2}\int_{V_4} d^4x\sqrt{-g}\left\{\mathcal{A}(\Phi)R - \mathcal{B}(\Phi)g^{\mu\nu}\nabla_\mu\Phi\nabla_\nu\Phi - 2\kappa^2 V(\Phi)\right\} + S_m\left(e^{2\alpha(\Phi)}g_{\mu\nu},\chi\right) \quad (2.1)$$

It contains four arbitrary functions of the dimensionless scalar field Φ: nonminimal coupling function $\mathcal{A}(\Phi)$, generic kinetic coupling of the scalar field $\mathcal{B}(\Phi)$, self-interaction potential of the scalar field $V(\Phi)$ and conformal coupling $e^{2\alpha(\Phi)}$ between the metric $g_{\mu\nu}$ and matter fields χ. Note that $\mathcal{A}(\Phi)$, $\mathcal{B}(\Phi)$ and $\alpha(\Phi)$ are dimensionless, but for the convenience of notation in cosmology the scalar potential is assumed to be of the dimension of energy density, $[V] = [\rho]$ in units $c = 1$. If we impose a physical condition that gravitational interaction is always finite and attractive, the nonminimal coupling function must satisfy $0 < \mathcal{A} < \infty$. We also assume from physical considerations that self-interaction potential is non-negative, $0 \leq V(\Phi) < \infty$.

As demonstrated by Flanagan [11], two of the four arbitrary functions can be fixed by transformations that contain two arbitrary functions $\bar{\gamma}(\bar{\Phi})$, $\bar{f}(\bar{\Phi})$ and leave the structure of action functional (2.1) invariant:

$$g_{\mu\nu} = e^{2\bar{\gamma}(\bar{\Phi})}\bar{g}_{\mu\nu}\,, \qquad \Phi = \bar{f}(\bar{\Phi}) \quad (2.2)$$

We will call the first transformation the change of the frame and the second one the reparametrization of the scalar field. The change of the frame is in fact a conformal rescaling of the metric and we assume that it is reasonable, i.e the function $\bar{\gamma}(\bar{\Phi})$ and its derivative $d\bar{\gamma}/d\bar{\Phi}$ do not diverge at any $\bar{\Phi}$.

The transformed action functional (2.1) retains its form

$$\bar{S} = \frac{1}{2\kappa^2}\int_{V_4} d^4x\sqrt{-\bar{g}}\{\bar{\mathcal{A}}(\bar{\Phi})\bar{R} - \bar{\mathcal{B}}(\bar{\Phi})\bar{g}^{\mu\nu}\bar{\nabla}_\mu\bar{\Phi}\bar{\nabla}_\nu\bar{\Phi} - 2\kappa^2\bar{V}(\bar{\Phi})\} + S_m\left(e^{2\bar{\alpha}(\bar{\Phi})}\bar{g}_{\mu\nu},\chi\right) \quad (2.3)$$

with transformed functions [11]

$$\begin{aligned}&\bar{\mathcal{A}}(\bar{\Phi}) = e^{2\bar{\gamma}(\bar{\Phi})}\mathcal{A}\left(\bar{f}(\bar{\Phi})\right), \quad \bar{V}(\bar{\Phi}) = e^{4\bar{\gamma}(\bar{\Phi})}V\left(\bar{f}(\bar{\Phi})\right), \quad \bar{\alpha}(\bar{\Phi}) = \alpha\left(\bar{f}(\bar{\Phi})\right) + \bar{\gamma}(\bar{\Phi})\\ &\bar{\mathcal{B}}(\bar{\Phi}) = e^{2\bar{\gamma}(\bar{\Phi})}\left(\mathcal{B}\left(\bar{f}(\bar{\Phi})\right)\left(\bar{f}'\right)^2 - 6\left(\bar{\gamma}'\right)^2\mathcal{A}\left(\bar{f}(\bar{\Phi})\right) - 6\bar{\gamma}'\bar{f}'\mathcal{A}'\right)\end{aligned} \quad (2.4)$$

where

$$\bar{f}' \equiv \frac{d\bar{f}(\bar{\Phi})}{d\bar{\Phi}}, \quad \mathcal{A}' \equiv \frac{d\mathcal{A}(\Phi)}{d\Phi} \quad \text{etc} \quad (2.5)$$

Note that at the conformal transformation (2.2) the energy-momentum tensor

$$T_{\mu\nu} = -\frac{2}{\sqrt{-g}}\frac{\delta S_m}{\delta g^{\mu\nu}} \quad (2.6)$$

transforms as $\bar{T}_{\mu\nu} = e^{2\bar{\gamma}}T_{\mu\nu}$ and its trace as $\bar{T} = e^{4\bar{\gamma}}T$.

From these transformation rules we can notice the following.

- The conditions on nonminimal coupling function $0 < \mathcal{A} < \infty$ and self-interaction potential $0 \leq V(\Phi) < \infty$ are preserved, i.e $0 < \bar{\mathcal{A}} < \infty$ and $0 \leq \bar{V}(\bar{\Phi}) < \infty$.

- If in some frame $\alpha = 0$, then in any other frame $|\bar{\alpha}| < \infty$.
- If we want to avoid ghosts, i.e if there is a frame where the tensorial and scalar part of the gravitational interaction are separated with $\mathcal{A} = 1$ and $\mathcal{B} > 0$, then in any related frame and parametrization it follows that

$$2\bar{\mathcal{A}}\bar{\mathcal{B}} + 3(\bar{\mathcal{A}}')^2 > 0 \tag{2.7}$$

We assume this relation to hold.

The field equations can be derived from the general action functional (2.1) by varying with respect to metric tensor $g^{\mu\nu}$ and scalar field Φ, respectively:

$$\mathcal{A}G_{\mu\nu}+\left[\frac{1}{2}\mathcal{B}+\mathcal{A}''\right]g_{\mu\nu}\nabla_\rho\Phi\nabla^\rho\Phi-\left[\mathcal{B}+\mathcal{A}''\right]\nabla_\mu\Phi\nabla_\nu\Phi+\mathcal{A}'\left[g_{\mu\nu}\Box\Phi-\nabla_\mu\nabla_\nu\Phi\right]+\kappa^2 g_{\mu\nu}V=\kappa^2 T_{\mu\nu} \tag{2.8}$$

$$\frac{1}{2}R\mathcal{A}' + \frac{1}{2}\mathcal{B}'g^{\mu\nu}\nabla_\mu\Phi\nabla_\nu\Phi + \mathcal{B}\Box\Phi - \kappa^2 V' = -\kappa^2\alpha' T \tag{2.9}$$

Upon substituting the scalar curvature R from the first equation into the second one and multiplying by $2\mathcal{A}$, the equation for the scalar field reads

$$\left(2\mathcal{A}\mathcal{B}+3(\mathcal{A}')^2\right)\Box\Phi+\frac{\left(2\mathcal{A}\mathcal{B}+3(\mathcal{A}')^2\right)'}{2}g^{\mu\nu}\nabla_\mu\Phi\nabla_\nu\Phi-2\kappa^2\left(\mathcal{A}V'-2\mathcal{A}'V\right)=\kappa^2\left(\mathcal{A}'-2\mathcal{A}\alpha'\right)T \tag{2.10}$$

A direct calculation demonstrates that upon the transformation (2.4) all terms in the equation of the metric tensor (2.8) acquire a common factor $e^{2\bar{\gamma}}$, which we have assumed to be regular. However, the transformed equation for the scalar field (2.10) gets a common factor $e^{6\bar{\gamma}}\bar{f}'$. If the transformation is regular, i.e $\bar{f}' \neq 0$, $\bar{f}' \neq \infty$, this equation should yield an equivalent account of the same physics in different parametrizations. What happens for the points where $\bar{f}'$ fails to be finite needs extra attention.

Finally, from the field equations (2.8) and (2.9) a continuity equation follows:

$$\nabla_\mu T^{\mu\nu} = \alpha' T\nabla^\nu\Phi \tag{2.11}$$

If $\alpha' = 0$ the right-hand side vanishes and the usual conservation of energy law holds; let us call the $\alpha = 0$ case the Jordan frame. Another well-known frame is the Einstein frame with $\mathcal{A} = 1$ and in general $\alpha' \neq 0$.

3. Some widely used action functionals

Sometimes in the literature one may encounter treatments which fix the frame (i.e fix $\alpha(\Phi)$, e.g $\alpha = 0$), but leave the parametrization of the scalar field unfixed, thus keeping three arbitrary functions in the STG action functional. But most often one meets a few distinct forms of the STG action functional obtained from the general action (2.1) by fixing two of the four arbitrary functions. These are the following.

1. The Jordan frame action in the Brans-Dicke-Bergmann-Wagoner parametrization (JF BDBW) [3, 4, 5] for the scalar field Ψ fixes $\mathcal{A} = \Psi$, $\alpha = 0$, while keeping $\mathcal{B} = \omega(\Psi)/\Psi$, $V = V(\Psi)$:

$$S = \frac{1}{2\kappa^2}\int d^4x\sqrt{-g}\left[\Psi R - \frac{\omega(\Psi)}{\Psi}\nabla^\rho\Psi\nabla_\rho\Psi - 2\kappa^2 V(\Psi)\right] + S_m\left(g_{\mu\nu}, \chi\right) \tag{3.1}$$

The original Brans-Dicke gravity (JF BD) [3] with a potential is a special case where $\omega = const.$,

$$S = \frac{1}{2\kappa^2} \int d^4x\sqrt{-g}\left[\Psi R - \frac{\omega}{\Psi}\nabla^\rho\Psi\nabla_\rho\Psi - 2\kappa^2 V(\Psi)\right] + S_m\left(g_{\mu\nu}, \chi\right) \tag{3.2}$$

2. The Jordan frame action in the parametrization used by Boisseau, Esposito-Farèse, Polarski and Starobinsky (JF BEPS) [12, 13] for the scalar field as ϕ is obtained by taking $\mathcal{B} = 1$, $\alpha = 0$, while having $\mathcal{A} = F(\phi)$, $V = V(\phi)$:

$$S = \frac{1}{2\kappa^2} \int d^4x\sqrt{-g}\left[F(\phi)R - \nabla^\rho\phi\nabla_\rho\phi - 2\kappa^2 V(\phi)\right] + S_m\left(g_{\mu\nu}, \chi\right) \tag{3.3}$$

In the so-called nonminimal coupling case (JF nm), the function F has a distinct form $F(\phi) = 1 - \xi\phi^2$, where ξ is a dimensionless parameter:

$$S = \frac{1}{2\kappa^2} \int d^4x\sqrt{-g}\left[\left(1 - \xi\phi^2\right)R - \nabla^\rho\phi\nabla_\rho\phi - 2\kappa^2 V(\phi)\right] + S_m\left(g_{\mu\nu}, \chi\right) \tag{3.4}$$

3. The Einstein frame action in canonical parametrization (EF can) [6, 4, 5] for the scalar field denoted as φ, fixes $\mathcal{A} = 1$, $\mathcal{B} = 2$, while keeping $\alpha = \alpha(\varphi)$ and $V = V(\varphi)$:

$$S = \frac{1}{2\kappa^2} \int d^4x\sqrt{-g}\left[R - 2g^{\mu\nu}\nabla_\mu\varphi\nabla_\nu\varphi - 2\kappa^2 V(\varphi)\right] + S_m\left(e^{2\alpha(\varphi)}g_{\mu\nu}, \chi\right) \tag{3.5}$$

The well known Einstein gravity with minimally coupled scalar field (EF min) can be viewed as a special case here with $\alpha = 0$,

$$S = \frac{1}{2\kappa^2} \int d^4x\sqrt{-g}\left[R - 2g^{\mu\nu}\nabla_\mu\varphi\nabla_\nu\varphi - 2\kappa^2 V(\varphi)\right] + S_m\left(g_{\mu\nu}, \chi\right) \tag{3.6}$$

However, in the latter case the scalar field equation (2.10) does not contain matter energy-momentum T as a source and strictly speaking the scalar field is not mediating the gravitational interaction any more.

The transformations between these most common frames and parametrizations are presented in Table 1. Note that the mutual derivatives of the scalar field in different parametrizations included in the Table 1 are in fact just $\bar{f}'$ which should satisfy the conditions $\bar{f}' \neq 0$, $\bar{f}' \neq \infty$ for a transformation to be regular.

4. Field equations and the limit of general relativity

Let us investigate the conditions under which a STG coincides with GR. Since the latter one does not involve a dynamical scalar field, a natural assumption is $\Phi = \text{const.}$, $\nabla_\mu\Phi = 0$. However, this condition should be made consistent by requiring that the source term for the scalar field also vanishes, otherwise constant Φ can not be maintained. Rewriting the scalar field equation (2.10) as

$$\Box\Phi + \frac{1}{2}\frac{\left(2\mathcal{A}\mathcal{B} + 3(\mathcal{A}')^2\right)'}{2\mathcal{A}\mathcal{B} + 3(\mathcal{A}')^2}g^{\mu\nu}\nabla_\mu\Phi\nabla_\nu\Phi = \kappa^2\frac{(\mathcal{A}' - 2\mathcal{A}\alpha')T + 2\left(\mathcal{A}V' - 2\mathcal{A}'V\right)}{2\mathcal{A}\mathcal{B} + 3(\mathcal{A}')^2} \tag{4.1}$$

it becomes clear that the STG equations can concur with those of GR at the values of Φ where the term on the RHS of (4.1) vanishes. Given that $\mathcal{A}$ is everywhere regular there are several possibilities.

The first and most obvious case is realized for $\Phi_\bullet$ which should simultaneously satisfy

$$\left(\mathcal{A}' - 2\alpha'\mathcal{A}\right)|_{\Phi_\bullet} T = 0 \tag{4.2a}$$

Table 1. Transformations between frames and parametrizations

	JF BDBW (Ψ)	JF BEPS (ϕ)	EF can (φ)
JF BDBW (Ψ)	Identity	$F(\phi)=\Psi$ $(\frac{d\phi}{d\Psi})^2=\frac{\omega(\Psi)}{\Psi}$ $(\frac{dF}{d\phi})^2=\frac{\Psi}{\omega(\Psi)}$	$\alpha(\varphi)=-\frac{1}{2}\ln\Psi$ $(\frac{d\varphi}{d\Psi})^2=\frac{2\omega(\Psi)+3}{4\Psi^2}$ $(\frac{d\alpha}{d\varphi})^2=\frac{1}{2\omega(\Psi)+3}$
JF BD (Ψ)	$\omega=$ const.	$F(\phi)=\Psi$ $(\frac{d\phi}{d\Psi})^2=\frac{\omega}{\Psi}$ $(\frac{dF}{d\phi})^2=\frac{\Psi}{\omega}$	$\alpha(\varphi)=-\frac{1}{2}\ln\Psi$ $(\frac{d\varphi}{d\Psi})^2=\frac{2\omega+3}{4\Psi^2}$ $(\frac{d\alpha}{d\varphi})^2=\frac{1}{2\omega+3}$
JF BEPS (ϕ)	$\Psi=F(\phi)$ $\frac{d\Psi}{d\phi}=\frac{dF}{d\phi}$ $\omega(\Psi)=F(\phi)\frac{1}{(\frac{dF}{d\phi})^2}$	Identity	$\alpha(\varphi)=-\frac{1}{2}\ln F(\phi)$ $(\frac{d\varphi}{d\phi})^2=\frac{3}{4}(\frac{d\ln F(\phi)}{d\phi})^2$ $+\frac{1}{2F(\phi)}$
JF nm (ϕ)	$\Psi=1-\xi\phi^2$ $\frac{d\Psi}{d\phi}=-2\xi\phi$ $\omega(\Psi)=\frac{\Psi}{4\xi(1-\Psi)}=\frac{1-\xi\phi^2}{4\xi^2\phi^2}$	$F(\phi)=1-\xi\phi^2$	$\alpha(\varphi)=\frac{1}{2}\ln\left(\frac{1}{1-\xi\phi^2}\right)$ $(\frac{d\varphi}{d\phi})^2=\frac{1-\xi\phi^2+6\xi^2\phi^2}{2(1-\xi\phi^2)^2}$ $\phi^2=\frac{1}{\xi}(1-e^{-2\alpha(\varphi)})$
EF can (φ)	$\Psi=e^{-2\alpha(\varphi)}$ $(\frac{d\Psi}{d\varphi})^2=4e^{-4\alpha(\varphi)}(\frac{d\alpha}{d\varphi})^2$ $\omega(\Psi)=\frac{1}{2}\left(\frac{1}{(\frac{d\alpha}{d\varphi})^2}-3\right)$	$F(\phi)=e^{-2\alpha(\varphi)}$ $\left(\frac{d\phi}{d\varphi}\right)^2=2e^{-2\alpha(\varphi)}\times$ $\times\left(1-3\left(\frac{d\alpha}{d\varphi}\right)^2\right)$ $(\frac{dF}{d\phi})^2=\frac{2e^{-2\alpha(\varphi)}(\frac{d\alpha}{d\varphi})^2}{1-3(\frac{d\alpha}{d\varphi})^2}$	Identity
EF min (φ)	$\Psi=1$	$F(\phi)=1$	$\alpha=0$

$$\left(\mathcal{A}V'-2\mathcal{A}'V\right)|_{\Phi_\bullet}=0 \tag{4.2b}$$

while $\mathcal{B}|_{\Phi_\bullet}$ is finite and nonvanishing. In addition, for the full compliance with GR the factors in front of the kinetic terms in the Einstein equation (2.8) and scalar field equation (4.1) should remain regular, hence $\mathcal{A}''|_{\Phi_\bullet}$ and $\mathcal{B}'|_{\Phi_\bullet}$ should not diverge. If the latter is not the case, then the STG does not allow a solution which behaves exactly as GR, but it may still be possible to have solutions which dynamically approach GR as a limiting process, provided

$$\mathcal{A}''\nabla_\mu\Phi\nabla_\nu\Phi\to 0 \quad \text{or/and} \quad \frac{\mathcal{B}'}{\mathcal{B}}\nabla_\mu\Phi\nabla_\nu\Phi\to 0 \quad \text{as} \quad \Phi\to\Phi_\bullet \tag{4.3}$$

The result is the GR Einstein equation with $V|_{\Phi_\bullet}$ effectively playing the role of the cosmological constant. It is instructive to observe that for JF BEPS parametrization the condition (4.2a), (4.2b) translates into $F'|_{\phi_\bullet} = 0$, $V'|_{\phi_\bullet} = 0$, for the nonminimal coupling case into $\phi_\bullet = 0$, $V'|_{\phi_\bullet} = 0$, and for the EF canonical parametrization into $\alpha'|_{\varphi_\bullet} = 0$, $V'|_{\varphi_\bullet} = 0$. But in the JF BDBW parametrization the condition (4.2a) can not be realized for general matter ($T \neq 0$) at all since $\mathcal{A}' \equiv 1$.

The second possibility to make the RHS of (4.1) to vanish is by having a value $\Phi_\star$ for which

$$\frac{1}{\mathcal{B}}\bigg|_{\Phi_\star} = 0 \tag{4.4}$$

while $\mathcal{A}'|_{\Phi_\star}$, $\alpha'|_{\Phi_\star}$, and $V'|_{\Phi_\star}$ do not diverge. An important difference with the previous case is that here we do not have a constant solution for the scalar field, but only a process of approaching to that value. For this process to correctly yield the GR, we must demand that $\mathcal{B}\nabla_\mu\Phi\nabla_\nu\Phi \to 0$ as $\Phi \to \Phi_\star$. In addition, if $\mathcal{B}'|_{\Phi_\star}$ or $\mathcal{A}''|_{\Phi_\star}$ happen to be singular as well, only the solutions with

$$\frac{\mathcal{B}'}{\mathcal{B}}\nabla_\mu\Phi\nabla_\nu\Phi \to 0, \quad \mathcal{A}''\nabla_\mu\Phi\nabla_\nu\Phi \to 0 \tag{4.5}$$

lead to GR as a limit. For a later remark we note that if the Einstein equation (2.8) and the scalar field equation (4.1) converge to the GR limit at the same "rate", i.e if

$$\mathcal{B}\nabla_\mu\Phi\nabla_\nu\Phi \propto \frac{\mathcal{B}'}{\mathcal{B}}\nabla_\mu\Phi\nabla_\nu\Phi \tag{4.6}$$

then $\frac{\mathcal{B}'}{\mathcal{B}^2}$ is finite. Among the particular forms of STG the condition (4.4) can be only realized in JF BDBW parametrization where it translates into $\frac{1}{\omega}|_{\Psi_\star} = 0$. In the JF BEPS and nonminimal, and EF canonical parametrizations the function $\mathcal{B}$ is fixed to a constant value which precludes (4.4).

At first there seems to be also a third option to make the RHS of (4.1) to vanish by letting $\frac{1}{\mathcal{A}'} = 0$. However by looking at the scalar field equation (2.9) this case turns out to be problematic. Namely, for GR it is well known that spacetime curvature and matter energy momentum are proportional to each other, $R \propto T$. But if $\mathcal{A}' \to \infty$ then finite T would correspond to vanishing R, unless α' also blows up. The latter would complicate the continuity equation (2.11). Hereby we restrict our attention to the cases where $\mathcal{A}'$ and α' are not singular at the same value of scalar field Φ, and therefore to achieve a GR-like behaviour we do not consider this possibility. Furthermore, for the sake of mathematical simplicity we also leave aside the rather fine-tuned theories where the conditions (4.2a), (4.2b) and (4.4) are realized together, or where both the numerator and denominator of the RHS term of (4.1) vanish simultaneously thus requiring a much more thorough analysis.

Now, given the rather different conditions (4.2a), (4.2b) and (4.4), one is entitled to ask whether there is any connection between them. It is interesting to note that there is. Under the transformations (2.4) the quantities

$$\frac{(\mathcal{A}' - 2\mathcal{A}\alpha')^2}{2\mathcal{A}\mathcal{B} + 3(\mathcal{A}')^2} \quad \text{and} \quad \frac{(\mathcal{A}V' - 2\mathcal{A}'V)^2}{\mathcal{A}^4\left(2\mathcal{A}\mathcal{B} + 3(\mathcal{A}')^2\right)} \tag{4.7}$$

retain their form, i.e do not acquire extra terms or common factors. Therefore, under a generic transformation the condition that the RHS of (4.1) vanishes remains invariant, i.e if some Φ in a certain frame and parametrization satisfies it, then the corresponding $\bar{\Phi}$ in another frame and parametrization will also satisfy it. Although, it is completely feasible that $\Phi_\bullet$ obeying (4.2a), (4.2b) may get translated into $\bar{\Phi}_\star$ obeying (4.4).

5. Parametrized post-Newtonian approximation

For being viable, the scalar-tensor theory of gravity must pass the tests on local scales, e.g, give a good account of the motions in our solar system. A natural framework for such a check is provided by the parametrized post-Newtonian (PPN) formalism adapted to slow motions in a weak field. To compare GR and STG there are two nonvanishing PPN parameters γ and β. They both have value 1 for Einstein's general relativity which is also favored by current observations. For an STG, the PPN parameters can deviate from unity as they depend on the spatially asymptotic background value of the scalar field [14, 15]. When STG has a potential, the PPN parameters cease to be constants as they also acquire an extra dependence on the distance r from the source [16, 17, 18, 19]. It is useful to express the result in units where the Newtonian potential $U_N = \frac{\kappa^2 M}{8\pi r}$ is dimensionless, while the dimensionless constant $G_{\text{eff}}(\Phi, r)$ modifies multiplicatively Newton's gravitational constant $G_N = \frac{\kappa^2}{8\pi}$ and determines the Cavendish force.

It is possible to translate the general results [19] from JF BDBW parametrization into a generic representation of $\mathcal{A}$, $\mathcal{B}$, V, α by using the transformations (2.2) and (2.4) where $\bar{\mathcal{A}} = \Psi$, $\bar{\mathcal{B}} = \frac{\omega(\Psi)}{\Psi}$, and $\bar{\alpha} = 0$. It follows that $\bar{\gamma} = -\alpha$, $\bar{\gamma}' = -\alpha' \bar{f}'$, while the other necessary quantity $\bar{f}'$ can be expressed by taking the derivative of the first line of Eq. (2.4). Since the PPN ansatz assumes flat Minkowski spacetime in spatial infinity, the internal consistency requires $V = 0$, $V' = 0$ (all values of the functions taken at the asymptotic value of the scalar field). In terms of the constant related to the scalar field effective mass,

$$m_\Phi = \kappa \sqrt{\frac{2\mathcal{A}}{e^{2\alpha}\left(2\mathcal{A}\mathcal{B} + 3(\mathcal{A}')^2\right)} \frac{d^2 V}{d\Phi^2}} \tag{5.1}$$

the results are

$$G_{\text{eff}} = \frac{e^{2\alpha}}{\mathcal{A}} \left(1 + \frac{\left(\mathcal{A}' - 2\mathcal{A}\alpha'\right)^2 e^{-m_\Phi r}}{2\mathcal{A}\mathcal{B} + 3(\mathcal{A}')^2}\right) \tag{5.2a}$$

$$\gamma - 1 = -\frac{2e^{2\alpha}\left(\mathcal{A}' - 2\mathcal{A}\alpha'\right)^2 e^{-m_\Phi r}}{G_{\text{eff}}\mathcal{A}\left(2\mathcal{A}\mathcal{B} + 3(\mathcal{A}')^2\right)} \tag{5.2b}$$

$$\begin{aligned} \beta - 1 = & \frac{\left(\mathcal{A}' - 2\mathcal{A}\alpha'\right)^2 \left[\left(2\mathcal{A}\mathcal{B} + 3(\mathcal{A}')^2\right)'\left(\mathcal{A}' - 2\mathcal{A}\alpha'\right) - 2\left(2\mathcal{A}\mathcal{B} + 3(\mathcal{A}')^2\right)\left(\mathcal{A}' - 2\mathcal{A}\alpha'\right)'\right] e^{-2m_\Phi r}}{2e^{-4\alpha} G_{\text{eff}}^2 \mathcal{A}\left(2\mathcal{A}\mathcal{B} + 3(\mathcal{A}')^2\right)^3} \\ & - \frac{\left(\mathcal{A}' - 2\mathcal{A}\alpha'\right)^2 m_\Phi r}{e^{-4\alpha} G_{\text{eff}}^2 \left(2\mathcal{A}\mathcal{B} + 3(\mathcal{A}')^2\right)} \left[e^{-m_\Phi r} \ln m_\Phi r + \ldots\right] \end{aligned} \tag{5.2c}$$

Among the number of r-dependent terms in the square brackets on the last line only the contribution that is leading for large $m_\Phi r$ is given. For different frames and parametrizations the corresponding expressions can be found in Table 2. These can be deduced by specifying the functions in the formulas above, or using the transformations (2.4) and the information in Table 1.

The conceptual difference with the previous section is that now we have a static configuration and the functions of the scalar field are taken at their spatially asymptotic values. However, the analysis of the limit where the STG PPN parameters coincide with those of general relativity, viz. $G_{\text{eff}} = 1$, $\gamma = 1$, $\beta = 1$ proceeds quite analogously. The first option is provided by the condition (4.2a), where in addition $\mathcal{B}$ is finite and

$$\left(\mathcal{A}' - 2\mathcal{A}\alpha'\right)^2 \left(\mathcal{A}' - 2\mathcal{A}\alpha'\right)' = 0 \tag{5.3}$$

Note that the twin condition (4.2b) is automatically satisfied due to the PPN ansatz. The second option would be given by (4.4), with α', α'' not infinite, and $\frac{\mathcal{B}'}{\mathcal{B}^3} = 0$. Comparing the latter with

Table 2. PPN parameters in different frames and parametrizations

JF BDBW	$G_{\text{eff}} = \frac{1}{\Psi}\left(1+\frac{e^{-m_\Psi r}}{2\omega+3}\right)$	$m_\Psi = \kappa\sqrt{\frac{2\Psi}{2\omega(\Psi)+3}\frac{d^2V}{d\Psi^2}}$
	$\gamma - 1 = -\frac{2e^{-m_\Psi r}}{G_{\text{eff}}\Psi(2\omega+3)}$	
	$\beta - 1 = \frac{\frac{d\omega}{d\Psi}e^{-2m_\Psi r}}{G^2_{\text{eff}}\Psi(2\omega+3)^3} - \frac{m_\Psi r}{G^2_{\text{eff}}\Psi^2(2\omega+3)}\left[e^{-m_\Psi r}\ln(m_\Psi r)+\ldots\right]$	
JF BEPS	$G_{\text{eff}} = \frac{1}{F}\left(1+\frac{\left(\frac{dF}{d\phi}\right)^2 e^{-m_\phi r}}{2F+3\left(\frac{dF}{d\phi}\right)^2}\right)$	$m_\phi = \kappa\sqrt{\frac{2F}{2F+3\left(\frac{dF}{d\phi}\right)^2}\frac{d^2V}{d\phi^2}}$
	$\gamma - 1 = -\frac{2\left(\frac{dF}{d\phi}\right)^2 e^{-m_\phi r}}{G_{\text{eff}}F\left(2F+3\left(\frac{dF}{d\phi}\right)^2\right)}$	
	$\beta - 1 = \frac{\left(\frac{dF}{d\phi}\right)^2\left(\left(\frac{dF}{d\phi}\right)^2 - 2F\frac{d^2F}{d\phi^2}\right)e^{-2m_\phi r}}{G^2_{\text{eff}}F\left(2F+3\left(\frac{dF}{d\phi}\right)^2\right)^3} - \frac{\left(\frac{dF}{d\phi}\right)^2 m_\phi r}{G^2_{\text{eff}}F^2\left(2F+3\left(\frac{dF}{d\phi}\right)^2\right)}\left[e^{-m_\phi r}\ln(m_\phi r)+\ldots\right]$	
JF nm	$G_{\text{eff}} = \frac{1}{1-\xi\phi^2}\left(1+\frac{2\xi^2\phi^2 e^{-m_\phi r}}{1-\xi\phi^2+6\xi^2\phi^2}\right)$	$m_\phi = \kappa\sqrt{\frac{2(1-\xi\phi^2)}{1-\xi\phi^2+6\xi^2\phi^2}\frac{d^2V}{d\phi^2}}$
	$\gamma - 1 = -\frac{4\xi^2\phi^2 e^{-m_\phi r}}{G_{\text{eff}}(1-\xi\phi^2)(1-\xi\phi^2+6\xi^2\phi^2)}$	
	$\beta - 1 = \frac{2\xi^3\phi^2 e^{-2m_\phi r}}{(1-\xi\phi^2)(1-\xi\phi^2+6\xi^2\phi^2)^3} - \frac{2\xi^2\phi^2 m_\phi r}{(1-\xi\phi^2)(1-\xi\phi^2+6\xi^2\phi^2)}\left[e^{-m_\phi r}\ln(m_\phi r)+\ldots\right]$	
EF can	$G_{\text{eff}} = e^{2\alpha}\left(1+\left(\frac{d\alpha}{d\varphi}\right)^2 e^{-m_\varphi r}\right)$	$m_\varphi = \kappa\sqrt{\frac{1}{2e^{2\alpha}}\frac{d^2V}{d\varphi^2}}$
	$\gamma - 1 = -\frac{2e^{2\alpha}\left(\frac{d\alpha}{d\varphi}\right)^2 e^{-m_\varphi r}}{G_{\text{eff}}}$	
	$\beta - 1 = \frac{e^{4\alpha}\left(\frac{d\alpha}{d\varphi}\right)^2\frac{d^2\alpha}{d\varphi^2}e^{-2m_\varphi r}}{2G^2_{\text{eff}}} - \frac{e^{4\alpha}\left(\frac{d\alpha}{d\varphi}\right)^2 m_\varphi r}{G^2_{\text{eff}}}\left[e^{-m_\varphi r}\ln(m_\varphi r)+\ldots\right]$	
EF min	$G_{\text{eff}} = 1$	$m_\varphi = \kappa\sqrt{\frac{1}{2}\frac{d^2V}{d\varphi^2}}$
	$\gamma - 1 = 0$	
	$\beta - 1 = 0$	

the discussion in the previous section we may note that the condition on $\mathcal{B}'$ to achieve the GR limit is marginally less strict in PPN than the one obtained from the equations of motion, i.e $\frac{\mathcal{B}'}{\mathcal{B}^2}$ finite. (A similar observation in the case of cosmology was made in Ref. [20]). The third option is realized by giving the scalar field a very large effective mass, i.e

$$\frac{1}{m_\Phi}\Big|_{\Phi_\blacklozenge} = \left(\frac{2\kappa^2\mathcal{A}}{e^{2\alpha}\left(2\mathcal{A}\mathcal{B}+3(\mathcal{A}')^2\right)}\frac{d^2V}{d\Phi^2}\right)^{-\frac{1}{2}}\Big|_{\Phi_\blacklozenge} = 0 \tag{5.4}$$

However, in that case it is not so obvious what the corresponding condition arising from the general equations of motion would be.

6. Example

To have an illustration let us take a look at a specific simple example. Let the JF BDBW functions be given by

$$\mathcal{A} = \Psi, \quad \mathcal{B} = \frac{\omega(\Psi)}{\Psi} = \frac{3}{2(1-\Psi)}, \quad V = \frac{1}{\left(\frac{1}{2}-\Psi\right)^2}, \quad \alpha = 0 \tag{6.1}$$

The attactive gravitation condition ($\mathcal{A} > 0$) and no ghosts condition (2.7) delimit $0 < \Psi \leq 1$. Recalling the discussions in Secs. 4 and 5 we may find that the field equations and PPN parameters reduce to those of general relativity in several different occasions.

- The first is realized when (4.4) holds, i.e $\Psi_\star = 1$, while $\frac{\mathcal{B}'}{\mathcal{B}^2} = \frac{2}{3}$ is finite. Here the PPN parameters also reduce to their general relativity values, as expected, since $\frac{\mathcal{B}'}{\mathcal{B}^3} = 0$.
- The second possibility to reduce the field equations to GR only occurs when the trace of matter energy-momentum tensor $T = 0$ and the condition (4.2a) does not apply. Then (4.2b) is satisfied at $\Psi_\bullet = \frac{1}{4}$. The PPN parameters, however, do not coincide with those of GR now.
- Finally, it is possible to draw the PPN parameters into the GR values by satisfying (5.4) with an extremely massive scalar field, $\Psi_\blacklozenge = \frac{1}{2}$. But now the field equations do not agree with those of general relativity.

We can transform the theory from the BDBW parametrization with Ψ into the BEPS parametrization with ϕ by using Table 1. Integrating

$$\mp\frac{d\phi}{d\Psi} = \sqrt{\frac{\omega(\Psi)}{\Psi}} = \sqrt{\frac{3}{2(1-\Psi)}} \tag{6.2}$$

gives (neglecting the additive integration constant)

$$\pm\phi = \sqrt{6(1-\Psi)}, \quad \Psi = 1 - \frac{1}{6}\phi^2 \tag{6.3}$$

and we see it is actually the nonminimal coupling subclass of BEPS, where

$$\mathcal{A} = F(\phi) = 1 - \frac{1}{6}\phi^2, \quad \mathcal{B} = 1, \quad V = \frac{1}{\left(\frac{1}{2}-\frac{\phi^2}{6}\right)^2}, \quad \alpha = 0 \tag{6.4}$$

Note that Ψ is mapped doubly to ϕ, as $\Psi \in (0,1]$ translates into $\phi \in (-\sqrt{6}, 0]$ and $\phi \in [0, \sqrt{6})$. Again, there are several possibilities to achive the general relativity limit of the field equations and PPN parameters.

- First, the field equations reduce to the ones of GR when Eq. (4.2a), given by $\mathcal{A}' = 0$, and (4.2b), given by $\mathcal{A}V' - 2\mathcal{A}'V = 0$, are satisfied. The only common solution is $\phi_\bullet = 0$. A glance to Table 2 reveals that the PPN parameters also trivially fall into their GR limit. By a direct comparison via (6.3) it becomes obvious that this value of ϕ corresponds to the first case in the BDBW case.
- If matter $T = 0$ and the condition (4.2a) is not enforced, the condition (4.2b) alone has also the solution $\pm\phi_\bullet = \frac{3}{\sqrt{2}}$. This does not lead the PPN parameters to their GR values. A direct check by (6.3) tells that the corresponding case in the BDBW parametrization was the second one.

- Last, when the scalar field acquires an extremely large mass by (5.4) at $\pm\phi_{\blacklozenge} = \sqrt{3}$ the PPN parameters reduce to those of GR, but the field equations do not. It corresponds to the third case above.

We may transform the same theory from JF BDBW parametrization into EF canonical parametrization by integrating

$$\mp\frac{d\varphi}{d\Psi} = \sqrt{\frac{2\omega(\Psi)+3}{4\Psi^2}} = \sqrt{\frac{3}{4\Psi^2(1-\Psi)}} \tag{6.5}$$

which gives (neglecting the additive integration constant)

$$\pm\varphi = \sqrt{3}\,\mathrm{arctanh}\sqrt{1-\Psi}\,, \quad \Psi = 1-\tanh^2\frac{\varphi}{\sqrt{3}} \tag{6.6}$$

The functions characterizing the frame and parametrization are

$$\mathcal{A}=1, \quad \mathcal{B}=2, \quad V = \frac{1}{\left(\frac{1}{2}-\tanh^2\frac{\varphi}{\sqrt{3}}\right)^2\left(1-\tanh^2\frac{\varphi}{\sqrt{3}}\right)^2}, \quad \alpha = -\frac{1}{2}\ln\left(1-\tanh^2\frac{\varphi}{\sqrt{3}}\right) \tag{6.7}$$

Alternatively, one may embark from the JF BEPS parametrization and integrate

$$\pm\frac{d\varphi}{d\phi} = \sqrt{\frac{3}{4}\left(\frac{d\ln F(\phi)}{d\phi}\right)^2 + \frac{1}{2F(\phi)}} = \frac{1}{\sqrt{2}\left(1-\frac{\phi^2}{6}\right)} \tag{6.8}$$

to obtain (again, neglecting the additive integration constant)

$$\pm\varphi = \sqrt{3}\,\mathrm{arctanh}\frac{\phi}{\sqrt{6}}, \quad \pm\phi = \sqrt{6}\tanh\frac{\varphi}{\sqrt{3}} \tag{6.9}$$

The mapping into EF canonical parametrization is again double for JF BDBW, as $\Psi \in (0,1]$ translates into $\phi \in (-\infty, 0]$ and $\phi \in [0,\infty)$, while JF BEPS $\phi \in (-\sqrt{6},\sqrt{6})$ translates into $\phi \in (-\infty,\infty)$ and equivalently into $-\phi \in (-\infty,\infty)$ according to the sign in Eq. (6.8). Analogously with the other parametrizations we can discuss the general relativity limit of the field equations and PPN parameters in three cases.

- When the conditions (4.2a) and (4.2b) both hold, i.e $\alpha' = 0$ and $V' = 0$, the value of the scalar field is $\varphi_{\bullet} = 0$. It takes the PPN parameters to their GR limit and by direct check using (6.6) and (6.9) one can conclude it corresponds to the first cases discussed above.
- For absent or radiative matter with $T = 0$ the condition (4.2a) does not apply and (4.2b) alone is also solved by $\pm\varphi_{\bullet} = \sqrt{3}\,\mathrm{arctanh}\left(\sqrt{3}/2\right)$. The PPN parameters differ from those of GR and it is straightforward to check that this value of φ corresponds to the second cases above.
- The scalar field mass diverges at $\pm\varphi_{\blacklozenge} = \sqrt{3}\,\mathrm{arctanh}\left(1/\sqrt{2}\right)$, satisfying (5.4) and reducing the PPN parameters to their GR values. The field equations still differ from those of GR and we recognize correspondence to the third cases described above.

We see that the derivative of the transformation function $\bar{f}'$ relating different scalar field parametrizations gets singular for different values of the field. For a transformation from JF BDBW to BEPS (6.2) it is singular at $\Psi = 1$, $\phi = 0$, for a transformation from JF BDBW to EF canonical (6.5) it is singular at $\Psi = 1$, $\varphi = 0$ and $\Psi = 0$, $\varphi = \pm\infty$, while for a transformation from JF BEPS to EF canonical (6.8) it is singular at $\phi = \pm\sqrt{6}$, $\varphi = \pm\infty$. The value $\Psi = 0$, $\phi = \pm\sqrt{6}$, $\varphi = \pm\infty$ also makes the the conformal transformation singular. But strictly speaking, this value is actually outside the range of the assumed validity of the theory, since for JF BDBW and BEPS it violates the attractive gravity assumption, while for EF canonical the infinite value of the field is arguably unphysical since α becomes singular.

So, it is only the singularity of transformation and the GR limit occuring at JF BDBW $\Psi = 1$ that is possibly problematic. However, we saw that despite the transformation becoming singular the general relativity limit in terms of the field equations and PPN parameters, namely the first cases discussed above, does also occur in the corresponding JF BEPS value $\phi = 0$ and EF canonical parametrization value $\varphi = 0$. It is interesting that in the JF BDBW parametrization this GR limit is realized by satisfying the condition (4.4), while in the JF BEPS and EF canonical parametrizations it comes from the conditions (4.2a), (4.2b). This confirms the discussion in Secs. 4 and 5 that the existence of the GR limit is invariant of the parametrization.

7. Conclusion

We studied general scalar-tensor gravity involving four free functions in different conformal frames and scalar field parametrizations. We investigated its general relativity limits in the sense of field equations and the values of PPN parameters coinciding with those of general relativity. Despite the transformation of the scalar field from one representation to another may possess a singularity, it turned out that the existence of general relativity limits is independent of the parametrization.

Acknowledgments

This work was supported by the Estonian Science Foundation Grant No. 8837, by the Estonian Ministry for Education and Science Institutional Research Support Project IUT02-27 and by the European Union through the European Regional Development Fund (Centre of Excellence TK114).

References

[1] Clifton T, Ferreira P G, Padilla A and Skordis C 2012 *Phys. Rep.* **513** 1
[2] Jordan P 1959 *Z. Phys.* **157** 112
[3] Brans C and Dicke R H 1961 *Phys. Rev.* **124** 925
[4] Bergmann P G 1968 *Int. J. Theor. Phys.* **1** 25
[5] Wagoner R V 1970 *Phys. Rev.* D **1** 3209
[6] Dicke R H 1962 *Phys. Rev.* **125** 2163
[7] Quiros I, Garcia-Salcedo R, Aguilar J E M and Matos T 2013 *Gen. Relativity Gravitation* **45** 489
[8] Tsamparlis M, Paliathanasis A, Basilakos S and Capozziello S 2013 *Gen. Relativity Gravitation* **45** 2003
[9] Chiba T and Yamaguchi M 2013 *JCAP* **1310** 040
[10] Järv L, Kuusk P and Saal M 2007 *Phys. Rev.* D **76** 103506
[11] Flanagan É É 2004 *Classical Quantum Gravity* **21** 3817
[12] Boisseau B, Esposito-Farèse G, Polarski D and Starobinsky A A 2000 *Phys. Rev. Lett.* **85** 2236
[13] Esposito-Farèse G and Polarski D 2001 *Phys. Rev.* D **63** 063504
[14] Nordtvedt K Jr 1970 *Astrophys. J.* **161** 1059
[15] Damour T and Esposito-Farese G 1992 *Classical Quantum Gravity* **9** 2093
[16] Olmo G J 2005 *Phys. Rev. Lett.* **95** 261102
[17] Olmo G J 2005 *Phys. Rev.* D **72** 083505
[18] Perivolaropoulos L 2010 *Phys. Rev.* D **81** 047501
[19] Hohmann M, Järv L, Kuusk P and Randla E 2013 *Phys. Rev.* D **88** 084054
[20] Järv L, Kuusk P and Saal M 2008 *Phys. Rev.* D **78** 083530

Deformed Richardson-Gaudin model

P Kulish[1], A Stolin[2] and L H Johannesson[2]

[1]St. Petersburg Department of Steklov Mathematical Institute, Russian Academy of Sciences
Fontanka 27, 191023 St. Petersburg, Russia

[2]University of Gothenburg, Box 100, SE-405 30 Gothenburg, Sweden

Abstract. The Richardson-Gaudin model describes strong pairing correlations of fermions confined to a finite chain. The integrability of the Hamiltonian allows the algebraic construction of its eigenstates. In this work we show that the quantum group theory provides a possibility to deform the Hamiltonian preserving integrability. More precisely, we use the so-called Jordanian r-matrix to deform the Hamiltonian of the Richardson-Gaudin model. In order to preserve its integrability, we need to insert a special nilpotent term into the auxiliary L-operator which generates integrals of motion of the system. Moreover, the quantum inverse scattering method enables us to construct the exact eigenstates of the deformed Hamiltonian. These states have a highly complex entanglement structure which require further investigation.

The Richardson-Gaudin model [1, 2] is an integrable spin-$\frac{1}{2}$ periodic chain with Hamiltonian

$$H = \sum_{j=1}^{N} \epsilon_j S_j^z + g \sum_{j,k=1}^{N} S_j^- S_k^+ \tag{1}$$

where g is a coupling constant and $S_l^{\pm} = S_l^x \pm i S_l^y$, with N copies of the Lie algebra $su(2)$ generators S_l^{α},

$$[S_l^{\alpha}, S_{l'}^{\beta}] = i\varepsilon^{\alpha\beta\gamma} S^{\gamma} \delta_{l\,l'}, \quad \alpha, \beta = x, y, z$$

As shown by Cambiaggio *et al* [3], by introducing the fermion operators $c_{lm}^{\dagger}$ and c_{lm} related to the $sl(2)$ generators by

$$S_l^z = 1/2 \sum_m c_{lm}^{\dagger} c_{lm} - 1/2, \quad S_l^+ = \frac{1}{2} \sum_m c_{lm}^{\dagger} c_{l\bar{m}}^{\dagger} = (S_l^-)^{\dagger}$$

the Richardson-Gaudin model in Eq. (1) gets mapped onto the pairing model Hamiltonian

$$H_P = \sum_l \epsilon_l \hat{n}_l + g/2 \sum_{l,l'} A_l^{\dagger} A_{l'} \tag{2}$$

Here $c_{lm}^{\dagger}$ (c_{lm}) creates (annihilates) a fermion in the state $|lm\rangle$ (with $|l\bar{m}\rangle$ in the time reversed state of $|lm\rangle$) and

$$n_l = \sum_m c_{lm}^{\dagger} c_{lm}, \quad A_l^{\dagger} = (A_l)^{\dagger} = \sum_m c_{lm}^{\dagger} c_{l\bar{m}}^{\dagger}$$

Journal of Physics: Conference Series **532** (2014) 012012 doi:10.1088/1742-6596/532/1/012012

are the corresponding number- and pair-creation operators. The pairing strengths $g_{ll'}$ are here approximated by a single constant g, with ϵ_l the single-particle level corresponding to the m-fold degenerate states $|\,lm\rangle$.

As it is well-known, the pairing model in Eq. (2) is central in the theory of superconductivity. Richardson's exact solution of the model [1], exploiting its integrability, has been important for applications in mesoscopic and nuclear physics where the small number of fermions prohibits the use of conventional BCS theory [4]. Moreover, its (pseudo)spin representation in the guise of the Richardson-Gaudin model, Eq. (1), provides a striking link between quantum magnetism and pairing phenomena, both central concepts in the physics of quantum matter.

The eigenstates of the Richardson-Gaudin Hamiltonian, eq. (1), can be constructed algebraically using the quantum inverse scattering method (QISM) [5, 6]. The main objects of this method are the classical r-matrix

$$r(\lambda,\,\mu) = \frac{4}{\lambda-\mu}\sum_{\alpha} S^{\alpha}\otimes S^{\alpha}\Big|_{s=\frac{1}{2}} \simeq \frac{1}{\lambda-\mu}\begin{pmatrix} 1 & 0 & 0 & 0\\ 0 & -1 & 2 & 0\\ 0 & 2 & -1 & 0\\ 0 & 0 & 0 & 1\end{pmatrix} \tag{3}$$

where $h(\lambda)$, $X^+(\lambda)$, $X^-(\lambda)$ are the generators of the loop algebra $\mathcal{L}(sl(2))$ whereas the L-matrix is

$$L(\lambda) = \begin{pmatrix} h(\lambda) & 2X^-(\lambda)\\ 2X^+(\lambda) & -h(\lambda)\end{pmatrix}$$

The commutation relations (CR) of loop algebra generators are given in compact matrix form

$$[L_1(\lambda), L_2(\mu)] = -[r_{12}(\lambda,\mu), L_1(\lambda) + L_2(\mu)]$$

where

$$L_1(\lambda) = L(\lambda)\otimes\mathbb{I},\quad L_2(\mu) = \mathbb{I}\otimes L(\mu)$$

and $r(\lambda,\,\mu)$ is the 4×4 c-number matrix in Eq. (3). A consequence of this form is the commutativity of transfer matrices,

$$t(\lambda) = \frac{1}{2}\mathrm{tr}_0(L^2(\lambda)) \quad \in \mathcal{L}(sl(2)),\quad [t(\lambda), t(\mu)] = 0 \tag{4}$$

The corresponding mutually commuting operators extracted from the decomposition of $t(\lambda)$ define a Gaudin model [2, 7]. However, to get Richardson Hamiltonian a mild change of the L-operator is necessary

$$L(\lambda) \to L(\lambda; c) := c\,h_0 + L(\lambda)$$

where $h_0 = \sigma_0^z$ in auxiliary space $\mathbb{C}_0^2$ of spin $1/2$. This transformation does not change the CR of matrix elements of this matrix $L(\lambda; c)$ due to the symmetry of the r-matrix (3):

$$[Y\otimes\mathbb{I} + \mathbb{I}\otimes Y, r(\lambda,\,\mu)] = 0,\quad Y\in sl(2)$$

The resulting transfer matrix obtains some extra terms

$$t(\lambda; c) = \frac{1}{2}\mathrm{tr}_0\,(L(\lambda; c))^2 = c^2\mathbf{1} + c\,h(\lambda) + h^2(\lambda) + 2\left(X^+(\lambda)X^-(\lambda) + X^-(\lambda)X^+(\lambda)\right)$$

Let us consider a spin-$\frac{1}{2}$ representation on auxiliary space $V_0 \simeq \mathbb{C}^2$ and spin ℓ_k representations on quantum spaces $V_k \simeq \mathbb{C}^{\ell_k+1}$ with extra parameters ϵ_k corresponding to site $k = 1, 2, \ldots, N$.

The whole space of quantum states is $\mathcal{H} = \otimes_1^N V_k$ and the highest weight vector (highest spin, "ferromagnetic state") $|\Omega_+\rangle$ satisfies

$$X^+(\lambda)\,|\Omega_+\rangle = 0, \quad h(\lambda)\,|\Omega_+\rangle = \rho(\lambda)\,|\Omega_+\rangle \tag{5}$$

where

$$\rho(\lambda) = \sum_{k=1}^{N} l_k/(\lambda - \epsilon_k)$$

It is useful to introduce notation for global operators of $sl(2)$-representation $Y_{gl} := \sum_{k=1}^{N} Y_k$. To find the eigenvectors and spectrum of $t(\lambda)$ on $\mathcal{H}$ one requires that vectors of the form

$$|\mu_1, \ldots, \mu_M\rangle = \prod_{j=1}^{M} X^-(\mu_j)\,|\Omega_+\rangle$$

are eigenvectors of $t(\lambda)$,

$$t(\lambda)\,|\{\mu_j\}_{j=1}^M\rangle = \Lambda\left(\lambda; \{\mu_j\}_{j=1}^M\right)|\{\mu_j\}_{j=1}^M\rangle$$

provided that the parameters μ_j satisfy the Bethe equations:

$$2c + \sum_{k=1}^{N} \ell_k/(\mu_i - \epsilon_k) - \sum_{j\neq i}^{M} 2/(\mu_i - \mu_j) = 0, \quad i = 1, \ldots, M \tag{6}$$

The realization of the loop algebra generators on the space $\mathcal{H}$ takes the form

$$h(\lambda) = \sum_{k=1}^{N} \frac{h_k}{\lambda - \epsilon_k}, \quad X^-(\lambda) = \sum_{k=1}^{N} \frac{X_k^-}{\lambda - \epsilon_k}, \quad X^+(\lambda) = \sum_{k=1}^{N} \frac{X_k^+}{\lambda - \epsilon_k} \tag{7}$$

The coupling constant g of (1) is connected with parameter $c = 1/g$ while the Hamiltonian (1) is obtained as operator coefficient of term $1/\lambda^2$ in the expansion of $t(\lambda; c)$ at $\lambda \to \infty$.

The quantum group theory provides a possibility to deform a Hamiltonian preserving integrability [8, 9]. Specifically, we use the so-called Jordanian r-matrix to quantum deform the Hamiltonian of Richardson-Gaudin model (1). We add to $sl(2)$ symmetric r-matrix (3) the Jordanian part

$$r^J(\lambda, \mu) = \frac{C_2^{\otimes}}{\lambda - \mu} + \xi\left(h \otimes X^+ - X^+ \otimes h\right)$$

with Casimir element $C_2^{\otimes}$ in the tensor product of two copies of $sl(2)$,

$$C_2^{\otimes} = h \otimes h + 2\left(X^+ \otimes X^- + X^- \otimes X^+\right)$$

After Jordanian twist the r-matrix (14) is commuting with the generator X_0^+ only

$$\left[X_0^+ \otimes \mathbb{I} + \mathbb{I} \otimes X_0^+, r^{(J)}(\lambda, \mu)\right] = 0$$

Hence, one can add the term $cX_0^+ + L(\lambda,\xi)$ to the L-operator. This yields the twisted transfer-matrix

$$t^{(J)}(\lambda) = \frac{1}{2}\mathrm{tr}_0(cX_0^+ + L(\lambda,\xi))^2 = cX^+(\lambda) + h(\lambda)^2 - 2h'(\lambda) + 2(2X^-(\lambda)+\xi)X^+(\lambda) \qquad (8)$$

The corresponding commutation relations between the generators of the twisted loop algebra are explicitly given by

$$\begin{aligned}
&[h(\lambda),h(\mu)] = 2\xi\left(X^+(\lambda) - X^+(\mu)\right), && [X^-(\lambda),X^-(\mu)] = -\xi\left(X^-(\lambda) - X^-(\mu)\right) \\
&[X^+(\lambda),X^-(\mu)] = -\frac{h(\lambda)-h(\mu)}{\lambda-\mu} + \xi X^+(\lambda), && [X^+(\lambda),X^+(\mu)] = 0 \\
&[h(\lambda),X^-(\mu)] = 2\frac{X^-(\lambda)-X^-(\mu)}{\lambda-\mu} + \xi h(\mu), && [h(\lambda),X^+(\mu)] = -2\frac{X^+(\lambda)-X^+(\mu)}{\lambda-\mu}
\end{aligned} \qquad (9)$$

The realization of the Jordanian twisted loop algebra $\mathcal{L}_J(sl(2))$ with CR (9) is given similar to (7) with extra terms proportional to the deformation parameter ξ

$$h(\lambda) = \sum_{k=1}^{N}\left(\frac{h_k}{\lambda-\epsilon_k} + \xi X_k^+\right), \quad X^-(\lambda) = \sum_{k=1}^{N}\left(\frac{X_k^-}{\lambda-\epsilon_k} - \frac{\xi}{2}h_k\right), \quad X^+(\lambda) = \sum_{k=1}^{N}\frac{X_k^+}{\lambda-\epsilon_k} \qquad (10)$$

To construct eigenstates for the twisted model one has to use operators of the form [9, 10]

$$B_M(\mu_1,\ldots,\mu_M) = X^-(\mu_1)\left(X^-(\mu_2)+\xi\right)\ldots\left(X^-(\mu_M)+\xi(M-1)\right)$$

acting by these operators on the ferromagnetic state $|\Omega_+\rangle$.

The deformed Richardson-Gaudin model Hamiltonian can now be extracted from the transfer-matrix $t^{(J)}(\lambda)$ as the operator coefficient in its expansion $\lambda\to\infty$.

According to (4) and (8) one can also extract quantum integrals of motion J_k using the realization (10). It would yield rather cumbersome expressions for J_k :

$$t^{(J)}(\lambda) = J_0 + \frac{1}{\lambda}J_1 + \frac{1}{\lambda^2}J_2 + \ldots$$

The corresponding quantum deformed Hamiltonian reads

$$H \simeq J_2 = c\sum_{j=1}^{N}\epsilon_j X_j^+ + 2\xi\left\{\left(\sum_{j=1}^{N}\epsilon_j h_j\right)X_{gl}^+ - h_{gl}\sum_{j=1}^{N}\epsilon_j X_j^+\right\} + \left(h_{gl}^2 + 2h_{gl} + 4X_{gl}^-X_{gl}^+\right)$$

It is instructive to write down a simplified case without the Jordanian twist: $\xi = 0$. One thus obtains

$$J_0 = 0, \quad J_1 = X_{gl}^+, \quad J_2 \simeq \sum_{k=1}^{N}\epsilon_k X_k^+ + g/2\left(h_{gl}^2 + 2h_{gl} + 4X_{gl}^-X_{gl}^+\right)$$

The case $\xi = 0$ can also be obtained by taken off from the inhomogeneous XXX spin chain. The model can be described by a 2×2 monodromy matrix [5]

$$T(\lambda) = \begin{pmatrix} A(\lambda) & B(\lambda) \\ C(\lambda) & D(\lambda)\end{pmatrix}$$

and entries of this matrix satisfy quadratic relations

$$R(\lambda,\mu)T(\lambda)\otimes T(\mu) = (I\otimes T(\mu))\,(T(\lambda)\otimes I)\,R(\lambda,\mu) \tag{11}$$

If we multiplay $T(\lambda)$ by a constant 2×2 matrix $M(\varepsilon)$ the resulting matrix $\widetilde{T}(\lambda) = M(\varepsilon)\cdot T(\lambda)$ will satisfy the same relation (11). Choosing a triangular matrix $M(\varepsilon) = \left(\begin{smallmatrix}1 & \varepsilon\\ 0 & 1\end{smallmatrix}\right)$the entries of monodromy matrices become simply related:

$$\widetilde{A} = A + \varepsilon C, \quad \widetilde{B} = B + \varepsilon D, \quad \widetilde{C} = C, \quad \widetilde{D} = D.$$

This choice of $M(\varepsilon)$ (of the same type as considered in [11]) permits us to use the same reference state $|\Omega_+\rangle \in \mathcal{H}$ (5) and $\widetilde{B}$ as a creation operator of the algebraic Bethe ansatz [5].

Bethe states are given by the same action of product operators $\widetilde{B}(\mu_j) = B(\mu_j) + \varepsilon D(\mu_j)$ although operators $B(\mu_j)$ do not commute with $D(\mu_j)$:

$$D(\lambda)B(\mu) = \alpha(\lambda,\mu)B(\mu)D(\lambda) + \beta(\lambda,\mu)B(\lambda)D(\mu)$$

where

$$\alpha(\lambda,\mu) = (\lambda-\mu+\eta)/(\lambda-\mu), \quad \beta(\lambda,\mu) = -\eta/(\lambda-\mu)$$

For a 3 magnon state one gets due to B-D ordering

$$\prod_{j=1}^{3}\widetilde{B}(\mu_j) = \prod_{j=1}^{3}B(\mu_j) + \varepsilon\sum_{s=1}^{3}\alpha(\mu_k,\mu_s)\alpha(\mu_s,\mu_l)B(\mu_k)B(\mu_l)D(\mu_s)$$
$$+\,\varepsilon^2\sum_{s=1}^{3}\alpha(\mu_k,\mu_s)\alpha(\mu_l,\mu_s)B(\mu_s)D(\mu_k)D(\mu_l) + \varepsilon^3\prod_{j=1}^{3}D(\mu_j)$$

Similar formula is valid for M-magnon state. Hence, acting on ferromagnet state $|\,\Omega_+\rangle$, we obtain filtration of states with eigenvalues of S^z : $\frac{N}{2}, \frac{N}{2}-1, \frac{N}{2}-2, \frac{N}{2}-3$.

More complicated deformations of the Richardson-Gaudin model can be obtained using r-matrices related to the higher rank Lie algebras [12]. The structure of the eigenstates of the transfer matrix and their entanglement properties [13] are under investigation.

Acknowledgments
This work was supported by RFBR grants 11-01-00570-a, 12-01-00207-a, 13-01-12405-ofi-M2 (P.K.), and by STINT grant IG2011-2028 (A.S. and H.J.). We would like to thank E. Damaskinsky for useful discussion.

References

[1] Richardson R W 1963 *J. Math. Phys.* **6** 1034
[2] Gaudin M 1983 *La fonction d'onde de Bethe* (Paris: Masson) chapter 13
[3] Cambiaggio M C, Rivas A M F and Saraceno M 1997 *Nucl. Phys.* A **624** 157
[4] Dukelsky J, Pittel S and Sierra G 2004 *Rev. Mod. Phys.* **76** 643
[5] Faddeev L D 1998 How algebraic Bethe ansatz works for integrable models *Quantum symmetries, Proceedings of the Les Houches summer school, session LXIV* eds A Connes, K Gawedzki and J Zinn-Justi (North-Holland) p 149
[6] Kulish P P and Sklyanin E K 1982 Quantum spectral transform method. Recent developments *Lecture Notes in Phys.* **151** (Springer-Verlag) p 61
[7] Sklyanin E K 1999 *Lett. Math. Phys.* **47** 275
[8] Kulish P P and Stolin A A 1997 *Czech. J. Phys.* **12** 207
[9] Kulish P P 2002 *Twisted sl(2) Gaudin model* Preprint PDMI 08/2002
[10] Antonio N C and Manojlovic N 2005 *J. Math. Phys.* **46** 102701
[11] Mukhin E, Tarasov V and Varchenko A 2010 *Contemp. Math.* **506** 187
[12] Stolin A A 1991 *Math. Scand.* **69** 57
[13] Dunning C, Links J and Zhou H-Q 2005 *Phys. Rev. Lett.* **94** 227002

The structure of $Ph^\star$, generalized de Rham, and entropy

O A Laudal

Matematisk institutt, University of Oslo, Pb. 1053, Blindern, N-0316 Oslo, Norway

Abstract. In this note I shall continue the study of the non-commutative phase space functor, Ph(A), defined for any associative algebra A, and its derived differential co-simplicial algebra, $Ph^\star(A)$. The main focus will be on its relationship to the classical de Rham complex, to the dynamics of finite dimensional $Ph^\infty(A)$-modules, and to the notion of Entropy. These subjects are treated within the set-up of my book [2011 *Geometry of Time-Spaces* (World Scientific)].

Keywords: Phase spaces, Dirac derivations, de Rham complexes, connections, entropy.

1. Introduction

If we want to study the *dynamics*, of some mathematical system, say composed of representations of an algebra A, we should, according to [5], introduce the notion of a (non-commutative) *Phase Space*-functor, $Ph(-) : Alg_k \to Alg_k$, k being a field.

This functor extends to the category of schemes, and its infinite iteration $Ph^\infty(-)$, is outfitted with a universal *Dirac derivation*, $\delta \in Der_k(Ph^\infty(-), Ph^\infty(-))$, with seemingly useful applications in mathematical models of quantum theory, see [5].

The Ph functor extends also, by iteration, in a natural way, to a functor of algebras into differential co-simplicial algebras, such that for any associative k-algebra A, the (graded) family of algebras, $Ph^n(A), n \geq 0$, is outfitted with a differential of degree 1, and a nicely related co-simplicial structure, see [3, 4] and also [5].

The main purpose of this note is, to show the relationship between this differential co-simplicial algebra $Ph^\star(A)$ and the different versions of the classical de Rham complex to be found in the literature.

This will force us to take a new look at the non-commutative versions of *connections* on modules of associative k-algebras, A, their reformulations as representations of a *dynamical system*, $A(\sigma_g)$, induced by the choice of a *metric* $g \in Ph(A)$, in the commutative case, see [5].

Extending classical notions of *Modular Suites*, see [6], to the case of representations of $Ph^\infty(A)$, and by considering finite dimensional *approximations* to vector fields, (as representations of the corresponding $Ph^\infty(A)$) we come close to a reasonable mathematical formulation of *Information*, *Entropy*, and classical thermodynamics, in terms of deformation theory.

2. Phase spaces

From now on, we shall be led to talk about an algebra, as a *space*, having in mind the relationship between an associative (geometric), see [5], algebra and the moduli space of finite dimensional simple representations, characterizing it.

Journal of Physics: Conference Series **532** (2014) 012013 doi:10.1088/1742-6596/532/1/012013

Given an associative k-algebra A, denote by $A/k - \underline{alg}$ the category where the objects are homomorphisms of k-algebras $\kappa : A \to R$, and the morphisms, $\psi : \kappa \to \kappa'$ are commutative diagrams

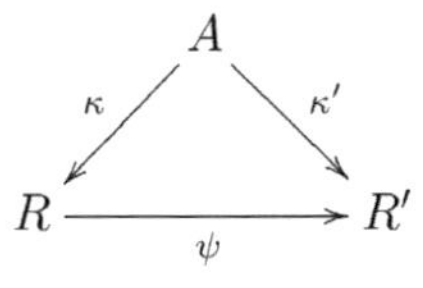

and consider the functor

$$Der_k(A, -) : A/k - \underline{alg} \longrightarrow \underline{Sets}$$

It is, see [4] and [5], representable by a k-algebra-morphism $\iota : A \longrightarrow Ph(A)$ with a *universal family* given by a universal derivation $d : A \longrightarrow Ph(A)$. Clearly we have the identities

$$d_* : Der_k(A, A) = Mor_A(Ph(A), A), \quad d^* : Der_k(A, Ph(A)) = End_A(Ph(A))$$

the last one associating d to the identity endomorphism of Ph.

2.1. *First properties*

Let now V be a right A-module, with structure morphism $\rho : A \to End_k(V)$. We obtain a universal derivation

$$c : A \longrightarrow Hom_k(V, V \otimes_A Ph(A))$$

defined by $c(a)(v) = v \otimes d(a)$. Using the long exact sequence of the Hochschild cohomology

$$0 \to Hom_A(V, V \otimes_A Ph(A)) \to Hom_k(V, V \otimes_A Ph(A)) \\ \overset{\iota}{\to} Der_k(A, Hom_k(V, V \otimes_A Ph(A))) \overset{\kappa}{\to} Ext^1_A(V, V \otimes_A Ph(A)) \to 0$$

we obtain the non-commutative *Kodaira-Spencer class*

$$c(V) := \kappa(c) \in Ext^1_A(V, V \otimes_A Ph(A))$$

inducing, via the identity d_*, the *Kodaira-Spencer morphism*

$$g : \Theta_A := Der_k(A, A) \longrightarrow Ext^1_A(V, V)$$

If $c(V) = 0$, then the exact sequence above proves that there exist an element, $\nabla \in Hom_k(V, V \otimes_A Ph(A))$ such that $c = \iota(\nabla)$. This is just another way of proving that if $c(V) = 0$, then c is given by a connection

$$\nabla : Der_k(A, A) \longrightarrow Hom_k(V, V)$$

In particular, we deduce, from the corresponding long exact sequence

$$0 \to Hom_A(V, V) \to Hom_k(V, V) \overset{\iota}{\to} Der_k(A, Hom_k(V, V)) \overset{\kappa}{\to} Ext^1_A(V, V) \to 0$$

the following elementary

Journal of Physics: Conference Series **532** (2014) 012013 doi:10.1088/1742-6596/532/1/012013

Lemma 2.1. *Let $\rho : A \to End_k(V)$, be an A-module, and let $\delta \in Der_k(A, Hom_k(V,V))$, map to 0 in $Ext^1_A(V,V)$, i.e assume $\kappa(\delta) = 0$, then there exist an Hamiltonian-element, $Q_\delta \in Hom_k(V,V)$, such that*

$$\rho(\delta(a)) = [Q_\delta, \tilde{\rho}(a)], \quad \forall a \in A$$

If V is a simple A-module then $ad(Q_\delta)$ is unique.

As is well known, in the commutative case, the Kodaira-Spencer class gives rise to a *Chern character* by putting

$$ch^i(V) := 1/i! \ \left(c(V)^i\right) \in Ext^i_A(V, V \otimes_A Ph(A))$$

and if $c(V) = 0$, the curvature $R(V)$ of ∇, induces a curvature class

$$R_\nabla \in H^2\left(k, A; \Theta_A, End_A(V)\right)$$

Any $Ph(A)$-module W, given by its structure map $\rho_W : Ph(A) \longrightarrow End_k(W)$ corresponds bijectively to an induced A-module structure on W, and a derivation $\delta_\rho \in Der_k(A, End_k(W))$, defining an element $[\delta_\rho] \in Ext^1_A(W,W)$.

Remark 2.2. Since $Ext^1_A(V,V)$ is the tangent space of the miniversal deformation space of V as an A-module, we see that the non-commutative space $Ph(A)$ also parametrizes the set of *generalized momenta*, i.e the set of pairs of an A- module V, and a tangent vector of the formal moduli of V, at that point.

$Ph(A)$ is relatively easy to compute. In particular, if $A = k[x_1,..,x_n] =: k[\underline{x}]$ is the polynomial algebra, we have

$$Ph(A) = k < x_1,..,x_n, dx_1,..,dx_n > /([x_i,x_j],[x_i,dx_j] + [dx_i,x_j])$$

Notice that, in this case, any rank 1 representation of $Ph(A)$ is represented by a pair, (q,p), of a closed point, q of $Spec(k[\underline{x}])$, and a tangent, p at that point.

We shall need the following formulas,

Theorem 2.3. *Given two such points, $(q_i,p_i), i = 1,2$, we find*

$$dim_k Ext^1_{Ph(A)}(k(q_1,p_1), k(q_2,p_2)) = 2n, \quad \textit{for } (q_1,p_1) = (q_2,p_2)$$
$$dim_k Ext^1_{Ph(A)}(k(q_1,p_1), k(q_2,p_2)) = n, \quad \textit{for } q_1 = q_2,, p_1 \neq p_2$$
$$dim_k Ext^1_{Ph(A)}(k(q_1,p_1), k(q_2,p_2)) = 1, \quad \textit{for } q_1 \neq q_2$$

Moreover, there is, in the last case, a generator of

$$Ext^1_{Ph(A)}(k(q_1,p_1), k(q_2,p_2)) = Der_k(Ph(A), Hom_k(k(q_1,p_1)), k(q_2,p_2)))/Triv$$

uniquely characterized by the tangent line defined by the vector $\overline{q_1q_2}$.

Proof. Assume for convenience that $n = 3$. Put

$$x_j(q_i,p_i) := q_{i,j}, \quad dx_j((q_i,p_i) := p_{i,j}, \quad \alpha_j = q_{1,j} - q_{2,j}, \quad \beta_j = p_{1,j} - p_{2,j}$$

See that for every element $\alpha \in Hom_k(k(q_1,p_1), k(q_2,p_2))$ we have

$$x_j\alpha = q_{1,j}\alpha, \quad \alpha x_j = q_{2,j}\alpha, \quad dx_j\alpha = p_{1,j}\alpha, \quad \alpha dx_j = p_{2,j}\alpha$$

with the obvious identification. Every derivation

$$\delta \in Der_k(Ph(A), Hom_k(k(q_1, p_1), k(q_2, p_2)))$$

must satisfy the relations

$$\delta([x_i, x_j]) = [\delta(x_i), x_j] + [x_i, \delta(x_j)] = 0$$
$$\delta([dx_i, x_j] + [x_i, dx_j]) = [\delta(dx_i), x_j] + [dx_i, \delta(x_j)] + [\delta(x_i), dx_j] + [x_i, \delta(dx_j)] = 0$$

Using the above left-right action-rules, the result follows from the long exact sequence computing Ext^1_{PhA}. The two families of relations above give us two systems of linear equations.

The first one, in the variables $\delta(x_1), \delta(x_2), \delta(x_3)$, with the matrix

$$\begin{pmatrix} -\alpha_2 & \alpha_1 & 0 \\ -\alpha_3 & 0 & \alpha_1 \\ 0 & -\alpha_3 & \alpha_2 \end{pmatrix}$$

and the second one, in the variables $\delta(x_1), \delta(x_2), \delta(x_3), \delta(dx_1), \delta(dx_2), \delta(dx_3)$, with the matrix

$$\begin{pmatrix} -\beta_2 & \beta_1 & 0 & -\alpha_2 & \alpha_1 & 0 \\ -\beta_3 & 0 & \beta_1 & -\alpha_3 & 0 & \alpha_1 \\ 0 & -\beta_3 & \beta_2 & 0 & -\alpha_3 & \alpha_2 \end{pmatrix}$$

In particular, we see that the *trivial* derivation given by

$$\delta(x_i) = \alpha_i, \quad \delta(dx_j) = \beta_j$$

satisfies the relations, and the generator of $Ext^1_{PhA}(k(q_1, p_1), k(q_2, p_2))$ is represented by

$$\delta(x_i) = 0, \quad \delta(dx_j) = \alpha_j$$

This is, in an obvious sense, the "tangent vector" $-\overline{q_1 q_2}$.

It is easy to extend this result from dimension 3 to any dimension n. □

Example 2.4. Let $A = M_2(k)$ and compute $Ph(A)$. Clearly the existence of the canonical homomorphism, $i : M_2(k) \to Ph(M_2(k))$ shows that $Ph(M_2(k))$ must be a matrix ring, generated, as an algebra, over $M_2(k)$ by $d\epsilon_{i,j}$, $i, j = 1, 2$, where $\epsilon_{i,j}$ is the elementary matrix. A little computation shows that we have the following relations

$$d\epsilon_{1,1} = \begin{pmatrix} 0 & (d\epsilon_{1,1})_{1,2} = -(d\epsilon_{2,2})_{1,2} \\ (d\epsilon_{1,1})_{2,1} = -(d\epsilon_{2,2})_{2,1} & 0 \end{pmatrix}$$
$$d\epsilon_{2,2} = \begin{pmatrix} 0 & (d\epsilon_{2,2})_{1,2} = -(d\epsilon_{1,1})_{1,2} \\ (d\epsilon_{2,2})_{2,1} = -(d\epsilon_{1,1})_{2,1} & 0 \end{pmatrix}$$
$$\epsilon_{1,2} = \begin{pmatrix} \epsilon_{1,2}(d\epsilon_{2,2})_{2,1} & (d\epsilon_{1,2})_{1,2} = -(d\epsilon_{2,1})_{2,1} \\ 0 & -(d\epsilon_{2,2})_{2,1}\epsilon_{1,2} \end{pmatrix}$$
$$\epsilon_{2,1} = \begin{pmatrix} (d\epsilon_{2,2})_{1,2}\epsilon_{2,1} & 0 \\ (d\epsilon_{2,1})_{2,1} = -(d\epsilon_{1,2})_{1,2} & \epsilon_{2,1}(d\epsilon_{1,1})_{1,2} \end{pmatrix}$$

From this follows that every cosection $\rho : Ph(M_2(k)) \to M_2(k)$ of $i : M_2(k) \to Ph(M_2(k))$ is given in terms of an element $\phi \in M_2(k)$, such that $\rho(da) = [\phi, a]$.

2.2. The structure of $Ph^\star(-)$ and the de Rham complex

The phase-space construction may, of course, be iterated. Given the k-algebra A we may form the sequence, $\{Ph^n(A)\}_{0\leq n}$, defined inductively by

$$Ph^0(A) = A, \quad Ph^1(A) = Ph(A), \ldots, \quad Ph^{n+1}(A) := Ph(Ph^n(A))$$

Let $i_0^n : Ph^n(A) \to Ph^{n+1}(A)$ be the canonical imbedding, and let $d_n : Ph^n(A) \to Ph^{n+1}(A)$ be the corresponding derivation. Since the composition of i_0^n and the derivation d_{n+1} is a derivation $Ph^n(A) \to Ph^{n+2}(A)$, there exist by universality a homomorphism $i_1^{n+1} : Ph^{n+1}(A) \to Ph^{n+2}(A)$, such that

$$d_n \circ i_1^{n+1} = i_0^n \circ d_{n+1}$$

Notice that we here compose functions and functors from left to right. Clearly we may continue this process constructing new homomorphisms

$$\{i_j^n : Ph^n(A) \to Ph^{n+1}(A)\}_{0\leq j\leq n}$$

with the property

$$d_n \circ i_{j+1}^{n+1} = i_j^n \circ d_{n+1}$$

We find, see [4], the following identities

$$\begin{aligned} i_p^n i_q^{n+1} &= i_{q-1}^n i_p^{n+1}, \quad p < q \\ i_p^n i_p^{n+1} &= i_p^n i_{p+1}^{n+1} \\ i_p^n i_q^{n+1} &= i_q^n i_{p+1}^{n+1}, \qquad q < p \end{aligned}$$

To see this, compose i_0^{n-1} with d_{n-1} and use induction. Thus, the $Ph^*(A)$ is a semi-cosimplicial k-algebra with a cosection onto A. Therefore, for any object

$$\kappa : A \to R \quad \in A/k - \underline{alg}$$

the semi-cosimplicial algebra above induces a semi-simplicial k-vectorspace $Der_k(Ph^\star(A), R)$ and one should be interested in its homology.

Consider now the diagram

$$\begin{array}{ccccccccc} A & \xrightarrow{i_0^0} & Ph(A) & \xrightarrow{i_p^1} & Ph^2(A) & \xrightarrow{i_p^2} & Ph^3(A) & \xrightarrow{i_p^3} & \\ & & \uparrow & & \uparrow & & \uparrow & & \\ & & \mathfrak{m}_1^1 & \xrightarrow{i_p^1} & \mathfrak{m}_2^1 & \xrightarrow{i_p^2} & \mathfrak{m}_3^1 & \xrightarrow{i_p^3} & \end{array}$$

where, for each integer n, the symbol i_p^n for $p = 0, 1, \ldots, n$ signifies the family of A-morphisms between $Ph^n(A)$ and $Ph^{n+1}(A)$ defined above, and where $\mathfrak{m}_n^1$ is the ideal of $Ph^n(A)$ generated by $im(d)$, which is the same as the ideal generated by the family

$$\{i_p^{n-1}(i_p^{n-2}(\ldots(i_p^1(d(A))\ldots))\}$$

Journal of Physics: Conference Series **532** (2014) 012013 doi:10.1088/1742-6596/532/1/012013

for all possible p. And, inductively, let $\mathfrak{m}_n^m$ be the ideal generated by $\mathfrak{m}_n^1\mathfrak{m}_n^{m-1}$. Then we find an extended diagram

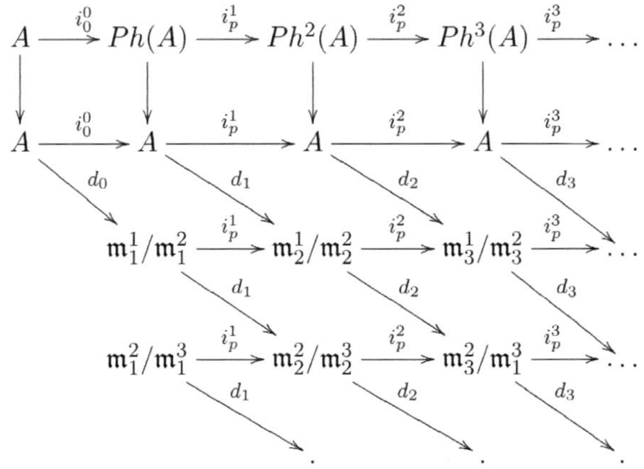

The diagonals are not necessarily complexes, but it suffices to kill d^2, to kill all d^n, $n \geq 2$, and for this it suffices to kill $d_1 d_0$, as one easily see, operating with the edge homomorphisms on the elements, $d_1(d_0(a))$ for $a \in A$. Therefore we shall, in this general situation, make the following definition.

Definition 2.5. The curvature $R(A)$ of the associative k algebra A is the k-linear map composition of d_0 and d_1

$$R(A) = d_0 d_1 : A \to \mathfrak{m}_2^2/\mathfrak{m}_2^3$$

Now, kill the curvature $R(A)$, and all the terms under the first diagonal, beginning with $\mathfrak{m}_1^2/\mathfrak{m}_1^3$, together with all terms generated by the actions of the edge homomorphisms on these terms, and let, $\mathbf{\Omega}_n^m$ be the resulting quotient of $\mathfrak{m}_n^m/\mathfrak{m}_n^{m+1}$, for $n \geq 0$. Clearly, $\Omega_n^0 = A$ for all $n \geq 0$, and we have got a graded semi co-simplicial A-module, with a k-differential d, such that $d^2 = 0$, looking like

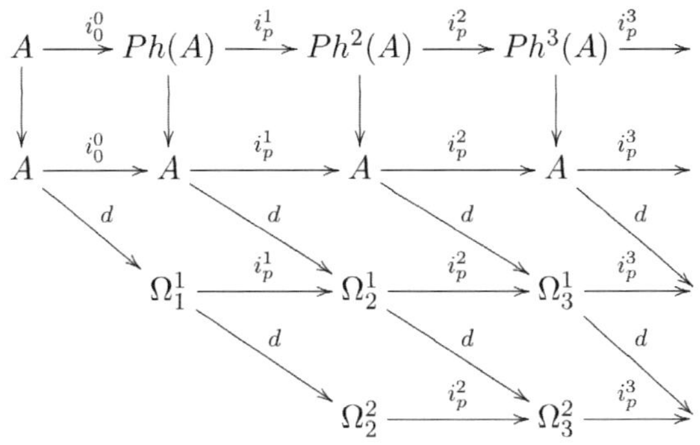

It is a graded complex, in two ways. First as a complex induced from the semi-cosimplicial structure, with differential of bidegree (1,0), and second, as complex with differential d, of bidegree (1,1).

Lemma 2.6. *Suppose A is commutative, then there is a natural morphism of complexes of A-modules $\Omega_A^\star \subset \Omega_\star^\star$ with $\Omega_A^n := \wedge^r \Omega_A \simeq \Omega_n^n$.*

Proof. Let $a_i \in A, i = 1, \ldots, r$, and compute in $\Omega^r_\star$ the value of, $d^r(a_1 a_2 \ldots a_r)$. It is clear that this gives the formula

$$\sum d_{i_1}(a_1) d_{i_2}(a_2) \ldots d_{i_r}(a_r) = 0$$

where the sum being over all permutation $(i_1, i_2, \ldots, i_r)$ of $(0, 1, \ldots, r-1)$. Here we consider A as a subalgebra of $Ph^n(A)$ via the unique compositions of the $i^s_0 : Ph^s(A) \subset Ph^{s+1}(A)$. In particular, we have

$$d_0(a_1)d_1(a_2) + d_1(a_1)d_0(a_2) = 0$$

for all $a_1, a_2 \in A$. This relation and the relation $d_0(a_2)d_1(a_1) = d_1(a_1)d_0(a_2)$, which follows from commutativity, $d(a_2)a_1 = a_1 d(a_2)$, forcing the left and right A-action on Ω_A to be equal, immediately give us, $d_0(a_1)d_1(a_2) = -d_0(a_2)d_1(a_1)$.

Consider now the diagram

$$\begin{array}{ccccccccc}
A & \xrightarrow{i^0_0} & Ph(A) & \xrightarrow{i^1_0} & Ph^2(A) & \xrightarrow{i^2_0} & Ph^3(A) & \xrightarrow{i^3_0} & \ldots \\
\downarrow & & \downarrow & & \downarrow & & \downarrow & & \\
A & \xrightarrow{i^0_0} & A \oplus \Omega^1_A & \xrightarrow{i^1_0} & A \oplus \Omega^1_A \oplus \Omega^2_A & \xrightarrow{i^2_0} & A \oplus \Omega^1_A \oplus \Omega^2_A \oplus \Omega^3_A & \xrightarrow{i^3_0} & \ldots
\end{array}$$

where the bottom line is a sequence of Nagata-extentions of the k-algebra A, and the vertical homomorphisms correspond to the natural derivations among these, defined by the derivations of the de Rham complex $\Omega^\star$ of A.

The universality of the two systems proves that there is a surjective map

$$\alpha : \Omega^n_n \to \Omega^n_A := \wedge^n \Omega_A$$

The map that sends the element $da_1 \wedge da_2 \wedge \ldots \wedge da_n \in \Omega^r_A$ to $d_0(a_1)d_1(a_2)..d_{n-1}(a_r) \in \Omega^n_n$ is an inverse, proving that α is an isomorphism. □

Let now V be a right A-module and assume $c(V) = 0$, such that there exist an element, $\nabla' \in Hom_k(V, V \otimes_A Ph(A))$ with $c = \iota(\nabla')$. This implies that for $a \in A$ and $v \in V$ we have $\nabla'(va) = \nabla'(v)a + v \otimes d_0(a)$. Composing ∇' with the projection, $o : Ph(A) \to A$, corresponding to the 0-derivation of A, we therefore obtain an A-linear homomorphism $P : V \to V$, *a potential.* Since $i^0_0 : A \to Ph(A)$ is a section of o, we find a k-linear map

$$\nabla_0 := \nabla' - P : V \to V \otimes \mathfrak{m}^1_1$$

Using the property

$$d_n \circ i^{n+1}_{j+1} = i^n_j \, d_{n+1}$$

we find well defined k-linear maps

$$\nabla_1 : V \to V \otimes \Omega^1_2, \quad \nabla_2 : V \to V \otimes \Omega^1_3 \quad \ldots \quad \nabla_n : V \to V \otimes \Omega^1_{n+1}, \quad \forall n \geq 0$$

given by

$$\nabla_{n+1} := \nabla_n \circ i^{n+1}_1, \quad n \geq 0$$

such that, for all, $v \in V, \omega \in \Omega_p^n$, the formula,

$$\nabla_n(v \otimes \omega) = \nabla_n(v)\omega + v \otimes d_n(\omega)$$

makes sense, and defines a sequence of *derivations*

$$\nabla_n : V \otimes \Omega_n^p \to V \otimes \Omega_{n+1}^{p+1}$$

sometimes just denoted d_n, and called a *connection* ∇ on the A-module V. We obtain a situation just like above,

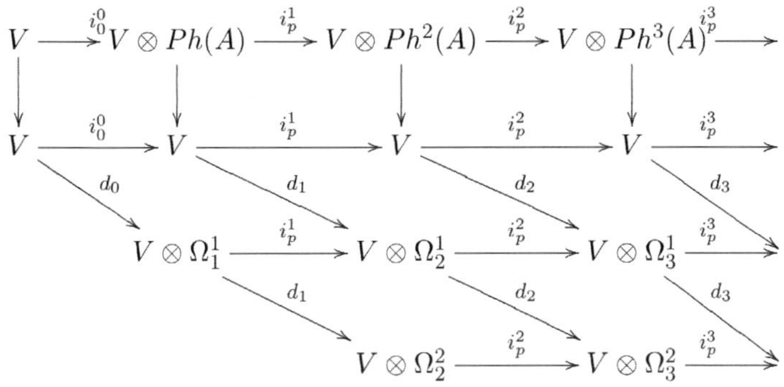

In general, there are no reasons for these derivations $d_n := \nabla_n,\ n \geq 0$ to define complexes and we make the following definition.

Definition 2.7. The curvature $R(V, \nabla)$ of the connection ∇, defined on the right A-module V, is the k-linear map, composition of d_0 and d_1,

$$R(V) = d_0 d_1 : V \to V \otimes \Omega_2^2$$

The following Lemma is then easily proved.

Lemma 2.8. *Suppose A is commutative, and assume $c(V) = 0$. Let $\nabla : \Theta_A \to End_k(V)$ be the classical connection corresponding to ∇_0. Suppose moreover that the curvature R of ∇ is 0, then $R(V) = 0$, implying that $d^2 = 0$, and so the diagonals in the diagram above, are all complexes.*

Proof. We may put

$$\nabla(v_i) = \sum_{j,k,} a_{i,j}^k v_j d_0(x_k)$$

and obtain

$$\nabla_1(\nabla_0(v_i)) = \sum_{j,k,l} \frac{\partial a_{i,j}^k}{\partial x_l} v_j d_1(x_l) d_0(x_k) + \sum_{j,k,l,m} a_{i,j}^k a_{j,m}^l v_m d_1(x_l) d_0(x_k).$$

Now the classical of curvature of ∇ may be defined as

$$R_{k,l}^i = \sum_j \frac{\partial a_{i,j}^k}{\partial x_l} v_j + \sum_{j,m} a_{i,j}^k a_{j,m}^l v_m - \sum_j \frac{\partial a_{i,j}^l}{\partial x_k} v_j - \sum_{j,m} a_{i,j}^l a_{j,m}^k v_m$$

so if $R = 0$ and $d_1(x_l)d_0(x_k) = -d_1(x_k)d_0(x_l)$, it follows that $\nabla_1(\nabla_0(v_i)) = 0$, from which it follows that $d^2 = 0$. □

2.3. Dirac derivation

The system of k-algebras and homomorphisms of k-algebras $\{Ph^n(A), i_j^n\}_{n,0\leq j\leq n}$ has an inductive (direct) limit, $Ph^\infty A$, together with homomorphisms, $i_n : Ph^n(A) \longrightarrow Ph^\infty(A)$ satisfying, see [5],

$$i_j^n \circ i_{n+1} = i_n, \quad j = 0, 1, \ldots, n$$

Moreover, the family of derivations $\{d_n\}_{0\leq n}$ defines a unique derivation, the *Dirac derivation*

$$\delta : Ph^\infty(A) \longrightarrow Ph^\infty(A)$$

such that

$$i_n \circ \delta = d_n \circ i_{n+1}$$

and it is easy to see that this is a universal construction, i.e for any pair of a morphism $i : A \longrightarrow B$ and a derivation $\xi \in Der_k(B)$, $\iota \circ \xi$ factorizes via $Ph^\infty(A)$ and δ. Put

$$Ph^{(n)}(A) := im\ i_n \subseteq Ph^\infty(A).$$

The k-algebra $Ph^\infty(A)$ has a descending filtration of two-sided ideals, $\{\mathbf{F}_n\}_{0\leq n}$ given inductively by

$$\mathbf{F}_1 = Ph^\infty(A) \cdot im(\delta) \cdot Ph^\infty(A)$$

and

$$\delta(\mathbf{F}_n) \subseteq \mathbf{F}_{n+1}, \ \mathbf{F}_{n_1}\mathbf{F}_{n_2}\ldots\mathbf{F}_{n_r} \subseteq \mathbf{F}_n, \quad \forall n_1 + \ldots + n_r = n$$

such that the derivation δ induces derivations $\delta_n : \mathbf{F}_n \longrightarrow \mathbf{F}_{n+1}$. Using the canonical homomorphism $i_n : Ph^n(A) \longrightarrow Ph^\infty(A)$ we pull the filtration $\{\mathbf{F}_p\}_{0\leq p}$ back to $Ph^n(A)$, not bothering to change the notation.

Definition 2.9. Let

$$\mathbf{D}(A) := \varprojlim_{n\geq 1} Ph^\infty(A)/\mathbf{F}_n$$

be the completion of $Ph^\infty(A)$ in the topology given by the filtration $\{\mathbf{F}_n\}_{0\leq n}$. The k-algebra $Ph^\infty(A)$ will be referred to as the k-algebra of higher differentials, and $\mathbf{D}(A)$ will be called the k-algebra of formalized higher differentials. Put

$$\mathbf{D}_n := \mathbf{D}_n(A) := Ph^\infty(A)/\mathbf{F}_{n+1}$$

Clearly δ defines a derivation on $\mathbf{D}(A)$, and an homomorphism of k-algebras,

$$\epsilon := \exp(\delta) : \mathbf{D}(A) \to \mathbf{D}(A)$$

and in particular, an algebra homomorphism

$$\tilde{\eta} := \exp(\delta) : A \to \mathbf{D}(A)$$

Recall now that for any associative k-algebra A, and for any right A-module V, any derivation $\delta \in Der_k(A, End_k(V))$ defines a class, $\xi(v) \in Ext^1_A(V, V)$, i.e a *tangent vector* of the *formal moduli* of the representation V, at the unique point.

Remark 2.10. The above implies that any representation $\rho : Ph^\infty(A) \to End_k(V)$ corresponds to a family of $Ph^n(A)$-module-structures on V for $n \geq 1$, i.e to an A-module $V_0 := V$, an element $\xi_0 \in Ext^1_A(V,V)$, i.e a tangent of the deformation functor of $V_0 := V$, as A-module, an element $\xi_1 \in Ext^1_{Ph(A)}(V,V)$, i.e a tangent of the deformation functor of $V_1 := V$ as $Ph(A)$-module, an element $\xi_2 \in Ext^1_{Ph^2(A)}(V,V)$, i.e a tangent of the deformation functor of $V_2 := V$ as $Ph^2(A)$-module, etc.

All this is just V, considered as an A-module, together with a sequence $\{\xi_n\}, 0 \leq n$, of a tangent, or a *momentum* ξ_0, an acceleration vector ξ_1, and any number of higher order *momenta* ξ_n. Thus, specifying a $Ph^\infty(A)$-representation V, implies specifying a *formal curve* through v_0, the base-point of the *miniversal deformation space* of the A-module V. Formally, this curve is given by the composition of the homomorphism $\epsilon(\tau) := exp(\tau\delta)$ and ρ. This is the *Time Evolution* of a representation ρ, to be treated later on.

3. Non-commutative algebraic geometry

The basic notions of affine non-commutative algebraic geometry related to a (not necessarily commutative) associative k-algebra, for k an arbitrary field, have been treated by many authors in several texts, see e.g [7, 8, 1, 2]. Given a finitely generated algebra A, we prove the existence of a non-commutative scheme-structure on the set of isomorphism classes of simple finite dimensional representations, i.e right modules, $Simp_{<\infty}(A)$.

We showed, in [2], that any finitely generated *geometric* k-algebra A, k algebraically closed, see definition in [5], including the commutative algebras, may be recovered from the (non-commutative) structure of $Simp_{<\infty}(A)$, and that there is an underlying quasi-affine (commutative) scheme-structure on each component $Simp_n(A) \subset Simp_{<\infty}(A)$, parametrizing the simple representations of dimension n. In fact, we have the following.

Theorem 3.1. *There is a commutative k-algebra $C(n)$ with an open subvariety $U(n) \subseteq Simp_1(C(n))$, an étale covering of $Simp_n(A)$, over which there exists a* versal representation $\tilde{V} \simeq C(n) \otimes_k V$, *a vector bundle of rank n defined on $Simp_1(C(n))$, and a* versal family, *i.e a morphism of algebras*

$$\tilde{\rho} : A \longrightarrow End_{C(n)}(\tilde{V}) \to End_{U(n)}(\tilde{V})$$

inducing all isoclasses of simple n-dimensional A-modules.

$End_{C(n)}(\tilde{V})$ induces also a bundle, of *operators*, on the étale covering $U(n)$ of $Simp_n(\mathbf{A})$. Assume given a derivation, $\gamma \in Der_k(A)$. Pick any $v \in Simp_n(A)$ corresponding to the right A-module V, with structure homomorphism $\rho_v : A \to End_k(V)$, then γ composed with ρ_v, gives us an element $\gamma_v \in Ext^1_A(V,V)$.

Therefore, γ defines a unique one-dimensional distribution in $\Theta_{Simp_n(A)}$, which, once we have fixed a versal family, defines a vector field $[\gamma] \in \Theta_{Simp_n(A)}$ and, in good cases, a (rational) derivation $[\gamma] \in Der_k(C(n))$. Notice also that we have the canonical isomorphism

$$Der_k(A,A) \simeq Mor_A(Ph(A),A)$$

Therefore the derivation γ, and the A-module V, correspond to a $Ph(A)$-module V_γ.

3.1. Dynamical structures and time

As we have seen, in the Remark (2.10), the dynamics of the space of representations of our algebra A, i.e the dynamics of the *space of measurements* of the family of *observables* that A is assumed to represent, can be encoded in the category of representations of the k-algebra $Ph^\infty(A)$, with the Dirac derivation as an infinitesimal time parameter. We would therefore like to use the tools developed above, for the k-algebra $Ph^\infty(A)$.

However, $Ph^\infty(A)$ is rarely of finite type, and so the space of simple modules does not have a classical algebraic geometric structure. We shall therefore introduce the notion of *dynamical structure*, to reduce the problem to a situation we can handle. This is also what physicists do. They invoke a parsimony principle, or an action principle, originally proposed by Fermat, and later by Maupertuis, with the purpose of reducing the preparation needed, to be able to see ahead.

Definition 3.2. A *dynamical structure*, σ, is a two-sided δ-stable ideal $(\sigma) \subset Ph^\infty(A)$, such that $A(\sigma) := Ph^\infty(A)/(\sigma)$ the corresponding, *dynamical system*, is of finite type. A dynamical structure, or system, is *of order n* if the canonical morphism $\sigma : Ph^{(n-1)}(A) \to A(\sigma)$ is surjective. If A is generated by the *coordinate functions*, $\{t_i\}_{i=1,2,...,d}$ a dynamical system of order n may be defined by a *force law*, i.e by a system of equations

$$\delta^n t_p = \Gamma^p(\underline{t}_i, \underline{d}t_j, \underline{d}^2 t_k, .., \underline{d}^{n-1} t_l), \quad p = 1, 2, \ldots, d$$

Put

$$A(\sigma) := Ph^\infty(A)/(\delta^n t_p - \Gamma^p)$$

where $\sigma := (\delta^n t_p - \Gamma^p)$ is the two-sided δ-ideal generated by the defining equations of σ. Obviously δ induces a derivation $\delta_\sigma \in Der_k(A(\sigma), A(\sigma))$, also called the Dirac derivation, and usually just denoted δ.

Notice that if σ_i, $i = 1, 2$, are two different order n dynamical systems, then we may well have

$$A(\sigma_1) \simeq A(\sigma_2) \simeq Ph^{(n-1)}(A)/(\sigma_*)$$

as k-algebras.

Assuming that the k-algebra A is finitely generated, and that the dynamical structure σ is such that also $A(\sigma)$ is finitely generated, we can now use the machinery above, with $\gamma = \delta$, the Dirac derivation. The following extension of Theorem (3.2), proved in [5], is important for the philosophy of this paper.

Theorem 3.3. *Formally, at any point $v \in U(n)$, with local ring $\hat{C}(n)_v$, there is a derivation $[\delta] \in Der_k(\hat{C(n)_v})$, and a Hamiltonian $Q \in End_{\hat{C(n)_v}}(\tilde{V}_v)$, such that, as operators on $\tilde{V}_v$, we have*

$$\delta = [\delta] + [Q, -]$$

This means that for every $a \in A(\sigma)$, considered as an operator $\tilde{\rho}(a) \in End_{C(n)_v}(\tilde{V}_v) \simeq M_n(C(n)_v)$, $\delta(a)$ acts formally on $\tilde{V}_v$ as

$$\tilde{\rho}(\delta(a)) = [\delta](\tilde{\rho}(a)) + [Q, \tilde{\rho}(a)]$$

Notice that, to make visible the action of $[\delta]$ on $\rho(a)$, we have included the obvious isomorphism, $End_{C(n)}(\tilde{V}) \simeq M_n(C(n))$, which will, in the sequel, be used without further warning.

There are local (and even global) extensions of this result, where $[\gamma]$ and Q may be assumed to be defined (rationally) on $C(n)$, see [5].

In line with our general philosophy, we shall therefore consider $[\delta]$ as measuring *time* in $Simp_n A(\sigma))$, respectively in $Spec(C(n))$. Assume for a while that $k = \mathbf{R}$, the real numbers, and that our constructions go through, as if k were algebraically closed. Let $v \in Simp_n(A(\sigma))$

Journal of Physics: Conference Series **532** (2014) 012013 doi:10.1088/1742-6596/532/1/012013

be an element, an *event*. Suppose there exist an integral curve $\mathbf{c}$ of $[\delta]$ through some element $v(\tau_0) \in Simp_1(C(n))$, equivalent to v, ending at $v(\tau_1) \in Simp_1(C(n))$, given by the automorphisms $e(\tau) := exp(\tau[\delta])$, for $\tau \in [\tau_0, \tau_1] \subset \mathbf{R}$. The supremum of τ for which the corresponding point, $v(\tau)$, of $\mathbf{c}$ is in $Simp_n(A(\sigma))$ should be called the *lifetime* of the *particle* $v(\tau)$. It is relatively easy to compute these lifetimes, and so be able to talk about *decay*, when the fundamental vector field $[\delta]$ has been computed. In [5], we have also proposed a mathematically sound way of treating interaction, purely in terms of non-commutative deformation theory.

Let $\phi(\tau_0) \in \tilde{V}(v_0) \simeq V$ be a (classically considered) state of our *quantum system*, at the time τ_0, and consider the (uni)versal family $\tilde{\rho} : A(\sigma) \longrightarrow End_{C(n)}(\tilde{V})$ restricted to $U(n) \subseteq Simp_1(C(n))$, the étale covering of $Simp_n(A(\sigma))$. We shall consider $A(\sigma)$ as our *ring of observables*. What happens to $\phi(\tau_0) \in V(0)$ when *time* passes from τ_0 to τ, along $\mathbf{c}$? This leads to the problem of finding a solution of the Schrödinger equation, *along* $\mathbf{c}$, given by the next result, proving that ϕ is completely determined, by the value of $\phi(\tau_0)$, for any $\tau_0 \in \mathbf{c}$. Here, we shall not go into the problem of *preparing* $\psi(\tau_0) \in V(\tau_0)$, i.e of how to exactly determin *where we are*, at some chosen clock-time, τ, see [5].

Theorem 3.4. *The* evolution operator $u(\tau_0, \tau_1)$ *that changes the state* $\phi(\tau_0) \in \tilde{V}(v_0)$ *into the state* $\phi(\tau_1) \in \tilde{V}(v_1)$, *where* $\tau_1 - \tau_0$ *is the length of the integral curve* $\mathbf{c}$ *connecting the two points* v_0 *and* v_1, *i.e the* time passed, *is given by*

$$\phi(\tau_1) = u(\tau_0, \tau_1)(\phi(\tau_0)) = \exp\left[\int_{\mathbf{c}} Q(\tau)d\tau\right](\phi(\tau_0))$$

where $\exp \int_{\mathbf{c}}$ *is the non-commutative version of the ordinary action integral, essentially defined by the equation*

$$\exp\left[\int_{\mathbf{c}} Q(\tau)dt\right] = \exp\left[\int_{\mathbf{c}_2} Q(\tau)d\tau\right] \circ \exp\left[\int_{\mathbf{c}_1} Q(\tau)d\tau\right]$$

where $\mathbf{c}$ *is* $\mathbf{c}_1$ *followed by* $\mathbf{c}_2$.

3.2. The generic dynamical structures associated to a metric

Now, let $C = k[t_1, \ldots, t_n]$, $k = \mathbf{R}$, the real numbers, be a commutative polynomial k-algebra, and let

$$g = \frac{1}{2} \sum_{i,j=1,..,r} g_{i,j} dt_i dt_j \quad \in Ph(C)$$

be a Riemannian metric. Recall the formula for the Levi-Civita connection

$$\sum_l g_{l,k}\Gamma^l_{j,i} = \frac{1}{2}\left(\frac{\partial g_{k,i}}{\partial t_j} + \frac{\partial g_{j,k}}{\partial t_i} - \frac{\partial g_{i,j}}{\partial t_k}\right)$$

Since in $Ph^\infty(C)$, we have

$$\delta(g) = \frac{1}{2} \sum_{i,j,k=1,..,r} \frac{\partial g_{i,j}}{\partial t_k} dt_k dt_i dt_j + \frac{1}{2} \sum_{i,j,=1,..,r} g_{i,j}(d^2t_i dt_j + dt_i d^2 t_j)$$

and we may plug in the formula

$$\delta^2 t_l = -\Gamma^l := -\sum \Gamma^l_{i,j} dt_i dt_j$$

on the right hand side, and see that we have got a solution of the Lagrange equation $\delta(g) = 0$ in the the commutativization of $Ph(C)$, i.e in the classical phase space. This solution has the form of a *force law*

$$d^2t_l = -\Gamma^l := -\sum \Gamma^l_{i,j} dt_i dt_j$$

generating a dynamical structure $(\sigma) := (\sigma(g))$ of order 2. The dynamical system is the algebra $\mathbf{C}(\sigma) = k[\underline{t}, \underline{\xi}]$, where ξ_j is the class of dt_j. The Dirac derivation now takes the form

$$\delta = \sum_l \left(\xi_l \frac{\partial}{\partial t_l} - \Gamma^l \frac{\partial}{\partial \xi_l} \right)$$

coinciding with the fundamental vector field $[\delta]$ in $Simp_1(\mathbf{C}(\sigma)) = Spec(k[t_i, \xi_j])$.

The equation $[\delta](g) = 0$ imply that g is constant along the integral curves of $[\delta]$ in $Simp_1(Ph(C))$, and these integral curves projects into $Simp_1(C)$ to give the geodesics of the metric g with equations

$$\ddot{t}_l = -\sum_{i,j} \Gamma^l_{i,j} \dot{t}_i \dot{t}_j$$

We may also look at this from another point of view. Suppose given any dynamical structure with Dirac derivation δ on $Ph(C)$. Consider $Simp_1(Ph(C))$. It is obviously represented by $C(1) := k[\underline{t}, \underline{\xi}]$, and the Dirac derivation induces a derivation $[\delta] \in Der_k(C(1))$, and the Hamiltonian must vanish. Therefore we have two options for *the same notion of time* in the picture, g and $[\delta]$. The last derivation should therefore be a *Killing vector field*, i.e we must have a solution of Lagrange equation $[\delta](g) = 0$ and we are left with the above solution for δ.

3.3. Connections as representations

As a contrast to this 1-dimensional representation case, let us consider infinite dimensional representations, for which the Dirac derivation $[\delta]$ vanishes, and the notion of time is taken care of by the Hamiltonian Q.

A non-degenerate metric $g = 1/2 \sum_{i=1}^d g_{i,j} dt_i dt_j \in Ph(C)$ induces an isomorphism of C-modules $\Theta_C = Hom_C(\Omega_C, C) \simeq \Omega_C$. Consider the bilateral ideal (σ_g) of $Ph(C)$ generated by

$$(\sigma_g) = ([dt_i, t_j] - g^{i,j})$$

and put $C(\sigma_g) := Ph(C)/(\sigma_g)$. Let, moreover,

$$T := -\frac{1}{2} \left(\sum_{k,l} \Gamma^k_{k,l} dt_l + \sum_{k,p,q} g^{k,q} \Gamma^p_{k,q} g_{p,l} dt_l \right)$$

and consider the inner derivation of $C(\sigma_g)$, defined by $\delta := ad(g - T)$. After a dull computation, we obtain, see [5], that in $C(\sigma_g)$, $\delta(t_i) = dt_i, i = 1, \dots, d$. Therefore we have a well-defined dynamical structure (σ_g), with Dirac derivation, $\delta = ad(g - T)$. In partiucular, there is a unique homomorphism, $Ph^\infty(C) \to C(\sigma_g)$, consistent with the Dirac derivation δ and $ad(g - T)$. It is also easy to see that (σ_g) is invariant w.r.t. isometries.

Example 3.5. The trivial metric, $g_0 = 1/2 \sum_{i=1}^d dt_i^2$ will give us the dynamic structure, $[dt_j, t_i] = \delta_{i,j}$, with the dynamic system $C(\sigma_{g_0}) = W(d)$, the Weyl space. Clearly $[g_0, t_i] = dt_i$, and, of course, $d^2t_i = [g_0, dt_i] = 0$ for all $i = 1, \dots, d$, so that this gives us the classical set-up for Quantum Mechanics.

Journal of Physics: Conference Series **532** (2014) 012013 doi:10.1088/1742-6596/532/1/012013

Any connection $\nabla : \Theta_C \to End_k(E)$, on a free C-module E is given in terms of the operators,

$$\delta_{t_i} := \frac{\partial}{\partial t_i}, \quad \nabla_{\delta_{t_i}} = \delta_{t_i} + \nabla_i$$

where $\nabla_i \in End_C(E)$ is now a representation

$$\rho_\nabla : C(\sigma_g) \to End_k(E)$$

mapping t_i to the obvious operator and mapping dt_i to $\sum_{j=1}^{d} g^{i,j} \nabla_{\delta_j}$. Put

$$\xi_i := \delta^i = \sum_{j=1}^{d} g^{i,j} \delta_j.$$

Conversaly, every C-representation $\rho : C(\sigma_g) \to End_k(E)$ will map the element, dt_i to a differential operator ξ_i, with $[\xi_i, t_j] = g^{i,j}$, i.e we have a connection on E.

Notice that the representation $\rho = \rho_\Theta$ of $C(\sigma_g)$ defined on Θ_C by the Levi-Civita connection has an Hamiltonian

$$Q := \rho(g - T) = \frac{1}{2} \sum_{i,j} g^{ij} \nabla_{\delta_i} \nabla_{\delta_j}$$

i.e the generalized Laplace-Beltrami operator, which is also invariant w.r.t. isometries, although the proof demands some algebra.

Notice also that for the Levi-Civita connection, we shall use the notations

$$D_{t_i} := \nabla_{\delta_{t_i}}, \quad D(\xi) := \nabla_\xi \quad \text{for } \delta_{t_i}, \quad \xi \in \Theta_{k[\underline{t}]}$$

As above, we now have two options for the notion of time, and time-development. In line with classical Quantum Theory, we would, for *a state* $\xi \in \Theta_C$, consider the Schrödinger equation

$$\frac{d\xi}{d\tau} = Q(\xi)$$

where τ would be an ad hoc chosen time parameter. In our case there are just two natural choices for τ, namely ξ itself, or the metric g, measuring the time t, see the discussion above. Since we have

$$D_\xi(\xi) = \mu \frac{d\xi}{dt}$$

where $\mu = g(\xi, \xi)^{1/2}$, it seems reasonable to replace the classical Schrødinger equation, in our situation, by the following one:

$$\frac{d\xi}{dt} = Q(\xi), \quad \xi \in \Theta_C$$

Notice that the left hand side of the Furniture Equation is

$$\frac{1}{\mu} D_\xi(\xi) = \frac{1}{\mu} \sum_{i,j} \xi_i \xi_j \Gamma^k_{i,j} \delta_l + \frac{1}{\mu} \sum_{i,l} \xi_l \frac{\partial \xi_l}{\partial t_i} \delta_l$$

If we go to our Toy Model, see [5], then $1/\mu \xi_i \xi_j$ would, in the physical interpretation of that model, have been written as $T^{i,j} = 1/E p_i p_j$, where $E := \mu$ is the energy, and (p_i) is the momentum of the *particles* that populate the universe. Classically T is called it the *Mass-stress Tensor*, T_{part}, in the classical relativity theory.

We shall return to this last equation, in the next section, where we, sometimes, will refer to it as the *Furniture Equation*.

But, first, notice the generic equation of formal motion for representations of $C(\sigma_g)$.

Journal of Physics: Conference Series **532** (2014) 012013 doi:10.1088/1742-6596/532/1/012013

Corollary 3.6 (the Generic Equation of Motion). *Consider a $C(\sigma_g)$-representation, $\rho : C(\sigma_g) \to End_k(V)$. Then the time development is, formally, given in terms of a parameter τ as*

$$\tilde{\rho}(\tau) = \rho(\epsilon(\tau)) : C(\sigma_g) \to End_k(V)$$

where

$$\epsilon(\tau) = \exp(\tau \cdot ad(g - T))$$

Now, to be able to handle this time development, we need to know formulas, in $C(\sigma_g)$, for $d^l t_i,\ l \geq 1,\ i = 1, \ldots, n$. To this end, put

$$\bar{\Gamma}^i_{p,q} := \sum_{l,r} g^{r,i}\Gamma^l_{r,p}g_{l,q}, \quad \bar{\nabla}_l := (\bar{\Gamma}^j_{i,l}), \quad T = \sum_l T_l dt_l$$

Then

$$T_l = -1/2(\sum_j (\Gamma^j_{j,l} + \bar{\Gamma}^j_{j,l}) = -1/2(\mathrm{tr}\,\nabla_l + \mathrm{tr}\,\bar{\nabla}_l),$$

$$\delta^2 t_i = [g - T, dt_i] = -1/2\sum_{p,q}(\bar{\Gamma}^i_{p,q} + \bar{\Gamma}^i_{q,p})dt_p dt_q + 1/2\sum_{p,q} g_{p,q}(R_{p,i}dt_q + dt_p R_{q,i}) + [dt_i, T]$$

where, as above, $R_{i,j} = [dt_i, dt_j]$. Put

$$\Gamma^{j,i}_p = \sum_k g^{j,k}\Gamma^i_{k,p}, \quad F_{i,j} := R_{i,j} - \sum_p (\Gamma^{j,i}_p - \Gamma^{i,j}_p)dt_p$$

and recall (see [5], p. 82) that for any connection ρ on a C-bundle E, the corresponding representation of $C(\sigma_g)$ will map $F_{i,j}$ to the ordinary curvature of the connection. In fact,

$$\rho(F_{i,j}) = [\rho(dt_i), \rho(dt_j)] - \sum_p (\Gamma^{j,i}_p - \Gamma^{i,j}_p)\rho(dt_p) = [\nabla_{\xi_i}, \nabla_{\xi_j}] - \nabla_{[\xi_i, \xi_j]}$$

Put, for short,

$$F(\xi_i, \xi_j) = F_{i,j} := \rho(F_{i,j}) \quad \in End_C(E)$$

Computing, we find, see [5], for a proof, the following.

Theorem 3.7 (the Generic Force Law). *In $C(\sigma_g)$ we have the following force laws*

$$(1)\ d^2 t_i = -\,1/2\sum_{p,q}(\bar{\Gamma}^i_{p,q} + \bar{\Gamma}^i_{q,p})dt_p dt_q + 1/2\sum_{p,q} g_{p,q}(R_{p,i}dt_q + dt_p R_{q,i}) + [dt_i, T]$$

$$(2)\ d^2 t_i = -\sum_{p,q}\Gamma^i_{p,q}dt_p dt_q - 1/2\sum_{p,q} g_{p,q}(F_{i,p}dt_q + dt_p F_{i,q})$$
$$+\,1/2\sum_{l,p,q} g_{p,q}[dt_p, (\Gamma^{i,q}_l - \Gamma^{q,i}_l)]dt_l + [dt_i, T]$$

Notice that considering the representation, ρ_Θ, corresponding to the Levi-Civita connection, the above translate into

$$\rho_\Theta(dt_i) = [Q, t_i], \quad \rho(d^2 t_i) = \sum_{j=1}^{d}[Q, \rho(dt_i)]$$

where Q is the Laplace-Beltrami operator.

Given any *observable* $f \in Ph(C)$, we would expect that the dynamics of the future *values* of f, measured via some representation, E, to be the spectrum of the operator

$$F(\tau) := \exp(\tau \cdot ad(Q))(f)$$

Notice again that any representation E of $C(\sigma_g)$, is of course a representation of $Ph^\infty(C)$. However, $Ph^\infty(C)$ is now infinitely generated, and the representation E of $C(\sigma_g)$, is never of finite dimension so we may not use the nice non-commutative algebraic geometry, and its quantum theoretical look-alike without some kind of adaptation, or at least, some kind of approximation, to which we shall return.

Remark 3.8. We shall consider the above formulas as general Force Laws, in $Ph(C)$, induced by the metric g. This means the following: Let $\mathfrak{c}$ be the δ- stable ideal generated by any one of these equation in $Ph^\infty(C)$. Since the force laws above holds in the dynamical system defined by (σ_g), we obviously have $\mathfrak{c} \subset (\sigma_g)$, and we might hope these new dynamical systems might lead to new Quantum Field Theories, as defined above, with equally new and interesting properties.

For a connection ∇ on a free C-module E, the second Force Law above will now take the form, in $End_C(E)$,

$$\begin{aligned}\rho_E(d^2t_i) + \sum_{p,q} \Gamma^i_{p,q} \nabla_{\xi_p} \nabla_{\xi_q} &= 1/2 \sum_p F_{p,i} \nabla_{\delta_p} + 1/2 \sum_p \nabla_{\delta_p} F_{p,i} \\ &+ 1/2 \sum_{l,q} \delta_q (\Gamma^{i,q}_l - \Gamma^{q,i}_l) \nabla_{\xi_l} + [\nabla_{\xi_i}, \rho_E(T)]\end{aligned}$$

where $\rho(dt_i) = \sum_j g^{i,j} \nabla_{\delta_j} =: \xi_i$.

4. Entropy

Consider an algebraic geometric object X, and let $aut(X)$ be the Lie algebra of automorphisms of X. The sub-Lie algebra $aut_0(X)$ that lifts to automorphisms of the formal moduli of X, is a Lie ideal. Put $\mathfrak{a}(X) := aut(X)/aut_0(X)$, then if $X(t)$ is a deformation of X along a parameter t, we find $dim_k \mathfrak{a}(X(t)) \leq dim_k \mathfrak{a}(X)$. One may phrase this saying that an object X can never gain *information* when deformed. Moreover, deformation is, obviously, not a reversible process, so information can get lost.

4.1. The classical commutative case

In [6], studying moduli problems of singularities in (classical) algebraic geometry, we were led to consider the notion of *Modular Suite*. This is a canonical partition $\{\mathbf{M}_\alpha\}$, of the versal base space, $\mathbf{M}$, of the deformation functor of an algebraic object, X. The different *rooms*, $\mathbf{M}_\alpha$, correspond to the subsets of equivalence classes of deformations in $\mathbf{M}$, along which the Lie algebra $\mathfrak{a} := aut/aut_0$ deforms as Lie-algebras, and therefore conserves its dimension. Working with Thermodynamics, see [Laudal O A *Cosmos and its Furniture* (Worls Sci.) to appear], it occurred to me that the notion of entropy has an interesting parallel in deformation theory. In fact I have proposed the following,

Definition 4.1. Let $X(\underline{t})$ corresponds to the point $\underline{t} \in \mathbf{M}_\alpha$, then we shall term *Entropy* of the *state* $\underline{t}$, the integer $S(\underline{t}) := dim_k(\mathbf{M}_\alpha)$.

In this classical situation, assuming that the field is algebraically closed, and that $\mathbf{M}$ is of finite Krull dimension, the modular suite $\{\mathbf{M}_\alpha\}$ is finite, with an inner room, the *modular substratum* and an *ambiant* (open) *maximal entropy* stratum. But the structure of the modular suite may be very complex, even for simple singularities X, see the example of the quasi homogenous plane curve singularity $x_1^5 + x_2^{11}$, in [6]. It is also clear that for any *algebraic dynamics* in $\mathbf{M}$, the entropy will always stay or grow, see again [6]. To be able to construct situations where the entropy is lowered, or the information goes up, we must leave classical algebraic geometry, and venture into non-commutative algebraic geometry. Here is where non-commutative deformation theory comes into play.

4.2. The general case
In the general situation, where our algebras of *observables* are associative but not necessarily commutative, the first interesting cases are deformations of associative algebras, A, or deformations of finite families of representations V_i of an associative algebra A. In [Laudal O A *Cosmos and its Furniture* (World Sci.) to appear] I will touch upon the first case. Here we shall look at the second case, based upon the technique of the previous sections.

There we worked with a polynomial algebra, $C = k[t_1, \dots, t_d]$, with a metric $g \in Ph(C)$, and the Levi-Civita connection, considered as a representation

$$A := Ph(C)/(\sigma_g) \to End_k(\Theta_C)$$

The Dirac derivation, [δ], in this case, vanish and the corresponding Hamiltonian turned out to be the Laplace-Beltrami operator, $Q \in End_k(\Theta_C)$. We were then, in analogy with Quantum theory, led to consider, for every *state* $\xi \in \Theta_C$, the time development, or Schrödinger equation, and the corresponding *Furniture Equation* [Laudal O A *Cosmos and its Furniture* (Worls Sci.) to appear] reads

$$\frac{d\xi}{dt} = Q(\xi)$$

We have shown in [Laudal O A *Cosmos and its Furniture* (Worls Sci.) to appear] that in the special case of our *toy model*, see [5] and *Cosmos and its Furniture*, this equation amounts to a combined Heat and Navier-Stokes equations. The corresponding notion of entropy, in the above sense, might be defined by the modular suite of the versal base of the deformation functor of the corresponding representation

$$\xi : Ph(C) \to C = End_C(C)$$

The formal moduli has a tangent space given by $Ext^1_{Ph(C)}(\xi, \xi)$, which is given by the classical long exact sequence

$$0 \to Hom_{Ph(C)}(\xi,\xi) \to Hom_C(\xi,\xi) \xrightarrow{\iota} Der_C(Ph(C), Hom_C(\xi,\xi)) \xrightarrow{\kappa} Ext^1_{Ph(C)}(\xi,\xi) \to 0$$

Here, $\iota = 0$ and $Hom_C(\xi, \xi) = C$, so that

$$Ext^1_{Ph(C)}(\xi,\xi) = C^d \simeq \Theta_C$$

as all relations $[dt_i, t_j] + [t_i, dt_j]$ in $Ph(C)$ are mapped to 0 by any C-derivation into C.

The miniversal base space of ξ, is therefore of infinite dimension, and since the Lie algebra of the automorphism group of C is the Lie algebra $Der_k(C) = \Theta_C$, the Lie algebra of the automorphism group of the representation ξ is

$$aut(\xi) := \{\eta \in \Theta_C | [\eta, \xi] = 0\}$$

We should now define the entropy of the state ξ as

$$S(\xi) := \dim\{\eta | aut(\xi) \vdash aut(\eta)\}$$

where $aut(\xi) \vdash aut(\eta)$ should mean that η is a deformation of ξ inducing a deformation of the Lie algebras of automorphisms. Obviously this is unrealistic, as most of the terms involved will be of infinite dimension.

We may try to overcome this difficulty by approximating the state ξ, as a representation, by a finite dimensional representation, defined for every point set object $\mathcal{P} = \{P_p\}$ of $Simp_1(C)$ by $\xi_{\mathcal{P}} : Ph(C) \to End_k(\mathcal{P})$ where

$$\xi_{\mathcal{P}}(t_i) = \alpha_i^0(p), \quad \xi_{\mathcal{P}}(dt_i) = \alpha_i^1(p)$$

where we, in anticipation of the treatment via the technique of the next section, have put

$$\alpha_i^0(p) = P_{p,i}, \quad \alpha_i^1(p) = \xi(P_p)_i$$

We may look at the object $\mathcal{P}$ as the set of molecules in our observatory (or in the Universe), and ξ as the combined state of these, at the outset maybe considered independently. Extending the representation $\xi_{\mathcal{P}}$ to a representation $\xi_{\mathcal{P}} : Ph^{\infty}(C) \to End_k(\mathcal{P})$, or cutting it down to a representation of some dynamical system,

$$\xi_{\mathcal{P},\sigma} : A(\sigma) := Ph^{\infty}(C)/(\sigma) \to End_k(\mathcal{P})$$

for some reasonable dynamical structure (σ) of $Ph^{\infty}(C)$, we can now use the technique of section (3), and look at the versal family $\mathbf{M}$ of the representation $\xi_{\mathcal{P},\sigma}$ for some fundamental *state* ξ, of the *Furniture* of the Universe. There will be a canonically defined *Moduli Suite* $\{\mathbf{M}_\alpha\}$ and for any deformation η of $\xi_{\mathcal{P},\sigma}$, there will be one α such that $\eta \in \mathbf{M}_\alpha$ and the resulting definition of the entropy would be $S(\eta) = \dim(\mathbf{M}_\alpha)$. The goal is to show that for reasonable classical cases, this should come out close to the Boltzmann's definition

$$S(\eta) := \log Vol(M_\alpha), \quad \eta \in M_\alpha$$

where M_α is the substratum of the corresponding *Coarse Graining* of the classical phase space of the situation, containing the *distribution* η, see again the very readable text of [9].

4.3. Representations of Ph^{∞}

Now let $A = k[t_1, \ldots, t_d]$, and consider a representation of $Ph^{\infty}(A)$ as a k-algebra,

$$\rho : Ph^{\infty}(A) \to End_k(V)$$

where V a k-vector space. Put

$$D_i^0 := \rho(t_i), \quad D_i^p := \rho(d^p t_i), \quad p \geq 1$$

The composition of $\exp(\tau\delta)$ and ρ is a homomorphisms of k-algebras

$$Ph^{\infty}(A) \overset{\rho[\tau]}{\to} End_k(V) \otimes_k k[[\tau]]$$

for which we have

$$X_i := \rho(\tau)(t_i) = \rho(\exp(\tau\delta)) = \sum_{p\geq 0} \tau^p/p! D_i^p$$

Since $[t_i, t_j] = 0$, we must have $[X_i, X_j] = 0$ and, since the relations in $Ph^{\infty}(A)$ are given by

$$\sum_{p+q=n\geq 0} \frac{1}{p!q!} [d^p t_i, d^q t_j] = 0$$

This is the condition:

$$\sum_{p+q=n\geq 0} \frac{1}{p!q!}\left[D_i^p, D_j^q\right] = 0$$

for the family of matrices $\{D_i^p,\ p \geq 0,\ i = 1, \ldots, d\}$ to define a homomorphism ρ of k-algebras.

Clearly if $\dim V = 1$ there are no conditions, and we may pick arbitrarily $D_i^p \in k$, and obtain formal power series,

$$X_i = \sum_n \frac{\tau^n}{n!} D_i^n$$

which, when convergent, gives the dynamics of the point.

Example 4.2. Assume $\dim_k V = 2$ and put

$$\rho(t_i) = D_i^0 = \begin{pmatrix} x_i(1) & 0 \\ 0 & x_i(2) \end{pmatrix} =: \begin{pmatrix} \alpha_i^0(1) & 0 \\ 0 & \alpha_i^0(2) \end{pmatrix}$$

and $\alpha_i^0(r,s) := x_i(r) - x_i(s),\ r,s = 1,2$. Let

$$D_i^q = \begin{pmatrix} \alpha_i^q(1) & r_i^q(1,2) \\ r_i^q(2,1) & \alpha_i^q(2) \end{pmatrix}, \quad q \geq 0$$

Put

$$\alpha_i^l(r,s) := \alpha_i^l(r) - \alpha_i^l(s), \quad r_i^k(r,s) = \sum_{l=0}^{k} \binom{k}{l} \sigma_{k-l}(r,s)\alpha_i^l(r,s), \qquad r,s = 1,2$$

where the sequence $\{\sigma_l(r,s)\},\ l = 0,1,\ldots$ is a sequence of arbitrary *coupling constants* with $\sigma_0(r,s) = 0$. Then $\rho(d^n t_i) := D_i^n$ defines a representation $\rho : Ph^\infty(A) \to End_k(V)$, if and only if

$$\sum_{p+q=n\geq 0} \frac{1}{p!q!}\left[D_i^p, D_j^q\right] = 0$$

which is exactly when

$$\sum_{p+q=n\geq 0} \frac{1}{p!q!}\left(\alpha_i^p(r,s) r_j^q(s,r) - r_i^p(r,s)\alpha_j^q(s,r)\right) = 0$$

Computing, we find the condition

$$\sum_{p+q=n\geq 0} \frac{1}{p!q!}\sigma_l(r,s)\left(\alpha_i^p(r,s)\alpha_j^{q-l}(s,r) - \alpha_i^{q-l}(r,s)\alpha_j^p(s,r)\right) = 0$$

for $r,s = 1,2$ and $l \geq 1$. The situation above arises when we consider two (different) points

$$P_1 = \left(\alpha_1^0(1), \ldots, \alpha_d^0(1)\right), \quad P_2 = \left(\alpha_1^0(2), \ldots, \alpha_d^0(2)\right)$$

in space, with pre-described tangents, $\xi_1 = (\alpha_1^1(1), \ldots, \alpha_d^1(1))$ and $\xi_2 = (\alpha_1^1(2), \ldots, \alpha_d^1(2))$. For these two points, considered as dimension 1 representations of $Ph(A)$, we saw that there is a

1-dimensional space of *tangents* between the points, i.e the $Ext^1_{Ph(A)}(k(P_1), k(P_2)) = k$. This leads to possibly non-zero elements $r_i^1(1,2)$, $r_i^1(2,1))$ in the matrix representation of the non-commutative deformation of the family $\{k(P_1), k(P_2)\}$ of $Ph(A)$-modules.

We now have a much more complete picture of the situation. The dynamics of the pair of points is described by the Dirac derivation. Assuming that for the time-moment $\tau = 0$ we know the position $\alpha^0(1), \alpha^0(2)$ and the momenta $\alpha^1(1), \alpha^1(2)$ of the two points, then the dynamics is described, in terms of the time τ by the matrices $X_i = \rho(exp(\tau\delta))(t_i)$. Putting

$$\alpha_i(r,s) = \sum_{n=0}^{\infty} \frac{\tau^n}{n!}\alpha_i^n(r,s), \quad \sigma(r,s) = \sum_{n=0}^{\infty} \frac{\tau^n}{n!}\sigma_n(r,s)$$

we find the explicit formulas

$$X_i = \begin{pmatrix} \alpha_i(1) & \sigma(1,2)\alpha_i(1,2) \\ \sigma(2,1)\alpha_i(2,1) & \alpha_i(2) \end{pmatrix}, \quad i = 1,\dots,d$$

The trace, and determinant are

$$\mathrm{tr}(X_i) = (\alpha_i(1) + \alpha_i(2)), \quad \det(X_i) = (\alpha_i(1)\alpha_i(2)) - \sigma(1,2))\alpha_i(1,2)\sigma(2,1))\alpha_i(2,1)$$

The *spectrum* of X_i, or the eigenvalues, are given as

$$X_i(r) = \frac{1}{2}\left(\mathrm{tr}(X_i) \pm \sqrt{\mathrm{tr}(X_i)^2 - 4\det(X_i)}\right), \quad r = 1,2$$

From this we see that if all coupling constants vanish, i.e if $\sigma(r,s) = 0$ $(r,s = 1,2)$ then we have undisturbed independent motions of the two points, $X_i(r) = \alpha_i(r)$, $r = 1,2$. If the coupling constants are nonzero the representation becomes simple, with trivial automorphism group, and so, according to the definition, of maximal entropy.

Moreover, assume $\sigma_l(1,2) = \sigma_l(2,1) = 0$ for $l = 0$ and $l \geq 2$, then the conditions above becomes for $r,s = 1,2$ and for all $n \geq 0$,

$$\sigma_l(r,s) \cdot \sum_{p+q=n} \frac{1}{p!q!}\left(\alpha_i^p(r,s)\alpha_j^{q-1}(s,r) - \alpha_i^{q-1}(r,s)\alpha_j^p(s,r)\right) = 0$$

If $d = 3$ this simplifies to,

$$\sigma_1(1,2) \cdot \sum_{p+q=n} \frac{1}{p!q!}(\alpha^p(1,2) \times \alpha^{q-1}(2,1)) = 0$$

where, of course

$$\alpha^m(r,s) = -\alpha^m(s,r) \quad r,s = 1,2, \quad m \geq 0$$

In general, considering an *Object*, $\mathcal{P}$ in space, consisting of r points $\{P_p\}_{p=1,\dots,r}$, we find that the dynamics is closely related to the interaction process, involving non-commutative deformation of families of representations, described in [5]. In fact, we easily obtain the following.

Theorem 4.3. *Given an* object, $\mathcal{P}$ *consisting of* r *points* $\{P_p\}_{p=1,\dots,r}$ *in* d*-space,* $Simp(A)$*, where* $A = k[t_1,\dots,t_d]$*. With the notations above, in particular,* $\alpha_i^n(p,q) = \alpha_i^n(p) - \alpha_i^n(q)$*,* $i = 1,\dots,d$*, consider the matrix*

$$D_i^n := \begin{pmatrix} \alpha_i^n(1) & r_i^n(1,2) & \dots & r_i^n(1,r) \\ r_i^n(2,1) & \alpha_i^n(2) & \dots & r_i^n(2,r) \\ \vdots & \vdots & \ddots & \vdots \\ r_i^n(r,1) & r_i^n(r,2) & \dots & \alpha_i^n(r)) \end{pmatrix}$$

Journal of Physics: Conference Series **532** (2014) 012013 doi:10.1088/1742-6596/532/1/012013

with

$$r_i^0(p,q)=0,\quad r_i^n(p,q)=\sum_{l=0}^{n}\binom{n}{l}\alpha_i^l(p,q)\sigma_{n-l}(p,q)$$

where $\sigma_m(p,q)\in k$ *are arbitrary* coupling constants, *with* $\sigma_0=0$. *Then these operators define a representation* $\rho: Ph^\infty(A)\to M_r(k)$ *with* $\rho(d^n t_i)=D_i^n$ *if and only if, for all* $n\geq 1$,

$$\sum_h\binom{n}{h}\sigma_{n-h}(p,q)\left(\alpha_i^h(p,q)\alpha_j^0(p,q)-\alpha_i^0(p,q)\alpha_j^h(p,q)\right)$$
$$=\sum_{k,l,m,s}\frac{n!\sigma_{n-k-m}(p,s)\sigma_{k-l}(s,q)}{l!m!(k-l)!(n-k-m)!}\left(\alpha_j^m(p,s)\alpha_i^l(s,q)-\alpha_i^l(p,s)\alpha_j^m(s,q)\right)$$

Proof. As above, consider the matrix

$$X_i=\rho(\exp(\tau\delta))(t_i)=\sum_{n\geq 0}\frac{\tau^n}{n!}D_i^n$$

and put

$$\alpha_i(r)=\sum_{n=0}^{\infty}\frac{\tau^n}{n!}\alpha_i^n(r),\quad \alpha_i(r,s)=\sum_{n=0}^{\infty}\frac{\tau^n}{n!}\alpha_i^n(r,s),\quad \sigma(r,s)=\sum_{n=0}^{\infty}\frac{\tau^n}{n!}\sigma_n(r,s)$$

we find the explicit formulas

$$X_i=\begin{pmatrix}\alpha_i(1) & \sigma(1,2)\alpha_i(1,2) & \dots & \sigma(1,r)\alpha_i(1,r)\\ \sigma(2,1)\alpha_i(2,1) & \alpha_i(2) & \dots & \sigma(2,r)\alpha_i(2,r)\\ \vdots & \vdots & \ddots & \vdots\\ \sigma(r,1)\alpha_i(r,1) & \sigma(r,2)\alpha_i(r,2) & \dots & \alpha_i(r)\end{pmatrix},\quad i=1,\dots,d$$

Now compute $[X_i,X_j]=0$ and see that the condition of the theorem emerges. □

Remark 4.4. We may consider the *space*

$$\mathbf{A}(r)=k[\alpha_i^n(p),\sigma_n(p,q)]/\mathfrak{a}$$

with coordinates

$$\{\alpha_i^n(p),\sigma_n(p,q),i=1,\dots,d,\ p,q=1,\dots,r,\ n\geq 0,\ \sigma_0(p,q)=0\}$$

and where the ideal $\mathfrak{a}$ is generated by the equations above, as the versal base space for the versal family of the non-commutative deformation theory applied to the family of $Ph^\infty(A)$ modules defined by the object $\mathcal{P}$, and with Dirac derivation, δ acting as, $[\delta]\left(\alpha_i^n(p)\right)=\alpha_i^{n+1}(p)$.

Since $Ph^\infty(A)$ is infinitely generated, there is, strictly speaking, no such thing, but we shall see that in special cases, we can overcome this difficulty by finding clever dynamical structures.

In (4.2) we saw that to overcome another difficulty related to non-existing moduli spaces, we opted for approximating a state $\xi\in\Theta_C$ by a finite dimensional representation, defined for every finite point set object $\mathcal{P}=\{P_p\}$ of $Simp_1(C)$ by $\xi_{\mathcal{P}}: Ph(C)\to End_k(\mathcal{P})$, where

$$\xi_{\mathcal{P}}(t_i)=\alpha_i^0(p),\quad \xi_{\mathcal{P}}(dt_i)=\alpha_i^1(p)$$

We wanted to look at the object $\mathcal{P}$ as the set of molecules in our observatory (or in the Universe), and at $\xi_{\mathcal{P}}$ as the combined state of these.

Extending the representation $\xi_{\mathcal{P}}$ to a representation $\xi_{\mathcal{P}} : Ph^{\infty}(C) \to End_k(\mathcal{P})$ or cutting it down to a representation of some dynamical system, $A(\sigma) := Ph^{\infty}(C)/(\sigma)$ $\xi_{\mathcal{P},\sigma} : A(\sigma) \to End_k(\mathcal{P})$ for some reasonable dynamical structure (σ) we might be able to compute the versal family $\mathbf{M}$ of the representation $\xi_{\mathcal{P},\sigma}$ for some fundamental *state* ξ, of the *Furniture* of the Universe, and also the *Moduli Suite*, $\{\mathbf{M}_{\alpha}\}$.

For any deformation η of $\xi_{\mathcal{P},\sigma}$ represented by an equivalence class $\tilde{\eta}$ in the versal base space $\mathbf{M}$ there will be one α such that $\tilde{\eta} \subset \mathbf{M}_{\alpha}$ and the entropy of η would be $S(\eta) = dim(\mathbf{M}_{\alpha})$.

From the classification above, it follows that the minimal entropy would correspond to all $\sigma_l(p,q) = 0$, i.e to some *dust-like* furniture of the space. And the maximal entropy would correspond to all $\sigma_l(p,q) \neq 0$, i.e to some *black hole-like* object $\mathcal{P}$ containing a huge number of gravitational and/or other parameters.

The goal is still to show that, for reasonable classical cases, this should come out close to the Boltzmann's definition

$$S(\eta) := \log Vol(M_{\alpha}), \quad \eta \in M_{\alpha}$$

where M_{α} is the substratum of the corresponding *Coarse Graining* of the classical phase space of the situation, containing the *distribution* η, see again [9], chapter 2. In particular, consider Penrose's need (for) some clear-cut way of saying that "the gravitational degrees of freedom were not activated", and his need to identify the mathematical quantity that actually measures "gravitational degrees of freedom". I suggest that the above, coupled with my "Toy Model", see [5], i.e working on $Hilb^{(2)}(\mathbf{A}^3)$ instead of the trivial affine space $\mathbf{A}^d = Spec(k[t_1, \ldots, t_d])$, may be of interest for this quest.

We have already looked at the case $r = 2$ and seen that the result makes physical sense. For $r = 3$ and $n = 2$ we find

$$\sigma_1(p,q)\big(\alpha_i^1(p,q)\alpha_j^0(p,q) - \alpha_i^0(p,q)\alpha_j^1(p,q)\big) = \sigma_1(p,s)\sigma_1(s,q)\big(\alpha_j^0(p,s)\alpha_i^0(s,q) - \alpha_i^0(p,s)\alpha_j^0(s,q)\big)$$

In dimension $d = 3$ this has a particularly nice interpretation. Let $\alpha^0(i,j)$ be the vector starting at P_i and ending at P_j, and let ξ_i be a tangent vector at P_i for $i = 1,2,3$. Put $\alpha^1(i,j) = \xi_i - \xi_j$, then the condition above reads

$$\sigma_1(p,q)\left(\alpha^1(p,q) \times \alpha^0(p,q)\right) = -\sigma_1(p,s)\sigma_1(s,q)\left(\alpha^0(p,s) \times \alpha^0(s,q)\right), \quad \forall p,s,q = 1,2,3$$

This says that for any two of the three points in space, the relative momentum must sit in the plane defined by the three points, the length being determined by the 3 coupling constants. Moreover, the sum of all three relative momenta must be 0. In fact, there are coefficients, $u, v \in k$ such that

$$\alpha^1(p,q) = u\alpha^0(p,s) + v\alpha^0(s,q)$$

Put this into the left side of the formula above, and find

$$\sigma_1(p,q)(v-u) = \sigma_1(p,s)\sigma_1(s,q)$$

assuming that the three points are not co-linear.

Assume now that the coupling constants are given as

$$\sigma_1(p,q) := m(p,q)|\alpha^0(p,q)|^{-2}$$

Then after some computation we find the following differential equations for the dynamics of the 3 points

$$\alpha^1(p,q) = -m(q,p)|\alpha^0(p,q)|^{-1}\epsilon(p,q) - m(p,s)|\alpha^0(p,s)|^{-1}\epsilon(p,s) + \rho\alpha^0(p,q)$$
$$\alpha^1(q,s) = -m(s,q)|\alpha^0(s,q)|^{-1}\epsilon(q,s) - m(q,p)|\alpha^0(q,p)|^{-1}\epsilon(q,p) + \rho\alpha^0(q,s)$$
$$\alpha^1(s,p) = -m(p,s)|\alpha^0(p,s)|^{-1}\epsilon(s,p) - m(s,q)|\alpha^0(s,q)|^{-1}\epsilon(s,q) + \rho\alpha^0(s,p)$$

Here $\epsilon(i,j)$ is the unit vector from the point P_i to the point P_j.

Notice that there is a different set-up, related to Grothendieck's generalized differential algebra. For A commutative, consider an A-module E, and an extension of this representation to $\rho : Ph^\infty(A) \to End_k(E)$. We have seen that ρ must be given in terms of operators,

$$D_i^p := \rho(d^p t_i) \quad \in End_k(E)$$

satisfying the conditions

$$\sum_{p+q=n} \frac{1}{p!q!}\left[D_i^p D_j^q\right] = 0, \quad \forall n \geq 0$$

There is an obvious family of solutions of these equations, given by any differential operator $Q \in Diff(E)$, with

$$D_i^0 = \rho(t_i) = t_i, \quad D_i^p := ad(Q)^p(t_i) \quad \in Diff(E), \quad p \geq 1$$

Looking at the case of finite dimensional representations, treated above, one sees the difference between the two set-ups, and the much greater generality obtained by considering the representations of Ph^∞, the way we do.

References

[1] Laudal O A 2002 *Homology, Homotopy, Appl.* **4** 357
[2] Laudal O A 2003 *Revista Matematica Iberoamericana* **9** 1
[3] Laudal O A 2005 Time-space and space-times *Noncommutative Geometry and Representatioon Theory in Mathematical Physics* eds J Fuchs *et al Contemp. Math.* **391** p 249
[4] Laudal O A 2008 *Acta Appl. Math.* **101** 191
[5] Laudal O A 2011 *Geometry of Time-Spaces* (World Sci.)
[6] Laudal O A and Pfister G 1988 *Local Moduli and Singularities* Lecture Notes in Math. **1310** (Springer-Verlag)
[7] Procesi C 1967 *Atti Accad. Naz. Lincei Rend. Cl. Sci. Fis. Mat. Natur.* **8** 239
[8] Procesi C 1973 *Rings with Polynomial Identities* (Marcel Dekker)
[9] Penrose R 2011 *What Came Before the Big Bang. Cycles of Time* (London: Vintage Books)

Journal of Physics: Conference Series **532** (2014) 012014 doi:10.1088/1742-6596/532/1/012014

On determinability of some classes of medial quasigroups by their endomorphisms

A Leibak and P Puusemp

Matemaatikainstituut, Tallinna Tehnikaülikool, Ehitajate tee 5, Tallinn 19086, Estonia

Abstract. We study the endomorphisms of idempotent medial quasigroups and determinability of some classes of medial quasigroups by their endomorphisms. We introduce the endomorphism algebra of idempotent medial quasigroup and prove that if the endomorphism algebras of quasigroups are isomorphic, then the corresponding quasigroups are isomorphic as well. In addition, we present a counterexample to demonstrate that the endomorphism algebras can not be replaced by the endomorphism monoids.

1. Introduction

Idempotent *medial* quasigroups [2] are distributive quasigroups with a transitive group of *middle* transformations. Recall that a *middle transformation* ϕ of a quasigroup Q is a permutation of Q, if there exists a permutation ϕ^*, such that $\phi(x)\cdot y = x\cdot\phi^*(y)$ for all $x, y \in Q$. The distributive quasigroups are related to the commutative Moufang loops [2]. It is known that the core of an arbitrary Moufang loop is a left distributive quasigroup. Analytic Moufang loops are used to study continuous symmetries [7]. The *left distributive* quasigroups with the identity $x(xy) = y$ are homogenous spaces, in the sense of Loos, see [6]. Also, recently several applications of quasigroups to cryptography (block ciphers, secret sharing schemes etc) were proposed.

In [4] the endomorphisms of commutative idempotent medial quasigroups were studied, therefore it was natural to study bigger class of quasigroups. Since Toyoda's theorem establishes a connection between medial quasigroups and Abelian groups, we restricted ourselves to idempotent medial quasigroups.Moreover, it was natural to ask, what we can get, if we combine Toyoda's theorem with the results known for the determinability of Abelian groups by their endomorphisms.

One can associate the endomorphism monoid $\mathrm{End}(A)$ to every algebraic structure. Depending on the underlying algebraic structure, the endomorphism monoid can be replaced by endomorphism semilattice, ring etc. Therefore it is natural to ask, if A and A' are algebraic structures of same type and that their endomorphism monoids $\mathrm{End}(A)$ and $\mathrm{End}(A')$ (ring etc) are isomorphic, are then A and A' isomorphic? Several mathematicians have studied this problem for several algebraic structures such as semigroups, groups, Abelian groups etc.

Looking the problem of the determinability of groups by their endomorphism semigroups, we know quite much. Then we have especially diverse information on endomorphisms of Abelian groups, because the endomorphisms of an Abelian group form a ring under the composition and the sum, and, therefore, it is possible to use methods of the theory of rings. An excellent overview of the present situation in the theory of endomorphism rings of Abelian groups is given in [3]. In general, for a given group G there exist many groups H such that the semigroups

End(G) and End(H) are isomorphic, and, for a given Abelian group G there exist many Abelian groups H such that their endomorphism rings End(G) and End(H) are isomorphic.

The class of all quasigroups contains the class of all groups, and, therefore, it is hopeless to describe a quasigroup by its endomorphism monoid in the class of all quasigroups. Probably, it is possible in some subclasses of quasigroups. Although, some authors tried to study endomorphisms and endomorphism monoids of quasigroups. Some aspects of automorphisms of quasigroups were studied by Shcherbacov [10]. His work [9] contains many facts on endomorphisms of quasigroups. Several observations on endomorphisms of linear and alinear quasigroups are made by Tabarov [11]. Nagy and Plaumann [5] characterized the quasigroups in which the left deviation, i.e, the map $e(x) = x\backslash x$ is an endomorphism.

We tried to characterize some classes of medial quasigroups by their endomorphism monoids. In [4], it was proved that if the endomorphism monoids of two finite idempotent medial commutative quasigroups are isomorphic, then these quasigroups are isomorphic. In this paper, we give some further results on endomorphism monoids of some classes of medial quasigroups.

2. Endomorphisms of medial quasigroups

Definition 2.1. A magma (groupoid) $\langle Q; \cdot\rangle$ is called a quasigroup, if each of the equations $ax = b$ and $ya = b$ has a unique solution for any $a, b \in Q$.

The solutions of these equations will be denoted by $x = a \backslash b$ and $y = b/a$, respectively. Since the solutions are unique, one can define the binary operations $\backslash$ and $/$ on Q as follows:

$$x = a \backslash b \Leftrightarrow ax = b \quad \text{(left division)}, \quad y = b/a \Leftrightarrow ya = b \quad \text{(right division)} \tag{2.1}$$

Proposition 2.2 ([2]). *If Q is a quasigroup, then the following identities hold*

$$x \backslash (xy) = x \cdot (x \backslash y) = y = (yx)/x = (y/x) \cdot x \tag{2.2}$$

where $\backslash$ and $/$ denote the left and right division, respectively.

It follows that the definition of quasigroup can be given in terms of binary operations and identities.

Definition 2.3. A set Q together with binary operations $\cdot$, $\backslash$, $/$, is called a quasigroup if it satisfies the identities (2.2)

Definitions 2.1 and 2.3 are equivalent [2].

It follows from Proposition 2.2 that the mappings $L_a, R_b \colon Q \to Q$, defined by $L_a(x) = ax$, $R_b(x) = xb$, are bijective. Therefore, to each quasigroup Q one can associate the subgroup Mlt(Q), generated by all mappings L_a and R_b ($a, b \in Q$) in the group of all bijections $Q \to Q$. The group Mlt(Q) is called *a multiplication group* of the quasigroup Q.

Definition 2.4. The mapping $\varphi\colon Q \to Q$ is called an endomorphism of a quasigroup Q, if φ preserves the binary operation $\cdot$, that is $\varphi(x \cdot y) = \varphi(x) \cdot \varphi(y)$ for all $x, y \in Q$.

Applying an endomorphism φ to $y = x \cdot (x \backslash y)$, we obtain $\varphi(y) = \varphi(x \cdot (x \backslash y)) = \varphi(x) \cdot \varphi(x \backslash y)$. Dividing by $\varphi(x)$ from the left, we have that $\varphi(x \backslash y) = \varphi(x) \backslash \varphi(y)$ i.e φ preserves the left division. Similary, one can show that φ preserves right division as well.

We denote by End(Q) and Aut(Q) the sets of all endomorphisms and all automorphisms (i.e, invertible endomorphisms) of a quasigroup Q, respectively. Obviously, the set End(Q) is a monoid with respect to the composition of mappings and Aut(Q) is a subgroup of End(Q).

Definition 2.5. A quasigroup Q is called medial (or entropic) if it satisfies the identity $(xy)(zw) = (xz)(yw)$.

Let Q be a quasigroup and $\varphi, \psi \in \mathrm{End}(Q)$. Let us define the map $\varphi \odot \psi : Q \longrightarrow Q$ as follows:

$$(\varphi \odot \psi)(x) = \varphi(x)\psi(x), \quad x \in Q$$

Endomorphisms φ and ψ are called *summable*, if $\varphi \odot \psi$ is an endomorphism, too. It is easy to check that each two endomorphisms of a medial quasigroup are summable, i.e, the situation is similar to the Abelian groups (recall that, all endomorphisms of Abelian groups are summable). We define *left* and *right division* on $\mathrm{End}(Q)$ elementwise, i.e,

$$(\varphi \oslash \psi)(x) = \varphi(x)/\psi(x), \quad (\varphi \otimes \psi)(x) = \varphi(x) \setminus \psi(x), \quad \forall \varphi, \psi \in \mathrm{End}(Q), \quad x \in Q$$

Proposition 2.6. *If Q is a medial quasigroup, then for all $\varphi, \psi \in \mathrm{End}(Q)$ the mappings $\varphi \oslash \psi$ and $\varphi \otimes \psi$ are endomorphisms of Q.*

Proof. Really, if $x, y \in Q$ and $\varphi, \psi \in \mathrm{End}(Q)$, we have, by (2.2) and mediality,

$$\begin{aligned}(\varphi \oslash \psi)(xy) &= \varphi(xy)/\psi(xy) = (\varphi(x)\varphi(y)) \,/\, (\psi(x)\psi(y)) \\ &= [(\varphi(x)/\psi(x) \cdot \psi(x))\,(\varphi(y)/\psi(y) \cdot \psi(y))] \,/\, (\psi(x)\psi(y)) \\ &= [(\varphi(x)/\psi(x) \cdot \varphi(y)/\psi(y))\,(\psi(x) \cdot \psi(y))] \,/\, (\psi(x)\psi(y)) \\ &= \varphi(x)/\psi(x) \cdot \varphi(y)/\psi(y) = (\varphi \oslash \psi)(x) \cdot (\varphi \oslash \psi)(y), \quad \varphi \oslash \psi \in \mathrm{End}(Q)\end{aligned}$$

Similarly, $\varphi \otimes \psi \in \mathrm{End}(Q)$. □

The connection between Abelian groups and medial quasigroups is given by *Toyoda's theorem* ([2], Theorem 2.10): *If Q is a medial quasigroup, then there exist an Abelian group $\langle Q; +, -, 0\rangle$, its commuting automorphisms φ and ψ, and an element $c \in Q$ such that*

$$x \cdot y = \varphi(x) + \psi(y) + c \tag{2.3}$$

This Abelian group $\langle Q; +, -, 0\rangle$ is called *an underlying Abelian group* of the medial quasigroup Q.

It is well-known that, if G is an Abelian group, then $\mathrm{End}(G)$ is a ring with respect to the addition and composition $\circ$ of endomorphisms. Next we demonstrate that the situation is somehow similar in the case of medial quasigroups.

Proposition 2.7. *Let Q be a medial quasigroup. Then*

(i) $\langle \mathrm{End}(Q); \odot, \oslash, \otimes \rangle$ *is a medial quasigroup,*

(ii) the following (distributivity) identities hold:

$$\begin{aligned}\varphi \circ (\psi \odot \eta) &= (\varphi \circ \psi) \odot (\varphi \circ \eta), & (\psi \odot \eta) \circ \varphi &= (\psi \circ \varphi) \odot (\eta \circ \varphi) \\ \varphi \circ (\psi \oslash \eta) &= (\varphi \circ \psi) \oslash (\varphi \circ \eta), & (\psi \oslash \eta) \circ \varphi &= (\psi \circ \varphi) \oslash (\eta \circ \varphi) \\ \varphi \circ (\psi \otimes \eta) &= (\varphi \circ \psi) \otimes (\varphi \circ \eta), & (\psi \otimes \eta) \circ \varphi &= (\psi \circ \varphi) \otimes (\eta \circ \varphi)\end{aligned}$$

Proof. To prove the first part, use Definition 2.3. Thus, it is sufficient to prove identities (2.2). Let $\varphi, \psi \in \mathrm{End}(Q)$ and $x \in Q$. Then

$$[(\varphi \oslash \psi) \odot \psi](x) = (\varphi \oslash \psi)(x) \cdot \psi(x) = \varphi(x)/\psi(x) \cdot \psi(x) = \varphi(x)$$

i.e we have $(\varphi \oslash \psi) \odot \psi = \varphi$. The rest of the identities can be proved analogously. Hence $\langle \mathrm{End}(Q); \odot, \oslash, \otimes \rangle$ is a quasigroup. Mediality follows from the immediate calculations:

$$[(\varphi \odot \psi) \odot (\chi \odot \pi)](x) = (\varphi \odot \psi)(x) \cdot (\chi \odot \pi)(x) = [\varphi(x)\psi(x)][\chi(x)\pi(x)]$$

$$= [\varphi(x)\chi(x)][\psi(x)\pi(x)] = (\varphi \odot \chi)(x) \cdot (\psi \odot \pi)(x)$$
$$= [(\varphi \odot \chi) \odot (\psi \odot \pi)](x), \quad \varphi, \psi, \chi, \pi \in \mathrm{End}(Q), \quad x \in Q$$

To prove the second part, check the distributivity identities. Let $\varphi, \psi, \chi, \pi \in \mathrm{End}(Q)$ and $x \in Q$. Then we have

$$[\varphi \circ (\psi \odot \eta)](x) = \varphi(\psi(x) \cdot \eta(x)) = [\varphi(\psi(x))] \cdot [\varphi(\eta(x))] = [(\varphi \circ \psi) \odot (\varphi \circ \eta)](x)$$

i.e, $\varphi \circ (\psi \odot \eta) = (\varphi \circ \psi) \odot (\varphi \circ \eta)$. The remaining five distributivity identities can be verified similarly. □

3. Endomorphisms of idempotent medial quasigroups

Let Q be a quasigroup. To each $x \in Q$ we associate the constant mapping $\pi_x \colon Q \longrightarrow \{x\}$. The map π_x is an endomorphism of Q. Let us denote by 0_Q the set of all constant mappings $\pi_x \colon Q \longrightarrow Q$, i.e, $0_Q = \{\pi_x \mid x \in Q\}$. Since $\pi_x \pi_y = \pi_x$ $(x, y \in Q)$, the set 0_Q is a subsemigroup of the monoid $\mathrm{End}(Q)$.

Definition 3.1. A quasigroup Q is called idempotent if the identity $xx = x$ holds in Q.

Proposition 3.2. *If Q is an idempotent quasigroup, then the subsemigroup $(0_Q; \circ)$ coincides with the semigroup of left-zeroes in* End Q.

Proof. Let $\varphi \in \mathrm{End}(Q)$ and $\pi_x \in 0_Q$. Then, for each $y \in Q$, we have

$$(\pi_x \circ \varphi)(y) = \pi_x(\varphi(y)) = x = \pi_x(y)$$

i.e, π_x is a left-zero. Therefore $(0_Q; \circ)$ is a subsemigroup of left-zeroes.

Conversely, suppose that an endomorphism ψ is left-zero in $\mathrm{End}(Q)$. Thus $\psi \circ \pi_x = \psi$ for each $x \in Q$. Let $x \in Q$ be fixed. If $y \in Q$, then we have

$$\psi(y) = (\psi \circ \pi_x)(y) = \psi(x)$$

i.e, ψ is a constant mapping. Therefore, each left-zero in $\mathrm{End}(Q)$ is a constant mapping. □

Proposition 3.2 implies that the mapping $\Psi \colon Q \longrightarrow 0_Q$, $\Psi(x) = \pi_x$, is bijective. Hence, we have the following corollary.

Corollary 3.3. *If the endomorphism monoid of an idempotent quasigroup Q is finite, then Q is finite, too.*

Recall that, for groups we have the similar result [1]: *If the endomorphism monoid of a group G is finite, then G is finite, too.*

Proposition 3.4. *Let Q be a medial idempotent quasigroup. Then*

(i) the quasigroup Q is distributive, i.e, the following identities hold in Q:

$$x(yz) = (xy)(xz), \quad (xy)z = (xz)(yz)$$

(ii) the multiplication group $\mathrm{Mlt}(Q)$ *is a subgroup in* $\mathrm{Aut}(Q)$,

(iii) the quasigroup $(0_Q; \odot)$ is isomorphic to $(Q; \cdot)$.

Proof. Assume that Q is a medial idempotent quasigroup. Since

$$x(yz) = (xx)(yz) = (xy)(xz), \quad (xy)z = (xy)(zz) = (xz)(yz), \quad \forall x, y, z \in Q$$

the quasigroup Q is distributive.

The identities $x(yz) = (xy)(xz)$ and $(xy)z = (xz)(yz)$ imply that the mappings $L_a(x) = ax$ and $R_b(x) = xb$ are endomorphisms of Q. Since $a\backslash(ax) = x$ and $(xb)/b = x$, we have that L_a and R_b are invertible, i.e, L_a and R_b are automorphisms. Therefore, $\mathrm{Mlt}(Q) \leqslant \mathrm{Aut}(Q)$, because the generators of $\mathrm{Mlt}(Q)$ are automorphisms of Q.

To prove that $(0_Q; \odot)$ is isomorphic to $(Q; \cdot)$, consider a map $\Psi\colon Q \longrightarrow 0_Q$ defined by $\Psi(x) = \pi_x$. Clearly, Ψ is injective. By Proposition 3.2, Ψ is surjective. Hence Ψ is bijective. If we show that Ψ is a homomorphism, then the proof is complete. Let $x, y \in Q$. Then

$$(\Psi(x) \odot \Psi(y))(z) = (\pi_x \odot \pi_y)(z) = \pi_x(z) \cdot \pi_y(z) = x \cdot y = \pi_{x \cdot y}(z) = (\Psi(x \cdot y))(z), \quad \forall z \in 0_Q \qquad \square$$

Corollary 3.5. *If Q is an idempotent medial quasigroup, then $(\mathrm{End}(Q); \odot)$ is also an idempotent medial quasigroup.*

Proof. Let $\varphi \in \mathrm{End}(Q)$ and $x \in Q$. Then

$$(\varphi \odot \varphi)(x) = \varphi(x) \cdot \varphi(x) = \varphi(x) \quad \Rightarrow \quad \varphi \odot \varphi = \varphi \quad \forall \varphi \in \mathrm{End}(Q)$$

Mediality of $(\mathrm{End}(Q); \odot)$ follows from Proposition 2.7. $\square$

Theorem 3.6. *Suppose that Q and Q' are idempotent medial quasigroups. If $\langle \mathrm{End}(Q); \circ, \odot, \oslash, \otimes \rangle$ and $\langle \mathrm{End}(Q'); \circ, \odot', \oslash', \otimes' \rangle$ are isomorphic, then Q and Q' are isomorphic, too.*

Proof. Let

$$\Phi : \langle \mathrm{End}(Q); \circ, \odot, \oslash, \otimes \rangle \longrightarrow \langle \mathrm{End}(Q'); \circ, \odot', \oslash', \otimes' \rangle$$

be an isomorphism. The isomorphism Φ induces an isomorphism $\Phi|_{0_Q}$ of the subsemigroups $(0_Q; \circ)$ and $(0_{Q'}; \circ)$ of left-zeroes of $\mathrm{End}(Q)$ and $\mathrm{End}(Q')$, respectively. By Proposition 3.4, we have isomorphisms of quasigroups $\Psi\colon Q \longrightarrow 0_Q$ and $\Psi'\colon Q' \longrightarrow 0_{Q'}$. Hence $\Psi'^{-1} \circ \Phi|_{0_Q} \circ \Psi$ is the isomorphism $Q \to Q'$. $\square$

4. Endomorphisms of commutative idempotent medial quasigroups

In this section we restrict ourselves to the commutative idempotent medial quasigroups (IMC-quasigroups for short).

Proposition 4.1. *A quasigroup Q is an IMC-quasigroup iff there exists an Abelian group $(Q, +)$ such that the mapping $\mathbb{2} : Q \longrightarrow Q$, where $\mathbb{2}(x) = x + x$, $x \in Q$, is an automorphism of $(Q, +)$ and $x \cdot y = \mathbb{2}^{-1}(x + y)$.*

Proof. Let Q be an IMC-quasigroup. By Toyoda's theorem, there exist an Abelian group $\langle Q; , +\rangle$, its commuting automorphisms ϕ and ψ, and an element $c \in Q$ such that

$$x \cdot y = \phi(x) + \psi(y) + c, \quad x, y \in Q$$

Since the group Q is commutative, we have

$$\phi(x) + c = \phi(x) + \psi(0) + c = x \cdot 0 = 0 \cdot x = \phi(0) + \psi(x) + c = \psi(x) + c$$

i.e, $\phi = \psi$. The idempotency implies

$$x = x \cdot x = \phi(x) + \phi(x) + c = (\mathbb{2} \circ \phi)(x) + c$$

Therefore,

$$0 = (\mathbb{2} \circ \phi)(0) + c = c, \quad 1_Q(x) = x = (\mathbb{2} \circ \phi)(x), \quad \mathbb{2} \circ \phi = 1_Q$$

Since ϕ is an automorphism of the group Q, the map $\mathbb{2}$ is an automorphism, too, and, we have $\phi = \mathbb{2}^{-1}$. Finally,

$$x \cdot y = \phi(x) + \psi(y) + c = \phi(x) + \phi(y) = \phi(x + y) = \mathbb{2}^{-1}(x + y)$$

Conversely, if $x \cdot y = \mathbb{2}^{-1}(x + y)$, then it is straightforward to check that Q is an IMC-quasigroup. □

Hence, any IMC-quasigroup is uniquely determined by its underlying Abelian group. Next corollary gives a complete description of finite Abelian groups which are the underlying Abelian groups of IMC-quasigroups.

It is well-known that the mapping $\mathbb{2}$ is an automorphism of a finite Abelian group $(Q, +)$ if and only if $(Q, +)$ is of odd order.

Hence, there does not exist finite IMC-quasigroups of even order. Combining the Proposition 4.1 with the properties of automorphism $\mathbb{2}$, we get the following corollary.

Corollary 4.2. *Let Q be an IMC-quasigroup and Q^+ the underlying Abelian group. Then each endomorphism of the group Q^+ is an endomorphism of the quasigroup Q.*

Proof. Really, if η is an endomorphism of the underlying Abelian group, then

$$(\mathbb{2} \circ \eta)(x) = \eta(x) + \eta(x) = \eta(x + x) = (\eta \circ \mathbb{2})(x), \quad \forall x \in Q$$

and, therefore, $\mathbb{2} \circ \eta = \eta \circ \mathbb{2}$. Since $\mathbb{2}$ is invertible, we obtain $\eta \circ \mathbb{2}^{-1} = \mathbb{2}^{-1} \circ \eta$. Hence

$$\eta(x) \cdot \eta(y) = \mathbb{2}^{-1}(\eta(x)) + \mathbb{2}^{-1}(\eta(y)) = \eta(\mathbb{2}^{-1}(x)) + \eta(\mathbb{2}^{-1}(x)) = \eta(x \cdot y)$$

for each $x, y \in Q$, i.e $\eta \in \mathrm{End}Q$. □

Let Q and Q' be finite IMC-quasigroups such that their endomorphism monoids are isomorphic. By Corollary 3.3, the corresponding underlying Abelian groups Q^+ and Q'^+ are finite, too. P. Puusemp proved (see [8, Theorem 4.2]) that if two finite Abelian groups have the isomorphic endomorphism monoids, then these groups are also isomorphic. Therefore, if we can show that the isomorphism between monoids $\mathrm{End}(Q^+)$ and $\mathrm{End}(Q'^+)$ follows from the isomorphism $\mathrm{End}(Q) \cong \mathrm{End}(Q')$, then the underlying Abelian groups are isomorphic and the quasigroups are isomorphic, too. The rest of the section is dedicated to prove that from $\mathrm{End}(Q) \cong \mathrm{End}(Q')$ implies $\mathrm{End}(Q^+) \cong \mathrm{End}(Q'^+)$.

Assume that Q is an IMC-quasigroup and Q^+ its underlying Abelian group. Let us consider the following binary operation $\oplus_k$ on Q:

$$x \oplus_k y = L_k^{-1}(x \cdot y) = k \backslash (xy), \quad \forall x, y \in Q, \quad k \in Q$$

Since $(Q, \cdot)$ is commutative, the magma $(Q, \oplus_k)$ is commutative, too.

Proposition 4.3. *The magma $(Q, \oplus_k)$ is an Abelian group with the identity element k and the inverse element of x is k/x.*

Proof. We start the proof by showing that k is the identity element. Let $x \in Q$. Then

$$x \oplus_k k = k \oplus_k x = k\diagdown(kx) = x$$

Next we prove the associativity of $\oplus_k$. Let $x, y, z \in Q$. Then

$$\begin{aligned} x \oplus_k (y \oplus_k z) &= (x \oplus_k k) \oplus_k (y \oplus_k z) = L_k^{-1}((x \oplus_k k) \cdot (y \oplus_k z)) \\ &= L_k^{-1}\left(L_k^{-1}(x \cdot k) \cdot L_k^{-1}(y \cdot z)\right) = L_k^{-1}\left(L_k^{-1}((x \cdot k) \cdot (y \cdot z))\right) \\ &= L_k^{-1}\left(L_k^{-1}((x \cdot y) \cdot (k \cdot z))\right) = L_k^{-1}\left(L_k^{-1}(x \cdot y) \cdot L_k^{-1}(k \cdot z)\right) \\ &= L_k^{-1}((x \oplus_k y) \cdot (k \oplus_k z)) = (x \oplus_k y) \oplus_k (k \oplus_k z) = (x \oplus_k y) \oplus_k z \end{aligned}$$

Finally, the inverse element of x is k/x. Indeed, due to the commutativity of $(Q, \cdot)$, we have $k/x = x\diagdown k$ and

$$x \oplus_k (k/x) = (k/x) \oplus_k x = k\diagdown(k/x \cdot x) = k\diagdown k = k$$ □

Corollary 4.4. *For each $k \in Q$, L_k is an automorphism of $(Q, \oplus_k)$ and $L_k^{-1}(x)=2_k(x)=x\oplus_k x$.*

Proof. Really,

$$L_k(x \oplus_k y) = x \cdot y = k\diagdown(k \cdot xy) = k\diagdown(k^2 \cdot xy) = k\diagdown(kx \cdot ky) = L_k(x) \oplus_k L_k(y)$$

Therefore, L_k is an automorphism of $(Q, \oplus_k)$. Since $(Q, \cdot)$ is an idempotent quasigroup, we have

$$x = x \cdot x = L_k(x \oplus_k x) = L_k(x) \oplus_k L_k(x) = (2_k \circ L_k)(x) \quad \Rightarrow \quad 2_k = L_k^{-1}$$ □

Proposition 4.5. *The Abelian groups $(Q, \oplus_k)$ and $(Q, \oplus_l)$ are isomorphic for each $k, l \in Q$.*

Proof. First we observe that the binary operation $\oplus_0$ coincides with the addition in the underlying Abelian group. Really, if $x, y \in Q$, then

$$x \oplus_0 y = 0\diagdown(xy) = 2(xy) - 0 = 2(2^{-1}(x + y)) = x + y$$

Let us define $\psi_k \colon Q \longrightarrow Q$ by $\psi_k(x) = x + k$. Clearly, ψ_k is bijective and

$$\psi_k(x) \oplus_k \psi_k(y) = (x + k) + (y + k) - k = (x + y) + k = \psi_k(x + y), \quad \forall x, y \in Q$$

Hence ψ_k is an isomorphism from $(Q, +)$ to $(Q, \oplus_k)$. □

Corollary 4.6. *Let $k \in Q$. Then each endomorphism of the group $(Q, \oplus_k)$ is an endomorphism of the quasigroup Q.*

The proof is similar to the proof of Corollary 4.2 and therefore is omitted.

We denote further the group $(Q, \oplus_k)$ by $Q^{\oplus_k}$. The submonoid $\mathrm{End}(Q^{\oplus_k})$ of $\mathrm{End}(Q)$ contains the identity mapping and only one constant map π_k. For each $k \in Q$, denote by $\mathcal{M}(k)$ the set of all submonoids M in $\mathrm{End}(Q)$ such that

$$M \in \mathcal{M}(k) \Leftrightarrow (\{\pi_k, 1_Q\} \subseteq M) \ \wedge \ (M \cap 0_Q = \{\pi_k\})$$

Thus $\mathcal{M}(k)$ is non-empty, because it contains $\mathrm{End}(Q^{\oplus_k})$. By Zorn's lemma, $\mathcal{M}(k)$ contains the maximal element, denoted by M_k.

Proposition 4.7. *The endomorphism monoid $\mathrm{End}(Q^{\oplus_k})$ coincides with the submonoid M_k of $\mathrm{End}(Q)$.*

Proof. Let $\varphi \in M_k$. Since $\varphi \circ \pi_k \in M_k$, we have

$$(\varphi \circ \pi_k)(x) = \varphi(k) = \pi_{\varphi(k)}(x), \quad x \in Q$$

Hence $\pi_{\varphi(k)} \in M_k \cap 0_Q$, i.e, $\pi_{\varphi(k)} = \pi_k$ and $\varphi(k) = k$. Next show that $M_k \subseteq \mathrm{End}(Q^{\oplus_k})$. Let $\varphi \in M_k$ and $x, y \in Q$. Then

$$\varphi(x \oplus_k y) = \varphi(k \backslash xy) = \varphi(k) \backslash (\varphi(x)\varphi(y)) = k \backslash (\varphi(x)\varphi(y)) = \varphi(x) \oplus_k \varphi(y).$$

Therefore, $\varphi \in \mathrm{End}(Q^{\oplus_k})$ and $M_k \subseteq \mathrm{End}(Q^{\oplus_k}) \subseteq \mathrm{End}\ Q$. Since $\mathrm{End}(Q^{\oplus_k}) \in \mathcal{M}(k)$, we have $M_k = \mathrm{End}(Q^{\oplus_k})$. □

Theorem 4.8. *Let Q and Q' be IMC-quasigroups and Q is finite. If the endomorphism monoids* $\mathrm{End}(Q)$ *and* $\mathrm{End}(Q')$ *are isomorphic, then the quasigroups Q and Q' are isomorphic as well.*

Proof. Assume that Q and Q' are IMC-quasigroups and Q is finite. Suppose that the monoids $\mathrm{End}(Q)$ and $\mathrm{End}(Q')$ are isomorphic, and let $\Gamma\colon \mathrm{End}(Q) \longrightarrow \mathrm{End}(Q')$ be the corresponding isomorphism. By Corollary 3.3, Q' is finite, too. We write $\boxdot$ and $\boxplus$ for the multiplication in Q' and addition in the underlying Abelian group of Q', respectively. Since both Q and Q' are IMC-quasigroups, we have, by Proposition 4.1,

$$x \cdot y = \mathfrak{2}^{-1}(x + y) \quad \text{and} \quad x' \cdot y' = \mathfrak{2}_{\boxplus}^{-1}(x' \boxplus y'), \quad \forall x, y \in Q, \quad x', y' \in Q'$$

The monoids $\mathrm{End}(Q^{+})$ and $\mathrm{End}(Q'^{\boxplus})$ are submonoids in $\mathrm{End}(Q)$ and $\mathrm{End}(Q')$, respectively.

Next we show that the monoids $\mathrm{End}(Q^{+})$ and $\mathrm{End}(Q'^{\boxplus})$ are isomorphic. Propositions 4.5 and 4.7 imply that $\mathrm{End}(Q^{+}) = M_0 \cong M_k$ for each $k \in Q$. Let M be the maximal submonoid of $\mathrm{End}(Q)$ such that $M \cap 0_Q = \{\pi_k\}$ for some $k \in Q$. Thus $M \cong \mathrm{End}(Q^{+})$. Therefore, the image of M under Γ is a maximal submonoid in $\mathrm{End}(Q')$ containing only one constant endomorphism. Hence, by Propositions 4.5 and 4.7, $\Gamma(M) \cong \mathrm{End}(Q'^{\boxplus})$, and, finally,

$$\mathrm{End}(Q^{+}) \cong M \overset{\Gamma}{\cong} \Gamma(M) \cong \mathrm{End}(Q'^{\boxplus}) \quad \Rightarrow \quad \mathrm{End}(Q^{+}) \cong \mathrm{End}(Q'^{\boxplus})$$

Since Q^{+} and $Q'^{\boxplus}$ are finite Abelian groups, the isomorphism $\mathrm{End}(Q^{+}) \cong \mathrm{End}(Q'^{\boxplus})$ implies an isomorphism $Q^{+} \cong Q'^{\boxplus}$ ([8], Theorem 4.2). Let $\xi\colon Q \longrightarrow Q'$ be the corresponding isomorphism. From immediate computations we obtain $\mathfrak{2}_{\boxplus}^{-1} \circ \xi = \xi \circ \mathfrak{2}^{-1}$. Let $x, y \in Q$. Then

$$\xi(x) \boxdot \xi(y) = \mathfrak{2}_{\boxplus}^{-1}(\xi(x) \boxplus \xi(y)) = (\mathfrak{2}_{\boxplus}^{-1} \circ \xi)(x + y) = (\xi \circ \mathfrak{2}^{-1})(x + y) = \xi(x \cdot y)$$ □

5. An example of a quasigroup not determined by its endomorphism monoid

In this section we present an example of idempotent medial quasigroup that is not determined by its endomorphism monoid. Let the underlying abelian group be C_5 which we regard as an additive group of integers modulo 5. Define two additional binary operations $\star$ and $\bullet$ on C_5:

$$x \star y \equiv 2x + 4y \pmod 5, \qquad x \bullet y \equiv 3x + 3y \pmod 5$$

Since the mappings $x \mapsto ax \pmod 5$ are automorphisms of C_5 for each $a \in \{2, 3, 4\}$, $\star$ and $\bullet$ are quasigroup operations on C_5. The quasigroups $(C_5, \bullet)$ and $(C_5, \star)$ are non-isomorphic, because the quasigroup $(C_5, \bullet)$ is commutative but $(C_5, \star)$ is not.

Next we show that $\mathrm{End}\,(C_5, \star) = \mathrm{End}\,(C_5, \bullet)$. Let $\varphi \in \mathrm{End}\,(C_5, \bullet)$. Since $\varphi(x \bullet 0) = \varphi(x) \bullet \varphi(0)$, we get

$$\varphi(3x) \equiv 3\varphi(x) + 3\varphi(0) \pmod 5$$

Hence,

$$\begin{aligned}
\varphi(x \star y) &= \varphi(2x + 4y \pmod 5) = \varphi(3(4x + 3y) \pmod 5) = \varphi(4x \bullet 3y) \\
&= \varphi(4x) \bullet \varphi(3y) \equiv 3[\varphi(4x) + \varphi(3y)] \pmod 5 \\
&\equiv 3[\varphi(3 \cdot 3x) + 3\varphi(y) + 3\varphi(0)] \pmod 5 \\
&\equiv 3[3\varphi(3x) + 3\varphi(0) + 3\varphi(y) + 3\varphi(0)] \pmod 5 \\
&\equiv 4\varphi(3x) + 4\varphi(y) + 3\varphi(0) \pmod 5 \\
&\equiv 4[3\varphi(x) + 3\varphi(0)] + 4\varphi(y) + 3\varphi(0) \pmod 5 \\
&\equiv 2\varphi(x) + 4\varphi(y) \pmod 5 = \varphi(x) \star \varphi(y)
\end{aligned}$$

i.e, $\mathrm{End}\,(C_5, \bullet) \subseteq \mathrm{End}\,(C_5, \star)$. Let us suppose now $\tau \in \mathrm{End}\,(C_5, \star)$. After expanding the equalities $\tau(x \star 0) = \tau(x) \star \tau(0)$ and $\tau(0 \star x) = \tau(0) \star \tau(x)$, we have

$$\tau(2x) \equiv 2\tau(x) + 4\tau(0) \pmod 5 \quad \text{and} \quad \tau(4x) \equiv 4\tau(x) + 2\tau(0) \pmod 5$$

Compute

$$\begin{aligned}
\tau(x \bullet y) &= \tau(3x + 3y \pmod 5) = \tau(2 \cdot 4x + 4 \cdot 2y \pmod 5) = \tau(4x \star 2y) \\
&= \tau(4x) \star \tau(2y) \equiv 2\tau(4x) + 4\tau(2y) \pmod 5 \\
&\equiv 2 \cdot 4\tau(x) + 2 \cdot 2\tau(0) + 4 \cdot 2\tau(y) + 4 \cdot 4\tau(0) \pmod 5 \\
&\equiv 3\tau(x) + 3\tau(y) \pmod 5 = \tau(x) \bullet \tau(y)
\end{aligned}$$

Hence $\tau \in \mathrm{End}\,(C_5, \bullet)$ and, therefore, $\mathrm{End}\,(C_5, \star) \subseteq \mathrm{End}\,(C_5, \bullet)$. We have proved that $\mathrm{End}\,(C_5, \star) = \mathrm{End}\,(C_5, \bullet)$. Therefore, the endomorphism monoids $\mathrm{End}\,(C_5, \star)$ and $\mathrm{End}\,(C_5, \bullet)$ are isomorphic, but the quasigroups $(C_5, \star)$ and $(C_5, \bullet)$ are not.

Therefore, the following quasigroups are not determined by their endomorphism monoid:

- the idempotent medial quasigroups in the class of idempotent medial quasigroups,
- medial quasigroups in the class of (medial) quasigroups,
- distributive quasigroups in the class of (distributive) quasigroups,
- left distributive quasigroups in the class of (left distributive) quasigroups,
- idempotent quasigroups in the class of (idempotent) quasigroups,
- commutative quasigroups in the class of quasigroups.

References

[1] Alperin J L 1962 *Pacific J. Math* **12** 1
[2] Belousov V D 1967 *Foundations of the Theory of Quasigroups and Loops* (Moscow: Nauka) (Russian)
[3] Krylov P A Mikhalev A V and Tuganbaev A A 2003 *Endomorphism Rings of Abelian Groups* (Kluwer)
[4] Leibak A and Puusemp P 2011 *Proc. Estonian Acad. Sci.* **60** 81
[5] Nagy P T and Plaumann P 2009 *J. Gen. Lie Theory Appl.* **3** 191
[6] Nagy P T and Strambach K 2002 *Loops in Group Theory and Lie Theory* (Walter de Gruyter)
[7] Paal E 2006 *Czechoslovak J. Phys.* **56** 1257
[8] Puusemp P 1975 *Acta Comment. Univ. Tartuensis* **366** 76 (in Russian)
[9] Shcherbacov V A 2009 *J. Gen. Lie Theory Appl.* **3** 197
[10] Shcherbacov V A 1991 *Mat. Issled.* **120** 104 (Russian)
[11] Tabarov A Kh 2007 *Diskret. Mat.* **19** 67 (Russian)

A short survey of Lie antialgebras

S Leidwanger[1] and S Morier-Genoud[2]

Institut de Mathématiques de Jussieu - Paris Rive Gauche UMR 7586
[1]Bâtiment Sophie Germain, case 7012, 75205 Paris Cedex 13, France
[2]Université Pierre et Marie Curie Paris 6, 4 place Jussieu, case 247, 75252 Paris Cedex 05, France

Abstract. This article is the writing notes of a talk on Lie Antialgebras given by the second author at the *QQQ* conference *3Quantum: Algebra Geometry Information* that held in Tallinn, July 2012. The aim of this note is to give a brief survey of the existing theory of Lie antialgebras and to suggest open questions.

1. Introduction

The notion of *Lie antialgebra* is quite recent and due to Valentin Ovsienko. Since the founding works [Ovsienko V arXiv:0705.1629], various directions of study on these algebras have been, and are still, investigated by Ovsienko [12] and others [11, 8, 9, 4], see also [arXiv:1012.3885]. This article is the writing notes of a talk on *Lie antialgebras* given by the second author at the *QQQ* conference *3Quantum: Algebra Geometry Information* that held in Tallinn, July 2012. The aim of this note is to give a brief survey of the existing theory and to suggest open questions.

What is a Lie antialgebra? The name is quite surprising as Lie antialgebras are neither Lie algebras nor antialgebras. The name was chosen to suggest a "dark face" of Lie algebras, a twin brother living in the shadow. The prefix *anti* also refers to parity inversion and changes of signs occuring in this theory in comparison to the theory of Lie algebras.

Ovsienko likes to illustrate the general idea behind Lie antialgebras with the diagram in Figure 1.1.

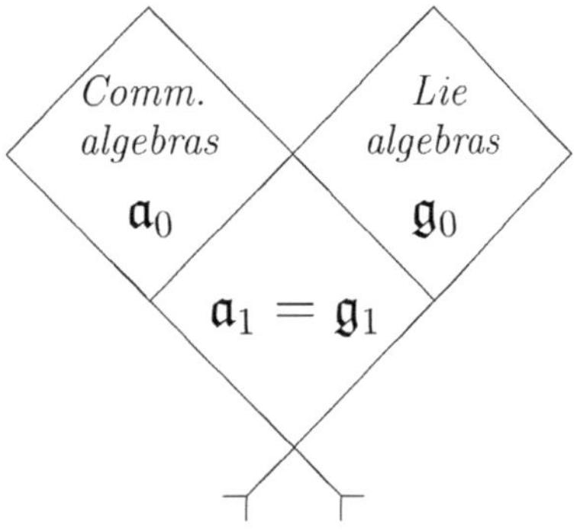

Figure 1.1. Lie antialgebras.

The left part consists of the class of commutative associative algebras, and the right one of the class of Lie algebras. These two classes are known to be Koszul dual. The idea is to relate the two classes in a concrete way, using $\mathbb{Z}_2$-graded structures. A commutative algebra $\mathfrak{a}_0$ is considered as the even part of a $\mathbb{Z}_2$-commutative superalgebra $\mathfrak{a} = \mathfrak{a}_0 \oplus \mathfrak{a}_1$, and a Lie algebra $\mathfrak{g}_0$ as the even part of a Lie superalgebra $\mathfrak{g} = \mathfrak{g}_0 \oplus \mathfrak{g}_1$, each of these even parts being generated by a common odd part $\mathfrak{a}_1 = \mathfrak{g}_1$.

2. Origin and first examples

2.1. A new invariant bivector fields

The origin of Lie antialgebras comes from the following observation. Consider the supervariety $M = \mathbb{R}^{2,1}$, on which the even coordinates are denoted p, q, and the odd coordinate is τ. The standard symplectic form on M is

$$\omega = dp \wedge dq + \frac{1}{2} d\tau \wedge d\tau$$

The form ω is preserved by the action of the Lie superalgebra $\mathbf{osp}(1,2)$.

A natural question is *what are the invariant bivectors fields with respect to the action of* $\mathbf{osp}(1,2)$?

One can immediately exhibit the following invariant bivector, which is the inverse of ω

$$\mathcal{P} = \frac{\partial}{\partial p} \wedge \frac{\partial}{\partial q} + \frac{1}{2} \frac{\partial}{\partial \tau} \wedge \frac{\partial}{\partial \tau}$$

It turns out that there is another invariant bivector, given by

$$\Lambda = \frac{\partial}{\partial \tau} \wedge \mathcal{E} + \tau \frac{\partial}{\partial p} \wedge \frac{\partial}{\partial q}$$

where

$$\mathcal{E} = p\frac{\partial}{\partial p} + q\frac{\partial}{\partial q} + \tau\frac{\partial}{\partial \tau}$$

is the Euler field.

It seems that the existence of this extra invariant bivector Λ had not been observed before Ovsienko's work. In addition, Ovsienko showed that there are no other invariant bivectors beside of $\mathcal{P}$ and Λ and their linear combinations.

What are the properties of Λ? One notes that the bivector Λ is odd with linear coefficients whereas $\mathcal{P}$ is even with constant coefficients. Let us briefly compare the properties of Λ and that of $\mathcal{P}$.

2.2. Algebraic structures associated to the bivectors $\mathcal{P}$ and Λ

The first thing one can do is to look at algebraic stuctures associated to these two bivectors. More precisely, given two homogeneous functions $F, G \in C^\infty(M)$, one can construct natural products using the duality between bivectors and 2-forms.

$$\{F, G\} := \langle \mathcal{P}, dF \wedge dG \rangle \longrightarrow \text{Lie (Poisson) superalgebra}$$

$$]F, G[:= \frac{-(-1)^{|F|}}{2} \langle \Lambda, dF \wedge dG \rangle \longrightarrow ??$$

The bracket $\{\,,\,\}$ associated to the even bivector $\mathcal{P}$ gives a structure of Lie superalgebra on $C^\infty(M)$. What about the bracket $]\,,\,[$ associated to Λ ? The sign $(-1)^{|F|}$, where $|F|$ stands for the parity of F, appearing in the definition, is added to make the operation $]F, G[$ supercommutative (the coefficient of $-\frac{1}{2}$ is just a normalization and is not essential). It is not easy to exhibit general properties of this bracket on the entire space of functions.

2.3. Examples

Let us restrict our attention to nice subspaces of $C^\infty(M)$. For instance, the space of quadratic polynomials is stable under the Lie bracket $\{,\}$ and forms a Lie superalgebra isomorphic to $\mathbf{osp}(1,2)$. Similarly, the space of linear functions is stable for the odd bracket $],[$, and forms (after parity inversion) a 3-dimensional algebra which is also well known: it is isomorphic to the *tiny Kaplansky algebra* K_3. Recall that K_3 has one even basis vector, say ε, and two odds, say a, b, subject to the following multiplication rules:

$$\varepsilon\varepsilon = \varepsilon, \quad \varepsilon a = \tfrac{1}{2}a, \quad \varepsilon b = \tfrac{1}{2}b, \quad ab = \tfrac{1}{2}\varepsilon$$

These relations may be obtained in naming ε, a, b the images of τ, q, p after parity inversion).

More generally, the space of homogeneous functions of degree 2 (here the homogeneity is defined with respect to the Euler field, i.e $\mathcal{E}(F) = 2F$) is stable under $\{\,,\,\}$. The subspace of rational functions with poles allowed only at points where $p = 0$ or $q = 0$, forms a Lie superalgebra isomorphic to the superconformal algebra $\mathcal{K}(1)$, also known as Neveu-Schwartz or super Virasoro algebra (see [12] for details). The parallel situation for the bracket $]\,,[$, is to consider the stable subspace of homogeneous rational functions of degree 1. It forms an infinite-dimensional algebra denoted $AK(1)$ and that can be described as follows: the even basis vectors ε_n, $n \in \mathbb{Z}$, and the odd basis vectors a_i, $i \in \mathbb{Z} + 1/2$, are subject to the multiplication rules

$$\varepsilon_n\varepsilon_m = \varepsilon_{n+m}, \quad \varepsilon_n a_i = \tfrac{1}{2}a_{n+i}, \quad a_i a_j = \tfrac{1}{2}(i-j)\varepsilon_{i+j}$$

These relations maybe obtained in naming ε_n and a_i the images of $\tau(\frac{q}{p})^n$ and $p(\frac{q}{p})^{i+\frac{1}{2}}$, after parity reversion.

The situation can be summarized as follows:

$$\begin{array}{lcll}
\{\text{quadratic polynomials}\} & = & <p^2, pq, q^2, p\tau, q\tau> & \xrightarrow{\{,\}} \quad \mathbf{osp}(1,2) \\
\{\text{linear polynomials}\} & = & <p, q, \tau> & \xrightarrow{],[} \quad K_3
\end{array}$$

$$\begin{array}{ll}
\{\text{rat. functions of deg 2}\} =< p^2(\frac{q}{p})^{n+1}, \tau p(\frac{q}{p})^{i+\frac{1}{2}}, n \in \mathbb{Z}, i \in \mathbb{Z}+\frac{1}{2} > & \xrightarrow{\{,\}} \quad \mathcal{K}(1) \\
\{\text{rat. functions of deg 1}\} =< p(\frac{q}{p})^{i+\frac{1}{2}}, \tau(\frac{q}{p})^{n}, n \in \mathbb{Z}, i \in \mathbb{Z}+\frac{1}{2} > & \xrightarrow{],[} \quad AK(1)
\end{array}$$

The algebras K_3 and $AK(1)$ are both known as Jordan superalgebras.

Question 1. What are the general properties of $]\,,[$? Can one construct other nice algebras out of this bracket? Do invariant odd bivector fields analogous to Λ exist in higher dimension?

3. General theory

3.1. Axioms

The above examples are quite encouraging to try to build out of them a general theory.

Definition ([12]). A Lie antialgebra $\mathfrak{a} = \mathfrak{a}_0 \oplus \mathfrak{a}_1$ is a supercommutative algebra satisfying

(i) $\mathfrak{a}_0$ is associative,

(ii) right multiplications $R_y : \mathfrak{a} \to \mathfrak{a}$, $a \mapsto ay$ are odd derivations for $y \in \mathfrak{a}_1$,

(iii) $\mathfrak{a}_0$ acts commutatively on $\mathfrak{a}_1$, i.e $x_1(x_2 y) = x_2(x_1 y)$ for $x_1, x_2 \in \mathfrak{a}_0$, $y \in \mathfrak{a}_1$.

Remarks on the axioms:

- $\mathfrak{a}_1$ is not a $\mathfrak{a}_0$-module, because of the axiom (ii), the axiom (iii) is equivalent to

$$\text{(iii')} \quad x_1(x_2y) = \frac{1}{2}(x_1x_2)y \quad \text{for} \quad x_1, x_2 \in \mathfrak{a}_0, \quad y \in \mathfrak{a}_1$$

and at a first sight this half-action may appear unnatural but it is actually essential to the theory,
- the defining identities of Lie antialgebras are all cubic,
- half-unital superalgebras satisfying (i) and (iii') are called Kaplansky algebras in [10],
- the axioms (i) and (ii) imply that $\mathfrak{a}$ is a Jordan superalgebra, (it is quite involved to go from these cubic identities to the quartic identites of Jordan superalgebra, one has to find the right sequence of transformations; this was done in details in [10] using the weaker requirement that $\mathfrak{a}_0$ is Jordan),
- if $\mathfrak{a}$ is generated by its odd part, then axioms (ii) and (iii) imply the axiom (i) (this was observed in [8]; axiom (iii) plays a crucial role in this property).

Lie antialgebras can be understood as a non trivial super analog to commutative associative algebras.

Example. The algebras K_3 and $AK(1)$ are Lie antialgebras. But, the entire space $C^\infty(M)$ equipped with the bracket] , [is not.

3.2. Adjoint Lie superalgebra

A nice feature of the theory of Lie antialgebras is a relationship to the theory of Lie superalgebras. The axiomatic definition of Lie antialgebra allows one to construct a formal Lie superalgebra sharing the same odd space.

To any Lie antialgebra $\mathfrak{a} = \mathfrak{a}_0 \oplus \mathfrak{a}_1$ associate the Lie superalgebra $\mathfrak{g}(\mathfrak{a}) = \mathfrak{g}_0 \oplus \mathfrak{g}_1$, where, as vector spaces, $\mathfrak{g}_1 = \mathfrak{a}_1$ and $\mathfrak{g}_0 = S^2_{\mathfrak{a}_0}\mathfrak{a}_1$ is the space of symmetric tensors of elements of $\mathfrak{a}_1$ over $\mathfrak{a}_0$. In other words, elements of $\mathfrak{g}_0$ are of the forms $y_1 \odot y_2$, where $y_1, y_2 \in \mathfrak{a}_1$ and $\odot$ is the symmetric tensor product over $\mathfrak{a}_0$, so that

$$\begin{cases} y_1 \odot y_2 &= y_2 \odot y_1 \\ y_1x \odot y_2 &= y_1 \odot y_2x, \qquad y_1, y_2 \in \mathfrak{a}_1, x \in \mathfrak{a}_0 \end{cases}$$

Now, define a bracket on $\mathfrak{g}(\mathfrak{a})$ as follows

$$\begin{aligned} [y_1, y_2] &= y_1 \odot y_2 \\ [y_1 \odot y_2, y_3] = -[y_3, y_1 \odot y_2] &= y_1(y_2y_3) + y_2(y_1y_3) \\ [y_1 \odot y_2, y_3 \odot y_4] &= [y_1 \odot y_2, y_3] \odot y_4 + [y_1 \odot y_2, y_4] \odot y_3 \end{aligned} \tag{3.1}$$

where y_1, y_2, y_3 and y_4 are elements of $\mathfrak{g}_1 = \mathfrak{a}_1$.

The above bracket is the most natural skew-symmetric bracket one may think of. It was announced to be a Lie superbracket in [12] and then the proof was completed in [8]. It is quite amazing how the axioms of Lie antialgebras combine to produce the Lie superalgebra structure. Once again here the axiom (iii) plays a crucial role. To convince the reader, we suggest as an exercice to show that the bracket on $\mathfrak{g}(\mathfrak{a})$ is well defined, i.e to show that $[y_1 \odot y_2x, y_3] = [y_1x \odot y_2, y_3]$.

Remark. Let us stress that the above construction is different from the well known Tits-Kantor-Koecher (TKK) construction. The adjoint Lie superalgebras associated to K_3 and $AK(1)$ are

$$\mathfrak{g}(K_3) = \mathbf{osp}(1,2), \quad \mathfrak{g}(AK(1)) = \mathcal{K}(1)$$

In these examples, the adjoint algebras coincide with the algebra of derivations. But, in general this is not true. One has only an embedding $\mathfrak{g}(\mathfrak{a}) \hookrightarrow \mathbf{Der}(\mathfrak{a})$.

Journal of Physics: Conference Series **532** (2014) 012015 doi:10.1088/1742-6596/532/1/012015

The following problems remain open.

Question 2. Is there a natural construction of a Lie antialgebra from a Lie superalgebra? Is there a direct link between the constructed adjoint Lie superalgebra $\mathfrak{g}(\mathfrak{a})$ and the one obtained through the TKK process?

Question 3. The relation $\mathbf{osp}(1,2) = \mathbf{Der}(K_3)$ is crucial in the construction of [2]. Can we expect similar construction from $\mathcal{K}(1) = \mathbf{Der}(AK(1))$, or more generally from $\mathfrak{a}$ and $\mathfrak{g}(\mathfrak{a})$?

3.3. Representation theory

Lie antialgebras are connected to three large classes of algebras: commutative associative, Jordan, and Lie. As a particular class of Jordan algebras, one may apply the classical representation theory of Jordan algebras to Lie antialgebras, but it seems important to adapt a little bit the classical definitions to take into account the specificity of Lie antialgebras.

Definition ([12, 8]). A linear map $\varrho : \mathfrak{a} \to \mathrm{End}(V)$, where V is a $\mathbb{Z}_2$-graded space, is a representation of Lie antialgebra if

(i) $\varrho(ab) = [\varrho(a)\,,\,\varrho(b)]_+$, for all $a, b \in \mathfrak{a}$, and
(ii) $\varrho(x_1x_2) = \varrho(x_1)\varrho(x_2)$, for all $x_1, x_2 \in \mathfrak{a}_0$.

The notation $[\,,\,]_+$ stands for the usual Jordan superbracket constructed out of an associative superalgebra, i.e on homogeneous elements

$$[X,Y]_+ = \frac{1}{2}(XY + (-1)^{|X||Y|}YX)$$

whereas the ususal commutator will be denoted $[\,,\,]$

$$[X,Y] = XY - (-1)^{|X||Y|}YX$$

In other words, a representation of Lie antialgebra is a Jordan representation with the extra requirement that the restriction to the even part is a morphism of associative algebra.

A nice feature of the theory is the following relationship between representations of $\mathfrak{a}$ and those of $\mathfrak{g}(\mathfrak{a})$.

Theorem ([8]). *Consider a Lie antialgebra $\mathfrak{a}$ and its adjoint Lie superalgebra $\mathfrak{g}(\mathfrak{a})$.*

(i) Every representation $\varrho : \mathfrak{a} \to \mathrm{End}(V)$ induces a unique representation $\tilde{\varrho} : \mathfrak{g}(\mathfrak{a}) \to \mathrm{End}(V)$ such that $\tilde{\varrho}(\mathfrak{g}_1) = \varrho(\mathfrak{a}_1)$.
(ii) There exists an ideal $\mathcal{I}_\mathfrak{a}$ in the universal enveloping algebra $U(\mathfrak{g}(\mathfrak{a}))$ such that every representation $\tilde{\varrho} : \mathfrak{g}(\mathfrak{a}) \to \mathrm{End}(V)$ vanishing on $\mathcal{I}_\mathfrak{a}$ induces a unique representation $\varrho : \mathfrak{a} \to \mathrm{End}(V)$ with $\tilde{\varrho}(\mathfrak{g}_1) = \varrho(\mathfrak{a}_1)$.

In other words, point (1) of the above theorem says that the images of odd elements of $\mathfrak{a}$ generate in $(\mathrm{End}(V), [\,,\,])$ a representation of $\mathfrak{g}(\mathfrak{a})$. This is quite surprising.

Consider for example an arbitrary representation of K_3, and denote by $A, B, \mathcal{E}$ the images of a, b, ε. The following relations hold

$$AB - BA = \mathcal{E}, \quad A\mathcal{E} + \mathcal{E}A = A, \quad B\mathcal{E} + \mathcal{E}B = B, \quad \mathcal{E}^2 = \mathcal{E}.$$

Now, denote

$$H := -(AB + BA), \quad E = A^2, \quad F := -B^2$$

These three elements automatically satisfy the relations of $\mathbf{sl}(2)$, i.e

$$HE - EH = 2E, \quad HF - FH = -2F, \quad EF - FE = H$$

and together with A, B they satisfy the relations of $\mathbf{osp}(1,2)$.

The converse is not true. The point (2) of the above theorem characterizes the representations of the Lie superalgebra for which the converse holds. In the case of $\mathfrak{a} = K_3$, the ideal $\mathcal{I}_\mathfrak{a}$ of $U(\mathbf{osp}(1,2))$ is the ideal generated by the Casimir element.

More details on the representation of K_3 viewed as a Lie antialgebra can be found in [11]. Elements of general representation theory for Lie antialgebras are developed in [8].

Question 4. In [8] the notion of universal enveloping algebra for a Lie antialgebra is studied and it is observed that $U(K_3)$ appears in a form similar to the generalized Weyl algebras (2.1) in [1]. Are there more relations between Lie antialgebras and generalized Weyl algebras?

4. More examples coming from geometry

In this last section, we describe a series of Lie antialgebras closely related to the Lie algebras of Krichever-Novikov. This series generalized $AK(1)$ in the same way as Krichever-Novikov algebras generalize the Witt algebra.

Consider a Riemann surface Σ of genus g with a set $\mathcal{M}$ of N marked points. The marked points are the points where poles of meromorphic functions on Σ are allowed.

Denote by $\mathcal{F}_\lambda$ the family of modules of meromorphic tensor densities of weight λ. In order to well define these modules for any complex parameter λ, one should fix a complex logarithm on Σ. In what follows we will only consider the cases where λ is integer or half integer.

The space $\mathcal{F}_0$ is well identified, it is simply the set of meromorphic functions on Σ, holomorphic outside of $\mathcal{M}$. The space $\mathcal{F}_{-1}$ is also well identified, it is the Lie algebra of meromorphic vector fields on Σ holomorphic outside of $\mathcal{M}$.

In the case of genus $g = 0$ and $N = 2$ marked points, the Lie algebra $\mathcal{F}_{-1}$ is the famous Witt algebra. In higher genus $g \geq 0$ and $N = 2$ marked points, the Lie algebras $\mathcal{F}_{-1}$ are known as Krichever-Novikov algebras, see [5, 6, 7]. The general case, $g \geq 0$ and $N \geq 2$, has been extensively studied by Schlichenmaier [13, 14, 15], see also [3].

One has natural actions of $\mathcal{F}_{-1}$ on itself (by commutator) and on $\mathcal{F}_0$ (by derivation). These actions can be unified and naturally deformed in a one-parameter family of actions, giving rise to the modules $\mathcal{F}_\lambda$. Using local coordinates, the actions of $\mathcal{F}_{-1}$ can be described as follows:

$$\begin{array}{ccc} \mathcal{F}_{-1} \times \mathcal{F}_\lambda & \longrightarrow & \mathcal{F}_\lambda \\ \big(f(z)dz^{-1}, g(z)dz^\lambda\big) & \mapsto & \big(f(z)g'(z) + \lambda f'(z)g(z)\big)dz^\lambda \end{array}$$

This operation can be generalized in a skew-symmetric operation between two density modules

$$\begin{array}{cccc} \{\ ,\ \}: & \mathcal{F}_\mu \times \mathcal{F}_\lambda & \longrightarrow & \mathcal{F}_{\lambda+\mu+1} \\ & \big(f(z)dz^\mu, g(z)dz^\lambda\big) & \mapsto & \big(-\mu f(z)g'(z) + \lambda f'(z)g(z)\big)dz^{\lambda+\mu+1} \end{array}$$

Together, with the natural multiplication of two tensor densities

$$\bullet : \mathcal{F}_\mu \times \mathcal{F}_\lambda \longrightarrow \mathcal{F}_{\lambda+\mu}, \quad \big(f(z)dz^\mu, g(z)dz^\lambda\big) \mapsto f(z)g(z)dz^{\lambda+\mu}$$

the latter provide a natural structure of Poisson algebra on the graded space of all tensor densities $\bigoplus_\lambda \mathcal{F}_\lambda$.

Theorem ([9]). *The $\mathbb{Z}_2$-graded space $\mathcal{F}_0 \oplus \mathcal{F}_{-1/2}$ has a natural structure of a Lie antialgebra, given by the product*

$$\begin{aligned} f(z) \circ g(z) &= f(z) \bullet g(z) \\ f(z) \circ \gamma(z)(dz)^{-\frac{1}{2}} &= \tfrac{1}{2} f(z) \bullet \gamma(z)(dz)^{-\frac{1}{2}} \\ \varphi(z)(dz)^{-\frac{1}{2}} \circ \gamma(z)(dz)^{-\frac{1}{2}} &= \left\{\varphi(z)(dz)^{-\frac{1}{2}}, \gamma(z)(dz)^{-\frac{1}{2}}\right\}. \end{aligned}$$

Its adjoint Lie superalgebra is the space $\mathcal{F}_{-1} \oplus \mathcal{F}_{-1/2}$, equipped with the bracket

$$\begin{aligned} \left[f(z)(dz)^{-1}, g(z)(dz)^{-1}\right] &= \left\{f(z)(dz)^{-1}, g(z)(dz)^{-1}\right\} \\ \left[f(z)(dz)^{-1}, \gamma(z)(dz)^{-\frac{1}{2}}\right] &= \left\{f(z)(dz)^{-1}, \gamma(z)(dz)^{-\frac{1}{2}}\right\} \\ \left[\varphi(z)(dz)^{-\frac{1}{2}}, \gamma(z)(dz)^{-\frac{1}{2}}\right] &= \frac{1}{2} \varphi(z)(dz)^{-\frac{1}{2}} \bullet \gamma(z)(dz)^{-\frac{1}{2}} \end{aligned}$$

In the case $g = 0$, $N = 2$, the Lie antialgebra is nothing but $AK(1)$. In the case $g = 0$, $N = 3$, the Lie antialgebra is closely related to a Jordan superalgebra algebraically constructed in [16] to provide a "new type" of Jordan superalgebra. Let us also mention computations of cocycles on the Lie antialgebras and Lie superalgebras of Krichever-Novikov type [4].

This construction gives a concrete realization of our initial picture, see Figure 4.2!

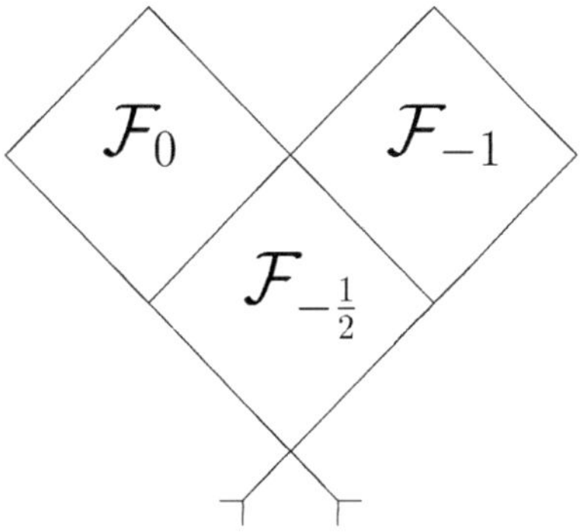

Figure 4.2. Realization of Lie antialgebras.

Question 5. Krichever-Novikov algebras carry a structure of almost-graded algebras, see [14]. How Lie antialgebras behave with respect to this almost-grading?

5. Conclusion

Lie antialgebras are friendly hybrid birds, suggesting a non-trivial super analog to commutative associative algebras, and offering more relations between the theory of Jordan algebras and Lie algebras. Many directions of the theory need to be explored. Questions 1-5 addressed throughout the paper are examples of precise questions naturally arising in the theory. More general open questions, as the use of Lie antialgebras in mathematical physics or link to integrable systems, are natural directions of investigations.

Acknowledgments

The authors are grateful to V. Ovsienko and M. Schlichenmaier for helpful remarks, references and suggestions.

References

[1] Bavula V and Van Oystaeyen F 2000 *J. Pure Appl. Algebra* **150** 41
[2] Benkart G and Elduque A 2002 *Proc. Amer. Math. Soc.* **130** 3209 (electronic)
[3] Dick R 1989 *Lett. Math. Phys.* **18** 255
[4] Kreusch M 2013 *Lett. Math. Phys.* **103** 1171
[5] Krichever I M and Novikov S P 1987 *Funktional Anal. Prilozhen.* **21** 46
[6] Krichever I M and Novikov S P 1987 *Funktional Anal. Prilozhen.* **21** 47
[7] Krichever I M and Novikov S P 1989 *Funkt. Anal. Prilozhen.* **23** 46
[8] Leidwanger S and Morier-Genoud S 2012 *Algebr. Represent. Theory* **15** 1
[9] Leidwanger S and Morier-Genoud S 2012 *Int. Math. Res. Not. IMRN* 4449
[10] McCrimmon K 1994 *J. Algebra***164** 656
[11] Morier-Genoud S 2009 *Int. Math. Res. Notices IMRN* 1838
[12] Ovsienko V 2011 *J. Algebra* **325** 216
[13] Schlichenmaier M 1990 *Lett. Math. Phys.* **19** 151
[14] Schlichenmaier M 1990 *Lett. Math. Phys.* **20** 33
[15] Schlichenmaier M 1993 *J. Math. Phys.* **34** 3809
[16] Zhelyabin V 2009 *Mat. Tr.* **12** 41 (Russian)

Supersymmetry: superfield equations of motion

R K Loide and P Suurvarik

Department of Physics, Tallinn University of Technology, Ehitajate tee 5, 19086 Tallinn, Estonia

Abstract. Supersymmetry and superfields are considered in connection with Poincaré superalgebras. A formalism of projection operators for deriving wave equations for ordinary fields and superfields is developed. Superfield equations of motion in the case of massive and massless fields are presented together with an application in linear supergravity.

1. Supersymmetry, superfields, Poincaré superalgebra

Supersymmetry has been known more than few decades, but it is still alive despite the absence of experimental verification. Supersymmetry offers several relevant theoretical solutions for modern physics.

1.1. Supersymmetry (Bose-Fermi symmetry)

Let us consider a general $N = 1$ superfield

$$\phi_i(x,\theta) = A_i(x) + \bar{\theta}\psi_i(x) + \bar{\theta}\theta F_i(x) + \bar{\theta}\gamma^5\theta G_i(x) + \bar{\theta}i\gamma^\mu\gamma^5\theta A_{\mu i} + \bar{\theta}\theta\bar{\theta}\chi_i(x) + (\bar{\theta}\theta)^2 D_i(x)$$

where θ_α is a four-component anticommuting Majorana spinor and i is a Lorentz index.

Now we introduce the most general Poincaré superalgebra [1]. The generators of the Poincaré group P^μ and $M^{\mu\nu}$, and n supergenerators S_α $(\alpha = 1, 2, \ldots, n)$ satisfy the following relations:

$$\begin{aligned}
\left[P^\mu, P^\nu\right] &= 0\\
\left[M^{\mu\nu}, P^\rho\right] &= \eta^{\nu\rho}P^\mu - \eta^{\mu\rho}P^\nu\\
\left[M^{\mu\nu}, M^{\rho\sigma}\right] &= \eta^{\nu\rho}M^{\mu\sigma} + \eta^{\mu\sigma}M^{\nu\rho} - \eta^{\mu\rho}M^{\nu\sigma} - \eta^{\nu\sigma}M^{\mu\rho}\\
\left[S_\alpha, P^\mu\right] &= 0\\
\left[S_\alpha, M^{\mu\nu}\right] &= B^{\mu\nu}_{\alpha\beta}S_\beta
\end{aligned}$$

where $\eta^{\mu\nu} = \text{diag}(+\,-\,-\,-)$.

The simple $N = 1$ supersymmetry algebra is related to the bispinor representation of the Lorentz group. We have four bispinor generators S_α, $B^{\mu\nu} = \frac{1}{2}\sigma^{\mu\nu}$ and $A_\mu = \gamma_\mu C$. The N-extended Poincaré superalgebra is related to the direct sum of N bispinor representations.

1.2. New possibilities [2]

- In the most general case the matrices A_μ are related to the β-matrices of an invariant first order wave equation

$$\left(i\partial_\mu\beta^\mu - m\right)\psi(x) = 0$$

in the following way: $A_\mu = \beta_\mu C$. Here $\psi(x)$ corresponds to a representation of the Lorentz group and $B^{\mu\nu}$ are the Lorentz generators of the corresponding representation.

- If we consider new possibilities for $N = 1$ superalgebras, we can take the Rarita-Schwinger equation for a vector-bispinor field and use the β-matrices of the Rarita-Schwinger equation

$$(\beta^\mu)^\rho_\sigma = \gamma^\mu \eta^\rho_\sigma + \left(\frac{a}{\sqrt{3}} - \frac{1}{2}\right)\eta^{\mu\rho}\gamma_\sigma + \left(\frac{b}{\sqrt{3}} - \frac{1}{2}\right)\gamma^\rho \eta^\mu_\sigma + \left(\frac{a+b}{4\sqrt{3}} + \frac{c}{4} + \frac{3}{8}\right)\gamma^\rho\gamma^\mu\gamma_\sigma$$

- If we take an irreducible representation for generators S_α (vector representation (1/2, 1/2), e.g), then there is no first order wave equation and we get $\{S_\alpha, S_\beta\} = 0$.
- If we consider the Kemmer-Duffin equation for spin 0, $\psi(x)$ is a direct sum of vector and scalar fields and therefore 5 generators S_α also correspond to vector and scalar representations. As a result we get the Bose superalgebra.

Let us turn to the ($N = 1$) Poincaré supergroup and algebra. It is well known that there are the following physically interesting irreducible representations: massive representations (m, Y), where superspin Y gives Poincaré spins $s = Y+1/2$, Y , $Y-1/2$, and the massless representations $(0, Y)$, where we have the helicities $s = Y + 1/2$ and Y.

The problem is: if we consider different Lorentz superfields $\varphi_i(x, \theta)$, we need additional conditions to guarantee that our field or combination of fields describe physical states with a given mass and superspin. Such additional conditions are relativistic wave equations. The most familiar of them for ordinary fields is the Dirac equation, which describes a fermion field with a given rest mass m and spin 1/2.

We have developed a quite powerful method of constructing and analyzing relativistically invariant wave equations for fields and superfields to describe states with a certain mass and spin (superspin), and also wave equations for massless fields. Our method is based on spin projection and superspin projection operators. In order to understand the main principles of our method, we first describe its ideas used in the case of ordinary fields and then generalize them to the case of superfields.

2. Ordinary Lorentz fields, wave equations, spin projection operators

We consider a finite dimensional Lorentz field $\psi(x)$ and demand, that $\psi(x)$ is a solution of the n-th order wave equation [3]

$$i\partial_{\mu_1} \dots i\partial_{\mu_n} \beta^{\mu_1 \dots \mu_n}\psi(x) = m^n \psi(x)$$

where the matrices $\beta^{\mu_1 \dots \mu_n}$ satisfy the commutation relations

$$\left[S^{\mu\nu}, \beta^{\mu_1 \dots \mu_n}\right] = \sum_l \left(\eta^{\nu\mu_l}\beta^{\mu_1 \dots \mu \dots \mu_n} - \eta^{\mu\mu_l}\beta^{\mu_1 \dots \nu \dots \mu_n}\right)$$

The most simple example of equations is the well-known Dirac equation.

It is easy to verify that the wave equation determines the mass of a given physical state. If we take the momentum representation and consider a $\hat{p}$-system with momentum $\hat{p}^\mu = (\epsilon m_i, \vec{0})$ (the rest system of a particle), the wave equation reduces to the eigenvalue problem of the matrix $\beta^{0 \dots 0}$

$$\beta^{0 \dots 0}\psi = \left(\frac{\epsilon m}{m_i}\right)^n \psi \quad \text{or} \quad \beta^{0 \dots 0}\psi = \lambda_i \psi$$

Now it follows that to each nonzero eigenvalue λ_i there corresponds a nontrivial solution ψ_i, which describes masses $m_i = \epsilon m (\lambda_i)^{-1/n}$.

When dealing with equations it is natural to suppose that no other subsidiary conditions are imposed, because it is well known that wave equations with additional restrictions on fields are in general inconsistent when interactions with other fields are introduced. Also it follows from the above given relation for masses that only the first and second order equations may have a physical (real) mass spectrum. Using relativistic wave equations we define the proper Poincaré basis. It follows that $\beta^{0\dots0}$ commutes with the Lorentz generators S^{kl}, $(k, l = 1, 2, 3)$ (spin generators). The Poincaré basis $\psi_{\mu_i s\sigma}$ at rest is defined as the eigenfunction of three operators $\beta^{0\dots0}$, $\vec{S}^2$ and S^3:

$$\beta^{0\dots0}\psi_{m_i s\sigma} = \lambda_i \psi_{m_i s\sigma}, \quad \vec{S}^2\psi_{m_i s\sigma} = s(s+1)\psi_{m_i s\sigma} \quad S^3\psi_{m_i s\sigma} = \sigma\psi_{m_i s\sigma}$$

The state of an arbitrary momentum $p = L\hat{p}$ is obtained from the rest state $\psi_{m_i s\sigma}$ via the boost transformation.

Dealing with irreducible Lorentz fields $\psi_i, \psi_j, \dots$, we introduce covariant spin projection operators P^s_{ij}, satisfying (*no sum* over j)

$$P^s_{ij} P^{s'}_{jk} = \delta_{ss'} P^s_{ik}$$

where P^s_{ii} are ordinary projection operators that extract spin s from ψ_i, i.e. $\psi^s_i = P^s_{ii}\psi_i$, and P^s_{ij} $(i \neq j)$ are spin transition operators, i.e. $\psi^s_i = P^s_{ij}\psi_j$. Now the n-th order wave equation previously introduced is presented in a general matrix form

$$(-\Box)^{n/2}\beta^{0\dots0}(\partial)\psi(x) = m^n\psi(x)$$

where

$$\beta^{0\dots0}(\partial) = \begin{pmatrix} a_{11}P_{11} & \dots & a_{1n}P_{1n} \\ \vdots & \ddots & \vdots \\ a_{n1}P_{n1} & \dots & a_{nn}P_{nn} \end{pmatrix}, \quad \psi(x) = \begin{pmatrix} \psi_1(x) \\ \vdots \\ \psi_n(x) \end{pmatrix}$$

Here a_{ij} are arbitrary numerical coefficients and operators P_{ij} are

$$P_{ij} = \sum_s a_{ij}(s) P^s_{ij}$$

In the first order equations case $a_{ij}(s)$ are uniquely determined, in the second order case there is some freedom of choice for $a_{ij}(s)$.

Example 2.1 (spin 3/2). Spin 3/2 equations for a vector-bispinor field $\psi_{\mu\alpha}(x)$ may be represented in the following *two-component* form:

$$i\,\Box^{1/2}\begin{pmatrix} P^{3/2}_{11} + \frac{1}{2}P^{1/2}_{11} & aP^{1/2}_{12} \\ bP^{1/2}_{21} & cP^{1/2}_{22} \end{pmatrix}\begin{pmatrix}\psi_1 \\ \psi_2\end{pmatrix} = m\begin{pmatrix}\psi_1 \\ \psi_2\end{pmatrix}$$

where 1 is a representation $1 = (1,\ 1/2)\oplus(1/2,\ 1)$ and $2 = (1/2,\ 0)\oplus(0,\ 1/2)$. Depending on the choice of free real parameters a, b and c we get a lot of different equations. For example, if $ab = -1/4$ and $c = -1/2$ we have a single spin 3/2 equation (Rarita-Schwinger equation). In the case of $ab = c/2$ we have an equation which describes fields of spin 3/2 with mass m and of spin 1/2 with mass $m' = m/|c + 1/2|$ (the latter equations are important in the massless case). There are also equations describing spin 3/2 and two spin 1/2 fields with different masses. The properties of spin 1/2 are determined by the eigenvalues of the 2×2 spin 1/2 matrix $\begin{pmatrix} 1/2 & a \\ b & c \end{pmatrix}$. The covariant form of the above given equation is

$$(i\partial - m)\psi^\mu + i\left(\frac{a}{\sqrt{3}} - \frac{1}{2}\right)\partial^\mu\gamma_\nu\psi^\nu + i\left(\frac{b}{\sqrt{3}} - \frac{1}{2}\right)\gamma^\mu\partial_\nu\psi^\nu + i\left(\frac{3}{8} + \frac{c}{4} - \frac{a+b}{4\sqrt{3}}\right)\gamma^\mu\partial\!\!\!/\gamma_\nu\psi^\nu = 0$$

At that the invariant bilinear form and Lagrangian are given, which means that one can develop the usual Lagrangian formalism for free and interacting fields, as well as derive Green's functions.

Example 2.2 (spin 2). In the spin 2 case the second order wave equation has a matrix form

$$\Box\begin{pmatrix} P^2_{11} - \frac{1}{2}P^0_{11} & aP^0_{12} \\ bP^0_{21} & cP^0_{00} \end{pmatrix}\begin{pmatrix}\psi_1 \\ \psi_2\end{pmatrix} + m^2\begin{pmatrix}\psi_1 \\ \psi_2\end{pmatrix} = 0$$

Its covariant form for a symmetric tensor field is

$$\begin{aligned}\Box h^{\mu\nu} &- \partial^\mu\partial_\rho h^{\rho\nu} - \partial^\nu\partial_\rho h^{\rho\mu} + \left(\frac{1}{2} - \frac{a}{\sqrt{3}}\right)\partial^\mu\partial^\nu h^\rho_\rho + \left(\frac{1}{2} - \frac{b}{\sqrt{3}}\right)\eta^{\mu\nu}\partial_\rho\partial_\sigma h^{\rho\sigma} \\ &+ \left(\frac{a+b}{4\sqrt{3}} + \frac{c}{4} - \frac{3}{8}\right)\eta^{\mu\nu}h^\rho_\rho + m^2h^{\mu\nu} = 0\end{aligned}$$

Depending on the choice of a, b and c we get different equations, describing either a single spin 2, or spin 2 and spin 0.

Similarly one can derive equations for higher spins: 5/2, 3 etc. These are of course more complex.

3. Massless gauge invariant equations

Let us consider the problem of massless fields [4]. It turns out that the same formalism of spin projection operators is useful to derive massless gauge invariant equations for a given helicity. Moreover, we also get additional restrictions on external sources when interactions with an external source are included.

There are different possibilities to describe massless fields, but the most common is to use gauge invariant equations. If we use equations for massive fields, treated above, it appears that some of them are invariant under gauge transformations and describe massless fields with a given helicity. If we write the relativistically invariant wave equation in the massless case ($m = 0$) as $\pi\psi(x) = 0$ and assume that there is a gauge transformation $\delta\psi = Q^g\epsilon$, where Q^g is an operator of gauge transformation, then for certain types of equations in addition to Q^g there also exists an operator Q^z, satisfying $Q^z\pi = 0$. The latter means that if we consider interactions with some external source $\pi\psi(x) = J$ we get a source constraint $Q^zJ = 0$.

Example 3.1 (helicity 3/2). If $ab = c/2$, the equation

$$i\partial\!\!\!/\psi^\mu + i\left(\frac{a}{\sqrt{3}} - \frac{1}{2}\right)\partial^\mu\gamma_\nu\psi^\nu + i\left(\frac{b}{\sqrt{3}} - \frac{1}{2}\right)\gamma^\mu\partial_\nu\psi^\nu + i\left(\frac{3}{8} + \frac{ab}{4} - \frac{a+b}{4\sqrt{3}}\right)\gamma^\mu\partial\!\!\!/\gamma_\nu\psi^\nu = 0$$

is gauge invariant under the following gauge transformation

$$\delta\psi^\mu = \partial^\mu\epsilon - \frac{1}{4}\left(1 + \frac{\sqrt{3}}{2a}\right)\gamma^\mu\partial\!\!\!/\epsilon$$

where ϵ is an arbitrary bispinor field. In our formalisms there are no additional conditions on ϵ. The source constraint is

$$\partial_\mu J^\mu - \frac{1}{4}\left(1 + \frac{\sqrt{3}}{2b}\right)\partial\!\!\!/\gamma_\mu J^\mu = 0$$

Example 3.2 (helicity 2). If $ab = -c/2$, the equation

$$\Box\, h^{\mu\nu} - \partial^\mu \partial_\rho h^{\rho\nu} - \partial^\nu \partial_\rho h^{\rho\mu} + \left(\frac{1}{2} - \frac{a}{\sqrt{3}}\right) \partial^\mu \partial^\nu h^\rho_\rho + \left(\frac{1}{2} - \frac{b}{\sqrt{3}}\right) \eta^{\mu\nu} \partial_\rho \partial_\sigma h^{\rho\sigma}$$
$$+ \left(\frac{a+b}{4\sqrt{3}} - \frac{ab}{2} - \frac{3}{8}\right) \eta^{\mu\nu} h^\rho_\rho = 0$$

is invariant under the gauge transformation

$$\delta h^{\mu\nu} = \partial^{(\mu}\epsilon^{\nu)} - \frac{1}{4}\left(1 + \frac{\sqrt{3}}{2a}\right) \eta^{\mu\nu} \partial_\rho \epsilon^\rho$$

The source constraint reads

$$\partial_\mu J^{\mu\nu} - \frac{1}{4}\left(1 + \frac{\sqrt{3}}{2b}\right) \partial^\nu J^\mu_\mu = 0$$

It should be mentioned that in our approach no other subsidiary conditions are needed. Usually the gauge transformation is written as $\delta h^{\mu\nu} = \partial^{(\mu}\epsilon^{\nu)}$ with the additional restriction $\partial_\rho \epsilon^\rho = 0$. Such additional restriction usually leads to inconsistencies when interactions are included.

It is interesting to note that the massive equations we obtain from gauge invariant equations adding the mass term are not single particle equations. In the $s = 3/2$ case, for example, spin $1/2$ is also present. The same is true in the $s = 2$ case, where spin 0 is also present.

4. Superfield equations of motion, superspin projection operators

Analogous considerations are applicable in the superfield case [5, 6, 7]. To consider irreducible Lorentz superfields $\psi_i(x,\theta), \psi_j(x,\theta), \ldots$, we introduce superspin projection operators E^Y_{ij}, satisfying (no sum over j)

$$E^Y_{ij} E^{Y'}_{jk} = \delta_{YY'} E^Y_{ik}$$

E^Y_{ii} are ordinary projection operators that separate superspin Y from ψ_i, i.e. $\psi^Y_i = E^Y_{ii}\psi_i$, and E^Y_{ij} $(i \neq j)$ are spin transition operators, i.e. $\psi^Y_i = E^Y_{ij}\psi_j$. However, the calculation of superprojectors is rather more complicated than in the ordinary field case. We have developed the formalism of calculation of projection operators in the cases of $N = 1$ and N-extended superfields.

Now the n-th order wave equation, previously introduced, is presented in a general matrix form

$$(-\Box)^{n/2} \Pi(\partial, D) \psi(x,\theta) = m^n \psi(x,\theta)$$

where

$$\Pi(\partial, D) = \begin{pmatrix} a_{11}E_{11} & \ldots & a_{1n}E_{1n} \\ \vdots & \ddots & \vdots \\ a_{n1}E_{n1} & \ldots & a_{nn}E_{nn} \end{pmatrix}, \quad \psi(x,\theta) \begin{pmatrix} \psi_i(x,\theta) \\ \vdots \\ \psi_n(x,\theta) \end{pmatrix}$$

D is a four component superderivative. Here a_{ij} are arbitrary numerical coefficients and operators E_{ij} are

$$E_{ij} = \sum_Y a_{ij}(Y) E^Y_{ij}$$

In most cases $a_{ij}(Y)$ are uniquely determined.

5. Massive superfield equation ($Y = 3/2$)

Here we consider, as an example, only the superspin 3/2 case which is mostly used in the $N = 1$ linear massive supergravity [8, 9]. It appears that if we use only the vector superfield $h^\mu(x, \theta)$ we obtain the equation

$$\Box \left(E_{11}^{3/2} - \frac{2}{3} E_{11}^0 \right)^\mu_\nu h^\nu(x,\theta) + m^2 h^\mu(x,\theta) = 0$$

which describes superspin 3/2 and also superspin 0 with nonphysical mass $im\sqrt{3/2}$. For that reason some other additional restrictions must be added to eliminate nonphysical superspin 0.

Therefore it is useful to add scalar superfield $\varphi(x,\theta)$ and use two superfields. Let us consider the following equation:

$$\Box \begin{pmatrix} E_{11}^{3/2} - \frac{2}{3} E_{11}^0 & aE_{12}^0 \\ bE_{21}^0 & cE_{00}^0 \end{pmatrix} \begin{pmatrix} \varphi_1 \\ \varphi_2 \end{pmatrix} + m^2 \begin{pmatrix} \varphi_1 \\ \varphi_2 \end{pmatrix} = 0$$

where φ_1 is a vector superfield and φ_2 is a scalar superfield.

The superspin 0 depends on the eigenvalues of the matrix $\left(\begin{smallmatrix} -2/3 & a \\ b & c \end{smallmatrix} \right)$. A single superspin 3/2 is described if $ab = -4/9$ and $c = 2/3$, in other cases one or two superspin 0 fields are also present. In the covariant form the equation reads

$$\frac{2}{3}\left[\left(\Box + \frac{d^2}{4}\right) h^\mu - \partial^\mu \partial_\nu h^\nu\right] - \frac{1}{6}\epsilon^\mu{}_{\nu\rho\sigma}\partial^\rho d^\sigma h^\nu + \frac{a}{2} d\partial^\mu \varphi + m^2 h^\mu = 0, \quad -\frac{b}{2} d\partial_\nu h^\nu - \frac{c}{4} d^2\varphi + m^2\varphi = 0$$

where

$$d = \bar{D}D, \quad d^\mu = i\bar{D}\gamma^\mu\gamma^5 D$$

Different approaches to massive equations are presented in Ref. [10].

6. Massless superfield equation for $Y = 3/2$

Let us consider the gauge invariant massless equations presented in Ref. [11]. If we substitute $m = 0$ in the massive superfield equation and write it as $\pi\psi(x,\theta) = 0$ we get massless equations only if there exists a gauge transformation $\delta\psi = Q^g\epsilon$, where Q^g is an operator and ϵ a superfield. Moreover, in addition to Q^g which satisfies $\pi Q^g = 0$ there also exists an operator Q^z, satisfying $Q^z\pi = 0$. The latter means that if we consider interactions with an external source

$$\pi\psi(x,\theta) = J(x,\theta)$$

we get a source constraint $Q^z J(x,\theta) = 0$. Gauge transformations also set conditions on free coefficients $a_{11}, a_{12}, \ldots$. As a matter of fact, the operators Q^g and Q^z are uniquely expressed via the superspin projection operators.

Example 6.1 (massless superspin 3/2)**.** The above mentioned superspin equation is gauge invariant only if $ab = -2c/3$. Take for simplicity $a = b = -2/3$, then

$$\frac{2}{3}\left[\left(\Box + \frac{d^2}{4}\right) h^\mu - \partial^\mu \partial_\nu h^\nu\right] - \frac{1}{6}\epsilon^\mu{}_{\nu\rho\sigma}\partial^\rho d^\sigma h^\nu - \frac{1}{3} d\partial^\mu \varphi = 0, \quad \frac{1}{3} d\partial_\nu h^\nu - \frac{1}{6} d^2\varphi = 0$$

The gauge transformation is now

$$\delta h^\mu = \bar{D}\gamma^\mu\epsilon(x,\theta), \quad \delta\varphi = \bar{D}\epsilon(x,\theta)$$

where $\epsilon(x,\theta)$ is a spinor superfield.

If we consider interaction with an external source

$$\frac{2}{3}\left[\left(\Box + \frac{d^2}{4}\right)h^\mu - \partial^\mu\partial_\nu h^\nu\right] - \frac{1}{6}\epsilon^\mu{}_{\nu\rho\sigma}\partial^\rho d^\sigma h^\nu - \frac{1}{3}d\partial^\mu\varphi = J^\mu(x,\theta)$$

$$\frac{1}{3}d\partial_\nu h^\nu - \frac{1}{6}d^2\varphi = J(x,\theta)$$

we have the source constraint

$$(\gamma_\mu D)_\alpha J^\mu(x,\theta) - D_\alpha J(x,\theta) = 0$$

From the last constraint it follows that

$$2i\partial_\mu J^\mu = dJ, \quad \bar{D}(d)DJ = 0$$

In $N = 1$ supergravity models $J^\mu(x,\theta)$ is connected with an axial source superfield (supercurrent), $J(x,\theta)$ is a scalar source superfield, and $h^\mu(x,\theta)$ is called the metric superfield. It should be mentioned that the latter conditions were posed out of some physical considerations. But adding to $h^\mu(x,\theta)$ a scalar superfield $\varphi(x,\theta)$, we obtain normal gauge invariant second order equation, which in the free field case describes massless superspin $3/2$. Including interaction with an external source we obtain a unique source constraint, which has previously been used in a more complicated and artifical form.

7. Conclusions

We have developed an interesting approach to fields and superfields of arbitrary spin and superspin. At present the field equations for higher spins and superfield equations may seem uninteresting, since they have no useful applications in modern field theory, but this need not remain so. For example, it is possible to analyze different string and superstring models via the fields and superfields. We have analyzed a few string models, writing down the equations for the component fields, and it appears that these equations are not consistent if interactions are introduced.

References

[1] Loide R-K 1984 *Phys. Lett.* B **141** 61
[2] Loide R-K and Suurvarik P 2010 *Proc. Estonian Acad. Sci.* **59** 332
[3] Loide R-K 1985 *J. Phys. A: Math. Gen.* **18** 2833
[4] Loide R-K, Ots I and Saar R 1992 *Proc. Estonian Acad. Sci.* **41** 270
[5] Loide R-K and Suurvarik P 1985 *Proc. Acad. Sci. of ESSR* **34** 248
[6] Loide R-K, Ots I, Saar R and Suurvarik P 2004 *Hadronic J.* **27** 151
[7] Suurvarik P and Loide R-K 2006 *Czech. J. Phys.* **56** 1299
[8] Hinterbichler K 2012 *Rev. Mod. Phys.* **84** 671
[9] Gregoire T, Schwartz M D and Shadmi Y 2004 *JHEP* **0407** 029
[10] Buchbinder I L, James Gates Jr S, Kuzenko S M and Phillips J 2005 *JHEP* **0502** 056
[11] Suurvarik P and Loide R-K 2005 *Czech. J. Phys.* **55** 1521

Journal of Physics: Conference Series **532** (2014) 012017 doi:10.1088/1742-6596/532/1/012017

Drinfeld twisting elements on Hom-bialgebras

A Makhlouf[1] and B Torrecillas[2]

[1]Université de Haute Alsace, Laboratoire de Mathématiques, Informatique et Applications 4 rue des Frères Lumière, F-68093 Mulhouse, France

[2]Department of Algebra and Analysis, University of Almeria, E-04071, Almeria, Spain

Abstract. The purpose of this paper is to introduce the concept of a twisting element based on a Hom-bialgebra and to use it to provide twists or deformations of Hom-associative algebras. Moreover we review the module theory in Hom-setting and show that a twisting element based on a bialgebra gives rise to a twisting element based on a Hom-bialgebra.

1. Introduction

Hom-type algebra structures appeared first in physics contexts. The study of q-deformations, based on discretized or deformed derivatives, of Heisenberg algebras (oscillator algebras), the Virasoro algebra and quantum conformal algebras reveals a generalized Lie algebra structure in which Jacobi identity is deformed by a linear map, see [1, 2, 5, 6, 7, 8, 11, 12, 13, 21, 22, 24]. This kind of algebras called Hom-Lie algebras where introduced and studied first by Hartwig, Larsson and Silvestrov in [17, 23]. The corresponding associative algebras, called Hom-associative algebras, where introduced in [27] and it was shown that a commutator of a Hom-associative algebra gives rise to a Hom-Lie algebra. The coalgebra counterpart and the related notions of Hom-bialgebra and Hom-Hopf algebra were introduced in [29, 30] and some of their properties, extending properties of bialgebras and Hopf algebras, were described. It has become an active area of research, see [3, 4, 14, 15, 16, 26, 28, 32, 35, 37] and [Yau D arXiv:0911.5402]. The corresponding module structure, comodule structure where studied deeply in [34, 36], see also [Yau D arXiv:0812.4695], and for recent developments see also [9, 10, 25, 31] and [Makhlouf A and Panaite F arXiv:1402.1893].

Twisting elements or Twists were introduced by Drinfeld [18] on quasi-Hopf algebras, in order to twist the coproduct without changing its product. They have become an important tool in the classification of finite-dimensional Hopf algebras. Inspired by this Drinfeld's theory of gauge transformations of quasi-Hopf algebras, Giaquinto and Zhang studied in [19] twisting elements based on actions of a bialgebra and related this to the theory of universal deformation formulas. A construction of a new associative algebra of a given associative algebra through twisting elements and universal deformation formulas, is provided.

Let $(B, \mu, \eta, \Delta, \varepsilon)$ be a bialgebra where B is $\mathbb{K}$-vector space, $\mu : B \otimes B \to B$ the multiplication, $\eta : \mathbb{K} \to B$ the unit, $\Delta : B \to B \otimes B$ the comultiplication and $\varepsilon : B \to \mathbb{K}$ the counit.

An element $F \in B \otimes B$ is a *twisting* element (based on B) if the two identities hold:

$$(\varepsilon \otimes \mathrm{id})F = 1 \otimes 1 = (\mathrm{id} \otimes \varepsilon)F, \quad [(\Delta \otimes \mathrm{id})(F)](F \otimes 1) = [(\mathrm{id} \otimes \Delta)(F)](1 \otimes F)$$

They were used to twist or deform the multiplication of any B-module algebra and the comultiplication of any B-module coalgebra.

Theorem 1.1 ([19]). *Let B be a bialgebra and $F \in B \otimes B$ be a twisting element.*

(i) If (A, μ) is a left B-module algebra, then $A_F = (A, \mu \circ F_l, 1_A)$ is an associative $\mathbb{K}$-algebra.

(ii) If (C, Δ, ε) is a right B-module coalgebra, then $C_F = (C, F_r \circ \Delta, \varepsilon)$ is a coassociative $\mathbb{K}$-coalgebra.

Our aim in this paper is to extend this theory and the result above to Hom-algebras. The paper is organized as follows. In the first section, we summarize the basics and properties of Hom-associative algebras, Hom-coassociative coalgebras and related Hom-bialgebras and Hom-Hopf algebras. Section 2 is devoted to module and module Hom-algebra structures over Hom-associative algebras and Hom-bialgebras; we review the theory and provide examples. In Section 3, we introduce the concept of twisting element based on Hom-bialgebra and provide a generalization of Giaquinto and Zhang's results in the Hom-setting. Moreover, we show that along linear maps and some assumptions, a twisting element based on bialgebra gives rise to a twisting element based on Hom-bialgebra.

Throughout this paper $\mathbb{K}$ denotes a field of characteristic 0 and A, B, M are $\mathbb{K}$-modules. In the sequel we denote by $\tau_{ij} : A^{\otimes n} \to A^{\otimes n}$ the map defined by

$$\tau_{ij}(x_1 \otimes \cdots \otimes x_i \otimes \cdots \otimes x_j \otimes \cdots \otimes x_n) = (x_1 \otimes \cdots \otimes x_j \otimes \cdots \otimes x_i \otimes \cdots \otimes x_n)$$

2. Hom-algebras, Hom-coalgebras and Hom-bialgebras

In this section, we review the definitions and properties of usual Hom-structures, provide some examples and recall the twisting principle giving a way to turn, in particular, a classical structure to Hom-structure using algebra morphisms.

A Hom-module is a pair (M, α_M) consisting of a $\mathbb{K}$-module M and a linear self-map $\alpha_M : M \to M$. A morphism of Hom-modules $f : (M, \alpha_M) \to (N, \alpha_N)$ is a morphism of the underlying $\mathbb{K}$-modules that is compatible with the twisting maps, in the sense that $\alpha_N \circ f = f \circ \alpha_M$. The tensor product of two Hom-modules M and N is the pair $(M \otimes N, \alpha_M \otimes \alpha_N)$.

We mean by a Hom-algebra a triple (A, μ, α) in which A is a $\mathbb{K}$-vector space, $\mu : A^{\otimes 2} \to A$ is a linear map and $\alpha : A \to A$ is a linear self, called a structure map. The linear map $\mu^{op} : A^{\otimes 2} \to A$ denotes the opposite map, i.e. $\mu^{op} = \mu \circ \tau_{12}$. A Hom-coalgebra is a triple (A, Δ, α) in which $\Delta : A \to A^{\otimes 2}$ is a linear map and $\alpha : A \to A$ is a linear self map. The linear map $\Delta^{op} : A \to A^{\otimes 2}$ denotes the opposite map, i.e. $\Delta^{op} = \tau_{12} \circ \Delta$. For a linear self-map $\alpha : A \to A$, we denote by α^n the n-fold composition of n copies of α, with $\alpha^0 \cong id$. A Hom-algebra (A, μ, α) (resp. a Hom-coalgebra (A, Δ, α)) is said to be *multiplicative* if $\alpha \circ \mu = \mu \circ \alpha^{\otimes 2}$ (resp. $\alpha^{\otimes 2} \circ \Delta = \Delta \circ \alpha$). The Hom-algebra is called *commutative* if $\mu = \mu^{op}$ and the Hom-coalgebra is called *cocommutative* if $\Delta = \Delta^{op}$. Classical algebras or coalgebras are also regarded as a Hom-algebras or Hom-coalgebras with identity twisting map. Given a Hom-algebra (A, μ, α), we often use the abbreviation $\mu(x, y) = xy$ for $x, y \in A$. Likewise, for a Hom-coalgebra (A, Δ, α), we will use Sweedler's notation $\Delta(x) = \sum_{(x)} x_{(1)} \otimes x_{(2)}$ but often omit the symbol of summation.

Definition 2.1. A *Hom-associative algebra* is a triple (A, μ, α) consisting of a vector space A, a bilinear map $\mu : A \times A \to A$ and a homomorphism $\alpha : A \to A$ satisfying for $x, y, z \in A$

$$\mu(\alpha(x) \otimes \mu(y \otimes z)) = \mu(\mu(x \otimes y) \otimes \alpha(z))$$

A Hom-associative algebra is said to be unital if there exists a unit $\mathbf{1}$ such that $\alpha(\mathbf{1}) = \mathbf{1}$ satisfying

$$\mu(x \otimes \mathbf{1}) = \mu(\mathbf{1} \otimes x) = \alpha(x)$$

We refer to a unital Hom-associative algebra with a quadruple $(A, \mu, \mathbf{1}, \alpha)$ or by (A, μ, η, α) where $\eta : \mathbb{K} \to A$ is a linear map such that $\eta(1) = \mathbf{1}$.

A linear map $\phi\ : A \to A'$ is a morphism of Hom-associative algebras if $\mu' \circ (\phi \otimes \phi) = \phi \circ \mu$ and $\phi \circ \alpha = \alpha' \circ \phi$. It is said to be weak if only the first condition is satisfied. For unital Hom-associative algebras, the morphism is, in addition, compatible with units.

Remark 2.2. Assume that $(A, \mu_A, \mathbf{1}_A, \alpha_A)$ and $(B, \mu_B, \mathbf{1}_B, \alpha_B)$ are two Hom-associative algebras; then $(A \otimes B, \mu_{A\otimes B}, \mathbf{1}_A \otimes \mathbf{1}_B, \alpha_A \otimes \alpha_B)$ is a Hom-associative algebra (called the tensor product of A and B), where $\mu_{A\otimes B}$ is the usual multiplication: $(a \otimes b)(a' \otimes b') = aa' \otimes bb'$.

Definition 2.3. A *Hom-coalgebra* is a triple (C, Δ, β) where C is a $\mathbb{K}$-vector space and $\Delta : C \to C \otimes C, \beta : C \to C$ are linear maps. A *Hom-coassociative coalgebra* is a Hom-coalgebra satisfying

$$(\beta \otimes \Delta) \circ \Delta = (\Delta \otimes \beta) \circ \Delta$$

A Hom-coassociative coalgebra is said to be *counital* if there exists a map $\varepsilon : C \to \mathbb{K}$ such that $\varepsilon \circ \beta = \varepsilon$ satisfying

$$(id \otimes \varepsilon) \circ \Delta = \beta \quad \text{and} \quad (\varepsilon \otimes id) \circ \Delta = \beta$$

Let (C, Δ, β) and (C', Δ', β') two Hom-coalgebras (resp. counital coalgebras). A linear map $\phi\ : C \to C'$ is a morphism of Hom-colgebras if $\Delta' \circ \phi = (\phi \otimes \phi) \circ \Delta$ and $\phi \circ \beta = \beta' \circ \phi$. It is said to be weak if only the first condition is satisfied. For counital Hom-coalgebras, the morphism is, in addition, compatible with the counits.

The original definitions of Hom-bialgebras and Hom-Hopf algebras in [29, 30] involve two different linear maps α and β, with α the structure map for the Hom-associativity condition and β the structure map for the Hom-coassociativity condition. In the sequel we refer to Hom-bialgebra (resp. Hom-Hopf algebras) for Hom-bialgebras (resp. Hom-Hopf algebras) in which $\beta = \alpha$. Notice that there is another direction in which $\beta = \alpha^{-1}$, they are called monoidal Hom-bialgebras and monoidal Hom-Hopf algebras [4].

Definition 2.4. A *Hom-bialgebra* is a 6-uple $(B, \mu, \eta, \Delta, \varepsilon, \alpha)$ where

(B1) (B, μ, η, α) is a unital Hom-associative algebra.

(B2) $(B, \Delta, \varepsilon, \alpha)$ is a counital Hom-coassociative coalgebra.

(B3) The linear maps Δ and ε are morphisms of Hom-algebras with respect to multiplication μ, that is

$$\begin{cases} \Delta(\mathbf{1}) = \mathbf{1} \otimes \mathbf{1}, \quad \text{where } \mathbf{1} = \eta(1) \\ \Delta(xy) = \Delta(x) \bullet \Delta(y) = \sum_{(x)(y)} (x_{(1)}y_{(1)}) \otimes (x_{(2)}y_{(2)}) \\ \varepsilon(\mathbf{1}) = 1 \\ \varepsilon(xy)) = \varepsilon(x)\varepsilon(y) \\ \Delta(\alpha(x)) = \sum_{(x)} \alpha(x_{(1)}) \otimes \alpha(x_{(2)}) \\ \varepsilon \circ \alpha(x) = \varepsilon(x) \end{cases}$$

A Hom-bialgebras morphism is a linear map which is both a morphism of Hom-associative algebras and Hom-coalgebras.

Following Loday's terminology on generalized bialgebras in which the unitality and counitality are removed, we also consider generalized Hom-bialgebras which are bialgebras without units and counits. We refer to them with quadruples (B, μ, Δ, α) and for simplicity we still call them Hom-bialgebras.

Given a Hom-bialgebra $(B, \mu, \eta, \Delta, \varepsilon, \alpha)$, the vector space $Hom(B, B)$ with the multiplication given by the convolution product carries a structure of Hom-associative algebra.

Proposition 2.5 ([29, 30]). *Let $(B, \mu, \eta, \Delta, \varepsilon, \alpha)$ be a Hom-bialgebra. Then the algebra $Hom(B, B)$ is a Hom-associative algebra with the multiplication given by the convolution product defined by*

$$f \star g = \mu \circ (f \otimes g) \circ \Delta$$

and the unit being $\eta \circ \epsilon$, the structure map γ is defined as $\gamma(f) = \alpha \circ f \circ \alpha$.

A Hom-Hopf algebra over a $\mathbb{K}$-vector space H is given by a tuple $(H, \mu, \eta, \Delta, \varepsilon, , \alpha, S)$, where $(H, \mu, \eta, \Delta, \varepsilon, \alpha)$ is a Hom-bialgebra and S is the antipode that is the inverse of the identity over H for the convolution product.

We have the following properties [29, 30]:

- The antipode S is unique,
- $S(\mathbf{1}) = \mathbf{1}$,
- $\varepsilon \circ S = \varepsilon$,
- Let x be a primitive element ($\Delta(x) = \mathbf{1} \otimes x + x \otimes \mathbf{1}$), then $\varepsilon(x) = 0$,
- If x and y are two primitive elements in H. Then we have $\varepsilon(x) = 0$ and the commutator $[x, y] = \mu(x \otimes y) - \mu(y \otimes x)$ is also a primitive element,
- The set of all primitive elements of H, denoted by $Prim(H)$, has a structure of Hom-Lie algebra.

The category of Hom-associative algebras is closed under twisting self-weak morphisms. The following Theorem gives an easy way to provide a new Hom-associative algebra with a given Hom-associative algebra and an algebra morphism. In particular, one may deform classical associative algebra structure to Hom-associative algebra structure. This construction was introduced first by D. Yau in [35] and then applied to several other algebraic structures.

Theorem 2.6 ([35] and Yau D arXiv:1010.3408). *Let $A = (A, \mu, \alpha, \eta)$ be a unital Hom-associative algebra and β be a weak morphism, then $A_\beta = (A, \mu_\beta = \beta \circ \mu, \beta \circ \alpha, \eta)$ is a unital Hom-associative algebra. In particular, let $A = (A, \mu, \eta)$ be a unital associative algebra and $\alpha : A \to A$ be an algebra morphism, i.e. $\alpha \circ \mu = \mu \circ \alpha^{\otimes 2}$. Then $A_\alpha = (A, \mu_\alpha = \alpha \circ \mu, \alpha, \eta)$ is a unital Hom-associative algebra.*

Similarly, the category of Hom-coassociative coalgebras is closed under weak Hom-coalgebra morphisms.

Theorem 2.7. *Let $(A, \Delta, \epsilon, \alpha)$ be a counital Hom-coassociative coalgebra and $\beta : A \to A$ be a weak Hom-coalgebra morphism. Then $(A, \Delta_\beta = \Delta \circ \beta, \alpha \circ \beta, \epsilon)$ is a counital Hom-coassociative coalgebra. In particular, let (A, Δ, ϵ) be a coalgebra and $\beta : A \to A$ be a coalgebra morphism. Then $(A, \Delta_\beta, \beta, \epsilon)$ is a counital Hom-coassociative coalgebra.*

The two previous Theorems lead to the following Proposition and Corollary.

Proposition 2.8. *Let (A, μ, Δ, α) be a Hom-bialgebra and $\beta : A \to A$ be a weak Hom-bialgebra morphism. Then $(A, \mu_\beta = \beta \circ \mu, \Delta_\beta = \Delta \circ \beta, \alpha \circ \beta)$ is a Hom-bialgebra. In particular, let (A, μ, Δ) be a bialgebra and $\beta : A \to A$ be a bialgebra morphism. Then $(A, \mu_\beta = \beta \circ \mu, \Delta_\beta = \Delta \circ \beta, \beta)$ is a Hom-bialgebra.*

Corollary 2.9. *Let (A, μ, α) be a multiplicative Hom-associative algebra, (C, Δ, α) be a Hom-coassociative coalgebra and (B, μ, Δ, α) be a Hom-bialgebra. Then for each integer $n \geq 0$*

(i) $A^n = (A, \mu^{(n)} = \alpha^{2^n - 1} \circ \mu, \alpha^{2^n})$ *is a Hom-associative algebra,*
(ii) $(C, \Delta^{(n)} = \Delta \circ \alpha^{2^n - 1}, \alpha^{2^n})$, *is a Hom-coassociative coalgebra,*
(iii) $(B, \mu^{(n)} = \alpha^{2^n - 1} \circ \mu, \Delta^{(n)} = \Delta \circ \alpha^{2^n - 1}, \alpha^{2^n})$ *is a Hom-bialgebra.*

Example 2.10. Let $\mathbb{K}G$ be the group-algebra over the group G. As a vector space, $\mathbb{K}G$ is generated by $\{e_g : g \in G\}$. If $\alpha : G \to G$ is a group homomorphism, then it can be extended to an algebra endomorphism of $\mathbb{K}G$ by setting

$$\alpha\left(\sum_{g\in G} a_g e_g\right) = \sum_{g\in G} a_g \alpha(e_g) = \sum_{g\in G} a_g e_{\alpha(g)}$$

Consider the usual bialgebra structure on $\mathbb{K}G$ and α a bialgebra morphism. Then, we define a generalized Hom-bialgebra $(\mathbb{K}G, \mu, \alpha, \Delta, \alpha)$ over $\mathbb{K}G$ by setting:

$$\mu(e_g \otimes e_{g'}) = \alpha(e_{g\cdot g'}), \quad \Delta(e_g) = \alpha(e_g) \otimes \alpha(e_g)$$

Example 2.11. Let X be a set and consider the set of non-commutative polynomials $\mathcal{A} = \mathbb{K}\langle X\rangle$. It carries a bialgebra structure with a comultiplication defined for $x \in X$ by

$$\delta(x) = 1 \otimes x + x \otimes 1, \quad \delta(1) = 1 \otimes 1$$

Let α be a bialgebra morphism. We define a Hom-bialgebra $(\mathcal{A}, \mu, \Delta, \alpha)$ by

$$\mu(f \otimes g) = f(\alpha(X))g(\alpha(X)), \quad \Delta(x) = \alpha(1) \otimes \alpha(x) + \alpha(x) \otimes \alpha(1), \quad \Delta(1) = \alpha(1) \otimes \alpha(1)$$

3. Modules and module Hom-algebras

In this section we summarize the definitions of modules over Hom-associative algebras and module Hom-algebra notion. Moreover, we provide a construction procedure leading to new examples through linear self maps.

Definition 3.1. Let (A, μ_A, α_A) be a Hom-associative algebra. A (left) *A-module* is a Hom-module (M, α_M) together with a linear map $\rho : A \otimes M \to M$, such that

$$\alpha_M \circ \rho = \rho \circ (\alpha_A \otimes \alpha_M), \quad \rho \circ (\alpha_A \otimes \rho) = \rho \circ (\mu_A \otimes \alpha_M) \tag{3.1}$$

A morphism $f : (M, \alpha_M) \to (N, \alpha_N)$ of A-modules is a morphism of the underlying Hom-modules that is compatible with the structure maps, in the sense that $f \circ \rho_M = \rho_N \circ (id_A \circ f)$.

The first identity in (3.1) is equivalent to ρ being a morphism of Hom-modules.

Theorem 3.2. *Let (A, μ_A, α_A) be a Hom-associative algebra and (M, α_M) be an A-module with structure map $\rho : A \otimes M \to M$. Let $\gamma_A : A \to A$ be a Hom-associative algebra morphism and $\gamma_M : M \to M$ be a linear map. Set for $s = 0, 1$*

$$\rho_\gamma^s = \gamma_M \circ \rho \circ (\gamma_A^s \otimes id) : A \otimes M \to M$$

Then ρ_γ^s gives the Hom-module $M = (M, \alpha_M \circ \gamma_M)$ the structure of an A_γ-module, where A_γ is the Hom-associative algebra $(A, \mu_\gamma = \gamma_A \circ \mu, \alpha_A \circ \gamma_A)$.

Proof. The conditions (3.1) are satisfied, respectively. Really, for $s = 0$ we have

$$\begin{aligned}(\gamma_M \circ \alpha_M) \circ (\gamma_M \circ \rho) &= (\gamma_M \circ \alpha_M) \circ (\rho \circ (\gamma_A \circ \gamma_M)) = \gamma_M \circ \rho \circ (\alpha_A \otimes \alpha_M) \circ (\gamma_A \circ \gamma_M) \\ &= (\gamma_M \circ \rho) \circ ((\alpha_A \circ \gamma_A) \otimes (\alpha_M \circ \gamma_M))\end{aligned}$$

and

$$(\gamma_M \circ \rho)((\alpha_A \circ \gamma_A) \otimes (\gamma_M \circ \rho)) = (\gamma_M \circ \rho) \circ ((\alpha_A \circ \gamma_A) \otimes (\rho \circ (\gamma_A \otimes \gamma_M)))$$

$$\begin{aligned}
&= (\gamma_M \circ \rho) \circ (\alpha_A \otimes \rho) \circ (\gamma_A \otimes \gamma_A \otimes \gamma_M) \\
&= (\gamma_M \circ \rho) \circ (\mu_A \otimes \alpha_M) \circ (\gamma_A \otimes \gamma_A \otimes \gamma_M) \\
&= (\gamma_M \circ \rho) \circ (\mu_A \circ (\gamma_A \otimes \gamma_A) \otimes (\alpha_M \circ \gamma_M)) \\
&= (\gamma_M \circ \rho) \circ (\gamma_A \circ \mu_A \otimes (\alpha_M \circ \gamma_M))
\end{aligned}$$

The proof is similar when $s = 1$. □

The Theorem above generalizes slightly the following results obtained in [arXiv:0911.5402].

Corollary 3.3. *Let (A, μ_A, α_A) be a multiplicative Hom-associative algebra and (M, α_M) be an A-module with structure map $\rho : A \otimes M \to M$. For each integer $n, k \geq 0$ define the map*

$$\rho^{n,k} = \alpha_M^{2^k-1} \circ \rho \circ (\alpha_A^n \otimes id_M) : A \otimes M \to M$$

Then each $\rho^{n,k}$ gives the Hom-module $M^k = (M, \alpha_M^{2^k})$ the structure of an A^k-module, where A^k is the k-derived Hom-associative algebra $(A, \mu_A^{(k)} = \alpha_A^{2^k-1} \circ \mu_A, \alpha_A^{2^k})$.

Note that

$$\rho^{0,0} = \rho,\ \rho^{1,0} = \rho \circ (\alpha_A \otimes id_M),\ \rho^{0,1} = \alpha_M \circ \rho,\ \rho^{n+1,0} = \rho^{n,0} \circ (\alpha_A \otimes id_M),\ \rho^{0,k+1} = \alpha_M^{2^k-1} \circ \rho$$

The A^k-module M^k with $\rho^{n,k}$ is called the (n,k)-derived module of M.

Corollary 3.4. *Let (A, μ_A) be an associative algebra and M be an A-module (classical sense) with structure map $\rho : A \otimes M \to M$. Suppose $\alpha_A : A \to A$ is an algebra morphism and $\alpha_M : M \to M$ is a linear self-map such that $\alpha_M \circ \rho = \rho \circ (\alpha_A \otimes \alpha_M)$. For any integer $n, k \geq 0$ define the map*

$$\rho_\alpha^{n,k} = \alpha_M^{2^k} \circ \rho \circ (\alpha_A^n \otimes id_M) : A \otimes M \to M$$

Then each $\rho^{n,k}$ gives the Hom-module $M_\alpha^k = (M, \alpha_M^{2^k})$ the structure of an A_γ-module, where $\gamma = \alpha_A^{2^k}$ and A_γ is the Hom-associative algebra $(A, \mu_\gamma = \gamma \circ \mu_A, \gamma)$.

Example 3.5 (Hom-quantum group $\mathcal{U}_q(\mathfrak{sl}_2)_\alpha$, see Yau D 2009 arXiv:0907.1880). Consider the quantum group $\mathcal{U}_q(\mathfrak{sl}_2)$ generated as usual by $E, F, K^{\pm 1}$ satisfying the relations

$$KK^{-1} = \mathbf{1} = K^{-1}K, \quad KE = q^2 EK,\ KF = q^{-2} FK, \quad EF - FE = \frac{K - K^{-1}}{q - q^{-1}}$$

and the bialgebra morphism $\alpha_\lambda : \mathcal{U}_q(\mathfrak{sl}_2) \to \mathcal{U}_q(\mathfrak{sl}_2)$ defined by

$$\alpha_\lambda(E) = \lambda E, \quad \alpha_\lambda(F) = \lambda^{-1} F, \quad \alpha_\lambda(K^{\pm 1}) = K^{\pm 1}$$

Then $\mathcal{U}_q(\mathfrak{sl}_2)_\alpha = (\mathcal{U}_q(\mathfrak{sl}_2), \mu_{\alpha_\lambda}, \Delta_{\alpha_\lambda}, \alpha_\lambda)$ is a Hom-bialgebra. One obtains the following finite-dimensional modules over $\mathcal{U}_q(\mathfrak{sl}_2)_\alpha$. Assume $q \in \mathbb{C} - \{0\}$ is not a root of unity. For each integer $n \geq 0$ and $\epsilon \in \{\pm 1\}$, there is an $(n+1)$-dimensional simple $\mathcal{U}_q(\mathfrak{sl}_2)$-module $V(\epsilon, n)$. Let $\{v_i\}_{1 \leq i \leq n}$ be a basis of $V(\epsilon, n)$. The action is defined by

$$Kv_i = \epsilon q^{n-2i} v_i\,, \quad Ev_i = \epsilon [n - i + 1]_q v_{i-1}\,, \quad Fv_i = [i + 1]_q v_{i+1}$$

where $v_{-1} = 0 = v_{n+1}$ and $[n]_q$ denotes the q-number of n. Pick any scalar $\xi \in \mathbb{C}$ and define $\alpha_\xi : V(\epsilon, n) \to V(\epsilon, n)$ by setting $\alpha_\xi(v_i) = \xi \lambda^{-i} v_i, \quad \forall i$. Then

$$\alpha_\xi(Uv) = \alpha_\lambda(U)\alpha_\xi(v), \quad \forall U \in \mathcal{U}_q(\mathfrak{sl}_2), \quad v \in V(\epsilon, n)$$

Applying the previous result, one constructs an uncountable, four parameter family $V(\epsilon, n)_\alpha^{r,k}$ of $(n+1)$-dimensional derived $\mathcal{U}_q(\mathfrak{sl}_2)_\beta$-module, where $\beta = \alpha_\lambda^{2^k} = \alpha_{\lambda^{2^k}}$ and the structure map is given for any $r, k \geq 0$ by

$$\rho_\alpha^{n,k}(K^{\pm 1} \otimes v_i) = (\epsilon q^{n-2i})^{\pm 1}(\xi\lambda^{-i})^{2^k} v_i\,, \quad \rho_\alpha^{n,k}(E \otimes v_i) = \epsilon[n-i+1]_q \xi^{2^k} \lambda^{r-2^k(i-1)} v_{i-1}$$
$$\rho_\alpha^{n,k}(F \otimes v_i) = \epsilon[i+1]_q \xi^{2^k} \lambda^{-r-2^k(i+1)} v_{i+1}$$

The classical case corresponds to $\lambda = \xi = 1$.

Now, we define the concept of B-module Hom-algebra, where B is a Hom-bialgebra, introduced and studied in [arXiv:0911.5402].

Definition 3.6. Let $(B, \mu, \Delta_B, \alpha_B)$ be a Hom-bialgebra and (A, μ_A, α_A) be a Hom-associative algebra. A *B-module Hom-algebra* structure on A consists of a B-module structure $\rho : B \otimes A \to A$ such that the module Hom-algebra axiom

$$\alpha_B^2(x)(ab) = (x_{(1)}a)(x_{(2)}b) \tag{3.2}$$

is satisfied for all $x \in B$ and $a, b \in A$, where $\rho(x \otimes a) = xa$.

In argument-free form the Hom-algebra axiom is

$$\rho \circ (\alpha_B^2 \otimes \mu_A) = \mu_A \circ \rho^{\otimes 2} \circ \tau_{23} \circ (\Delta_B \otimes id_A \otimes id_A)$$

Remark 3.7. Let $(B, \mu, \Delta_B, \alpha_B)$ be a Hom-bialgebra and let (M, α_M) and (N, α_N) be B-modules with structure maps ρ_M and ρ_N. Then $M \otimes N$ is a B-module with structure map

$$\rho_{MN} = (\rho_M \otimes \rho_N) \circ \tau_{23} \circ (\Delta \otimes id_M \otimes id_N) : \quad B \otimes M \otimes N \to M \otimes N$$

Theorem 3.8 (arXiv:0911.5402). *Let $(B, \mu_B, \Delta_B, \alpha_B)$ be a Hom-bialgebra, (A, μ_A, α_A) be a Hom-associative algebra and $\rho : B \otimes A \to A$ be a B-module structure on A.*

Then the module Hom-algebra axiom is satisfied if and only if $\mu_A : A \otimes A \to A$ is a morphism of B-modules, in which $A \otimes A$ and A are given the B-module structure maps ρ_{AA} and $\rho^{2,0}$, respectively.

The following Theorem generalizes slightly the construction of a derived double-sequence of module Hom-algebras from a given module Hom-algebra obtained in [arXiv:0911.5402] (see the Corollary).

Theorem 3.9. *Let $(B, \mu_B, \Delta_B, \alpha_B)$ be a Hom-bialgebra, (A, μ_A, α_A) be a Hom-associative algebra and $\rho : B \otimes A \to A$ be a B-module structure on A. Suppose $\gamma_B : B \to B$ is a Hom-bialgebra morphism and $\gamma_A : A \to A$ is a Hom-associative algebra morphism such that $\gamma_A \circ \rho = \rho \circ (\gamma_B \otimes \gamma_A)$. Set, for $s = 0, 1$,*

$$\rho_\gamma^s = \gamma_B \circ \rho \circ (\gamma_B^s \otimes id_A) : B \otimes A \to A$$

Then each ρ_γ^s gives A_γ the structure of an B_γ-module Hom-algebra, where A_γ is the Hom-associative algebra $(A, \mu_\gamma = \gamma_A \circ \mu_A, \alpha_A \circ \gamma_A)$ and B_γ is the Hom-bialgebra

$$(B, \mu_\gamma = \gamma_B \circ \mu_B, \Delta_\gamma = \Delta \circ \gamma_B, \alpha_B \circ \gamma_B)$$

Proof. The Hom-associative algebra structure of A_γ and the Hom-bialgebra structure of B_γ are ensured by Theorems 2.6 and 2.8. We check now the compatibility condition.

$$\begin{aligned}
\rho_\gamma^s \circ ((\alpha_B \circ \gamma_B)^2 \otimes \gamma_A \circ \mu_A) &= \gamma_B \circ \rho \circ (\gamma_B^s \otimes id_A) \circ (\alpha_B^2 \circ \gamma_B^2 \otimes \gamma_A \circ \mu_A) \\
&= \gamma_B \circ \rho \circ (\alpha_B^2 \otimes \mu_A) \circ (\gamma_B^{s+2} \otimes \gamma_A \otimes \gamma_A) \\
&= \gamma_B \circ \mu_A \circ (\rho \otimes \rho) \circ \tau_{23} \circ (\Delta_B \otimes id_A \otimes id_A) \circ (\gamma_B^{s+2} \otimes \gamma_A \otimes \gamma_A) \\
&= \gamma_B \circ \mu_A \circ (\rho \otimes \rho) \circ \tau_{23} \circ (\gamma_B^{s+1} \otimes \gamma_B^{s+1} \otimes \gamma_A \otimes \gamma_A) \circ ((\Delta_B \circ \gamma_B) \otimes id_A \otimes id_A) \\
&= \gamma_B \circ \mu_A \circ (\gamma_B \circ \rho \circ (\gamma_B^s \otimes id_A))^{\otimes 2} \circ \tau_{23} \circ ((\Delta_B \circ \gamma_B) \otimes id_A \otimes id_A) \\
&= (\gamma_B \circ \mu_A) \circ (\rho_\gamma^s)^{\otimes 2} \circ \tau_{23} \circ ((\Delta_B \circ \gamma_B) \otimes id_A \otimes id_A)
\end{aligned}$$ □

Corollary 3.10 (arXiv:0911.5402). *Let (B, μ_B, Δ_B) be a bialgebra, (A, μ_A) be an associative algebra and $\rho : B \otimes A \to A$ be a B-module structure on A. Suppose $\alpha_B : B \to B$ is a bialgebra morphism and $\alpha_A : A \to A$ is an algebra morphism such that $\alpha_A \circ \rho = \rho \circ (\alpha_B \otimes \alpha_A)$. For any integer $n, k \geq 0$ define the map*

$$\rho_\alpha^{n,k} = \alpha_A^{2^k} \circ \rho \circ (\alpha_B^n \otimes id_A) : B \otimes A \to A.$$

Then each $\rho^{n,k}$ gives A_β the structure of an B_γ-module Hom-algebra, where $\beta = \alpha_A^{2^k}$, $\gamma = \alpha_B^{2^k}$, A_β is the Hom-associative algebra $(A, \mu_\beta = \beta \circ \mu_A, \beta)$ and B_γ is the Hom-bialgebra

$$(B, \mu_\gamma = \gamma \circ \mu_B, \Delta_\gamma = \Delta \circ \gamma, \gamma)$$

Example 3.11 (arXiv:0911.5402). Assume $q \in \mathbb{C} - \{0\}$ is not a root of unity. The $\mathcal{U}_q(\mathfrak{sl}_2)$-module algebra structure on the quantum plane

$$\mathbb{A}_q^{2|0} = \mathbb{K} < x, y > /(yx - qxy)$$

is defined using the following quantum partial derivatives

$$\partial_{q,x}(x^n y^m) = [m]_q x^{m-1} y^n, \quad \partial_{q,y}(x^n y^m) = [n]_q x^m y^{n-1}$$

and for $P = P(x, y) \in \mathbb{A}_q^{2|0}$ we define

$$\sigma_x(P) = P(qx, y), \quad \sigma_y(P) = P(x, qy)$$

The action $\rho : \mathcal{U}_q(\mathfrak{sl}_2) \otimes \mathbb{A}_q^{2|0} \to \mathbb{A}_q^{2|0}$ is determined by

$$\begin{aligned}
&EP = x(\partial_{q,y}P), \quad FP = (\partial_{q,x}P)y \\
&KP = \sigma_x\sigma_y^{-1}(P) = P(qx, q^{-1}y), \quad K^{-1}P = \sigma_y\sigma_x^{-1}(P) = P(q^{-1}x, qy)
\end{aligned}$$

The bialgebra morphism $\alpha_\lambda : \mathcal{U}_q(\mathfrak{sl}_2) \to \mathcal{U}_q(\mathfrak{sl}_2)$ defined by

$$\alpha_\lambda(E) = \lambda E, \quad \alpha_\lambda(F) = \lambda^{-1}F, \quad \alpha_\lambda(K^{\pm 1}) = K^{\pm 1}$$

and the algebra morphism $\alpha : \mathbb{A}_q^{2|0} \to \mathbb{A}_q^{2|0}$ defined by $\alpha(x) = \xi x$ and $\alpha(y) = \xi\lambda^{-1}y$ satisfies $\alpha \circ \rho = \rho \circ (\alpha_\lambda \otimes \alpha)$.

By the Theorem, for any integer $l, k \geq 0$ the map

$$\rho_\alpha^{l,k} = \alpha^{2^k} \circ \rho \circ (\alpha_\lambda^l \otimes id) : \mathcal{U}_q(\mathfrak{sl}_2) \otimes \mathbb{A}_q^{2|0} \to \mathbb{A}_q^{2|0}$$

gives the Hom-quantum-plane $(\mathbb{A}_q^{2|0})_\beta$ the structure of a $(\mathcal{U}_q(\mathfrak{sl}_2))_\gamma$-module. We have

$$\begin{aligned}
&\rho_\alpha^{l,k}(E \otimes x^m y^n) = [n]_q \xi^{2^k(m+n)} \lambda^{l-2^k(n-1)} x^{m+1} y^{n-1} \\
&\rho_\alpha^{l,k}(F \otimes x^m y^n) = [m]_q \xi^{2^k(m+n)} \lambda^{-l-2^k(n+1)} x^{m-1} y^{n+1} \\
&\rho_\alpha^{l,k}(K^{\pm 1} \otimes P) = P(q^{\pm 1}\xi^{2^k} x, q^{\mp 1}(\xi\lambda^{-1})^{2^k} y)
\end{aligned}$$

4. Hom-type Drinfeld twisting elements

We define now twisting elements based on Hom-bialgebras and discuss deformations of Hom-associative algebras which are B-module Hom-algebras using these twisting elements.

Definition 4.1. Let $(B, \mu_B, \Delta_B, \mathbf{1}_B, \varepsilon_B, \alpha_B)$ be a Hom-bialgebra where α_B is invertible. An element $F \in B \otimes B$ is a twisting element (based on B) if

$$[(\Delta \otimes \alpha_B)(F)][(\alpha_B^2 \otimes \alpha_B^2)(F) \otimes \mathbf{1}] = [(\alpha_B \otimes \Delta)(F)][\mathbf{1} \otimes (\alpha_B^2 \otimes \alpha_B^2)(F)]. \tag{4.1}$$

Remark 4.2. A second axiom needed to handle unitality is expressed by the following identity

$$(\varepsilon \otimes id) \circ (\alpha_B^2 \otimes \alpha_B^2)(F) = 1_B \otimes 1_B = (id \otimes \varepsilon) \circ (\alpha_B^2 \otimes \alpha_B^2)(F). \tag{4.2}$$

We recover the classical notion when the structure map equals the identity.

Universal deformation formulas are particular cases of twistings. A universal deformation formula (UDF) based on a Hom-bialgebra B is a twisting element F of the form

$$F = 1_B \otimes 1_B + tF_1 + t^2F_2 + \cdots + t^nF_n + \cdots \tag{4.3}$$

where each $F_i \in B \otimes B$ and t is a parameter.

Remark 4.3. Since $\Delta(\mathbf{1}) = \mathbf{1} \otimes \mathbf{1}$ and $\alpha_B(\mathbf{1}) = \mathbf{1}$ then $F = \mathbf{1} \otimes \mathbf{1}$ is always a twisting element for any Hom-bialgebra (unital).

We show that under some assumptions the twisting element may remain when the Hom-bialgebra is deformed by a Hom-bialgebra morphism.

Proposition 4.4. *Let $(B, \mu_B, \mathbf{1}_B, \Delta_B, \varepsilon_B, \alpha_B)$ be a Hom-bialgebra and $F \in B \otimes B$ be a twisting element based on B. Suppose $\gamma_B : B \to B$ is a Hom-bialgebra morphism (with $\gamma_B(\mathbf{1}_B) = \mathbf{1}_B$) such that α_B and γ_B commute and $(\gamma_B \otimes \gamma_B)(F) = F$.*

Then F is a twisting element based on the Hom-bialgebra B_γ, where B_γ is defined as

$$(B, \gamma_B \circ \mu, \Delta \circ \gamma_B, \mathbf{1}_B, \varepsilon_B, \alpha_B \circ \gamma_B)$$

Proof. First, B_γ is a Hom-bialgebra according to Proposition 2.8. We verify the identity (4.1):

$$\begin{aligned}
&[(\Delta \circ \gamma_B \otimes \alpha_B \circ \gamma_B)(F)] \cdot_\gamma [((\alpha_B \circ \gamma_B)^2 \otimes (\alpha_B^2 \circ \gamma_B)^2)(F) \otimes \mathbf{1}] \\
&\quad = \gamma_B^{\otimes 3}([(\Delta \otimes \alpha_B)(F)][(\alpha_B^2 \otimes \alpha_B^2)(F) \otimes \mathbf{1}]) \\
&\quad = \gamma_B^{\otimes 3}([(\alpha_B \otimes \Delta)(F)][\mathbf{1} \otimes (\alpha_B^2 \otimes \alpha_B^2)(F)]) \\
&\quad = [(\alpha_B \otimes \Delta)(F)] \cdot_\gamma [\mathbf{1} \otimes (\alpha_B^2 \otimes \alpha_B^2)(F)] \\
&\quad = [(\alpha_B \otimes \Delta) \circ (\gamma_B \otimes \gamma_B)(F)] \cdot_\gamma [\mathbf{1} \otimes (\alpha_B^2 \otimes \alpha_B^2) \circ (\gamma_B^2 \otimes \gamma_B^2)(F)] \\
&\quad = [(\alpha_B \circ \gamma_B \otimes \Delta \circ \gamma_B)(F)] \cdot_\gamma [\mathbf{1} \otimes (\alpha_B^2 \circ \gamma_B^2 \otimes \alpha_B^2 \circ \gamma_B^2)(F)] \\
&\quad = [(\alpha_B \circ \gamma_B \otimes \Delta \circ \gamma_B)(F)] \cdot_\gamma [\mathbf{1} \otimes ((\alpha_B \circ \gamma_B)^2 \otimes (\alpha_B \circ \gamma_B)^2)(F)]
\end{aligned}$$

□

Now, we generalize Theorem 1.1 to Hom-setting and show that a twisting element on a Hom-bialgebra B gives rise to a deformed Hom-associative algebra when it carries a structure of B-module Hom-algebra.

Theorem 4.5. *Let $(B, \mu_B, \mathbf{1}_B, \Delta_B, \varepsilon_B, \alpha_B)$ be a Hom-bialgebra, $F \in B \otimes B$ be a twisting element and $(A, \mu_A, \mathbf{1}_A, \alpha_A)$ be a unital Hom-associative algebra.*

If A is a left B-module Hom-algebra, where $\mathbf{1}_B a = a$ for all $a \in A$, then

$$A_F = (A, \mu_A \circ [(\alpha_B^2 \otimes \alpha_B^2)F]_l, \alpha_A)$$

is a Hom-associative algebra, where the twisted multiplication is the composite defined as

$$[(\alpha_B^2 \otimes \alpha_B^2)F]_l(a \otimes b) = \rho(\alpha_B^2(F_1) \otimes a) \otimes \rho(\alpha_B^2(F_2) \otimes b)$$

with ρ denoting the action of B on A, $a, b \in A$ and $F = F_1 \otimes F_2$.

If in addition, the axiom (4.2) holds and $\rho(x \otimes 1_A) = \varepsilon(x)1_A$ for all $x \in B$, then the Hom-associative algebra A_F is unital.

Proof. The Hom-associativity of $\mu_A \circ [(\alpha_B^2 \otimes \alpha_B^2)F]_l$ can be determined by considering the following diagram, it is a consequence of the commutativity of the outer square:

$$\begin{array}{ccccc}
A \otimes A \otimes A & \xrightarrow{[(\alpha_B^2\otimes\alpha_B^2)F]_l\otimes id} & A \otimes A \otimes A & \xrightarrow{\mu_A\otimes\alpha_A} & A \otimes A \\
\downarrow {\scriptstyle id\otimes[(\alpha_B^2\otimes\alpha_B^2)F]_l} & & \downarrow {\scriptstyle [(\Delta_B\otimes\alpha_B)F]_l} & & \downarrow {\scriptstyle [(\alpha_B^2\otimes\alpha_B^2)F]_l} \\
A \otimes A \otimes A & \xrightarrow{[(\alpha_B\otimes\Delta_B)F]_l} & A \otimes A \otimes A & \xrightarrow{\mu_A\otimes\alpha_A} & A \otimes A \\
\downarrow {\scriptstyle \alpha_A\otimes\mu_A} & & \downarrow {\scriptstyle \alpha_A\otimes\mu_A} & & \downarrow {\scriptstyle \mu_A} \\
A \otimes A & \xrightarrow{[(\alpha_B^2\otimes\alpha_B^2)F]_l} & A \otimes A & \xrightarrow{\mu_A} & A
\end{array}$$

We show that each of the four inner squares commutes. The lower right square expresses the Hom-associativity of μ. The top left square corresponds to the condition (4.1) since we are assuming that F is a twisting element. The commutativity of the two "off-diagonal" squares follows from the module Hom-algebra structure. Indeed, let $a_1, a_2, a_3 \in A$. In one side we have

$$\begin{aligned} P &= [(\alpha_B^2 \otimes \alpha_B^2)F]_l \circ (\mu_A \otimes \alpha_A)(a_1 \otimes a_2 \otimes a_3) = [\alpha_B^2(F_1) \otimes \alpha_B^2(F_2)]_l(\mu_A(a_1 \otimes a_2) \otimes \alpha_A(a_3)) \\ &= (\alpha_B^2(F_1)\mu_A(a_1 \otimes a_2)) \otimes (\alpha_B^2(F_2)\alpha_A(a_3)) \end{aligned}$$

and in the other side we have

$$\begin{aligned} Q &= (\mu_A \otimes \alpha_A) \circ [(\Delta_B \otimes \alpha_B)F]_l(a_1 \otimes a_2 \otimes a_3) = (\mu_A \otimes \alpha_A)(F_{1(1)}a_1 \otimes F_{1(2)}a_2 \otimes \alpha_B(F_2)a_3) \\ &= (\mu_A \left(F_{1(1)}a_1 \otimes F_{1(2)}a_2\right) \otimes \alpha_A(\alpha_B(F_2)a_3) \end{aligned}$$

Since

$$(\alpha_B^2(F_1)\mu_A(a_1 \otimes a_2)) = (\mu_A(F_{1(1)}a_1 \otimes F_{1(2)}a_2), \quad \alpha_B^2(F_2)\alpha_A(a_3) = \alpha_A(\alpha_B(F_2)a_3)$$

then $P = Q$.

The proof is similar for the lower left square. Now, we prove that 1_A remains the unit.

$$\begin{aligned} \mu_A \circ ((\alpha_B^2 \otimes \alpha_B^2)F)_l(a \otimes 1_A) &= \mu_A(\alpha_B^2(F_1)a \otimes \alpha_B^2(F_2)1_A) = \mu_A(\alpha_B^2(F_1)a \otimes \varepsilon(\alpha_B^2(F_2))1_A) \\ &= \mu_A((id \otimes \varepsilon) \circ (\alpha_B^2 \otimes \alpha_B^2)(F)(a \otimes 1_A)) \\ &= \mu_A((1_B \otimes 1_B)(a \otimes 1_A)) = \mu_A(a \otimes 1_A) = \alpha_A(a) \end{aligned}$$ □

Notice that if one use universal deformation formula, that is F is based on $B[[t]]$ then the previous produces clearly on $A[[t]]$ a one parameter formal deformation.

Proposition 4.6. *Let $(B, \mu_B, \Delta_B, \mathbf{1}_B, \varepsilon_B, \alpha_B)$ be a Hom-bialgebra and $(A, \mu_A, \mathbf{1}_A, \alpha_A)$ be a left B-module Hom-algebra. Suppose $\gamma_B : B \to B$ is a Hom-bialgebra morphism and $\gamma_A : A \to A$ is a Hom-associative algebra morphism such that $\gamma_A \circ \rho = \rho \circ (\gamma_B \otimes \gamma_A)$.*

If $F \in B \otimes B$ is a twisting element based on B such that $(\gamma_B \otimes \gamma_B)(F) = F$ then F is a twisting element of the Hom-bialgebra B_{α_B} and the Hom-associative algebra A_{γ_A} deforms to a Hom-associative $(A_{\gamma_A})_F$.

Proof. The proof follows from Theorem 3.9 and Proposition 4.4. □

In particular, when $\alpha_B = id_B$ and $\alpha_A = id_A$, this proposition provides a procedure to obtain a Hom-associative algebra from an associative algebra, which is a B-module algebra where B is a bialgebra and using a twisting element. It is done along linear maps on A and B that deform both the structure of bialgebra and B-module algebra.

Example 4.7 (Hom-Type Taft-Sweedler bialgebra). We consider T_2, the 4-dimensional unital Taft-Sweedler algebra generated by g, x and the relations $(g^2 = 1, x^2 = 0, xg = -gx)$. The comultiplication is defined by $\Delta(g) = g \otimes g$ and $\Delta(x) = x \otimes 1 + g \otimes x$, the counit is given by $\varepsilon(g) = 1,\ \varepsilon(x) = 0$. Set $\{u_1 = 1, u_2 = g, u_3 = x, u_4 = gx\}$ be a basis. The only twisting elements of T_2 of the form

$$F = u_1 \otimes u_1 + \sum_{i,j=1}^{4} \alpha_{i,j} u_i \otimes u_j$$

are given by

$$F = u_1 \otimes u_1 + p_1(u_1 \otimes u_3 - u_2 \otimes u_3) + p_2 u_4 \otimes u_3$$

where p_1, p_2 are parameters in $\mathbb{K}$.

Pick any $\lambda \in \mathbb{K}$, the map $\alpha : T_2 \to T_2$ defined by

$$\alpha(1) = 1, \quad \alpha(g) = g, \quad \alpha(x) = \lambda x, \quad \alpha(gx) = \lambda gx$$

is a bialgebra morphism. Therefore, we obtain a Hom-bialgebra $(T_2)_\lambda$ which is defined by the following table that describes multiplying the ith row elements by the jth column elements:

	u_1	u_2	u_3	u_4
u_1	u_1	u_2	λu_3	λu_4
u_2	u_2	u_1	λu_4	λu_3
u_3	λu_3	$-\lambda u_4$	0	0
u_4	λu_4	$-\lambda u_3$	0	0

with

$$\Delta(u_1) = u_1 \otimes u_1,\ \Delta(u_2) = u_2 \otimes u_2,\ \Delta(u_3) = \lambda(u_3 \otimes u_1 + u_2 \otimes u_3),\ \Delta(u_4) = \lambda(u_4 \otimes u_2 + u_1 \otimes u_4)$$

and $\varepsilon(u_1) = \varepsilon(u_2) = 1, \quad \varepsilon(u_3) = \varepsilon(u_4) = 0$. For $\lambda = -1$, the elements $F = u_1 \otimes u_1 + p_2 u_4 \otimes u_3$, where p_2 is a parameter in $\mathbb{K}$, are twisting elements of the Hom-bialgebra $(T_2)_{\lambda=-1}$.

Remark 4.8. Likewise, all previous definitions and constructions works similarly when considering Hom-coalgebras which are right B-module Hom-coalgebras.

Given a Hom-bialgebra $(B, \mu, \Delta_B, \alpha_B)$, a *$B$-module Hom-coalgebra* structure on Hom-coassociative coalgebra (C, Δ_C, α_C) consists of a (right) B-module structure $\rho : C \otimes B \to C$ such that the following identity holds

$$\Delta_C(\rho(a \otimes \alpha_B^2(x))) = \rho(a_{(1)} \otimes x_{(1)}) \otimes \rho(a_{(2)} \otimes x_{(2)}), \quad \forall x \in B, \quad \forall a \in C$$

The proof of the following result is similar to that of Theorem 4.5. One has to reverse the arrows in the square.

Proposition 4.9. *Let $(B, \mu_B, \mathbf{1}_B, \Delta_B, \varepsilon_B, \alpha_B)$ be a Hom-bialgebra, $F \in B \otimes B$ be a twisting element and $(C, \Delta_C, \varepsilon_C, \alpha_C)$ be a counital Hom-coassociative coalgebra. If C is a right B-module Hom-coalgebra, where $\rho(a \otimes \mathbf{1}_B) = a$ for all $a \in C$, then*

$$C_F = (A, [(\alpha_B^2 \otimes \alpha_B^2)F]_r \circ \Delta_C, \alpha_C)$$

is a Hom-coassociative coalgebra, where

$$[(\alpha_B^2 \otimes \alpha_B^2)F]_r(a \otimes a') = \rho(a \otimes \alpha_B^2(F_1)) \otimes \rho(a' \otimes \alpha_B^2(F_2))$$

with ρ denoting the action of B on C and $a, a' \in C$.

Acknowledgments
The first author thanks the Department of Algebra and Geometry of the Almeria University for their kind hospitality during his visit. The second author was partially supported by MTM2011-27090 from MCI.

References
[1] Alvarez-Gaumé L, Gomez C and Sierra G 1989 *Phys. Lett.* B **220** 142
[2] Aizawa N and Sato H 1990 *Phys. Lett.* B **256** 185
[3] Ammar A, Ejbehi Z and Makhlouf 2011 *J. Lie Theory* **21** 813
[4] Caenepeel S and Goyvaerts I 2011 *Comm. Algebra* **39** 2216
[5] Chaichian M, Ellinas D and Popowicz Z 1990 *Phys. Lett.* B **248** 95
[6] Chaichian M, Isaev A P, Lukierski L, Popowicz Z and Prešnajder P 1991 *Phys.Lett.* B **262** 32
[7] Chaichian M, Kulish P and Lukierski J 1990 *Phys. Lett.* B **237** 401
[8] Chaichian M, Popowicz Z and Prešnajder P 1990 *Phys. Lett.* B **249** 63
[9] Chen Y and Zhang L 2014 *J. Math. Phys.* **55** 031702
[10] Chen Y Wang Z and Zhang L 2013 *J. Math. Phys.* **54** 073515
[11] Curtright T L and Zachos C K 1990 *Phys. Lett.* B **243** 237
[12] Damaskinsky E V and Kulish P 1992 *Zap. Nauch. Semin. LOMI* **189** 37 (Russian); 1992 *J. Soviet Math.* **62** 2963
[13] Daskaloyannis C 1992 *Mod. Phys. Lett.* A **7** 809
[14] Elhamdadi M and Makhlouf A 2013 Hom-quasi-bialgebras *Contemp. Math.* **585** eds N Andruskiewitch, J Cuadra and B Torrecillas (Providence, RI: AMS)
[15] Fregier Y, Gohr A and Silvestrov S 2009 *J. Gen. Lie Theory Appl.* **3** 285
[16] Gohr A 2010 *J. Algebra* **324**, 1483
[17] Hartwig J T, Larsson D and Silvestrov S 2006 *J. Algebra* **295** 314
[18] Drinfeld V.G 1989 *Algebra i Analiz* **1** 114 (Russian); 1990 *Leningrad Math. J.* **1** 1419
[19] Gianquinto T and Zhang J J 1998 *J. Pure Appl. Algebra* 128 133
[20] Guillot P and Kassel C 2010 *Int. Math. Res. Notices IMRN* **10** 1894
[21] Hu N 1999 *Algebra Colloq.* **6** 51
[22] Kassel C 1992 *Comm. Math. Phys.* **146** 343
[23] Larsson D and Silvestrov S 2005 *J. Algebra* **288** 321
[24] Liu K Q 1992 *Lett. Math. Phys.* **24** 257
[25] Liu L and Shen B 2014 *J. Math. Phys.* **55** 031701
[26] Makhlouf A 2010 Paradigm of nonassociative Hom-algebras and Hom-superalgebras, *Jordan Structures in Algebra and Analysis'* eds J Carmona Tapia *et al* (Circulo Rojo: Publishing House) p 145
[27] Makhlouf A and Silvestrov S 2008 *J. Gen. Lie Theory Appl.* **2** 51
[28] Makhlouf A and Silvestrov S 2010 *Forum Math.* **22** 715
[29] Makhlouf A and Silvestrov S 2008 Hom-Lie admissible Hom-coalgebras and Hom-Hopf algebras *Generalized Lie theory in Mathematics, Physics and Beyond* eds S Silvestrov, E Paal, V Abramov and A Stolin (Springer-Verlag) p 189
[30] Makhlouf A and Silvestrov S 2010 *J. Algebra Appl.* **9** 553
[31] Makhlouf A and Panaite F 2014 *J. Math. Phys.* **55** 013501
[32] Sheng Y 2012 *Algebr. Represent. Theory* **15** 1081
[33] Whiterspoon S 2006 *Comm. Algebra* **34** 4187
[34] Yau D 2010 *Int. Electron. J. Algebra.* **8** 45
[35] Yau D 2009 *J. Lie Theory* **19** 409
[36] Yau D 2012 *J. Phys.* A **45** 065203; 2009 **42** 165202
[37] Yau D 2009 *J. Phys.* A **42** 165202

Complexity vs energy: theory of computation and theoretical physics

Y I Manin

Max-Planck-Institut für Mathematik, Vivatsgasse 7, 53111 Bonn, Germany

Abstract. This paper is a survey based upon the talk at the satellite *QQQ* conference to ECM6, *3Quantum: Algebra Geometry Information*, Tallinn, July 2012. It is dedicated to the analogy between the notions of *complexity* in theoretical computer science and *energy* in physics. This analogy is not metaphorical: I describe three precise mathematical contexts, suggested recently, in which mathematics related to (un)computability is inspired by and to a degree reproduces formalisms of statistical physics and quantum field theory.

Contents

1. Introduction and summary

This survey of several recent papers [15, 17, 18, 20, 21] and [19], the latter can also be read in my preprint [arXiv:1301.0081], is dedicated to a deep analogy between the notions of *complexity* in theoretical computer science and *energy* in physics. The analogy is not metaphorical: we describe several precise mathematical contexts, suggested recently, in which mathematics related to (un)computability is inspired by (and to a degree reproduces) formalisms of statistical physics and quantum field theory.

Namely, after recalling basics of the classical computability theory in Sec. 2, we turn to three main subjects:

a) *The problem of (un)computability of the asymptotic bound for error-correcting codes over a fixed finite alphabet (Sec. 3).*

Here M. Marcolli and the author have shown in [21] based upon [15] that the asymptotic bound arises as a phase transition curve between different thermodynamic phases. The relevant partition function is a sum over the ensemble of all codes in which the role of energy is played by the Kolmogorov complexity of the code.

b) *The problem of mathematical foundation for the empirical Zipf's law (describing e.g. the frequency distributions of words in natural languages).*

   It was suggested that this distribution reflects minimization of certain "effort". I show that (in certain contexts) if this effort is defined as complexity, Zipf's law emerges from L. Levin's *a priori* distributions, mathematical theory of which was founded in the 1970's: see Sec. 4 and more detailed argumentation in [20].

c) *The problem of uncomputable in the computability theory (Sec. 5-7).*

It is well-known that the theory of computability unavoidably leads to effects of uncomputability in its own realm: basically, it may be impossible to decide in finite time whether a partial recursive function is defined at a given point.

I draw an analogy between this and problems of infinities in perturbative Quantum Field Theory (QFT). Moreover, I suggest that renormalization schemes from QFT involving first a deformation of the problem and then "subtraction of infinities" can be fruitfully applied in computation theory. This procedure as well involves Kolmogorov complexity. The basic common elements of the two formalisms are graphs appearing as Feynman diagrams in QFT and as flowcharts in computation theory. For more details and related results cf [17, 18, 16].

There are no proofs in this report: we focus on the presentation of basic ideas. The paper is a survey based upon the talk at the satellite *QQQ* conference to ECM6, *3Quantum: Algebra Geometry Information*, Tallinn, July 2012.

2. A brief guide to computability: operadic and categorical perspective

Any single approach to mathematical notion of computability - Turing's machines, Church's lambda calculus, Markov's algorithms - by necessity bypasses rich intuitions governing other approaches. But since this is unavoidable, and since our goal here is to pave the shortest way to Kolmogorov's complexity, for us computability theory here will be based on the theory of (partial) recursive functions.

2.1. Three descriptions of partial recursive functions

A "function", say, $f : X \to Y$, below always means a pair $(f, D(f))$, where $D(f) \subset X$ and $f : D(f) \to Y$ a set-theoretic map. The definition domain is not always mentioned explicitly. If $D(f) = X$, the function might be called "total"; generally it may be called "partial" one. The other extremal case is that of "empty function", with $D(f) = \emptyset$. We put $\mathbf{Z}_+ := \{1, 2, 3, \dots\}$.

(i) *Intuitive description.* A function $f : \mathbf{Z}_+^m \to \mathbf{Z}_+^n$ is (partial) recursive iff it is "semi-computable" in the following sense: there exists an algorithm F accepting as inputs vectors $x = (x_1, \dots, x_m) \in \mathbf{Z}_+$ with the following properties:
   - if $x \in D(f)$, F produces as output $f(x)$.
   - if $x \notin D(f)$, F either produces the output "NO", or works indefinitely long without producing any output.

(ii) *Formal description (sketch).* It starts with two lists:
   - An explicit list of "obviously" semi-computable *basic functions* such as constant functions, projections onto i-th coordinate etc.
   - An explicit list of *elementary operations*, performed over functions, such as composition, inductive definition, and implicit definition by equation, that can be applied to several semi-computable functions and "obviously" produces from them a new semi-computable function.

After that, the set of *partial recursive functions* is defined as the minimal set of functions $f : \mathbf{Z}_+^m \to \mathbf{Z}_+^n$, with all $m, n \geq 0$, containing all basic functions and closed w.r.t. all elementary operations. For details, see e.g. [14], Ch. V.

(iii) *Diophantine description (a difficult theorem).* A function $f : \mathbf{Z}_+^m \to \mathbf{Z}_+^n$ is partial recursive iff there is a polynomial

$$P(x_1, \ldots, x_m; y_1, \ldots, y_n; t_1, \ldots, t_q) \quad \in \mathbf{Z}[x, y, t]$$

such that the graph

$$\Gamma_f := \{(x, f(x))\} \quad \subset \mathbf{Z}_+^m \times \mathbf{Z}_+^n$$

is the projection of the subset $P = 0$ in $\mathbf{Z}_+^m \times \mathbf{Z}_+^n \times \mathbf{Z}_+^q$. For references and a proof, see e.g. [14], Ch. VI.

2.2. Constructive worlds
An (infinite) *constructive world* is a countable set X (usually of some finite Bourbaki structures, such as the set of all error-correcting codes in a fixed alphabet, cf sec. 3.1 below) given together with a class of *structural numberings:* intuitively computable bijections $\nu : \mathbf{Z}_+ \to X$ which form a principal homogeneous space over the group of totally recursive permutations of $\mathbf{Z}_+$. A finite constructive world is any finite set.

Categorical Church's thesis, Part I. Let X, Y be two infinite constructive worlds, $\nu_X : \mathbf{Z}_+ \to X$ and $\nu_Y : \mathbf{Z}_+ \to X$ be their structural numberings, and F be an (intuitive) algorithm that takes as input an object $x \in X$ and produces an object $F(x) \in Y$ whenever x lies in the domain of definition of F; otherwise it outputs "NO" or works indefinitely.

Then $f := \nu_Y^{-1} \circ F \circ \nu_X : \mathbf{Z}_+ \to \mathbf{Z}_+$ is a partial recursive function.

Categorical Church's thesis, Part II. Let $\mathcal{C}$ be a category, whose objects are some infinite constructive worlds, and some finite constructive worlds of all finite cardinalities. Define the set of morphisms $\mathcal{C}(X, Y)$ to be the set partial maps that can be algorithmically computed.

Then $\mathcal{C}$ is equivalent to the category having one infinite object $\mathbf{Z}_+$, one finite object $\{1, \ldots, n\}$ of each cardinality, and partial recursive functions as morphisms. If X is finite, then $\mathcal{C}(X, Y)$ consists of all partial maps.

2.3. Kolmogorov complexity and Kolmogorov order
Let X be a constructive world. For any (semi)-computable function $u : \mathbf{Z}_+ \to X$, the (exponential) complexity of an object $x \in X$ relative to u is

$$K_u(x) := \min \{m \in \mathbf{Z}_+ \,|\, u(m) = x\}$$

If such m does not exist, we put $K_u(x) = \infty$.

Claim 2.1. *There exists such u ("an optimal Kolmogorov numbering", or "decompressor") that for each other (semi)-computable $v : \mathbf{Z}_+ \to X$, some constant $c_{u,v} > 0$, and all $x \in X$, one has*

$$K_u(x) \leq c_{u,v} K_v(x)$$

This $K_u(x)$ is called *Kolmogorov complexity* of x.

A Kolmogorov order of a constructive world X is a bijection $\mathbf{K} = \mathbf{K}_u : X \to \mathbf{Z}_+$ arranging elements of X in the increasing order of their complexities K_u.

Notice that any optimal numbering is only partial function, and its definition domain is not decidable. Moreover, the Kolmogorov complexity K_u itself is *not computable*: it is the lower bound of a sequence of computable functions.

The same can be said about the Kolmogorov order. Moreover, on $\mathbf{Z}_+$ it cardinally differs from the natural order in the following sense: it puts in the initial segments very large numbers that can be at the same time Kolmogorov simple. For example, let $a_n := n^{n^{\cdot^{\cdot^{n}}}}$ (n times). Then $K_u(a_n) \le cn$ for some $c > 0$.

In Sec. 4 below we will discuss other remarkable properties of complexity, in particular, its self-similar fractal properties.

Finally, the indeterminacy of the complexity related to different choices of optimal functions u, v is multiplicatively $\exp(O(1))$. The same is true for the Kolmogorov order.

For a thorough treatment of Kolmogorov complexity, cf [10]. Notice that in the literature one often uses the *logarithmic* Kolmogorov complexity which is defined as the length of the binary presentation of $K_u(x)$. It is interpreted as the length of the *maximally compressed description* of x. For our purposes, exponential version is more convenient, in particular, because it allows us to define an unambiguous Kolmogorov order on $\mathbf{Z}_+$ or any infinite constructive world.

2.4. Oracle assisted computations

The formal description of partial recursive functions in Sec. 2.1 (ii) allows one to define larger classes of partial functions that can be obtained by *oracle assisted* computations. The point is that the standard elementary operations can be applied to *arbitrary* partial functions. Therefore we can add any uncomputable (not partial recursive) functions to the list of basic functions and consider the minimal subset of partial functions containing this expanded list and closed wrt elementary operations.

This option was used in [20] in order to define the respective extensions of the notion of complexity and apply them to the explanation of Zipf's law in the situations, related to oracle assisted computations and library reuse, cf Sec. 4 below.

Formally, we are considering the (pro)perad generated by the elementary operations on partial functions and various algebras over it. It would be important to understand all relations between elementary operations. For the first steps in this direction, cf [26]; for a general formalism, cf [1].

3. Error-correcting codes and their asymptotic bounds

3.1. Basic notations

Choose an alphabet A, a finite set of cardinality $q \ge 2$. A *code* $C \subset A^n$ is a subset of words of length n. *Hamming distance* between two words of the same length is defined as

$$d((a_i),(b_i)) := \mathrm{card}\{i \in (1,\dots,n) \,|\, a_i \ne b_i\}$$

Code parameters are the cardinality of the alphabet q and the numbers $n(C), k(C), d(C)$ defined by:

$$n(C) := n, \quad k(C) := k := [\log_q \mathrm{card}(C)], \quad d(C) := d = \min\{d(a,b) \,|\, a, b \in C, a \ne b\}$$

Briefly, C is an $[n,k,d]_q$-code. Its *code point* is the point

$$x(C) := \left(\frac{k(C)}{n(C)}, \frac{d(C)}{n(C)}\right) \in [0,1]^2$$

Coordinates of $x(C) = (R(C), \delta(C))$ are called *transmission rate* and *relative distance*, respectively.

The idealized scheme of using error-correcting codes for information transmission can be described as follows. Some source data are encoded by a sequence of code words. After transmission through a noisy channel at the receiving end we will get a sequence of possibly corrupted words. If we know probability of corruption of a single letter, we can calculate, how many corrupted letters in a word we may allow for safe transmission; pairs of code words must be then separated by a larger Hamming distance. This necessity puts an upper bound on the achievable transmission rate.

A good code must maximize minimal relative distance when the transmission rate is chosen.

Our discussion up to now was restricted to *unstructured* codes: arbitrary subsets of words. Arguably, one more property of good codes is the existence of efficient algorithms of encoding and decoding. This can be achieved by introduction of *structured* codes. A typical choice is represented by *linear codes*: for them, A is a finite field of q elements, and C is a linear subspace of $\mathbf{F}_q^n$.

3.2. Asymptotic bound

Call *the multiplicity* of a code point the number of codes that project onto it.

Theorem 3.1 (Yu. Manin, 1981 and 2011). *There exists a continuous function $\alpha_q(\delta)$, $\delta \in [0,1]$, with the following properties:*

(i) The set of code points of infinite multiplicity is exactly the set of rational points $(R,\delta) \in [0,1]^2$ satisfying $R \leq \alpha_q(\delta)$. The curve $R = \alpha_q(\delta)$ is called the asymptotic bound.

(ii) Code points x of finite multiplicity all lie strictly above the asymptotic bound and are called isolated ones: for each such point there is an open neighborhood containing x as the only code point.

(iii) The same statements are true for linear codes, with a possibly different asymptotic bound $R = \alpha_q^{lin}(\delta)$.

3.3. Can one compute an asymptotic bound?

During the thirty years since the discovery of the asymptotic bounds, many upper and lower estimates were established for them, especially for the linear case: see the monograph [25]. Upper bounds helped to pinpoint a number of isolated codes.

However, the following most natural problems remain unsolved:

- To find an explicit formula for α_q or α_q^{lin}.
- To find any single value of $\alpha_q(\delta)$ or $\alpha_q^{lin}(\delta)$ for $0 < \delta < 1-q^{-1}$ (at the end segment $[1-q^{-1}, 1]$ these function vanish).
- To find any method of approximate computation of $\alpha_q(\delta)$ or $\alpha_q^{lin}(\delta)$.
- Clearly, $\alpha_q^{lin} \leq \alpha_q$. Is this inequaliy *strict* somewhere?

3.4. A brief survey of some known results

(i) One can count the number of codes of bounded block length n and plot their code points. The standard probabilistic methods then give the following *Gilbert-Varshamov bounds.*

Most unstructured q-ary codes lie lower or only slightly above the Hamming curve

$$R = 1 - H_q(\delta/2), \quad H_q(\delta) = \delta \log_q(q-1) - \delta \log_q \delta - (1-\delta)\log_q(1-\delta)$$

Most linear q-ary codes lie near or only slightly above the Gilbert-Varshamov bound

$$R = 1 - H_q(\delta)$$

In particular,

$$\alpha_q(R) \geq \alpha_q^{lin}(R) \geq 1 - H_q(\delta)$$

(ii) A useful combinatorial upper estimate is the Singleton bound:

$$R(C) + \delta(C) \leq 1 + 1/n(C)$$

Hence $\alpha_q(\delta) \leq 1 - \delta$. It follows that code points lying above this bound are isolated. The following Reed-Solomon (linear) codes $C \subset \mathbf{F}_q^n$ belong to this group.

Choose parameters $1 \leq k \leq n \leq q, d = n + 1 - k$. Choose pairwise distinct $x_1, \ldots, x_n \in \mathbf{F}_q$, Embed the space of polynomials $f(x) \in \mathbf{F}_q[x]$ of degree $\leq k - 1$ into $\mathbf{F}_q^n$ by

$$f \mapsto (f(x_1), \ldots, f(x_n)) \in \mathbf{F}_q^n$$

After works of Goppa, this construction was generalized. Points $x_1, \ldots, x_n \in \mathbf{F}_q$ were replaced by rational points of any smooth algebraic curve over $\mathbf{F}_q$, and polynomials by sections of an invertible sheaf. This allowed one to construct non-isolated linear codes lying partly strictly above the Gilbert-Varshamov bound.

This implies that we cannot "see" the asymptotic bound, plotting the set of (linear) code points of bounded size: we will see a cloud of points, whose upper bound concentrates near the Hamming or Varshamov-Gilbert bounds.

3.5. Partition function for codes involving complexity

The situation drastically changes, at least theoretically, if we allow ourselves to rearrange the codes in the order of growing Kolmogorov complexity.

In order to state our principal theorem, notice that the function $\alpha_q(\delta)$ is continuous and strictly decreasing for $\delta \in [1, 1 - q^{-1})$. Hence the limit points domain $R \leq \alpha_q(\delta)$ can be equally well described by the inequality $\delta \leq \beta_q(R)$ where β_q is the function inverse to α_q.

Fix an $R \in \mathbf{Q} \cap (0, 1)$. For $\Delta \in \mathbf{Q} \cap (0, 1)$, put

$$Z(R, \Delta; \beta) := \sum_{C: R(C) = R,\, \Delta \leq \delta(C) \leq 1} K_u(C)^{-\beta + \delta(C) - 1}$$

where K_u is an (exponential) Kolmogorov complexity on the constructive world of all codes in a given alphabet of cardinality q.

Theorem 3.2. *(i) If $\Delta > \beta_q(R)$, then $Z(R, \Delta; \beta)$ is a real analytic function of β.*

(ii) If $\Delta < \beta_q(R)$, then $Z(R, \Delta; \beta)$ is a real analytic function of β for $\beta > \beta_q(R)$ such that its limit for $\beta - \beta_q(R) \to +0$ does not exist.

The following thermodynamical analogies justify our interpretation of the asymptotic bound a phase transition curve.

a) The argument β of the partition function corresponds to the inverse temperature.
b) The transmission rate R corresponds to the density ρ.
c) Our asymptotic bound transported into $(T = \beta^{-1}, R)$-plane as $T = \beta_q^{-1}(R)$ becomes the phase transition boundary in the (temperature, density)-plane.

4. Zipf's law and Kolmogorov order

4.1. Zipf's law

G. Zipf studied the frequencies with which words of a natural language are used in various texts. He found a remarkably stable pattern [27, 28]: if all words w_k of a language are ranked according to decreasing frequency of their appearance in a representative corpus of texts, then the frequency p_k of w_k is approximately inversely proportional to its rank k: see e.g. Fig. 1 in [12] based upon a corpus containing $4 \cdot 10^7$ Russian words.

Zipf himself has suggested that this distribution "minimizes effort". Mandelbrot in [11] has shown that if we postulate and denote by C_k a certain "cost" (of producing, using etc.) of the word of rank k, then the frequency distribution $p_k \sim 2^{-h^{-1}C_k}$ minimizes the ratio $h = C/H$, where $C := \sum_k p_k C_k$ is the average cost per word, and $H := -\sum_k p_k \log_2 p_k$ is the average entropy, see [13].

We get from this a power law, if $C_k \sim \log k$. An additional problem, what is so special about power -1, must be addressed separately.

In all such discussions, it is more or less implicitly assumed that empirically observed distributions concern fragments of a potential countable infinity of objects. In the mathematical model suggested in [20] it is assumed that these objects form an infinite constructive world in the sense of 2.2 above. Below I will survey this model.

4.2. How minimization of complexity leads to Zipf's law

A mathematical model of Zipf's law is based upon two postulates:

(A) *Rank ordering coincides with a Kolmogorov ordering (up to a factor* $exp(O(1))$*), cf 2.3 above.*

(B) *The probability distribution producing Zipf's law (with exponent* -1*) is (an approximation to) the L. Levin maximal computable from below distribution: see [29, 8, 9, 10].*

If we accept (A) and (B), then Zipf's law follows from two basic properties of Kolmogorov complexity:

(a) *rank of* w *defined according to (A) is* $exp(O(1)) \cdot K(w)$.

(b) *Levin's distribution assigns to an object* w *probability* $\sim KP(w)^{-1}$ *where* KP *is the exponentiated prefix Kolmogorov complexity (cf [10, 2]), and we have, up to* $exp(O(1))$*-factors,*

$$K(w) \preceq KP(w) \preceq K(w) \cdot \log^{1+\varepsilon} K(w)$$

with arbitrary $\varepsilon > 0$.

There is a slight discrepancy between the growth orders of K and KP. This discrepancy ensures the convergence of the series $\sum_w KP(w)^{-1}$. On finite sets of data this small discrepancy is additionally masked by the dependence of both K and KP on the choice of an optimal encoding.

"Minimization of effort" is thus achieved if effort itself is interpreted as the length of the maximally compressed prefix free description of an object.

Such a picture makes sense especially if the objects satisfying Zipf's distribution, are *generated* rather than simply *observed.*

This matches very well the results of the previous section on asymptotic bounds for error-correcting codes: if one produces codes in the order of their Kolmogorov complexity rather than size, their code points will well approximate the picture of the whole domain under the asymptotic bound. Moreover, Levin's distribution very naturally leads to the thermodynamic partition function on the set of codes, and to the interpretation of asymptotic bound as a phase transition curve. In sec. 2, we have written it in the form the form $\sum_C K(C)^{-s(C)}$ where $s(C)$ is a certain function defined on codes and including as parameters analogs of temperature and density. We could replace K with KP, and freely choose the optimal family defining complexity: this would have no influence at all on the form of the phase curve/asymptotic bound.

It is interesting to observe that the mathematical problem of generating good error-correcting codes historically made a great progress in the 1980's with the discovery of algebraic geometric

Goppa codes, that is precisely with the discovery of greatly compressed descriptions of large combinatorial objects.

To summarize, the class of a priori probability distributions that we are considering here is *qualitatively distinct* from those that form now a common stock of sociological and sometimes scientific analysis: cf a beautiful synopsis of the latter by Terence Tao in [23] who also stresses that "mathematicians do not have a fully satisfactory and convincing explanation for how the Zipf's law comes about and why it is universal".

What arguments could furnish such an explanation? Ubiquity of Gaussian distribution, for example, is often explained away by appealing to the central limit theorem: average of many independent random (equally distributed) variables tends to be Gaussian for whatever initial distribution. Below I will argue that universality of Zipf's law is similarly based on the surprisingly self-similar nature of Kolmogorov complexity.

4.3. Fractal landscape and self-similarity of the Kolmogorov complexity

In [10] (pp 103, 105, 178), one can find a schematic graph of logarithmic complexity of naturals. The visible "continuity" of this graph reflects the fact that complexity of $k+1$ in any reasonable encoding is almost the same as complexity of k. It looks as follows: most of the time it follows closely the graph of $\log k$, but infinitely often it drops down, lower than any given computable function:

One does not see or suspect self-similarity. But it is there: if one restricts this graph onto any infinite decidable subset of $\mathbf{Z}_+$ in increasing order, one will get the same complexity relief as for the whole $\mathbf{Z}_+$: in fact, for any recursive bijection f of $\mathbf{Z}_+$ with a subset of $\mathbf{Z}_+$ we have $K(f(x)) = \exp(O(1)) \cdot K(x)$.

If we pass from complexity to a Levin's distribution, that is, basically, invert the values of complexity, these fractal properties survive.

This property can be read as the extreme stability of such a distribution with respect to the passage to various sub-universes of objects, computable renumbering of objects etc., in the same way as the picture of random noise in a stable background is held responsible for universality of normal distribution.

4.4. Complexity on the background of oracle assisted computations and library reuse

In [24] and [arXiv:cs/0508023] T. Veldhuizen considers Zipf's law in an unusual context that did not exist in the days when Kolmogorov, Solomonov and Chaitin made their ground-breaking discoveries, but which provides, in a sense, landscape for an industrial incarnation of complexity. Namely, Veldhuizen studies actual software and software libraries and analyzes possible profits from software reuse. Metaphorically, this is a picture of human culture whose everyday existence depends on a continuous reuse of treasures created by researchers, poets, philosophers, cf [Manin Y I arXiv:1301.0081].

Mathematically, reuse furnishes new tools of compression: roughly speaking, a function f may have a very large Kolmogorov complexity, but the length of the library address of its program may be short, and only the latter counts if one can simply copy the program from the library.

In order to create a mathematical model of reuse and its Zipf's landscape, the notion of an admissible set of partial functions note, I need to define the mathematical notion of *relative Kolmogorov complexity* $K(f|\Phi)$.

4.5. Admissible sets of functions

Consider a set Φ of partial functions $f : (\mathbf{Z}_+^+)^m \to (\mathbf{Z}_+^+)^n$, $m, n \geq 0$. We will call Φ *an admissible set*, if it is countable and satisfies the following conditions.

(i) Φ is closed under composition and contains all projections (forget some coordinates), and embeddings (permute and/or add some constant coordinates).
Any $(m+1,n)$-function can be considered as a family of (m,n)-functions

$$(u_k) : u_k(x_1,\dots,x_m) := u(x_1,\dots,x_m,k)$$

From (i) it follows that for any $u \in \Phi$ and $k \in \mathbf{Z}_+^+$, also $u_k \in \Phi$. Similarly, if $u(x_1,\dots,x_m)$ is in Φ, then

$$U(x_1,\dots,x_m,x_{m+1},\dots,x_{m+n}) \equiv u(x_1,\dots,x_m)$$

is in Φ.

(ii) For any (m,n), there exists such an $(m+1,n)$-function $u \in \Phi$ that the family of functions $u_k : (\mathbf{Z}_+^+)^m \to (\mathbf{Z}_+^+)^n$, contains all (m,n)-functions belonging to Φ. We will say that such a function u (or family (u_k)) is ample.

(iii) Let f be a total recursive function f whose image is decidable, and f defines a bijection between $D(f)$ and image of f. Then Φ contains both f and f^{-1}.

It is shown in [20] that one can define analog of complexity with respect to such a set, $K(x \mid \Phi)$ and, moreover, that such sets can be obtained as "algebras" over a (pro)perad generated by standard operations that usually are applied only to partially recursive functions.

There are many instances of empiric Zipf's laws where our picture might be applicable: cf [6, 5, 4] and also [Murtra B C and Solé R 2010 *On the Universality of Zipf's Law*, Santa Fe Institute, unpublished, available online]. Such a reduction of the Zipf law for natural languages might require for its justification some neurobiological data: cf [12], appendix A in [arXiv:0710.0105].

5. Feynman graphs and perturbation series in quantum physics

5.1. A toy model

Feynman path integral is an heuristic expression of the form

$$\frac{\int_{\mathcal{P}} e^{S(\varphi)} D(\varphi)}{\int_{\mathcal{P}} e^{S_0(\varphi)} D(\varphi)} \tag{5.1}$$

or, more generally, a similar heuristic expression for *correlation functions.*

In the expression (5.1), $\mathcal{P}$ is imagined as a functional space of *classical fields* φ on a *space-time manifold* M; $S : \mathcal{P} \to \mathbf{C}$ is a functional of *classical action* measured in Planck's units. S_0 is its *quadratic part*, or *"free field action"*.

Usually $S(\varphi)$ itself is an integral over M of a local density on M called *Lagrangian.* In our notation

$$S(\varphi) = -\int_M L(\varphi(x))dx$$

The Lagrangian density may depend on derivatives, include distributions etc.

Finally, the integration measure $D(\varphi)$ and the integral itself $\int_{\mathcal{P}}$ should be considered as symbolic constituents of the total expression (4.1) conveying a vague but powerful idea of *"summing quantum amplitudes over virtual classical trajectories"*.

In our toy model, we will replace $\mathcal{P}$ by a finite-dimensional real space. We endow it with a basis indexed by a finite set of "colors" A, and an Euclidean metric g encoded by the symmetric tensor $(g^{ab})_{a,b\in A}$. We put $(g^{ab}) = (g_{ab})^{-1}$.

Journal of Physics: Conference Series **532** (2014) 012018 doi:10.1088/1742-6596/532/1/012018

The action functional $S(\varphi)$ is a formal series in linear coordinates (φ^a) on $\mathcal{P}$, of the form

$$S(\varphi) = S_0(\varphi) + S_1(\varphi)$$

where

$$S_0(\varphi) := -\frac{1}{2}\sum_{a,b} g_{ab}\varphi^a\varphi^b \quad \text{and} \quad S_1(\varphi) := \sum_{k=1}^{\infty}\frac{1}{k!}\sum_{a_1,\dots,a_k\in A} C_{a_1,\dots,a_k}\varphi^{a_1}\dots\varphi^{a_k} \tag{5.2}$$

where $(C_{a_1,\dots,a_n})$ are certain symmetric tensors.

Below we will consider (g_{ab}) and $(C_{a_1,\dots,a_n})$ as independent formal variables, *"formal coordinates on the space of theories"*.

We will express the toy version of (5.1) as a formal series over (isomorphism classes of) graphs.

A (combinatorial) graph τ, by definition, consists of two finite sets: flags F_τ and vertices V_τ. Besides, an involution j_τ of F_τ is given, showing which pairs of flags form halves of edges, and which are not (tails). Finally, the map $\partial_\tau : F_\tau \to V_\tau$ shows to which vertex each graph is incident. The geometric realization of τ is a topological space whose structure is suggested by the choice of words in the definition given by Figure 1:

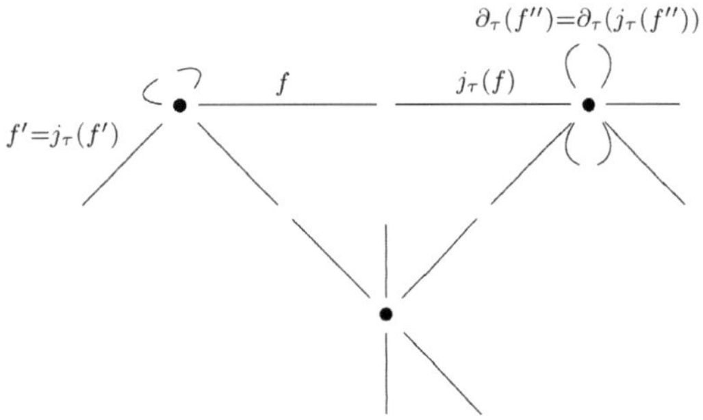

Figure 1. Geometric realization of τ

Each edge e consists of a pair of flags denoted ∂e, and each vertex v determines the set of flags incident to it denoted $F_\tau(v)$. By $\chi(\tau)$ we denote the Euler characteristic of the geometric realization of τ.

Theorem 5.1. *Let λ be a formal parameter. Then*

$$\frac{\int_{\mathcal{P}} e^{\lambda^{-1}S(\varphi)}D(\varphi)}{\int_{\mathcal{P}} e^{\lambda^{-1}S_0(\varphi)}D(\varphi)} = \sum_{\tau\in\Gamma}\frac{\lambda^{-\chi(\tau)}}{|\mathrm{Aut}\,\tau|}w(\tau) \tag{5.3}$$

where τ runs over isomorphism classes of all finite graphs τ. The weight $w(\tau)$ of such a graph is determined by the action functional (1.2) as follows:

$$w(\tau) := \sum_{u:F_\tau\to A}\prod_{e\in E_\tau} g^{u(\partial e)}\prod_{v\in V_\tau} C_{u(F_\tau(v))} \tag{5.4}$$

More precisely, the identity (5.2) is obtained by first interpreting the integrands in the numerator of (5.2) as formal series in $(g_{ab}, C_{a_1,\dots,a_k})$, and then integrating term-wise by using the well known formulas for Gaussian integrals.

6. Graphs as flowcharts, and Hopf algebras

6.1. Graphs as flowcharts

Feynman diagrams of more realistic models and graphs used in the computation theory can be considered as *flowcharts* describing the flow of information from a part of tails playing role of *inputs* to another part, playing role of *outputs*. At vertices, the information gets processed.

In order to make such an interpretation workable, we need pay more attention to orientation. *Orientation* of a graph τ is the decoration $F_\tau \to L_F = \{in, out\}$ such that halves of any edge are decorated by different labels.

Tails of τ oriented *in* (resp. *out*) are called *(global) inputs* T_τ^{in} (resp. *(global) outputs* T_τ^{out}) of τ. Similarly, $F_\tau(v)$ is partitioned into inputs and outputs of the vertex v.

An oriented graph τ is called *directed* if it satisfies the following condition:

On each connected component one can define a continuous real valued function ("time") in such a way that moving in the direction of orientation along each flag inreases the value of this function.

In particular, oriented trees and forests are always directed, and physical Feynman diagrams without loops as well.

An abstract *flowchart* is a directed graph endowed with the decoration of its vertices by a set *Op* of (names of) operations that can be performed on certain inputs producing certain outputs. Generally, flags are also labeled by *types* of the arguments.

To be more precise, flowcharts in theoretical computer science form a natural hierarchy.

At the lower level of this hierarchy, *histories of computations* are situated. For example, the sequence of the states of a Turing machine, performing a concrete computation, may be encoded by a flowchart, in which inputs of all vertices are decorated by 0 or 1, and vertices themselves carry either the name of identical operation or the name of the internal state of the head, reading the respective site. Such a history may well be *infinite.*

At higher levels flowcharts may serve as *descriptions*: programs represented as compositions of some subprograms, but not specifying concrete values of arguments and thus hiding the actual computation process and/or compressing the notation.

We omit here a formal definition of admissible sets of decorated flowcharts: cf [17, 18] for further details. Briefly, an admissible set must be closed wrt finite disjoint unions and *cuts* that will be defined below.

For another version of flowcharts see [22].

6.2. Connes-Kreimer bialgebras of flowcharts [3]

Let Fl be an admissible set of decorated graphs, $k :=$ a commutative ring. We denote by $H = H_{Fl}$ the k-linear span of isomorphism classes $[\tau]$ of graphs τ in Fl and define multiplication by

$$m : H \otimes H \to H, \quad m([\sigma] \otimes [\tau]) := \left[\sigma \coprod \tau\right]$$

We pass now to cuts and comultiplication.

Let τ be an oriented graph. Call *a proper cut* C of τ any partition of V_τ into a disjoint union of two non-empty subsets V_τ^C (upper vertices) and $V_{\tau,C}$ (lower vertices) satisfying the following conditions:

(i) For each oriented wheel in τ, all its vertices belong either to V_τ^C, or to $V_{\tau,C}$.

(ii) If an edge e connects a vertex $v_1 \in V_\tau^C$ to $v_2 \in V_{\tau,C}$, then it is oriented from v_1 to v_2 ("information flows only from past to future").

(iii) Two improper cuts: $\tau^C := \tau$ or $\tau_C = \tau$.

Denote by τ^C (resp. τ_C) the subgraphs of τ consisting of vertices V_τ^C (resp. $V_{\tau,C}$) and incident flags. Put

$$\Delta : H \to H \otimes H, \quad \Delta([\tau]) := \sum_C [\tau^C] \otimes [\tau_C]$$

sum being taken over all cuts of τ.

Claim 6.1. *(i) m defines on H the structure of a commutative k-algebra with unit $[\emptyset]$. Set $\eta : k \to H$ and $1_k \mapsto [\emptyset]$.*

(ii) Δ is a coassociative comultiplication on H with counit

$$\varepsilon : H \to k, \quad \sum_{\tau \in Fl} a_{[\tau]}[\tau] \mapsto a_{[\emptyset]}$$

(iii) $(H, m, \Delta, \varepsilon, \eta)$ is a commutative bialgebra with unit and counit.

Theorem 6.2 (K. Ebrahimi-Fard, D. Manchon [7])**.** *H is a Hopf algebra (i. e. has a unique antipode) if one can introduce an grading on H such that*

$$m(H_p \otimes H_q) \subset H_{p+q}, \quad \Delta(H_n) \subset \oplus_{p+q=n} H_p \otimes H_q$$

and moreover, $H_0 = k[\emptyset]$ is one-dimensional, so that H is connected.

A possible choice of such grading:

$H_n :=$ the k-submodule of H spanned by $[\tau] \in Fl$ with $|F_\tau| = n$.

7. Regularization and renormalization

7.1. Regularization by "minimal subtraction"

Generally, by regularization we mean "producing a finite answer from infinite one". A typical example is this.

Consider the ring $\mathcal{A}$ of germs of meromorphic functions of z at $z = 0$. Put $\mathcal{A}_- := z^{-1}\mathbf{C}[z^{-1}]$, and denote by $\mathcal{A}_+$ the ring of germs of regular functions at $z = 0$. The value of regular function at zero is $\varepsilon_{\mathcal{A}}(f) := f(0)$. Any germ is unique sum of regular one and one belonging to $\mathcal{A}_-$.

If a function is not necessarily regular, the regularized value of f at 0 is $\varepsilon_{\mathcal{A}}(f_+) = f_+(0)$, where

$f_+(z) := f(z) -$ the polar part of f.

Generally, a "minimal subtraction algebra" is a commutative associative K-algebra $\mathcal{A}$ represented as the direct sum of two linear subspaces $\mathcal{A} = \mathcal{A}_+ \oplus \mathcal{A}_-$, each being a subalgebra. Usually $\mathcal{A}$ is unital and $1 \in \mathcal{A}$; besides, we have an augmentation homomorphism $\varepsilon_{\mathcal{A}} : \mathcal{A}_+ \to K$.

7.2. Connes-Kreimer renormalization

This is a version of regularization that:

(i) is performed simultaneously for an infinite family of functions indexeded by flowcharts;

(ii) uses the "division by the collective pole part" in a noncommutative group in place of subtraction of an individual pole.

More precisely, consider a Hopf K-algebra $\mathcal{H}$, and a minimal subtraction unital algebra $\mathcal{A}_+, \mathcal{A}_- \subset \mathcal{A}$, $\varepsilon_\mathcal{A} : \mathcal{A} \to K$.

Denote by $G(\mathcal{A})$ the group of K-linear maps $\varphi : \mathcal{H} \to \mathcal{A}$ such that $\varphi(1_\mathcal{H}) = 1_\mathcal{A}$, with the convolution product

$$\varphi * \psi(x) := m_\mathcal{A}(\varphi \otimes \psi)\Delta(x) = \varphi(x) + \psi(x) + \sum_{(x)} \varphi(x')\psi(x'')'$$

identity $e(x) := u_\mathcal{A} \circ \varepsilon(x)$, and inversion

$$\varphi^{*-1}(x) = e(x) + \sum_{m=1}^{\infty} (e - \varphi)^{*m}(x)$$

In situations that we will consider, for any $x \in \ker \varepsilon$ the latter sum contains only finitely many non-zero summands.

We will say that φ is *a character* if it is a homomorphism of algebras.

Following Birkhoff, we may now define "collective pole" and "collective regular part" of φ. More precisely, if $\mathcal{A}$ is a minimal subtraction algebra, each $\varphi \in G(\mathcal{A})$ admits a unique decomposition of the form

$$\varphi = \varphi_-^{*-1} * \varphi_+, \quad \varphi_-(1) = 1_\mathcal{A}, \quad \varphi_-(\ker \varepsilon) \subset \mathcal{A}_-, \quad \varphi_+(\mathcal{H}) \subset \mathcal{A}_+$$

Values of renormalized polar (resp. regular) parts φ_- (resp. φ_+) on $\ker \varepsilon$ are given by the inductive formulas

$$\varphi_-(x) = -\pi \left(\varphi(x) + \sum_{(x)} \varphi_-(x')\varphi(x'') \right), \quad \varphi_+(x) = (\mathrm{id} - \pi) \left(\varphi(x) + \sum_{(x)} \varphi_-(x')\varphi(x'') \right)$$

Here $\pi : \mathcal{A} \to \mathcal{A}_-$ is the polar part projection in the algebra $\mathcal{A}$.

Physicists invented these inductive formulas: they are known as BPZH-renormalization, for Bogolyubov-Parasyuk-Zimmermann-Hepp.

7.3. Deforming the Halting Problem

Let f be a partial recursive function.The Halting Problem for f is that of recognizing whether a number $k \in \mathbf{Z}_+$ belongs to its definition domain $D(f)$. In this subsection, we will translate it into the problem, whether an analytic function $\Phi(k, f; z)$ of a complex parameter z has a pole at $z = 1$.

The relevant minimal subtraction algebra will be a version of our example from 6.1. Let $\mathcal{A}_+$ be the algebra of analytic functions in $|z| < 1$, continuous at $|z| = 1$, $\varepsilon_\mathcal{A} : \Phi(z) \mapsto \Phi(1)$. Put

$$\mathcal{A}_- := (1-z)^{-1}\mathbf{C}[(1-z)^{-1}], \quad \mathcal{A} := \mathcal{A}_+ \oplus \mathcal{A}_-$$

We now choose an appropriate programming method P and construct its Hopf algebra. Basically, $\mathcal{H} = \mathcal{H}_P$ is the symmetric algebra, spanned by isomorphism classes $[p]$ of certain descriptions. Comultiplication in $\mathcal{H}_P$ is dual to the composition of descriptions.

The main choice is that of characters, corresponding to the halting problem.

The character $\varphi_k : \mathcal{H}_P \to \mathcal{A}$ corresponding to the halting problem at a point $k \in \mathbf{Z}_+$ for the partial recursive function computable with the help of a description $p \in P(\mathbf{Z}_+, \mathbf{Z}_+)$, will be defined as $\varphi_k([p]) := \Phi(k, f; z) \in \mathcal{A}$ where the function Φ is described below.

Using the trick used in the theory of quantum computation (usually applied in the context of finite automata) we will first reduce the general halting problem to the recognition of fixed points of permutations.

Start with a partial recursive function $f: X \to X$, where X is a constructive world. Extend X by one point, i. e. form $X\coprod\{*_X\}$. Choose a total recursive structure of an additive group without torsion on $X\coprod\{*_X\}$ with zero $*_X$. Extend f to the everywhere defined function

$$g: X\coprod\{*_X\} \to X\coprod\{*_X\}, \quad g(y) := *_X \text{ if } y \notin D(f)$$

Define

$$\tau_f : \left(X\coprod\{*_X\}\right)^2 \to \left(X\coprod\{*_X\}\right)^2, \quad \tau_f(x,y) := (x+g(y), y)$$

It is a permutation. Since $(X\coprod\{*_X\}, +)$ has no torsion, the only finite orbits of $\tau_f^{\mathbf{Z}}$ are fixed points.

Moreover, the restriction of τ_f upon the recursive enumerable subset $D(\sigma_f) := (X\coprod\{*_X\}) \times D(f)$ induces a partial recursive permutation σ_f of this subset. Since $g(y)$ never takes the zero value $*_X$ on $y \in D(f)$, but always is zero outside it, the complement to $D(\sigma_f)$ in Y consists entirely of fixed points of τ_f. Thus, the halting problem for f reduces to the fixed point recognition for τ_f.

7.4. The halting problem renormalization character

Define a Kolmogorov numbering on a constructive world X as a bijection $\mathbf{K} = \mathbf{K}_u : X \to \mathbf{Z}_+$ arranging elements of X in the increasing order of their complexities K_u.

Let $\sigma : X \to X$ be a partial recursive map, such that σ maps $D(\sigma)$ to $D(\sigma)$ and induces a permutation of this set. Put $\sigma_{\mathbf{K}} := \mathbf{K}\circ\sigma\circ\mathbf{K}^{-1}$ and consider this as a permutation of the subset

$$D(\sigma_{\mathbf{K}}) := \mathbf{K}\left(D(\sigma)\right) \quad \subset \mathbf{Z}_+$$

consisting of numbers of elements of $D(\sigma)$ in the Kolmogorov order.

If $x \in D(\sigma)$ and if the orbit $\sigma^{\mathbf{Z}}(x)$ is infinite, then there exist such constants $c_1, c_2 > 0$ that for $k := \mathbf{K}(x)$ and all $n \in \mathbf{Z}$ we have

$$c_1 \cdot \mathbf{K}(n) \le \sigma_{\mathbf{K}}^n(k) \le c_2 \cdot \mathbf{K}(n)$$

Now let $X = \mathbf{Z}_+$ and let σ be a partial recursive map, inducing a permutation on its definition domain. Put

$$\Phi(k,\sigma;z) := \frac{1}{k^2} + \sum_{n=1}^{\infty} \frac{z^{\mathbf{K}(n)}}{\left(\sigma_{\mathbf{K}}^n(k)\right)^2}$$

Then we have:

Theorem 7.1. *(i) If σ-orbit of x is finite, then $\Phi(x,\sigma;z)$ is a rational function in z whose all poles are of the first order and lie at roots of unity.*

(ii) If this orbit is infinite, then $\Phi(x,\sigma;z)$ is the Taylor series of a function analytic at $|z|<1$ and continuous at the boundary $|z|=1$.

References

[1] Borisov D and Manin Y 2008 Generalized operads and their inner cohomomorhisms In *Geometry and Dynamics of Groups and Spaces* (In memory of Aleksander Reznikov) eds M Kapranov *et al* Progress in Math **265** (Birkhäuser) p 247
[2] Calude Ch S and Staiger L 2009 *Math. Struct. Comput. Sci.* **19** 45
[3] Connes A and Kreimer D 2000 *Comm. Math. Phys.* **210** 249
[4] Dehaene S 1997 *The Number Sense. How the Mind Creates Mathematics* (Oxford Univ. Press)
[5] Dehaene S and Mehler J 1992 *Cognition* **43** 1
[6] Delahaye J-P 2012 *Pour la Science* **421** 80
[7] Ebrahimi-Fard K and Manchon D 2009 *IRMA Lect. Math. Theor. Phys.* **15** 179 (Zürich: EMS)
[8] Levin L A 1976 *Soviet Math. Dokl.* **17** 522
[9] Levin L A 1984 *Information and Control* **61** 15
[10] Ming Li and Vitányi P 1993 *An Introduction to Kolmogorov Complexity and its Applications* (Springer-Verlag)
[11] Mandelbrot B 1953 An Information Theory of the Statistical Structure of Languages ed W Jackson *Communication Theory* (Woburn, MA: Butterworth) p 486
[12] Manin D Yu 2008 *Cognitive Science* **32** 1075
[13] Manin D Yu 2009 *J. Quantitative Linguistics* **16** 274
[14] Manin Yu I 2010 *A Course in Mathematical Logic for Mathematicians* 2nd edition Graduate Texts in Math. **53** (Springer-Verlag)
[15] Manin Yu 2012 *WTCS 2012 (Calude Festschrift)* **LNCS 7160** p 174
[16] Manin Yu 1999 Classical computing, quantum computing, and Shor's factoring algorithm *Séminaire Bourbaki* **862** (June 1999); 2000 *Astérisque* **266** 375
[17] Manin Yu 2012 Renormalization and computation I. Motivation and background eds J-L Loday and B Vallette *Proceedings OPERADS 2009 Séminaires et Congrès 26* Soc Math de France p 181
[18] Manin Yu 2012 *Math. Struct. Comp. Sci.* **22** Special issue (Cambridge Univ. Press) 729
[19] Manin Yu 2013 Kolmogorov complexity as a hidden factor of scientific discourse: from Newton's law to data mining Talk at the Plenary Session of the Pontifical Academy of Sciences on *Complexity and Analogy in Science: Theoretical, Methodological and Epistemological Aspects* Vatican 5-7 November 2012
[20] Manin Yu 2014 *Funct. Anal. Appl.* **48** 116
[21] Manin Yu and Marcolli M 2014 *J. Differential Geom.* **97** 91
[22] Scott D 1971 The lattice of flow diagrams *Symposium on Semantics of Algorithmic Languages* Lecture Notes in Math. **188** p 311
[23] Tao T 2012 *Daedalus* **141** 23
[24] Veldhuizen T L 2005 Software libraries and their reuse: entropy, Kolmogorov complexity, and Zipf's law *Library-Centric Software Design* (LCSD 2005) an OOPSLA 2005 workshop
[25] Vladut S G, Nogin D Yu and Tsfasman M A 2007 *Algebraic Geometric Codes: Basic Notions* Mathematical Surveys and Monographs **139** (Providence, RI: AMS)
[26] Yanofsky N S 2011 *J. Logic Comput.* **21** 253
[27] Zipf G K 1936 *The Psycho-Biology of Language* (London: Routledge)
[28] Zipf G K 1949 *Human Behavior and the Principle of Least Effort* (Addison-Wesley)
[29] Zvonkin A K and Levin L A 1970 *Uspehi Mat. Nauk* **25** 8

Journal of Physics: Conference Series **532** (2014) 012019 doi:10.1088/1742-6596/532/1/012019

Melikyan algebra is a deformation of a Poisson algebra

H Melikyan[1] and P Zusmanovich[2]

[1]Department of Mathematics and Physics, North Carolina Central University
Durham, NC 27707, U.S.A

[2]Institute of Mathematics and Statistics, University of São Paulo
Rua do Matão 1010, São Paulo 05508-090, Brazil

Abstract. We prove, using computer, that the restricted Melikyan algebra of dimension 125 is a deformation of a Poisson algebra.

1. Introduction

The classification of finite-dimensional simple Lie algebras over an algebraically close field of characteristic $p > 3$ spanned the period of more than 50 years, and was achieved relatively recently, see [6] and [7]. Roughly, each simple algebra is either of classical type or of Cartan type. As it is customary in many branches of mathematics, small characteristics provide additional difficulties: in characteristic $p = 5$, there is a series of simple Lie algebras, discovered by the first author [5], and dubbed thereafter as Melikyan algebras, see [7, Vol. I, §4.3], which do not have analogs in higher characteristics. They are, however, intimately related to Lie algebras of Cartan type, and the present note provides an additional evidence of such a connection.

It was observed long time ago that simple Lie algebras of Cartan type can be deformed into each other (see, for example, [1] and references therein), and such kind of relationship, being interesting for its own sake, is also proved to be useful in structure theory. A similar connection exists between Melikyan algebras and Lie algebras of Cartan type: though they are different, Melikyan algebras are deformations of certain kind of Poisson algebras, the latter being extensions of Hamiltonian algebras. This was stated without proof in [4, §5]. Unfortunately, the proof was never published, and more than decade later, the details seem to be lost.

In the present note we give a proof of this statement in the case of the smallest (and also the only restricted one) algebra in the series of dimension 125, $\mathcal{M}(1,1)$. The strategy of our proof is the following. We construct on computer the deformed algebra $\mathcal{D}$ in question. Then we verify, using computer, that the constructed Lie algebra is simple. Over an algebraically closed field of characteristic 5, there are only three simple Lie algebras of dimension 125 - the algebra of general type $\mathcal{W}_1(3)$, the contact algebra $\mathcal{K}_3(1,1,1)$, and the Melikyan algebra $\mathcal{M}(1,1)$, the latter has a symmetric invariant bilinear form, and the formers do not. We check that the constructed algebra $\mathcal{D}$ has a symmetric invariant bilinear form, and hence it is isomorphic to the Melikyan algebra. After giving all necessary definitions and constructions in Sec. 2, the computer calculations are explained in detail in Sec. 3. In the last Sec. 4 we outline some further questions and possible lines of development.

2. Poisson algebras, cocycles and deformations

We remind some standard definitions and facts concerning Poisson algebras, in a slightly more general, than usual, setting which will be useful for our purposes.

Let A be a commutative associative algebra with unit 1. Let us call a linear map $D : A \to A$ a *generalized derivation* of A, if it satisfies the identity

$$D(ab) = D(a)b + aD(b) - abD(1)$$

for any $a, b \in A$. Note that this class of maps includes ordinary derivations (when $D(1) = 0$), and multiplications by an element of A. It is straightforward to check that the set of all generalized derivations of an algebra A forms a Lie algebra under the operation of commutator (which contains the usual derivations as a subalgebra).

Given a set $\{D_1, \dots, D_n, F_1, \dots F_n\}$ of pairwise commuting generalized derivations of A, the bracket

$$[a, b] = \sum_{i=1}^{n} D_i(a)F_i(b) - F_i(a)D_i(b) \tag{2.1}$$

defines a Lie algebra structure on A which will be denoted as $A_{\{D_i,F_i\}}$. In the particular case when A is a polynomial algebra in even number of variables $x_1, \dots, x_n, y_1, \dots, y_n$ (or, more generally, an algebra of smooth functions on a symplectic manifold), and generalized derivations are the usual ones, the partial derivatives with respect to x_i, y_i, $i = 1, \dots, n$, the so obtained algebra is the ubiquitous Poisson algebra.

In fact, every summand in (2.1) defines a Lie algebra structure on A. In other words, they form *compatible Poisson structures.* This means that first, for any two subsets $\{D_i, F_i \mid i \in I\}$, $\{D_i, F_i | i \in J\}$ of the initial set of commuting generalized derivations, the bracket defining the Lie algebra structure on $A_{\{D_i,F_i \mid i \in J\}}$, is a 2-cocycle on the Lie algebra $A_{\{D_i,F_i \mid i \in I\}}$, and second, the Massey square of this cocycle is zero (or, in other words, the infinitesimal deformation defined by this cocycle is prolonged trivially).

Now, following [4], consider a certain Poisson algebra structure on a 2-variable divided powers algebra. Recall that the divided powers algebra $\mathcal{O}_n(m_1, \dots, m_n)$, corresponding to n-tuple of positive integers $m_1, \dots, m_n$, is a commutative associative algebra defined over a field of characteristic $p > 0$, with the basis

$$\{x_1^{(i_1)} \cdots x_n^{(i_n)} \mid 0 \le i_k < p^{m_k},\ k = 1, \dots, n\}$$

and multiplication

$$x_1^{(i_1)} \cdots x_n^{(i_n)} \cdot x_1^{(j_1)} \cdots x_n^{(j_n)} = \binom{i_1 + j_1}{i_1} \cdots \binom{i_n + j_n}{i_n} x_1^{(i_1+j_1)} \cdots x_n^{(i_n+j_n)}$$

where the standing assumption is that $x_k^{(0)} = 1$, the unit of the algebra, and $x_k^{(i)} = 0$ if i is outside the allowed range, i.e. $i < 0$ or $i \ge p^{m_k}$, $k = 1, \dots, n$.

Obviously,

$$\mathcal{O}_n(m_1, \dots, m_n) \simeq \mathcal{O}_1(m_1) \otimes \cdots \otimes \mathcal{O}_1(m_n) \tag{2.2}$$

The algebra $\mathcal{O}_n(m_1, \dots, m_n)$ possess the following n derivations, lowering the power of each variable:

$$\partial_{x_k} : x_1^{(i_1)} \cdots x_k^{(i_k)} \cdots x_n^{(i_n)} \mapsto x_1^{(i_1)} \cdots x_k^{(i_k-1)} \cdots x_n^{(i_n)}, \quad k = 1, \dots, n$$

Obviously, these derivations pairwise commute. Of course, any map of $\mathcal{O}_n(m_1, \dots, m_n)$ of the form $a\partial_{x_k}$, where $a \in \mathcal{O}_n(m_1, \dots, m_n)$, is a derivation too (in fact, the derivation algebra of $\mathcal{O}_n(m_1, \dots, m_n)$ is linearly spanned by derivations of such form, and constitute the general Lie algebra of Cartan type $\mathcal{W}_n(m_1, \dots, m_n)$).

Let specialize the bracket (2.1) to the following situation: $A = \mathcal{O}_2(2,1)$ (it will be more convenient to denote the generating variables in this case as x and y instead of x_1 and x_2), $p = 5$, $n = 1$, and $D_1 = \partial_x$, $F_1 = \partial_y$. The resulting Poisson Lie algebra $\mathcal{P}$ of dimension 125 has one-dimensional center $K1$, and the commutator of the quotient $[\mathcal{P}/K1, \mathcal{P}/K1]$ is isomorphic to the simple 123-dimensional Lie algebra $\mathcal{H}_2(2,1)$ of Hamiltonian type.

Now define the following 2-cochains $\varphi, \psi : \mathcal{P} \times \mathcal{P} \to \mathcal{P}$:

$$\varphi(a,b) = \partial_x^2(a)\partial_x^3(b) - \partial_x^3(a)\partial_x^2(b, \quad \psi(a,b) = (\mathrm{id} - x\partial_x)(a)\partial_x^5(b) - \partial_x^5(a)(\mathrm{id} - x\partial_x)(b)$$

where $a, b \in \mathcal{O}_2(2,1)$ and id denotes the identity map. Note that both these cochains are of the form (2.1) for $n = 1$, but while ψ is formed by commuting generalized derivation $\mathrm{id} - x\partial_x$ and derivation ∂_x^5, and hence is automatically a cocycle on $\mathcal{P}$ by the discussion above, the map φ is formed by the maps ∂_x^2 and ∂_x^3 which are not derivations. However, straightforward computations (which may be performed on a computer, see below) show that φ is a cocycle too.

Consider the cocycle $\varphi + 2\psi$. Direct computations (which, again, can be performed with the aid of computer) show that this cocycle can be prolonged trivially, and hence define a deformation of the algebra $\mathcal{P}$. Let us denote this deformation by $\mathcal{D}$.

Theorem (Kostrikin-Dzhumadil'daev). $\mathcal{D} \simeq \mathcal{M}(1,1)$.

This theorem is proved on computer, as explained in the next section.

3. Computer calculations

We use the GAP version 4.7.2 [`http://www.gap-system.org/`] which already contains a good deal of Lie-algebraic structures and functions for computation of various invariants of Lie algebras, and augment them by the necessary stuff. All GAP routines described in this section can be found at [`http://justpasha.org/math/poisson-melikyan/`].

The base field in Theorem is algebraically closed, of characteristic 5. However, on computer we naturally work over the finite field $GF(5)$. While this difference is almost immaterial (all algebras in question are definable over $GF(5)$), in order to be rigorous, we distinguish between algebra $\mathcal{X}$, where $\mathcal{X} \in \{\mathcal{P}, \mathcal{D}\}$, defined over an algebraically closed field K, and its form $\mathcal{X}'$ defined over the prime subfield $GF(5)$ (so $\mathcal{X}' \otimes_{GF(5)} K \simeq \mathcal{X}$ as K-algebras).

Step 1. Construction of the algebra $\mathcal{P}'$

First, we provide two routines `PoissonAlgebra` and `SumOfAlgebraStructures` to construct Poisson algebras of kind (2.1). The routine `PoissonAlgebra` takes as arguments an algebra A, and two linear maps $D, F : A \to A$, and returns the bracket (2.1) in the particular case $n = 1$:

$$[a,b] = D(a)F(b) - F(a)D(b)$$

for $a, b \in A$. The routine `SumOfAlgebraStructures` takes as arguments several algebra structures defined on the same vector space V, and returns the algebra structure on V which is the sum of the supplied algebra structures.

Second, we provide two routines `DividedPowersAlgebra` and `TensorProductOfAlgebras` to construct divided powers algebras. The routine `DividedPowersAlgebra` takes as arguments a field K of positive characteristic, and a positive integer n, and returns the algebra $\mathcal{O}_1(n)$ over the field K. The routine `TensorProductOfAlgebras` takes as arguments several algebras defined over the same field, and returns their tensor product. Due to isomorphism (2.2), this enables us to construct an arbitrary divided powers algebra.

All this, together with a few other auxiliary routines implementing the derivations ∂_x and ∂_y as linear operators acting on $\mathcal{O}_2(2,1)$, enables us to construct the algebra $\mathcal{P}'$.

Step 2. Construction of the algebra $\mathcal{D}'$

Using the already employed routines `PoissonAlgebra` and `SumOfAlgebraStructures`, we construct cocycles φ and ψ, and then the bracket

$$\{a,b\} = [a,b] + \varphi(a,b) + 2\psi(a,b)$$

where $a,b \in \mathcal{O}_2(2,1)$, what gives us the algebra $\mathcal{D}'$.

3.1. Step 3. Identifying algebras $\mathcal{D}$ and $\mathcal{M}(1,1)$

Generally, to establish on computer whether two given Lie algebras are isomorphic or not, amounts to solving a system of quadratic equations, and as such, is a difficult task. There are several algorithms for doing that, some of them quite sophisticated (see, for example, [2]), but neither of them, as of time of this writing, would work for algebras of dimension 125 on a reasonable computer within a reasonable amount of time.

However, to establish whether two modules over the *same* algebra are isomorphic or not, is a linear problem, and as such, is much more tractable. Some little theory enables us to take advantage of the latter fact.

In dealing with modules over algebras, we use a highly efficient Meataxe suit of algorithms (see [3, §7.4]). Initially, Meataxe was developed for dealing with finite groups representations, but it works on the level of associative matrix algebras (via group algebras), so it can be utilized for study of representations of Lie algebras (via associative envelopes of Lie algebras) equally well. Using Meataxe routines available in GAP, we can compute whether a given Lie algebra is simple (i.e., the adjoint representation is irreducible), central simple (i.e., the adjoint representation is absolutely irreducible), and possess a symmetric invariant bilinear form (i.e., the adjoint representation is equivalent to its dual). We borrowed the idea to use Meataxe in the Lie-algebraic context from [2].

First, we establish that the algebra $\mathcal{D}'$ is central simple (so $\mathcal{D}$ is simple). According to the classification of simple modular Lie algebras, over an algebraically closed field of characteristic 5 there are only three simple Lie algebras of dimension 125 - the algebra of general type $\mathcal{W}_1(3)$, the algebra of contact type $\mathcal{K}_3(1,1,1)$, and the Melikyan algebra $\mathcal{M}(1,1)$ (see, for example, [7, Vol. I, §4.2] for the dimensions of Lie algebras of Cartan type). In order to distinguish between them, we have to choose some invariant, and the suitable invariant in our case is the presence of a nonzero symmetric invariant bilinear form.

Recall that a symmetric bilinear form $\omega : L \times L \to K$ on a Lie algebra L is called *invariant*, if $\omega([x,z],y) + \omega(x,[y,z]) = 0$ for any $x,y,z \in L$. On simple Lie algebras, this form, if exists, is unique up to a scalar due to Schur lemma, and the existence of nonzero form is equivalent to isomorphism of L-modules $L \simeq L^*$. The algebras $\mathcal{W}_1(3)$ and $\mathcal{K}_3(1,1,1)$ do not have such a form (see [8, §4.6, Theorems 6.3 and 6.6], and also [10] for a short alternative proof and further references), while the Melikyan algebra does ([6, Proposition 6.1]).

As the last step, we establish that $\mathcal{D}'$ has a symmetric invariant bilinear form, and hence so does $\mathcal{D}$. To summarize: $\mathcal{D}$ is a simple Lie algebra of dimension 125 having nonzero symmetric invariant bilinear form, and hence have to be isomorphic to the Melikyan algebra, thus proving the Theorem.

The whole computation takes about 1.5 minutes on a machine with 2.40 GHz CPU.

4. What next?

It goes without saying that a more satisfactory proof of the Theorem would cover the whole series of Melikyan algebras, will not involve computer, and will provide an explicit isomorphism. We are planning to systematically compute deformations of Poisson Lie algebras of the kind appearing in this note, and identify among them those which are isomorphic to Melikyan algebras and other interesting Lie algebras in small characteristics. As a first step, we envisage a linear-algebraic approach to computation of (low-degree) cohomology of a more general class of Poisson algebras, in which the underlying commutative associative algebra A is decomposed as the tensor product of two algebras. This should be somewhat similar to computation of cohomology of current Lie algebras in [9].

Another interesting question is how the p-map comes into play. Note that the Poisson algebra $\mathcal{P}$ is not restricted, while its deformation $\mathcal{M}(1,1)$ is. This phenomenon, where a restricted Lie algebra is a deformation of a not restricted one, and, generally, how the p-map behaves under deformations, deserves further investigation.

Acknowledgements

Thanks are due to Askar Dzhumadil'daev for turning our attention to [4] in the first place, and for subsequent useful discussion. This work was carried out while Zusmanovich was visiting North Carolina Central University, and was supported by the grants HRD-0833184 (NSF), NNX09AV07A (NASA), and ETF9038 (Estonian Science Foundation).

References

[1] Dzhumadil'daev A S 1989 *Mat. Sbornik* **180** 168 1990 Math. USSR Sbornik **66** 169
[2] Eick B 2010 *J. Symb. Comput.* **45** 943
[3] Holt D F, B. Eick B and O'Brien E A 2005 *Handbook of Computational Group Theory* (Chapman & Hall/CRC)
[4] Kostrikin A I and Dzhumadil'daev A S 2000 Modular Lie algebras: new trends *Algebra. Proceeding of the International Algebraic Conference on the Occasion of the 90th Birthday of A.G. Kurosh* ed Yu Bahturin (de Gruyter) p 181
[5] Melikyan H 1980 *Uspekhi Mat. Nauk* **35** N1 203 (Russian); *Russ. Math. Surv.* **35** N1 219 (English translation)
[6] Premet A and Strade H 2008 *J. Algebra* **320** 3559
[7] Strade H 2004 2009 2012 *Simple Lie Algebras over Fields of Positive Characteristic* Vol I-III (de Gruyter)
[8] Strade H and Farnsteiner R 1988 *Modular Lie Algebras and Their Representations* (Marcel Dekker)
[9] Zusmanovich P 2005 *Lin. Algebra Appl.* **407** 71
[10] Zusmanovich P 2011 *Comm. Algebra* **39** 548

Journal of Physics: Conference Series **532** (2014) 012020 doi:10.1088/1742-6596/532/1/012020

Categorical generalization of spinfoam models

A Miković[1,2] and M Vojinović[2]

[1]Departamento de Matemática, Universidade Lusófona de Humanidades e Tecnologias
Av. do Campo Grande, 376, 1749-024 Lisboa, Portugal

[2]Grupo de Fisica Matemática, Universidade de Lisboa
Av. Prof. Gama Pinto, 2, 1649-003 Lisboa, Portugal

Abstract. We give a brief review of the problem of quantum gravity. After the discussion of the nonrenormalizability of general relativity, we briefly mention the main research directions which aim to resolve this problem. Our attention then focuses on the approach of Loop Quantum Gravity, specifically spinfoam models. These models have some issues concerning the semiclassical limit and coupling of matter fields. The recent developments in category theory provide us with the necessary formalism to introduce a new action for general relativity and perform covariant quantization so that the issues of spinfoam models are successfully resolved.

1. Introduction

It is well known that Einstein's theory of General Relativity is not straightforward to quantize. This is easily seen from the fact that GR is not perturbatively renormalizable. Simply put, one can attempt to quantize GR as an ordinary spin-two field in flat Minkowski spacetime, in the following way, for a nice review see [Hamber H W arXiv:0704.2895]. Starting from the usual Einstein-Hilbert action

$$S_{EH} = \int d^4x\, \sqrt{-g}R$$

one rewrites the metric tensor $g_{\mu\nu}$ as the flat Minkowski metric $\eta_{\mu\nu}$ and the spin-two field $h_{\mu\nu}$ as

$$g_{\mu\nu} = \eta_{\mu\nu} + h_{\mu\nu}$$

and substitutes it into the action, rewriting it in terms of the new variable $h_{\mu\nu}$. Thereby one obtains

$$S_{EH} = \int d^4x\, h_{\mu\nu}\Box h^{\mu\nu} + (\text{gauge fixing terms}) + (\text{self-interaction terms})$$

The D'Alambertian operator is defined in the usual way, in the flat Minkowski space,

$$\Box \equiv \eta^{\mu\nu}\partial_\mu\partial_\nu$$

From here one can proceed to perform the standard field theory quantization in the naive way — first formulate the free quantum field theory, and then introduce interactions perturbatively.

However, very soon one is bound to face the difficulty of nonrenormalizability of this theory. The tree-level Feynman diagrams are finite, the one-loop divergences can be removed by wavefunction renormalization, but at the two-loop level a Lagrangian counterterm of the form

$$\mathcal{L}_2 = \frac{const}{\varepsilon^2} R^{\alpha\beta}{}_{\mu\nu} R^{\mu\nu}{}_{\rho\sigma} R^{\rho\sigma}{}_{\alpha\beta} \quad (\varepsilon \to 0)$$

appears [1], which is nonzero on-shell. Here $\varepsilon = 4 - D$ is the cutoff parameter from dimensional regularization scheme. At higher loop levels similar terms involving R^4, R^5, etc. terms are also expected to appear, rendering the theory perturbatively nonrenormalizable. This means that in order to remove all divergences one needs to introduce at least one additional coupling constant for each loop level. The infinite number of these coupling constants implies the loss of predictive power of the theory, since all experiments doable in principle can only ever fix a finite number of coupling constants.

This property of General Relativity has been known for quite some time, and there are various research directions which attempt to address this issue. They can be broadly separated into two classes, by the methodology.

The first class of approaches considers modifying or substituting GR by another theory, which should preferably be renormalizable. Such attempts have evolved into vast research directions such as supergravity, string field theory, noncommutative geometry, and so on. The goal of each proposed model is to have a renormalizable theory that looks like GR at least on the length scales which can be tested experimentally, while at the same time have only a finite number of coupling constants. These coupling constants could then in principle be used to predict the values of the infinite set of coupling constants appearing in the perturbative quantum gravity approach.

The second class of approaches is based on the point of view that abandons the renormalization paradigm, and essentially gives physical meaning to the cutoff parameters of some particular regularization scheme. In other words, the assumption is that at some scale (typically expected to be near the Planck scale) expectation values of the physical observables will start to depend explicitly on cutoff parameters. This dependence is assumed to be measurable (in principle), rather than being removed by renormalization. These attempts have also evolved into vast research directions such as loop quantum gravity, causal dynamical triangulations, causal set theory, etc. The goal of all proposed models is exactly the same as before — predict some definite values for the infinite number of coupling constants present in the perturbative quantum gravity.

All these research directions have had limited success, and in the absence of any experimental data relevant at the Planck scale, none of these directions can be preferred over the others.

In what follows, we shall be mainly concerned with the approach of loop quantum gravity (for a review see [2]), more specifically spin foam models, and we shall propose one novel particular model that addresses some serious issues present in all other spin foam models so far.

In section 2 we shall give a short overview of the status of LQG in general and spin foam models in particular. We will argue that the main drawbacks of all 4D spin foam models stem from the fact that tetrad fields are not basic variables of the theory. Section 3 deals with the categorical generalization of the Poincaré group, called the Poincaré 2-group. This will give us the necessary mathematical tools to reformulate the GR action in a convenient way which includes tetrad fields as basic variables. The analysis of this new action is then given in section 4, with a sketch of a quantization procedure giving rise to the so-called spincube model. Section 5 contains conclusions and discussion of the results.

2. Loop quantum gravity and spin foam models

A detailed review of the Loop Quantum Gravity approach can be found in [2]. Here we just give some basic properties at an informal level.

The basic idea of LQG is to choose diffeomorphism-invariant quantities as basic degrees of freedom for the gravitational field, and then perform a canonical nonperturbative quantization of gravity in terms of these quantities. The natural candidates for basic variables turned out to be Wilson loops, and subsequently their generalizations called spin networks. This choice of variables introduces a natural diffeomorphism-invariant cutoff at the Planck length scale l_P, thereby rendering the theory UV-finite. The quantization is performed in the Schrödinger picture, and provides one with a mathematically well-defined constructions of the kinematical Hilbert space for the theory and some basic operators for geometric observables such as lengths, areas and volumes of space. Evolution in time is embodied in the Hamiltonian constraint, corresponding to the Wheeler-de Witt equation in the LQG setting.

The main features of such canonical approach to quantization are as follows. The theory represents a nonperturbative quantization of GR, and can in principle be applied to the study of physical systems where gravity is the dominant factor at short distances — such systems include the black hole and cosmological singularities. It gives one a mathematical handle on a well-defined Hilbert space of states for the gravitational field, thereby giving some insight into the quantum mechanical features of gravity. The natural basis for the Hilbert space is the set of the *spin network states*, combinatorial graphs colored by the irreducible representations of the $SU(2)$ group, and corresponding intertwiners. Finally, the study of the geometric observables — the length, area and volume operators — reveals that each of them has a discrete spectrum, giving rise to the geometric interpretation of the gravitational field wavefunctional, as well as the discrete character of space.

The theory also has some drawbacks. First, the Hamiltonian constraint is not uniquely defined, due to the usual ordering problems present in quantum mechanics. Second, even if one chooses some particular ordering, the Hamiltonian constraint is extremely complicated and impossible to solve in practice. This severely limits the possibility for any practical calculations and the study of the dynamics of the theory. As the main obstacle, the proof of the correct semiclassical limit of the theory is still missing, as well as any attempt to predict the coupling constants from the perturbative gravity approach.

A way to resolve these drawbacks has been found in the spin foam approach [Rovelli C arXiv:1102.3660]. The idea is to give up canonical quantization, but instead attempt a covariant, path-integral quantization of the theory. Building on the results of the canonical approach, one wants to define the gravitational path-integral

$$Z = \int \mathcal{D}g_{\mu\nu} \exp\left(iS_{EH}[g_{\mu\nu}]\right)$$

in some way, in order to be able to calculate expectation values of observables, both in deep quantum regime and the semiclassical regime. This approach tends to give one a good handle on the dynamics of the theory, in addition to all features of the canonical approach.

The basic procedure of defining Z goes as follows. One starts from the Plebanski action for General Relativity,

$$S = \int B_{ab} \wedge R^{ab} + \phi^{abcd} B_{ab} \wedge B_{cd}$$

The first part of this action represents the topological BF theory for the $SO(3,1)$ group. The R^{ab} is the curvature 2-form, a field strength "F" for the $SO(3,1)$ connection 1-form ω^{ab}. The B_{ab} is the Lagrange multiplier 2-form. The second part of the action is the Plebanski constraint, featuring B_{ab} and the 0-form Lagrange multiplier ϕ^{abcd}. The purpose of the constraint is to enforce the B_{ab} to be a simple 2-form (i.e. an exterior product of two 1-forms). This constraint is therefore called "simplicity constraint", and it can be shown that the simplicity requirement of the B_{ab} field is enough to convert the topological BF theory into General Relativity. The fact that B_{ab} is simple gives rise to nontrivial degrees of freedom in the theory, reducing the equation of motion for ω^{ab} from Riemann-flat to Ricci-flat.

The second step is the quantization of the topological BF theory. This can be done in a rigorous way by employing the methods of topological quantum field theory. One first discretizes spacetime into 4-simplices, motivated by the structure of space in the canonical LQG, and rewrites the BF action in the form

$$\int B_{ab} \wedge R^{ab} \stackrel{\text{discr.}}{\longrightarrow} \sum_{\triangle} B_{\triangle} R_{\triangle}$$

where the sum goes over all triangles in the triangulation. Then one defines a topological invariant

$$Z \equiv \int \mathcal{D}\omega \int \mathcal{D}B \exp\Big(i \sum_{\triangle} B_{\triangle} R_{\triangle} \Big) = \sum_{\Lambda} \prod_{f} A_2(\Lambda_f) \prod_{v} A_4(\Lambda_v)$$

Here Λ are the irreducible representations of $SO(3,1)$, labeling the faces f, edges e and vertices v of the Poincaré dual lattice corresponding to the triangulation. The colored 2-complex dual to the spacetime triangulation is called a *spin foam*. The amplitudes $A_2(\Lambda)$ and $A_4(\Lambda)$ are determined such that Z is in fact a topological invariant — the total expression must not depend on the particular choice of the spacetime triangulation. In that way one arrives at the TQFT corresponding to the BF theory for the $SO(3,1)$ group, commonly called the *Ooguri spin foam model*. Of course, the invariant Z may be (and actually is) badly divergent, but that is not important at this stage, since we are only interested in the structure of the path integral.

The last step in the quantization procedure is to enforce the simplicity constraint on the BF path integral at the quantum level. The exact technique for this is quite involved [3, 4], but the bottom-line is that one projects the $SO(3,1)$ irreducible representations Λ to the $SU(2)$ representations present in the canonical LQG formalism, in order to obtain the same structure of the Hilbert space on the spin foam boundary. The resulting theory is not topologically invariant, but represents one possible rigorous definition for the theory of quantum gravity. The most advanced spin foam model in this respect is the EPRL/FK model, developed independently by two research groups [3, 4].

The main feature of spin foam models is that they correct some drawbacks of the canonical theory, primarily the dynamical sector is more under control. In addition, there remains a certain ambiguity in the choice of the amplitudes A_2 and A_4. This can be conveniently utilized to redefine the model such that it becomes IR-finite and to have a correct semiclassical limit [5, 6]. One can also employ standard QFT methods to calculate the effective action for the model in the semiclassical limit, which opens a possibility to explicitly determine the coupling constants from perturbative quantum gravity.

Unfortunately, the spin foam models introduce their own set of problems. Aside from the "unusual" properties like fuzziness of geometry at the Planck scale, all spin foam models suffer from two major handicaps. The first is related to the fact that, in addition to the good semiclassical limit, all models have *additional semiclassical limits*, which do not give rise to the standard GR, but to the so-called *area-Regge geometry*. Since these different classical limits are not observed in experiments, one needs some additional mechanism to suppress such solutions. However, so far no mechanism could be constructed to deal with this problem.

The second handicap is related to the inability of the spin foam models to couple matter fields to gravity. Namely, the basic geometric variables which are employed in description of spacetime geometry are areas and volumes of space, but not lengths. This situation makes it extremely complicated (and in the case of massive fermionic matter even impossible) to incorporate matter fields into the spin foam model. Even if doable (see [7] for the massless fermion coupling), the resulting theory would be too complicated to be useful for any calculation.

As it turns out, both of these handicaps have a common origin — the edge lengths in the triangulation are not well-defined at the quantum level. This is itself a consequence of the choice of spin network states as basic degrees of freedom in the canonical LQG — the choice which emphasizes the spin connection ω^{ab}, while entirely ignoring the tetrad fields e^a. At the level of spin foam models, it is easy to see that the Plebanski constraint was purposefully designed to require the simplicity of B_{ab}, while avoiding to explicitly state that (the dual of) B_{ab} is the product of two tetrad 1-forms. The reason for this is that the tetrad fields do not appear as variables in the topological BF sector of the theory, which is being used for the definition of the path integral.

In the remainder of this paper we will present a novel way to address this main difficulty, and to introduce tetrad fields explicitly in the topological sector of the theory. However, in order to do this, it is important to introduce some mathematical concepts which provide the background formalism for the new model.

3. Poincaré 2-group

We begin by giving a very brief review of the so-called *categorification ladder*, an important and active research topic in category theory. We shall not attempt at any rigor, leaving out most of the details, which can be found for example in [8] and references therein.

In the branch of mathematics called *category theory*, one defines a structure called a *category* as a set of *objects* and a set of *morphisms* between those objects, satisfying some basic axioms. Such a structure is fairly general and does not have many interesting properties itself. However, this generality allows one to use it for all sorts of purposes. For example, one can define the usual structure of a *group* as a category which has only one object, while all morphisms (mapping the object onto itself) are invertible. The composition rules for the morphisms can be chosen to be the group multiplication, thereby providing an isomorphism between a given group and the corresponding category with one element.

The first step in the categorification ladder is to introduce the concept of a *2-category*. A 2-category consists of a set of objects, a set of morphisms and a set of *2-morphisms*, maps between morphisms. Intuitively, if a category can be represented by a linear graph of dots (objects) and arrows connecting them (morphisms), a 2-category can be represented by a planar graph, consisting of dots (objects), arrows connecting them (morphisms) and "surface arrows" mapping one arrow into another (see [8] for details and pictures). The main point is that the dimensionality of the graph has been raised by one. The categorification ladder can continue by introducing a 3-category (or in general an n-category) by a similar process, leading to 3-dimensional (in general n-dimensional) graphs.

In analogy with a group, one can then define a 2-group, as a 2-category which has only one element, while all morphisms and 2-morphisms are invertible. A 2-group is a categorical generalization of a group, and is not a group itself. One can prove that any 2-group is equivalent to a *crossed module*, a structure that has been studied independently by mathematicians before the idea of the categorification ladder has even been introduced. A crossed module is a quadruple $(G, H, \partial, \triangleright)$. This is a pair of groups G and H, such that $\partial : H \to G$ is a homomorphism and $\triangleright : G \times H \to H$ is an action of G on H such that certain axioms are satisfied, which turn out to be directly related to the structure of a 2-category, see [8]. The elements of G represent the 1-morphisms, while the elements of the semidirect product $G \ltimes H$ represent the 2-morphisms. The canonical example of a 2-group relevant for physics is the Poincaré 2-group, where $G = SO(3,1)$, $H = \mathbb{R}^4$, ∂ is a trivial homomorphism and $\triangleright$ is the usual action of the Lorentz transformations on the $\mathbb{R}^4$ space. The Lorentz group is the group of morphisms, while the usual Poincaré group is the group of 2-morphisms.

The main feature of the whole 2-group formalism is that one can generalize the concept of a *holonomy* along a line to its two-dimensional analog — a *surface holonomy.* The initial interest in this came from string theory. A point-particle travels along a world line in spacetime, and one is naturally led to the concept of a parallel transport along a given line. String theory promotes the point particle into a one-dimensional object — a string — which then travels along a world surface in spacetime. Thus one would like to have a concept of a *parallel transport along a given surface.* One of the main aims of the 2-category and 2-group formalism is to introduce and formalize this concept.

Given the strong categorical relationship between groups and 2-groups, one can construct a gauge theory on a 4-manifold $\mathcal{M}$ based on a crossed module $(G, H, \partial, \triangleright)$ of Lie groups by using 1-forms A, which take values in the Lie algebra $\mathfrak{g}$ of G, and 2-forms β, which take values in the Lie algebra $\mathfrak{h}$ of H [9, 10]. The forms A and β transform under the usual gauge transformations $g : \mathcal{M} \to G$ as

$$A \to g^{-1}Ag + g^{-1}dg\,, \quad \beta \to g^{-1} \triangleright \beta$$

while the gauge transformations generated by H are given by

$$A \to A + \partial\eta\,, \quad \beta \to \beta + d\eta + A \wedge^{\triangleright} \eta + \eta \wedge \eta$$

where η is a one-form taking values in $\mathfrak{h}$, see [10]. When the group H is Abelian, which happens in the Poincaré 2-group case, then the $\eta \wedge \eta$ term vanishes, and one obtains the gauge transformations given in [9].

The pair (A, β) represents a 2-connection on a 2-fiber bundle associated to the 2-Lie group (G, H) and the manifold $\mathcal{M}$. The corresponding curvature forms are given by

$$\mathcal{F} = dA + A \wedge A - \partial\beta\,, \quad \mathcal{G} = d\beta + A \wedge^{\triangleright} \beta$$

and they transform as

$$\mathcal{F} \to g^{-1}\mathcal{F}g\,, \quad \mathcal{G} \to g^{-1} \triangleright \mathcal{G}$$

under the usual gauge transformations, while

$$\mathcal{F} \to \mathcal{F}\,, \quad \mathcal{G} \to \mathcal{G} + \mathcal{F} \wedge^{\triangleright} \eta$$

under the H-gauge transformations.

One can introduce a natural topological gauge theory determined by the vanishing of the 2-curvature $\mathcal{F} = 0$, $\mathcal{G} = 0$. These equations can be obtained from the action

$$S = \int \langle B \wedge \mathcal{F} \rangle_{\mathfrak{g}} + \langle C \wedge \mathcal{G} \rangle_{\mathfrak{h}}$$

where B is a Lagrange multiplier 2-form taking values in $\mathfrak{g}$, C is a Lagrange multiplier 1-form taking values in $\mathfrak{h}$, $\langle\ ,\ \rangle_{\mathfrak{g}}$ is a G-invariant nondegenerate bilinear form in $\mathfrak{g}$ and $\langle\ ,\ \rangle_{\mathfrak{h}}$ is a G-invariant nondegenerate bilinear form in $\mathbf{h}$. This action is called $BFCG$ action, in analogy with the BF theory action. The gauge transformations of the Lagrange multiplier fields are given by

$$B \to g^{-1}Bg\,, \quad C \mapsto g^{-1} \triangleright C$$

for the usual gauge transformations, while

$$B \to B - [C, \eta]\,, \quad C \mapsto C$$

for the H-gauge transformations.

Let us now examine the case of the Poincaré 2-group. In this case $A = \omega^{ab} J_{ab}$, $\beta = \beta^a P_a$, where $a, b \in \{0,1,2,3\}$, J_{ab} are the generators of the Lorentz group while P_a are the generators of the translation group $\mathbb{R}^4$. Consequently

$$\begin{array}{rclcl} \mathcal{F} & = & (d\omega^{ab} + \omega^a{}_c \wedge \omega^{cb}) J_{ab} & = & R^{ab} J_{ab} \\ \mathcal{G} & = & \left(d\beta^a + \omega^a{}_b \wedge \beta^b \right) P_a & = & (\nabla \beta^a) P_a \end{array}$$

The G-gauge transformations are the local Lorentz rotations

$$\omega \to g^{-1} \omega g + g^{-1} dg \,, \qquad \beta \to g^{-1} \triangleright \beta$$

while the H-gauge transformations are the local translations

$$\delta_\varepsilon \omega^{ab} = 0 \,, \qquad \delta_\varepsilon \beta^a = d\varepsilon^a + \omega^a{}_b \wedge \varepsilon^b$$

where $\eta = \varepsilon^a P_a$.

The $BFCG$ action then becomes

$$S = \int_{\mathcal{M}} \left(B^{ab} \wedge R_{ab} + C_a \wedge \nabla \beta^a \right)$$

where

$$\delta_\varepsilon B = 0 \,, \qquad \delta_\varepsilon C = 0$$

At this point a very important observation is in order. The transformation properties of the 1-form C^a are the same as the transformation properties of the tetrad 1-form e^a under the local Lorentz and the diffeomorphism transformations. In addition, the equation of motion for C^a is $\nabla C^a = 0$, just like the no-torsion equation for the tetrad, $\nabla e^a = 0$. Based on this, *we identify the Lagrange multiplier* C^a *with the tetrad field* e^a, and write the action in the form

$$S = \int_{\mathcal{M}} \left(B^{ab} \wedge R_{ab} + e^a \wedge \nabla \beta_a \right)$$

In this way one can construct a categorical generalization of the topological BF action. The new action is again topological, but more rich in structure, since the tetrad fields are explicitly present. In addition, the 2-group formalism provides a framework to construct a topological quantum field theory from this action, in analogy with the BF case. This provides us with the necessary tools to construct a categorical generalization of a spin foam model, based on the $BFCG$ action instead of the BF action. The explicit presence of the tetrads should help us resolve the two handicaps of spin foam models discussed in section 2.

4. The spincube model

The first step in the construction of the new model is to write the action for General Relativity, starting from the $BFCG$ action. In order to do this, all we need is the simplicity constraint,

$$B_{ab} = \varepsilon_{abcd} \, e^c \wedge e^d$$

which can now be added into the action as it stands, as opposed to the BF case where the Plebanski constraint had to be introduced due to the absence of the tetrads e^a in the BF action. Therefore, one can write the *constrained BFCG action* in the form

$$S = \int_{\mathcal{M}} \left[B^{ab} \wedge R_{ab} + e^a \wedge \nabla \beta_a - \phi_{ab} \wedge \left(B^{ab} - \varepsilon^{abcd} e_c \wedge e_d \right) \right] \qquad (4.1)$$

where ϕ_{ab} is an additional Lagrange multiplier 2-form field, introduced in order to enforce the simplicity constraint.

The equations of motion are obtained by varying S with respect to B, e, ω, β and ϕ, respectively, to give:

$$R_{ab} - \phi_{ab} = 0, \; \nabla\beta_a + 2\varepsilon_{abcd}\phi^{bc} \wedge e^d = 0, \; \nabla B_{ab} - e_{[a} \wedge \beta_{b]} = 0, \; \nabla e_a = 0, \; B_{ab} - \varepsilon_{abcd}e^c \wedge e^d = 0$$

With the usual assumption that the tetrad fields are nondegenerate, these equations can be reworked into an equivalent form:

$$\phi^{ab} = R^{ab}, \quad B_{ab} = \varepsilon_{abcd}e^c \wedge e^d, \quad \beta^a = 0, \quad \nabla e^a = 0, \quad \varepsilon_{abcd}R^{bc} \wedge e^d = 0$$

The first three equations determine β^a and the multipliers B_{ab} and ϕ_{ab} in terms of e^a and ω^{ab}. The fourth equation is the no-torsion equation, which determines the connection ω^{ab} to be the Levi-Civita connection (a function of the tetrads e^a). The last equation is nothing but the Einstein field equation for the only remaining field e^a. Thus we see that the action (4.1) is classically equivalent to General Relativity. More precisely, it is equivalent to the Einstein-Cartan theory,

$$S_{EC} = \int_{\mathcal{M}} \varepsilon_{abcd}e^a \wedge e^b \wedge R^{cd}$$

since the torsion is equal to zero as an equation of motion rather than by definition.

Given the new action for General Relativity, we can proceed with the covariant quantization in analogy with the spin foam models. The action has the topological term and the constraint term, so as a first step we construct a topological quantum field theory by defining the path integral for the $BFCG$ part of the action. In the second step, we enforce the constraint term by requiring a suitable restriction in the path integral of the topological theory.

One begins by triangulating spacetime into 4-simplices, and rewriting the topological part of the action in the form

$$\sum_{\triangle} B_{\triangle}R_{\triangle} + \sum_{l} e_l(\nabla\beta)_l$$

where the first sum goes over all triangles and the second goes over all edges in the triangulation of the spacetime manifold. Then one constructs a topologically invariant path integral in the form (see [11] for the details of the construction)

$$\begin{aligned} Z &\equiv \int \mathcal{D}\omega \int \mathcal{D}B \int \mathcal{D}e \int \mathcal{D}\beta \exp\Big(i\sum_{\triangle} B_{\triangle}R_{\triangle} + i\sum_{l} e_l(\nabla\beta)_l\Big) \\ &= \sum_{\Lambda}\prod_{p} A_1(\Lambda_p)\prod_{f} A_2(\Lambda_f)\prod_{v} A_4(\Lambda_v) \end{aligned} \tag{4.2}$$

The labels $\Lambda = (L_p, m_f)$, where $L_p \in \mathbb{R}_0^+$ and $m_f \in \mathbb{Z}$, are now irreducible representations of the Poincaré 2-group, and in addition to vertices v and faces f of the Poincaré dual lattice, we also take the product over all the polyhedra p, since they are dual to the edges of the triangulation and naturally appear in the construction due to the presence of the $e \wedge \nabla\beta$ term in the $BFCG$ action. The amplitudes $A_1(\Lambda)$, $A_2(\Lambda)$ and $A_4(\Lambda)$ are chosen so that Z does not change under the action of the Pachner moves, which guarantees its independence of the triangulation. The polyhedra are colored with L_p, which have the interpretation as lengths of triangulation edges, while faces are colored with m_f, which have the interpretation as areas of the triangles in the triangulation. In the topological theory, edge lengths and triangle areas are independent of each other.

Note that the path integral is not defined over a colored 2-complex (the spinfoam), but rather over a colored 3-complex (called *spincube*).

Finally, we can impose the simplicity constraint, in order to turn the topological path integral into a realistic model for quantum gravity. Based on the geometric interpretation of the variables, the constraint actually says that a very natural requirement should be enforced — the triangle areas must be compatible with the corresponding edge lengths. This can be formalized in the requirement

$$|m_f| l_P^2 = A_f(L), \quad \forall f$$

where $A_f(L)$ is the Heron formula for the triangle area in terms of its edges. The Planck length appears naturally in order to balance the dimensions of the two sides of the equation. As a last step, one redefines the amplitudes A_1, A_2 and A_4 in order to render the theory IR-finite, as well as to enforce the correct semiclassical limit, in a way similar to the spinfoam models.

Note that imposing this constraint leaves only edge lengths as independent variables in the theory, so that the "area-Regge" problem present in spinfoam models is resolved automatically. In addition, the edge length variables allow for a completely straightforward coupling of matter fields to the spincube model. Namely, at the level of the classical theory, one can introduce fermions via the action

$$\begin{aligned} S \;=\; & \int \left[B^{ab} \wedge R_{ab} + e^a \wedge \nabla \beta_a - \phi_{ab} \wedge \left(B^{ab} - \varepsilon^{abcd} e_c \wedge e_d \right) \right] \\ & + i\kappa_1 \int \varepsilon_{abcd}\, e^a \wedge e^b \wedge e^c \wedge \bar{\psi} \left[\gamma^d \overleftrightarrow{d} + \{\omega, \gamma^d\} + \frac{im}{2} e^d \right] \psi \\ & + i\kappa_2 \int \varepsilon_{abcd} e^a \wedge e^b \wedge \beta^c \, \bar{\psi} \gamma_5 \gamma^d \psi \end{aligned} \tag{4.3}$$

where

$$\omega = \omega_{ab}[\gamma^a, \gamma^b]/8, \quad \kappa_1 = 8\pi l_P^2/3, \quad \kappa_2 = -2\pi l_P^2$$

The first term is the constrained $BFCG$ action, while the second and third terms introduce fermion coupling which results in the same equations of motion as in the ordinary Einstein-Cartan theory with fermions.

The quantization procedure of the action (4.3) is essentially the same as the one without fermions. The only difference is in the fact that the vertex amplitude A_4 will change to reflect the presence of the fermionic matter, as

$$A_4 \to A_4 \exp\left[i S_R^{(\text{ferm})}(L, \psi) \right]$$

where $S_R^{(\text{ferm})}$ is the Regge discretized action of a fermion field ψ coupled to gravity. The expressions which appear in $S_R^{(\text{ferm})}$ can be easily obtained, in contrast to the EPRL/FK model case, where the expression for the 4-simplex volume is impossible to define uniquely in terms of the spin foam variables [7].

Similarly to (4.3), one can also couple other matter fields to (4.1) in a completely straightforward way, including gauge and scalar fields, the cosmological constant, the Holst term, and so on.

5. Conclusions

The proposed 2-group reformulation of GR can be used to obtain a categorical ladder generalization of Loop Quantum Gravity. The advantage of this generalization is that the edge lengths of a triangulation become the basic dynamical variables. This will facilitate the construction of the path integral such that the classical limit of the corresponding quantum theory is GR and the coupling of matter will be much easier to accomplish.

The categorical nature of the theory implies that the edge labels of a spacetime triangulation should be the 2-group irreducible representations on a 2-Hilbert space. Note that this is not unique, since one can also use the category of chain complexes of vector spaces in order to define the representations, see [10, 12]. The structure of the chain-complex representations is different from the 2-Hilbert space representations, which means that chain-complex representation theory defines an alternative quantization of GR. Hence it would be interesting to develop the chain-complex representation theory of the Poincaré 2-group.

The physical significance of 2-Hilbert space representations could be better understood by performing a canonical quantization of the action (4.1).

As far as the construction of 4-manifold invariants based on the $BFCG$ state sum is concerned, one would have to regularize the topological state sum/integral based on the amplitude (4.2) such that the triangulation independence is preserved. One way to do it is to try to implement a gauge-fixing procedure, see [13]. Another way is to find a quantum group regularization, since there are strong indications that categorified quantum groups and their representations will be important for the construction of 4-manifold invariants [14]. Hence one can try to find a crossed module of Hopf algebras which is a deformation of the Poincaré 2-group, and then try to find an appropriate 2-category of representations which will give a finite topological state sum.

Acknowledgments

This work has been partially supported by FCT project PTDC/MAT/099880/2008. MV was also supported by the FCT grant SFRH/BPD/46376/2008.

References

[1] Goroff M H and Sagnotti A 1986 *Phys. Lett.* B **160** 81; *Nucl. Phys.* **B266** 709
[2] Rovelli C 2004 *Quantum Gravity* (Cambridge Univ. Press)
[3] Engle J, Livine E, Pereira R and Rovelli C 2008 *Nucl. Phys.* B **799** 136
[4] Freidel L and Krasnov K 2008 *Classical Quantum Gravity* **25** 125018
[5] Miković A and Vojinović M 2011 *Classical Quantum Gravity* **28** 225004
[6] Miković A and Vojinović M 2013 *Classical Quantum Gravity* **30** 035001
[7] Bianchi E, Han M, Magliaro E, Perini C, Rovelli C and Wieland W 2013 *Classical Quantum Gravity* **30** 235023
[8] Baez J C and Huerta J 2011 *Gen. Relativity Gravitation* **43** 2335
[9] Girelli F, Pfeiffer H and Popescu E M 2008 *J. Math. Phys.* **49** 032503
[10] Martins J F and Miković A 2011 *Adv. Theor. Math. Phys.* **15** 1059
[11] Miković A and Vojinović M 2012 *Classical Quantum Gravity* **29** 165003
[12] Cirio L S and Martins J F 2012 *Diff. Geom. App.* **30** 238
[13] Baratin A and Friedel L 2007 *Classical Quantum Gravity* **24** 2027
[14] Crane L and Frenkel I B 1994 *J. Math. Phys.* **35** 5136

Centre-of-mass for the finite speed of light

Z Oziewicz

Universidad Nacional Autonoma de México, Facultad de Estudios Superiores
C.P. 54714 Cuautitlan Izcalli, Estado de México

Abstract. In 1632, Galilei was aware of relativity of velocity and that this implies relativity of spaces-of-locations. During centuries the relativity of spaces-of-locations was ignored. Professor Harald Keres considered the space-of-locations as a congruence of world-lines, and there is no universal absolute three-dimensional space-of-locations. In applications, velocities relative to centre-of-mass are important. But the concept of centre-of-mass is impossible within relativity theory postulating that each pair of reference systems is related by the Lorentz isometry group transformation. We show that centre-of-mass of many-body interacting (bound) system for the case of finite light-speed is a well defined concept within the group-free approach using algebra epimorphisms as splits. We consider the Keres space-of-locations as the Grassmann factor-algebra of differential forms where a material body with a positive mass is interpreted as idempotent algebra epimorphism of the Grassmann algebra of spacetime onto the Grassmann factor-algebra of corresponding space-of-locations of that material body. A material body as a reference system is a group-free split, and this allows us to express all motions, velocities, accelerations and rotations, as *relative* with respect to the choice of variable reference system.

2010 Mathematics Subject Classification: 51B20, 53A17, 53A35, 53B30, 83A05.
2010 Physics and Astronomy Classification Scheme: 03.30.+p, 97.80.-d.
Keywords: relative velocity, split-velocity, time-like centre-of-mass, total mass.

Dedicated *im memoriam* of *Professor Harald Keres* (1912-2010)
who insisted on relativity of all kinds of motion
and initiated investigations of Einstein's gravity in Estonia

1. Harald Keres (1912-2010)

I am familiar with Professor Harald Keres publications since 40 years. I was immediately strongly convinced by the Keres crystal clear concept of *relativity* of spaces of locations, relativity of diachronic, relativity of 'simul-localidad' (relativity to be in the same location). I like his group-free definition of arbitrary (not necessarily constant) *relative* velocity (for infinite light-speed) of the Keres x-space relative to y-space [8, §2.6 p 68 formula (34)]. Moreover, during all these years I was permanently impressed by the Keres insistence that *every* motion must be relative motion, that relativity theory must not end with relativity of arbitrary velocity, but must include also relativity of acceleration and relativity of rotation (general relativity) [9]. The Keres group-free approach strongly motivated my research on group-free relativity for finite light-speed, i.e the Lorentz-isometry-group-free relativity theory [15, Oziewicz Z arXiv:math-ph/0608062], and [16, 17, 18, 19, 20, 21].

The Keres lecture in 1970 at Kiev's Symposium on *Philosophical Questions of Relativistic Physics and Cosmology* [9] is of fundamental importance. In this lecture Keres states that the concept of inertial reference system is a convention, because every reference system can be considered as being inertial. Keres stresses that every motion, including acceleration and rotation must be considered to be relative, to be dependent on the free choice of the variable reference system, identified with a material body of a positive mass.

Journal of Physics: Conference Series **532** (2014) 012021 doi:10.1088/1742-6596/532/1/012021

> There does not exist a universal (unique) absolute three-dimensional space (of locations), common for all phenomena, and therefore there does not exist universal absolute acceleration and rotation. Every motion is relative.
>
> Harald Keres [9] p 142

The above opinion is contrasted with the long standing in history of physics opinion by Langevin in 1911 [11, 12] that acceleration and rotation must have an absolute meaning, reference-system-free, independent of the choice of the three-dimensional space-of-locations.

> There is no absolute space (of locations), and we only conceive of relative motion; and yet in most cases mechanical facts are enunciated as if there is an absolute space (of locations) to which they can be referred.
>
> Henri Poincaré (1854-1912), Science and Hypothesis
> Chapter 6: Classical Mechanics 1902

Galileo Galilei (1564-1642) was the first who observed in 1632 relativity of velocity [7]. This implies immediately relativity of location: are we changing the location or do we stay in the same place? – this is relative. There does not exist a concept of location in Nature. Location is a mathematical convention and not a physical concept. In reality what counts are events (space-time of events) and the location of an event in some space-of-locations is an irrelevant convention [19, 20].

Einstein [4, 5] defined acceleration as the Christoffel derivative applied to one-body system, acceleration $=\nabla_X X$ is a geodesic vector field (see also [25] §14, [2, 3]). No reference body is involved, so it is acceleration with respect to what? According to Keres one must specify another material body, another the Keres x-space, with respect to which relative velocity of one-body y-space are to be defined (velocity as a change of location, but location of what reference body?), and then relative acceleration is expected to be given by a derivative of relative velocity for two-body system. All this is beautifully explained in the Keres publications ([8] §2.8, [9, 10]).

Unfortunately, instead of generalizing *relativity* of velocity (supposed to be done by special relativity in terms of an isometry group), by including the *relativity* of acceleration, the Christoffel derivative applied to one-body system was like throwing the baby out with the bathwater: it made acceleration as defined by Einstein, Weyl, and later on by Ehlers, not relative but an absolute concept, neither the second reference material body nor the centre-of-mass were involved.

Keres considered the Galilean relativity of three-dimensional spaces-of-locations. Galilean relativity is due to the relative velocity among two material bodies with positive masses. A material body is identified with a congruence of time-like curves as 'the Lagrangian description' of the motion of a fluid material particle in terms of a congruence of curves, in fact, due to Euler and not to Lagrange. Such congruence is called by Keres to be x-space, and for a pair of such congruences, for x-space and y-space, Keres defined arbitrary relative velocity between them (for infinite light-speed). In the Keres notation x denotes the conserved zero-grade scalar fields [8, 10].

We present the Keres group-free concept of arbitrary relative velocity as Eulerian in terms of derivations of commutative *algebra* $\mathcal{F}$ of 'scalar fields'. An algebra derivation corresponds to a geometric vector field, but we avoid geometric interpretation considering an algebra as a primordial concept within manifold-free approach. The idea to represent material bodies in terms of a Lie algebra of derivations goes back to Leonhard Euler in 1757 (for velocity of a fluid) [6], and in the case of the finite light-speed to Hermann Minkowski in 1908.

Relativity of velocity needs at least two-body system, and each such system possesses a mass-centre, therefore the genuine relativity of velocity should include the concept of velocity *relative* to mass-centre rather than between bodies. Section 7 is devoted to the problem of a mass-centre for the finite light-speed.

2. Universal property of the differential and other notations

In what follows $\mathcal{F}$ denotes an (unital, associative, commutative) algebra of zero-grade scalar fields. We need two Grassmann algebras, the Grassmann algebra of differential multiforms $\mathcal{F}^{\wedge}$, and the Grassmann algebra $(\operatorname{der}\mathcal{F})^{\wedge}$ of 'multivector fields',

$$\mathcal{F}^{\wedge 0} \equiv (\operatorname{der}\mathcal{F})^{\wedge 0} \equiv \mathcal{F}, \quad (\operatorname{der}\mathcal{F})^{\wedge 1} \equiv \operatorname{der}\mathcal{F} \equiv \operatorname{der}(\mathcal{F},\mathcal{F}) \tag{2.1}$$

For X and $Y \in (\operatorname{der}\mathcal{F})^{\wedge}$ we use the left Grassmann multiplication

$$\wedge_X Y \equiv X \wedge Y, \quad (\wedge_X) \circ (\wedge_Y) = \wedge_{X\wedge Y}, \quad (\wedge_X)^2 = 0 \quad \text{for grade } X = \text{odd} \tag{2.2}$$

2.1 Definition (evaluation)**.** The left *evaluation*, denoted by ev, is defined as an involutive pull-back of the left Grassmann wedge product

$$\operatorname{ev}_X \equiv (\wedge_X)^*, \quad (\wedge_X) \circ (\wedge_Y) = \wedge_{X\wedge Y} \quad \Longleftrightarrow \quad (\operatorname{ev}_Y) \circ (\operatorname{ev}_X) = \operatorname{ev}_{X\wedge Y} \tag{2.3a}$$

$$\text{grade } X = \text{odd} \quad \Longrightarrow \quad (\operatorname{ev}_X)^2 = 0 \tag{2.3b}$$

2.2 Exercise. Let for $X \in \operatorname{der}\mathcal{F}$ and $S \in \mathcal{F}^{\wedge 1}$, $(\operatorname{ev}_X)S = 1$. Then

$$(\operatorname{ev}_X) \circ \wedge_S \circ (\operatorname{ev}_X) = \operatorname{ev}_X \quad \text{and} \quad \wedge_S \circ (\operatorname{ev}_X) \circ \wedge_S = \wedge_S \tag{2.4}$$

2.3 Clarification (universal property of the differential)**.** The differential $d \in \operatorname{der}(\mathcal{F},\mathcal{F}^{\wedge 1})$ (extended to graded differential $d \in \operatorname{der}(\mathcal{F}^{\wedge},\mathcal{F}^{\wedge})$ with $d^2 = d \circ d = 0$), possesses a universal property: each derivation of an algebra $\mathcal{F}$, $X \in \operatorname{der}(\mathcal{F},\mathcal{F})$, is a composition of a differential with evaluation. Thanks to the Grassmann condition (2.3b) and because the evaluation by a derivation-vector ev_X diminish grade by one, we can write for grade $X = 1$,

$$(\operatorname{ev}_X)|\mathcal{F} \equiv 0, \quad X = (\operatorname{ev}_X) \circ d + d \circ (\operatorname{ev}_X) = (d + \operatorname{ev}_X)^2 \quad \in \operatorname{der}(\mathcal{F},\mathcal{F}) \tag{2.5}$$

2.4 Definition (Christoffel 1869)**.** Elwin Bruno Christoffel (1829-1900) introduced a zero-grade derivation of all tensor fields, for $X \in \operatorname{der}\mathcal{F}$ and $f \in \mathcal{F}$, by two conditions [1]

$$\nabla_X \circ d \neq d \circ \nabla_X, \quad \nabla_{fX} = f\,\nabla_X \quad \text{and} \quad \operatorname{ev} \circ \nabla_X = \nabla_X \circ \operatorname{ev} \tag{2.6}$$

2.5 Definition (Slebodzinski 1931)**.** Władysław Ślebodziński (1884-1972) introduced in his PhD . Thesis [23] a zero-grade derivation of all tensor fields by two conditions, for $f \in \mathcal{F}$ and $X \in \operatorname{der}\mathcal{F}$,

$$\mathcal{L}_X \circ d = d \circ \mathcal{L}_X, \quad \mathcal{L}_{fX} \neq f\mathcal{L}_X \quad \text{and} \quad \operatorname{ev} \circ \mathcal{L}_X = \mathcal{L}_X \circ \operatorname{ev} \tag{2.7}$$

In 1932, Van Dantzig called the Ślebodziński derivation (2.7) the Lie derivation and introduced a notation $\mathcal{L}$, however, Sophus Lie passed away in 1899 and has nothing to do with this concept of derivation of a tensor field. We keep the name the Ślebodziński derivation (or the Ślebodziński-Lie derivation). From definition (2.7) one can prove the following

$$\mathcal{L}_X = (d + \operatorname{ev}_X)^2 \quad \in \operatorname{der}(\mathcal{F}^{\wedge}), \quad \text{and} \quad \forall Y \in \operatorname{der}\mathcal{F}, \quad \mathcal{S}_X Y = [X, Y] \tag{2.8}$$

The Ślebodziński conditions (2.7) determine the unique derivation $\mathcal{L}$. The Christoffel conditions (2.6) left a lot of freedom, and an extra condition put on by Christoffel, $\nabla g = 0$, for g being the metric tensor, do not determine the Christoffel derivatation ∇ uniquely as was observed by Cartan in 1922.

2.6 Theorem (Cartan 1922). *Élie Cartan (1869-1951) observed in 1922 that a derivation $X \in \operatorname{der}\mathcal{F}$, extends to a graded derivation of the Grassmann algebra of differential multiforms, $\mathrm{ev}_X \in \operatorname{der}(\mathcal{F}^\wedge)$. For any differential multiforms α and β we have the Cartan graded derivation with the following Cartan theorem,*

$$(\mathrm{ev}_X)(\alpha \wedge \beta) = (\mathrm{ev}_X\,\alpha) \wedge \beta + (-)^\alpha \alpha \wedge (\mathrm{ev}_X\,\beta), \quad \mathrm{ev}_X \in \operatorname{der}(\mathcal{F}^\wedge) \tag{2.9}$$

Cartan denoted the above derivation by i_X, and called *inner product* or inner derivation, however the name *evaluation* is more appropriate.

2.7 Notation (algebra map versus derivation of an algebra). In the sequel we use the following notation for grade $X = 0$,

$$\begin{aligned} X \in \operatorname{alg}(\mathcal{F}^\wedge) \quad &\Longleftrightarrow \quad X\,(\alpha \wedge \beta) = (X\alpha) \wedge (X\beta) \\ X^p \in \operatorname{der}(\mathcal{F}^\wedge) \quad &\Longleftrightarrow \quad X^p(\alpha \wedge \beta) = (X^p\alpha) \wedge \beta + \alpha \wedge (X^p\beta) \end{aligned} \tag{2.10}$$

2.8 Definition (split). The Grassmann algebra map X as a $(1,1)$-tensor field in terms a vector field X and a differential one-form S_X such that $(\mathrm{ev}_X)S_X \equiv 1$, given as follows, is said to be a X-split

$$\begin{aligned} X &\equiv (\mathrm{ev}_X) \circ (\wedge_{S_X}) \in \operatorname{alg}(\mathcal{F}^\wedge) &&\text{and} \quad X^p \equiv (\wedge_{S_X}) \circ (\mathrm{ev}_X) \in \operatorname{der}(\mathcal{F}^\wedge) \\ X^* &\equiv (\mathrm{ev}_{S_X}) \circ (\wedge_X) \in \operatorname{alg}((\operatorname{der}\mathcal{F})^\wedge) &&\text{and} \quad X^{p*} \equiv (\wedge_X) \circ (\mathrm{ev}_{S_X}) \in \operatorname{der}((\operatorname{der}\mathcal{F})^\wedge) \end{aligned} \tag{2.11}$$

The Grassmann algebra derivation X^p is an involute permutation of an algebra map X.

2.9 Exercise (split is idempotent). For X-split and X^*-split the following holds

$$X \circ X = X, \quad X^p \circ X^p = X^p, \quad X^p \circ X = 0 = X \circ X^p \tag{2.12a}$$

$$X^p + X = (\mathrm{ev}_X + \wedge_{S_X})^2 = \mathrm{id}_{\mathcal{F}^\wedge} \tag{2.12b}$$

$$(X^p + X)^* = X^{*p} + X^* = (\mathrm{ev}_{S_X} + \wedge_X)^2 = \mathrm{id}_{(\operatorname{der}\mathcal{F})^\wedge} \tag{2.12c}$$

3. From the Keres space-of-locations to algebra derivation

Keres introduced three-dimensional space-of-locations as a congruence of time-like world-curves in four-dimensional space-time manifold [8, §2.1], [10, §3]. Each congruence corresponds to a material particle-body with a positive mass. Having explicitly a pair of two congruences, *two* spaces-of-locations, called x-space and y-space, Keres defined conceptually an *arbitrary velocity* of one space *relative* to another space, i.e an arbitrary velocity explicitly as *relative* velocity, without invoking the group theory concept, i.e group-free relative velocity. This definition allows us to define also acceleration and rotation of the Keres y-space relative to x-space, i.e acceleration and rotation explicitly as *relative* acceleration and *relative* rotation ([8] §2.8).

One can equivalently understand each congruence of world-lines in Eulerian way as integral curves of the corresponding vector field or synonymously as the integral curves of an ordinary differential equation. This allows us to replace congruence by an algebra derivation, and thus a material particle is the same as a derivation of an algebra $\mathcal{F}$, where $\{x^1, x^2, x^3\}$ describe the corresponding conserved scalar 'integrals of motion' of a particle. With analogous conserved scalar 'integrals of motion' of the second particle in 'adopted coordinate system' we have the following two-body free-system as a pair of derivations of an algebra $\mathcal{F}$,

$$t, x^1, x^2, x^3, y^1, y^2, y^3 \quad \in \mathcal{F} \tag{3.1a}$$

$$X = \left(\frac{\partial}{\partial t}\right)_{x^1,x^2,x^3} \quad \text{and} \quad Y = \left(\frac{\partial}{\partial t}\right)_{y^1,y^2,y^3} \quad \in \operatorname{der}\mathcal{F} \tag{3.1b}$$

$$(\mathrm{ev}_X)(dt) = X\,t \equiv 1 = (\mathrm{ev}_Y)dt, \quad (\mathrm{ev}_X)(dx^i) \equiv 0, \quad (\mathrm{ev}_Y)(dy^i) \equiv 0 \tag{3.1c}$$

Every material body in (3.1b) is basis-free because derivation of an algebra $\mathcal{F}$ is an example of a basis-free tensor field.

In the present section for the Keres two-body system (3.1b), we set the Galilean simultaneity

$$S_X = S_Y = dt \quad \Longrightarrow \quad X^{p*}\,Y = X \tag{3.2}$$

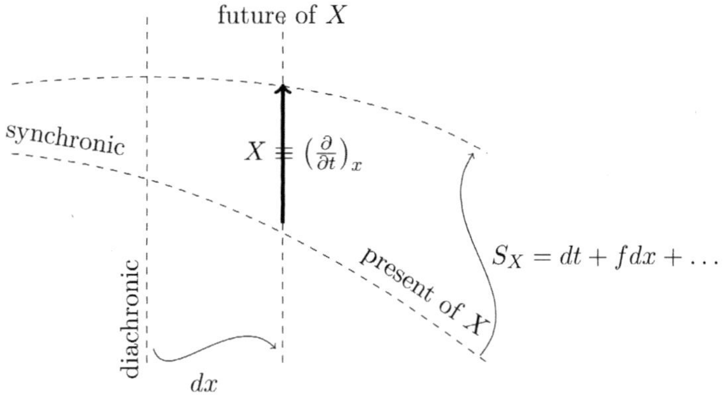

Figure 1. The reference system X is a material body (with a positive mass) as a permanent diachronic process-derivation $X \in \operatorname{der}\mathcal{F}$ from past to near future. Process X is going alongside of the crests of a wave of X-locations $\{dx\}$, $(\mathrm{ev}_X)(dx) = X\,x = 0$. Synchronic X-simultaneity of a reference system X is a differential one-form $S_X = dt + f dx + \ldots$, such that $(\mathrm{ev}_X)S_X \equiv 1$.

3.1 Definition (*relative*-velocity, [8] §2.6 formula (34))**.** An arbitrary (not necessarily constant) velocity v_{XY} of a material body Y (the Keres y-space) relative to a reference system X (relative to the Keres observer x-space) is defined as an X^*-split of Y,

$$v_{XY} \equiv X^*\,Y, \quad Y = (X^{*p} + X^*)Y = X + v_{XY}, \quad v_{XY} \equiv Y - X, \quad (dt)v_{XY} = 0 \tag{3.3}$$

Above definition is valid only for absolute simultaneity dt (3.1c), and only in this particular case the relative velocity is reciprocal, $v_{XY} = -v_{YX}$. The above concept of relative velocity is a group-free concept.

3.2 Corollary. *The scalar components, $v_{XY}\,x^i$ and $v_{XY}\,y^i$, are exactly the Keres expressions for the velocity of y-space relative to x-space, and vice versa, see* [8, §2.6] *and* [10, p 351 before formula (10)],

$$v_{XY}\,x^i = Y\,x^i = \left(\frac{\partial x^i}{\partial t}\right)_{y^1,y^2,y^3} \quad \text{and} \quad v_{XY}\,y^i = -X\,y^i = -\left(\frac{\partial y^i}{\partial t}\right)_{x^1,x^2,x^3} \tag{3.4}$$

4. Galileo Galilei 1632 and Leonhard Euler 1757

The Keres definition of relative split-velocity [8, §2.6 formula (34)], Definition 3.1, expressed the observation made by Galileo Galilei in 1632 about the relativity of velocity. Note that the same material body Y possesses infinite many velocities with respect to many different reference systems, possesses many distinct splits

$$Y = X + v_{XY} = Z + v_{ZY} = \ldots \tag{4.1}$$

If $v_{XY}x = v = \mathrm{const}$, then $x' = x - vt$ does the job, however the definition (3.3) is not restricted to constant velocities.

4.1 Comment. A velocity (relative or not) seen as a vector field must be an algebra derivation $v \in \operatorname{der}\mathcal{F}$. An acceleration is a derivation of a velocity vector field, a derivation of a derivation. However before invention by Christoffel in 1869, the concept of a derivation of a vector field was not known, and therefore Newton in 1687, and Euler in 1757, defined acceleration not as a derivation of a vector field, but as a second derivation of an algebra of scalar fields. Newtonian definition of acceleration is still in University Physics textbooks as a composition of two algebra derivations. Let $Y \in \operatorname{der}\mathcal{F}$ and $c^* \in \operatorname{alg}(\mathcal{F}^\wedge, \ldots)$ be an algebra epimorphism to algebra on integral curve of Y. Then acceleration is commonly defined as a composition of derivations,

$$\left(\frac{d}{dt} \circ \frac{d}{dt}\right) \circ c^* = c^* \circ (Y \circ Y) \quad \notin \operatorname{der}\mathcal{F} \tag{4.2}$$

A derivation of an algebra of scalar fields is a basis-free vector field, but composition of derivations is not. Thus Newton's and Euler's acceleration is not a tensor field. Einstein replaced the composition of derivations by the Christoffel derivation. The Christoffel derivation of a vector is a vector, and Albert Einstein [4, 5] defined acceleration as the Christoffel derivative of a velocity vector (giving the geodesic vector field).

The Keres relative velocity (3.3)-(4.1) is the same as the Euler definition in 1757 of a fluid, and fluid's *relative* velocity. The Euler derivative as a sum of vector fields, (4.3) below, is known as substantial or material, or barycentric, or hydrodynamic derivative. In the fluid context a material body is a *vector* field on space-time $\in \operatorname{der}\mathcal{F}$. The Euler derivative is interpreted in Fluid Mechanics textbooks in the following way: "the total rate of change is a sum of the local rate of change plus the convective rate of change"

$$Y \equiv \frac{\partial}{\partial t} + \sum v^i \frac{\partial}{\partial x^i} = X + v_{XY}, \quad \frac{\partial}{\partial t} \equiv \left(\frac{\partial}{\partial t}\right)_{x^1,x^2,x^3} \tag{4.3}$$

In fact, the Euler derivative should be interpreted as a definition of a velocity v of a fluid Y relative to a reference system X in adopted coordinates $X = (\partial_t)_x$. Given vector fields X and v, then a fundamental theorem of ordinary differential equations assure that for a sum $Y = X + v$ there exist local adopted coordinates that Y takes a form as in (3.1b).

For an observer X being the reference system, $X\,x = 0$, if x is in the Keres x-space, cf (3.1b), $v_{XX} = 0_X$. Therefore, the velocity of Y relative to X is

$$v_{XY}\,x = (Y - X)\,x = Y\,x \tag{4.4}$$

Euler defined acceleration of a fluid following Newton as the second derivative of scalar fields, i.e not as a basis-free tensor,

$$a_{\text{Newton}}|_x = Y \circ Y, \quad a_{\text{Euler}} = Y \circ (Y - X) = (X + v) \circ v \quad \notin \operatorname{der}\mathcal{F} \tag{4.5a}$$

$$a_{\text{Euler}}|_{x^i} = \{(X + v) \circ v\}|_{x^i} = (X + v)\,v^i = \frac{\partial v^i}{\partial t} + \sum v^j \frac{\partial v^i}{\partial x^j} \tag{4.5b}$$

5. Simultaneity differential form

The Riemannian metric tensor g can be considered as an algebra morphism from the Grassmann algebra of multivector fields to the Grassmann algebra of differential multiforms, and his inverse g^{-1}, if it exists, as an algebra morphism from the Grassmann algebra of differential forms to multivector fields. Metric tensor must not be interpreted as a distance between events. Distance between points of a manifold is an *integral* over curve.

5.1 Notation. We found convenient to use the following notation for the Grassmann algebra map with volume dependent scalar field det S. The 'reciprocity' symmetry (5.1b) is postulated

$$S \equiv \frac{1}{\det g} g \in \operatorname{alg}((\operatorname{der}\mathcal{F})^{\wedge}, \mathcal{F}^{\wedge}), \quad (\det S)(\det g)^3 = 1, \quad \text{with} \quad \det g = -c^2 \tag{5.1a}$$

$$S_{X\wedge Y} = S_X \wedge S_Y, \quad \gamma_{XY} \equiv S_{X\otimes Y} \equiv (\mathrm{ev}_Y)S_X = (\mathrm{ev}_X)S_Y = S_{Y\otimes X} \equiv \gamma_{YX} \tag{5.1b}$$

5.2 Axiom (absence of privileged reference material body)**.** The relativity postulate of the absence of a privileged reference body is expressed in the following way. Each time-like material body X (with a positive mass) as a traveller-process-derivation to his near future, possesses its own simultaneity differential Pfaff form, g_X or S_X, that is ambient-dependent, i.e metric-tensor g-dependent, such that for all *time-like* material bodies X the following axiom holds (i.e det g is the reference-body-free),

$$X \xrightarrow{\;g\;} g_X, \quad (g_X)X = \det g, \quad (\mathrm{ev}_X)S_X \equiv S_X X = S_{X\otimes X} \equiv 1 \tag{5.2}$$

6. Arbitrary relative split-velocity for the finite light-speed

In this section we drop the Galilean condition of absolute simultaneity and we will consider the case when $dt' \wedge dt \neq 0$.

$$X = \left(\frac{\partial}{\partial t}\right)_{x^1,x^2,x^3} \quad \text{and} \quad Y = \left(\frac{\partial}{\partial t'}\right)_{y^1,y^2,y^3} \quad \in \operatorname{der}\mathcal{F} \equiv \operatorname{der}(\mathcal{F},\mathcal{F}) \tag{6.1}$$

6.1 Example. In adopted coordinates in (6.1) we have

$$S_X = dt + f_i dx^i, \quad S_Y = dt' + g_i dy^i \tag{6.2}$$

The adopted dual bases of vector fields associative to simultaneity differential forms are as follows

$$\begin{aligned} X_i &\equiv \frac{\partial}{\partial x^i} - \left(S_X \frac{\partial}{\partial x^i}\right) X, \quad S_X X_i \equiv 0, \quad (dx^i)X_i = \delta^i_j \\ Y_i &\equiv \frac{\partial}{\partial y^i} - \left(S_Y \frac{\partial}{\partial y^i}\right) Y, \quad S_Y Y_i \equiv 0, \quad (dy^i)Y_j = \delta^i_j \end{aligned} \tag{6.3}$$

$$\mathrm{id} = X \otimes S_X + X_i \otimes dx^i = Y \otimes S_Y + Y_i \otimes dy^i \tag{6.4}$$

Each material body as in (6.1) is coordinate-free and basis-free because derivation of an algebra $\mathcal{F}$ is an example of a coordinate-free tensor field.

One can express the derivation X in a basis adopted for Y with S_Y and vice versa,

$$X = (S_Y X)\{Y + (X\,y^i)Y_i\}, \quad Y = (S_X Y)\{X + (Y\,x^i)X_i\}. \tag{6.5}$$

The scalar components, $X\,y^i$ (and $Y\,x^i$), are exactly the Keres expressions for the velocity of x-space relative to y-space ([10] p 351 before formula (10))

$$X\,y^i = \left(\frac{\partial y^i}{\partial t}\right)_{x^1,x^2,x^3} \quad \text{and} \quad Y\,x^i = \left(\frac{\partial x^i}{\partial t'}\right)_{y^1,y^2,y^3} \tag{6.6}$$

6.2 Definition (basis-free split-velocity for finite-light-speed). The above expressions (6.5)-(6.6) lead to the following basis-free definition of the velocity of the x-space *relative* to the observer y-space in the case of the finite light-speed. This definition explicitly depends on the simultaneity map S that can be identified with an environment,

$$\begin{aligned} &\text{Velocity of } X \text{ relative to } Y \otimes S_Y \text{ is} \quad v_{YX} \equiv \frac{1}{\gamma} X - Y, \quad S_Y \, v_{YX} = 0 \\ &\text{Velocity of } Y \text{ relative to } X \otimes S_X \text{ is} \quad v_{XY} \equiv \frac{1}{\gamma} Y - X, \quad S_X \, v_{XY} = 0 \end{aligned} \tag{6.7}$$

6.3 Proposition. *Speed of relative velocities is the same,* $(v_{YX})^2 = (v_{XY})^2$.

Proof. We have

$$\begin{aligned} (v_{YX})^2/\det g &\equiv S_v v = \left(\gamma^{-1} S_X - S_Y\right) v = \gamma^{-1} S_X \, v = \gamma^{-1} S_X \left(\gamma_{-1} X - Y\right) \\ &= -\left(1 - 1/\gamma^2\right) = (v_{XY})^2/\det g \end{aligned} \tag{6.8}$$

□

If simultaneity forms are not parallel, $S_X \wedge S_Y \neq 0$, then the relative velocities cannot be reciprocal,

$$\gamma = \left(1 + v^2/\det g\right)^{-\frac{1}{2}}, \quad v_{YX} \wedge v_{XY} = -\left(1 - 1/\gamma^2\right) X \wedge Y \tag{6.9a}$$

$$(S_X \wedge S_Y)(X \wedge Y) = (S_Y Y)(S_X X) - (S_Y X)(S_X Y) = -(\gamma^2 - 1) \tag{6.9b}$$

7. Centre-of-mass

Relativity of velocity is not complete without of the concept of centre-of-mass. A material body possesses many different velocities relative to choices of different reference bodies, and this fact is crystal clear in the Keres publications. But many-body interacting system possesses a centre-of-mass and more important are velocities relative to the centre-of-mass. Therefore a genuine special relativity would be complete only with the centre-of-mass concept. We consider that all phenomena considered in physics, the Doppler shift, stellar aberration discovered by Bradley, absorptions, scatterings, etc., should be analyzed in terms of velocities relative to centres-of-mass only. This means that the only important boost should be the boost from the centre-of-mass. In our opinion the concept of the centre-of-mass with its own simultaneity is of vital importance for special relativity theory.

However it is well known, see e.g [22, 13, 14] and any textbook on special or general relativity [Landau and Lifshitz *The Classical Theory of Fields*], that within the Lorentz isometry group approach, the concept of intrinsic centre-of-mass does not exist for finite light-speed, and whenever centre-of-mass is used in practice of cosmology the Newtonian limit is accepted. We are showing in the present section that the group-free concept of relative split-velocity, in the spirit of the Keres approach, extended to finite light-speed allows the perfect concept of centre-of-mass and split-velocities relative to centre-of-mass.

It is stressed in many publications that the real problem facing any attempt to construct a relativistic many-particle theory (theory for the finite light-speed) is the lack of a clear notion of simultaneity. Here, within the group-free approach, split-velocities (7.1) offer perfect solution of this problem: one must forget Lorentz isometry group, in the spirit of the group-free approach by Keres.

For a given a priori two-body system as the Keres x-space and y-space, a mass-centre is in another the Keres z-space with a total mass $M = \sum m_i$ – binding energy, with nontrivial mass-dependence and variable velocity dependence in case of the finite light-speed. One-body problem will stay always as one-body, however two-body problem is in fact always three-body, because it must involve a set of three the Keres spaces: x-, y- and mass-centre z-space.

We consider a mass-centre with a total mass M, as a time-like derivation (a vector field) Z with its own simultaneity differential one-form S_Z, for a free two-body system $\{X, Y\}$ with corresponding masses $\{m_x, m_y\}$. A mass-centre Z is a bound system and in order to be ionized – decoupled into free-system there must be an external massless gravity field G (or a massless electromagnetic field), in terms of energy-momenta. To write down the mathematical expression, we need the Minkowski scalar factors in terms of simultaneity differential forms. For a two-body time-like system, X with a positive mass m_x, and Y with a positive mass m_y, following Minkowski [1908] we set for a group-free relative split-velocity v_{XY},

$$\gamma_{XY} \equiv S_X Y = S_Y X = \gamma_{YX} = 1/\sqrt{1 - (v_{XY})^2/c^2} \tag{7.1}$$

The energy of the massless radiation G (the gravity or electromagnetic radiation with $S_G G \equiv 0$) relative to time-like material body X with a positive mass, $S_X X \equiv 1$, was defined by Schrödinger in 1956 as the following evaluation

$$E_{XG} \equiv S_G X = S_X G \tag{7.2}$$

With these definitions we define mass-centre Z with its own simultaneity differential one-form S_Z, $S_Z Z \equiv 1$, by the following energy-momenta conservation law, rotations and spins are disregarded not only for simplicity, but more for pedagogical reasons.

7.1 Definition (mass-centre)**.** Many-body interacting material system, as a set of time-like vectors-derivations $\{X_i\}$, possesses a time-like mass-centre $Z \in \operatorname{der}\mathcal{F}$ with a total mass M according to the following definition (equivalent to the law of conservation of energy-momenta). On the right there is the free decoupled many-body system, whereas on the left, Z is a mass-centre describing the bound system, if radiation is not absent $G \neq 0$,

$$G + M\,Z = \sum m_i\,X_i, \quad S_Z = \frac{1}{M}\left(\sum m_i\,S_{X_i} - S_G\right) \tag{7.3}$$

The above definition gives immediately two expressions for the total mass for finite light-speed, the first expression in terms of split-velocities and radiation energy relative to the mass-centre, whereas the second expresses the same total mass in terms of split-velocities between constitutive particles of the given many-body system.

We denote by $\gamma_i \equiv \gamma_{i\,Z}$ the Minkowski factor relative to mass-centre Z.

$$\text{Because:} \quad S_{\sum m_i X_i} \sum m_i X_i = \sum m_i\,m_j\,\gamma_{ij} \quad \text{and} \quad S_Z \sum m_i X_i = \sum m_i\,\gamma_i \quad \text{then:} \tag{7.4a}$$

$$M = \sum m_i\,\gamma_i - E_{ZG}, \quad M^2 = (M\,S_Z)(M\,Z) = \sum m_i\,m_j\,\gamma_{ij} - 2\sum m_i\,E_{iG} \tag{7.4b}$$

7.2 Lemma (gravity ionization energy)**.** *The gravity (or alias electromagnetic) ionization energy E_{ZG} in terms of Minkowski factors is*

$$E_{ZG} = \sqrt{\sum m_i\,m_j\,(\gamma_i\gamma_j - \gamma_{ij})} \tag{7.5a}$$

Proof. We have

$$M\,E_{ZG} = S_G \sum m_i X_i = \sum m_i\,S_i G = \sum m_i\,S_i\left(\sum m_j\,X_j - M\,Z\right) \tag{7.6a}$$

$$= \sum m_i\,m_j\,\gamma_{ij} - M \sum m_i\,\gamma_i \tag{7.6b}$$

and

$$\sum m_i\,m_j\,\gamma_{ij} = M\left(\sum m_i\,\gamma_i + E_{ZG}\right) = \left(\sum m_j\,\gamma_j - E_{ZG}\right)\left(\sum m_i\,\gamma_i + E_{ZG}\right) \quad \square \tag{7.7}$$

In the Newtonian limit of the infinite light-speed we have

$$\begin{array}{cccc} \sum m_i \gamma_i \xrightarrow{c\to\infty} \sum m_i, & \sum m_i m_i \gamma_{ij} \xrightarrow{c\to\infty} \left(\sum m_i\right)^2 \\ E_{ZG} \xrightarrow{c\to\infty} 0, & M \xrightarrow{c\to\infty} \sum m_i \end{array} \tag{7.8}$$

Lemma 7.2 with expression (7.5a) contains also mutual relative velocities between bodies of the system, γ_{ij}, and it would be better to express it in terms of the velocities relative to mass-centre γ_i. This will not reduce the number of independent variables, because γ_{ij} will be expressed in terms of γ_i and cosines of angles $\upsilon_{Zi}\cdot\upsilon_{Zj}$. The following identity we derived in [15] that holds for non-reciprocal split-velocities and is analogous (but not identical) to the Sommerfeld identity [24] for scalar speed of composed isometric-velocities (reciprocal) within the Lorentz isometry group,

$$\gamma_{ij} = \gamma_i \gamma_j \left(1 - (\upsilon_{Z\,i}\cdot\upsilon_{Z\,j})/c^2\right) \implies (E_{ZG})^2 = \sum m_i m_j \sqrt{(\gamma_i^2 - 1)(\gamma_j^2 - 1)}\,\cos(ij) \tag{7.9}$$

In the Sommerfeld identity in the last factor in (7.9) there is plus sign, because isometric velocity is reciprocal, $\upsilon_{Z\,i} = -\upsilon_{i\,Z}$, but this is not possible within the group-free approach, where by definition, $S_Z\,\upsilon_{Z\,X} \equiv 0$ and $S_X\,\upsilon_{X\,Z} \equiv 0$, and simultaneity is relative, $S_Z \wedge S_X \neq 0$ [15, Oziewicz Z arXiv:math-ph/0608062], and [16, 17, 18, 19, 20, 21].

For two-body system $\{m_x, m_y\}$ with mass-centre $\{m_x.m_y, M\}$ we have the following system of scalar equations for total mass M,

$$G + MZ = m_x X + m_y Y, \qquad M E_{ZG} = m_x E_{XG} + m_y E_{YG} \tag{7.10a}$$
$$E_{ZG} + M = m_x \gamma_{ZX} + m_y \gamma_{ZY}, \quad E_{XG} + M\gamma_{ZX} = m_x m + m_y \gamma_{XY} \tag{7.10b}$$
$$E_{YG} + M\gamma_{ZY} = m_x \gamma_{XY} + m_y \tag{7.10c}$$

The total mass of this bound system for finite light-speed is given in terms of three Minkowski factors (three Minkowski factors for two-body system), because we are obliged to consider in fact three-body system $\{m_x, m_y, M\}$ in space-time.

We will end this section with a comment on maximal binding energy within the above group-free theory. The binding energy is

$$E_b := \text{Binding energy} = \sum m_i - M = E_{ZG} - \sum m_i(\gamma_i - 1) \tag{7.11}$$

Assuming $\cos(ij) = 1$ in (7.9), we get the the maximal binding energy,

$$E_{ZG}^{\max} = \sum m_i \sqrt{\gamma_i^2 - 1} \implies \text{Maximal } E_b = \sum m_i \left(\sqrt{\gamma_i^2 - 1} - (\gamma_i - 1)\right) < m_i \tag{7.12a}$$

where $0 \leq \sqrt{\gamma^2 - 1} - (\gamma - 1) < 1$.

8. Universe as a bunch of Grassmann factor-algebras of relative spaces-of-locations

The Cartan theorem (2.9) implies that the kernel of an algebra derivation, $\ker(\text{ev}_X)$, is a Grassmann *factor-algebra*, or quotient-algebra, that we baptize as the Grassmann X-factor-algebra of relative X-space, shortly 'X-space' is a $\mathcal{F}$-algebra,

$$\alpha \in \ker(\text{ev}_X) \quad \text{and} \quad \beta \in \ker(\text{ev}_X) \quad \implies \quad \alpha \wedge \beta \quad \in \ker(\text{ev}_X) \tag{8.1}$$

An algebra map (morphism from one algebra to another algebra) possesses a kernel, and this kernel is two-sided ideal. Therefore each factor-algebra can be presented as an image of some algebra map (instead of as a kernel of a derivation as we started above), so moreover we have

$$\text{Grassmann } X\text{-factor-algebra-of-locations} \quad = \quad \frac{\text{Grassmann algebra}}{X\text{-simultaneity ideal}} \tag{8.2}$$

For each material body X of a positive mass, i.e of a positive internal energy,

$$\dim_{\mathcal{F}}(\text{Grassmann algebra}) = 2^4 = 16 \quad \text{and} \quad \dim_{\mathcal{F}}(X\text{-factor-algebra}) = 2^3 = 8 \tag{8.3}$$

8.1 Example. One can list some members of the Grassmann X-factor-algebra of X-space-of-locations for the Keres congruence in terms of a derivation X given in (6.1).

$$\begin{aligned}&\text{If } dx \text{ is in } \ker(\mathrm{ev}_X), \text{ then} \quad 0 = (\mathrm{ev}_X)dx = \{(\mathrm{ev}_X)\circ d\}x = X\,x\\ &\text{and } x \text{ is conserved scalar 'integral' of } X.\end{aligned} \tag{8.4}$$

Let $f \in \mathcal{F}$ denotes any scalar of a zero-grade, then

$$\{\mathcal{F}, f, t, t', x^1, x^2, x^3, y^1, f dx^1, \ldots, f dx^1 \wedge dx^2 \wedge dx^3\} \in \ker(\mathrm{ev}_X) \equiv X\text{-factor-algebra} \tag{8.5a}$$

$$X\text{-simultaneity ideal} \quad \ni \quad \{f dt, f dt \wedge dx^1, \ldots\} \quad \notin \ker(\mathrm{ev}_X) \tag{8.5b}$$

In particular, the Grassmann X-factor-algebra contains all X-electric and X-magnetic relative fields.

8.2 Warning. We must stress very important peculiarities. Each material body X is owner of the Grassmann X-factor-algebra of his own X-factor-space-of-locations. It is good even simply identify each material body with such Grassmann X-factor-algebra of his space-of-locations. And when we say in what follows, X-factor-algebra(-of-locations), this means the *Grassmann* X-factor-algebra of the Grassmann $\mathcal{F}$-algebra (of a space-time). Now, for each material body X ('descendant' of the Keres congruence as in (6.1)) being the Grassmann X-factor-algebra, the following two warnings hold.

- Every X-factor-algebra-of-locations is over the same entire algebra $\mathcal{F}$ of zero-grade scalars, i.e it is a different $\mathcal{F}$-algebra for each body X, but always over the *same* algebra $\mathcal{F}$ of scalars.
- Not each two-sided simultaneity-ideal I_X (two-sided ideal in the Grassmann algebra generated by a simultaneity Pfaffian differential form assigned to a material body X) possesses a factor-derivation d/I_X that we will denote by d_X, and thus a factor-differential $(d_X)^2 = 0$. The necessary and sufficient condition for the existence of d_X is $dI_X \subset I_X$, and then an ideal is said to be closed. This fact is important for the concept of an inertial or non-inertial material bodies. If factor-differential does exist, then and only then the Grassmann X-factor-algebra-of-locations is invariant (stable) with respect to the factor-differential

$$(d_X)\{\text{Grassmann } X\text{-factor-algebra}\} \quad \subset \quad \{\text{Grassmann } X\text{-factor-algebra}\} \tag{8.6}$$

An ideal X-simultaneity-ideal in the Grassmann $\mathcal{F}$-algebra of the differential multiforms is also known as an exterior differential system in the terminology of Élie Cartan.

So how we can see the Universe? The Universe is 'matter in relative motion', and in absolute motion from past to future. We can consider the Universe as a collection of material bodies 'in relative motions' and each material body as a Grassmann factor-algebra of differential forms, always over the same entire algebra $\mathcal{F}$ of scalar fields. Universe is a bunch of the Grassmann factor-algebras.

8.1. Recapitulation

We formulated the Keres concept of an x-space-of-locations as a congruence of world-lines in a space-time manifold [10] in terms of a basis-free and a manifold-free Grassmann X-factor-algebra over an algebra $\mathcal{F}$. We stress that each the Keres space-of-locations must include entire algebra $\mathcal{F}$ of zero-grade scalars. In this realm space-of-locations can be considered as the Grassmann factor-algebra of the Grassmann $\mathcal{F}$-algebra – no change of algebra of scalars! The Keres x-space include *all* coordinates, also $t \in \{x\text{-space}\}$, however simultaneity one-form, $S_X = dt$, is not in the Keres x-space,

$$(d_X)f = df - (X\,f)\,S_X \quad \in \{x\text{-space}\} \equiv \ker(\mathrm{ev}_X) \tag{8.7}$$

Moreover we warn that the existence of a factor-differential d_X is not granted a priori.

I learn from Professor the Keres publications [8, 9, 10] two important lessons.

- Three-dimensional space-of-locations is not a certain instant of time as majority of present-day textbooks claim. This wrong claim of what it is 'the physical space' dominate in all present-day scientific journals and in scientific conferences on gravity and cosmology. Three-dimensional space-of-locations is not a submanifold of a space-time manifold. Keres defined a space-of-locations as a congruence of world-lines. Therefore the Keres space-of-locations is the *quotient* manifold: entire world-line is just only one single *location*. No definite instant of time has a 'conscience' of a location, instead instance tells us about relative simultaneity. An event, as an element of a cross-section of a space-time for a given instant of time, can belong to infinite many different congruences of world-lines, and, as Professor Keres was clearly aware, there is no way to assign to such single event some location in a 'real physical space-of-locations'. I explained on several illustrative examples in another publication [19], that the space-of-locations must be defined as a **quotient**-manifold and never as a sub-manifold, and of course Professor Keres was clearly aware of this fact,

$$\text{Space-of-locations is a Grassmann factor-algebra} \equiv \frac{\text{Grassmann } \mathcal{F}\text{-algebra}}{\text{Simultaneity ideal}} \tag{8.8}$$

- Keres is explaining that there does *not* exist just one 'physical space-of-locations' (as a factor-manifold; in the present paper re-interpreted as a Grassmann factor-algebra). His original enunciation is in terms of different congruences. the Keres clear idea of *many* distinct 'physical' spaces-of-locations can be traced and attributed also to observations by Galileo Galilei and by Minkowski, but this idea of relativity of 'simul-localidad' was never elucidated so explicitly as in the Keres publications.

9. Relativity vs gravity

It is good, at least in the beginning for clear pedagogical reasons, distinguish conceptually the theory of gravitational interaction from the theory of relativity. This is because relativity theory (relativity of locations, relativity of velocities, etc.) is fundamental for *all* kinds of interactions including electromagnetic interactions, weak and nuclear interactions. For historical reasons, relativity theory is divided into two parts, relativity of the concept of location and of velocity, and the second part deals with relativity of acceleration and with relativity of rotation.

To be relative means to be dependent on the irrelevant choice of the reference system. There are still different non-equivalent understandings of the mathematical description of the physical reference system, when the reference system is defined as a material body with a positive mass. Thus within relativity theory, the velocity, acceleration, and rotation of one material body depend on the free choice of another reference body. Often 'general relativity' is understood as the theory of gravity only (excluding electromagnetic field), however I think it is more adequate not to mix 'relativity' with 'gravity', and interpret 'general relativity' rather as

Journal of Physics: Conference Series **532** (2014) 012021 doi:10.1088/1742-6596/532/1/012021

relativity of *all* motions, i.e relativity of location, relativity of velocity, relativity of acceleration and relativity of rotations, without selecting any specific kind of interaction. Thus consideration of the gravitational interaction would be under the name 'gravitation', avoiding the double terminology in some journals 'general relativity and gravitation'.

10. Reference system as a monad within manifold-free and group-free approach

The idea to define a material body as a time-like vector field is due to Hermann Minkowski in 1908. Einstein in 1905 adopted alternative and not equivalent definition of a material body as a coordinate system alias a coordinate basis, and this is not used in the present paper; we follow an approach by Minkowski, and the Keres ideas of congruence. There are two advantages to define a material body, and in this way every reference system, as a time-like vector field or as the Keres congruence of world-lines, instead of Einstein's coordinate system. Theory of reference systems in terms of algebra derivations is known as the *monad* theory.

(i) Each derivation in (6.1) is an example of a tensor field, thus it is coordinate-free and basis-free, and in this way we can forget irrelevant coordinates and bases.

(ii) The second advantage to define a material body as a derivation is that this description is manifold-free in the spirit of the non-commutative point-free geometry. The commutative geometry deals with the commutative *algebra* of scalar fields and for historical reasons the name 'geometry' dominate over more adequate 'algebra'. The recent birth (at the end of XX century) of non-commutative 'geometry' opens the way to manifold-free algebraic approach. The term 'geometry' we are able to exchange for 'algebra', and for this ideological exchange non-commutativity or commutativity of the primary algebra of 'scalars' is not so important: one can develop manifold-free approach also for commutative differential algebra as it was developed in the recent decades in the realm of manifold-free non-commutative 'geometry'.

What advantage we see in abandoning, or never using, the space-time *manifold* of events, the world of events, so popular in the present-day textbooks on relativity, on gravity, and on cosmology?

(i) Each manifold of higher than one dimension must possesses by definition one-dimensional sub-manifolds, and in particular must possess closed curves. Closed curve is natural in the three-dimensional space-of-locations, however it is hard to imagine a real closed curve in a manifold of events.

(ii) In the present paper a 'space-time algebra' is the Grassmann algebra of differential multiforms and the Grassmann algebra of 'multiderivations' (alias multivector fields). Such 'space-time' *algebra* with the Grassmann functor, $\mathcal{F} \mapsto \mathcal{F}^{\wedge}$, allows us to avoid to think about universe as a four-dimensional manifold of events that has existed since eternity, exists now in this very moment, and implies that the future exists now. In the manifold-free approach universe is a collection of material bodies (and massless radiation) á la Kant and each material body is a process-derivation of a commutative algebra as in (6.1) – reinstating the Keres idea of congruences in terms of algebra derivations.

The manifold-free algebraic approach assumes the commutative algebra $\mathcal{F}$ as the primary concept, and the 'coordinates' in (6.1) are elements of a commutative algebra without of necessity to use the concept of an event, i.e this is event-free and point-free approach to relativity theory. We avoid the name 'vector field' and replace it by adequate 'derivation of algebra of scalars' that we use as a synonym of a material body, as it is clear in expressions (6.1),

$$\operatorname{der}\mathcal{F} \quad \ni X \quad \longrightarrow \quad \operatorname{ev}_X \quad \in \operatorname{der}(\mathcal{F}^{\wedge}) \tag{10.1}$$

Within the Riemannian geometry, with non-singular or with singular metric tensor g, it is important to consider also the Clifford algebra of differential forms and tensor algebra together with the Grassmann algebra, however for simplicity we restricted ourselves in the present paper to consider the Grassmann algebra alone.

11. Singular metric

During 1957-1964 Professor Harald Keres published four papers in the Transactions of the Institute of Physics and Astronomy of the Estonian Academy of Sciences on the concept of inertial reference system within the Einsteinian gravity. In 1965 he demonstrated that the infinite light-speed limit, $c \to \infty$, of the Einstein gravity contains rotational non-Newtonian fields.

In 1976 Keres reformulated the Newtonian gravity with rotation within singular metric [8, 10]. The way in which Keres introduced singular metric, $\det\, g = 0$, deserves attention. We ask: what vector-derivation should be chosen to be in the kernel $v \in (\ker\, g)$? Still there is not crystal clear, at least for me, what in good mathematics means the Newtonian limit.

The Galilean-Newtonian limit is considered to be the infinite light-speed limit, $c \to \infty$, however how exactly? Maybe mathematically more precise must be the limit of a concept of simultaneity in terms of the Minkowski factor, $\gamma \to 1$? But the limit $\gamma \to 1$ could have also another interpretation as a limit from bound to a free-system, i.e disconnection of an interaction. Maybe it would be convenient to consider $\det g$ as a free variable parameter with a limit $\det\, g \to 0$, but what kernel of a metric tensor, $g \in \text{alg}((\text{der}\,\mathcal{F})^\wedge, \mathcal{F}^\wedge)$ defined to be an algebra map, would be the best choice for such Galilean limit?

If a simultaneity algebra map S is an algebra isomorphism then

$$X \wedge Y \neq 0 \quad \Longleftrightarrow \quad S_{X\wedge Y} = S_X \wedge S_Y \neq 0 \tag{11.1}$$

One can expect that the Galilean-Newtonian limit would mean that for a two-body system as in (6.1), simultaneities of these bodies coincide in this limit with simultaneity of mass-centre

$$S_X, S_Y \quad \xrightarrow{\text{Newtonian limit}} \quad S_Z \tag{11.2}$$

The expected kernel of a singular metric could be the Keres relative velocity, $Y - X \in \ker\, g$. An algebra map to be a singular algebra map do not mean to be algebra isomorphism possessing nontrivial kernel as two-sided ideal in the Grassmann algebra. Kernel of an algebra map, $g \in \text{alg}((\text{der}\,\mathcal{F})^\wedge, \mathcal{F}^\wedge)$, is an ideal, and is a factor-space = algebra/sub-algebra.

It is not complicated to define such singular metric, and develop, following the Keres idea, an analogy of Einstein's gravity theory for the singular metric, calculate the Christoffel derivative for the singular metric, $\nabla g = 0$. However the choice of the singular metric tensor made by professor Keres for two-body system (6.1) was different from our proposal, the Keres choice was, $X = Y + v_{XY} \in \ker g$, [10, §3 p 351 formula (10)].

With the singular metric tensor Keres was able to develop the complete analogy of the Einstein gravity theory for the Newtonian limit in terms of the Christoffel derivative [10].

12. The Keres publications are rich in many more fundamental ideas

In the present paper we reminded only the concept of relativity of spaces of locations, the main introductory but of fundamental importance the Keres idea of x-space as different from y-space. This allows Professor Keres to define a general (not constant) relative velocity in a group-free way. We pointed that this group-free way can be extended to the case of finite light-speed that is the source of relativity of simultaneities. We also left for another publication the problem of the reduced mass for finite light-speed.

Keres developed in full clarity the concepts of relative acceleration, and relative rotation (of x-space relative to y-space) [8, 10], and very deep insight into gravity theory that is outside of the scope of the present paper. I am sure many other fundamental concepts developed by Professor Keres deserve separate papers.

I am convinced that it would be most desirable to edit the (selected) Collected Works of Professor Harald Keres, with translating them from Russian into English to allow these important publications be available to wideer number of students and researches. All known to me publications of Professor Keres are in Russian, they were published during the Estonia Soviet period after 1945.

Acknowledgments

I am deeply grateful, *palju tänu*, to Professor Piret Kuusk, a daughter of *Professor Harald Keres*, for inviting this talk on her session *Gravitation and Cosmology* of the *QQQ* conference *3Quantum: Algebra Geometry Informations* (Tallinn, July 2012), the former was dedicated to centenary of Professor Harald Keres (1912-2010) who initiated investigations of Einstein's gravity in Estonia. During my first visit to Tartu University in October 1992 I wished much to meet personally Professor Keres, and Professor Piret Kuusk was so kind and arranged a very delightful personal conversation with Professor Harald Keres. I am thankful to Professor Eugen Paal, the Chair of the Tallinn meeting, for invitation and for many interesting discussions.

References

[1] Christoffel E B 1869 *J. Reine Angew. Math.* **70** 46
[2] Ehlers J 1961 *Akad. Wiss. Lit. Mainz Abh. Math.-Nat. Kl.* **11** 793
[3] Ehlers J 1962 *Colloques Int. C.N.R.S.* **91** *Les théories relativistes de la gravitation* eds A Lichnerowicz and M A Tonnelat 275
[4] Einstein A 1914 *Sitzungsber. der preuss. Akademie d. Wiss. Phys.-Math. Kl.* **II** 1030; 1915 **XLVII** 831
[5] Einstein A 1916 *Annalen der Physik* **49** 769
[6] Euler L 1757 *Mémoires de l'académie des sciences de Berlin* **11** 274
[7] Galilei G 1632 *Dialogo sopra i due massimi sistemi del mondo: Tolemaico e Copernico* ed G B Landini (Florence, Italy)
[8] Keres H 1972 Representations of Newtonian theory *Gravitation Problems Prospects* eds V P Shelest and K A Pyragas (Kiev: Naukova Dumka) p 62
[9] Keres H 1973 General relativity principle as the principle of universal relativity of mechanical motion *Methodological Analysis of Theoretical and Experimental Foundations of Gravity Physics* eds A Z Petrov and P S Dyshlevyi (Kiev: Naukova Dumka) p 133
[10] Keres H 1976 *Eesti NSV Teaduste Akadeemia Toimetised: Füüsika*Matemaatika* **25** 349
[11] Langevin P 1911 *Scientia* **10** 31
[12] Langevin P 1921 *Compt. Rend.* **173** 831
[13] Møller C 1949 *Ann. Inst. Henri Poincaré* **11** 251
[14] Møller C 1969 *The Theory of Relativity* (Oxford: Clarendon Press)
[15] Oziewicz Z 2005 How do you add relative velocities? *Group Theoretical Methods in Physics* eds G S Pogosyan, L E Vicent and K B Wolf (Bristol: Institute of Physics, Conference Series No 185)
[16] Oziewicz Z 2007 *Int. J. Geom. Methods Mod. Phys.* **4** 1
[17] Oziewicz Z 2007 Ternary relative velocity. Astonishing conflict of the Lorentz group with relativity *Physical Interpretations of Relativity Theory* ed V Gladyshev (Moscow: Bauman Moscow State Technical University) p 292
[18] Oziewicz Z 2009 *Adv. Appl. Clifford Algebr.* **19** 793
[19] Oziewicz Z 2010 Relative velocity: a dichotomy *Proceedings of the Natural Philosophy Alliance 17th Conference at California State University Long Beach* **7** No 2 (Arlington MA USA: Space Time Analyses) p 748
[20] Oziewicz Z and Page W S 2012 The many relative spaces-of-locations of Galileo and Poincaré *Proceedings of the Natural Philosophy Alliance* **9** (Eagan MN USA: NPA) p 406
[21] Oziewicz Z 2013 *Studium Vilnense* A **10** 3
[22] Pryce M H L 1948 *Proc. Roy. Soc. London* A **195** 62
[23] Ślebodziński W 1931 *Acad. Roy. Belg. Bul.* **17** 864
[24] Sommerfeld A 1909 *Physikalische Zeitschrift* **10** 826
[25] Weyl H 1921 *Space-Time-Matter* (London: Methuen)

Note on homological modeling of the electric circuits

E Paal and M Umbleja

Tallinn University of Technology, Ehitajate tee 5, 19086 Tallinn, Estonia

Abstract. Based on a simple example, it is explained how the homological analysis may be used for modeling of the electric circuits. The homological *branch*, *mesh* and *nodal* analyses are presented. Geometrical interpretations are given.

1. Introduction and outline of the paper

The classical electric circuit analysis is based on the 2 Kirchhoff Laws [1]:

(i) [KCL] *Kirchhoff's current law* says that: At any instant in a circuit the algebraic sum of the currents entering a node equals the algebraic sum of those leaving.

(ii) [KVL] *Kirchhoff's voltage law* says that: At any instant around a loop, in either a clockwise or counterclockwise direction, the algebraic sum of the voltage drops equals the algebraic sum of the voltage rises.

The homological analysis of the electric circuits is based on its geometric elements - nodes, contours (edges, branches), meshes (simple closed loops), also called the chains, and using the geometric boundary operator of the circuit. The latter depends only on the geometry (topology) of the circuit. Then, both of the Kirchhoff laws can be presented in a compact algebraic form called the homological Kirchhoff Laws (HKL).

In the present note, based on a simple example, it is explained how the homological analysis may be used for modeling of the electric circuits. The homological *branch*, *mesh* and *nodal* analyses are presented. With slight modifications, we follow [2, 3, 4, 5], where the reader can find more involved theoretical details and related references as well. Geometrical interpretations are given. For simplicity, the cohomological aspects are not exposed.

2. Notations

Consider a simple DC electric circuit C on Fig 2.1. It has the following basic geometric spanning spaces:

- *Node space* $C_0 := \langle v_1 v_2 \rangle_{\mathbb{R}}$
- *Contour space* $C_1 := \langle e_1 e_2 e_3 \rangle_{\mathbb{R}}$
- *Mesh space* $C_2 := \langle m_1 m_2 \rangle_{\mathbb{R}}$

Elements of C_n are called n-chains and we denote $C := (C_n)_{n=0,1,2}$. Denote the algebraic electrical parameters (with values from the coefficient field $\mathbb{R}$) as follows:

- ϕ_1, ϕ_2 - the node potentials,
- i_1, i_2, i_3 - the contour currents,

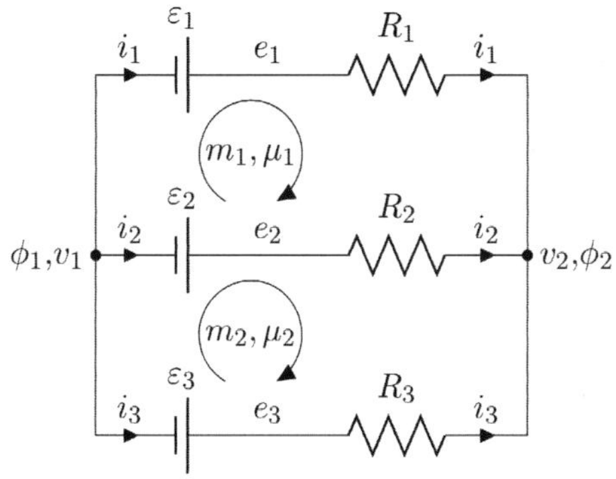

Figure 2.1. DC circuit

- $\varepsilon_1, \varepsilon_2, \varepsilon_3$ - the voltages,
- μ_1, μ_2 - the mesh currents.
- $\mathbb{R}_+ \ni R_1, R_2, R_3$ - the resistors,
- $\mathbb{R}_+ \ni G_1, G_2, G_3$ - the conductances, defined by $G_n R_n := 1$ $(n = 1, 2, 3)$.

To denote the physical variables, it is convenient to use the Dirac *bra-ket* notations. Thus, denote the rows by *bra*-vectors, e.g,

$$\langle \phi| := \langle \phi_1 \phi_2| := [\phi_1 \phi_1] \tag{2.1a}$$
$$\langle i| := \langle i_1 i_2 i_3| := [i_1 i_2 i_3] \tag{2.1b}$$
$$\langle \mu| := \langle \mu_1 \mu_2| := [\mu_1 \mu_2] \tag{2.1c}$$
$$\langle \varepsilon| := \langle \varepsilon_1 \varepsilon_2 \varepsilon_3| := [\varepsilon_1 \varepsilon_2 \varepsilon_3] \tag{2.1d}$$
$$\langle R| := \langle R_1 R_2 R_3| := [R_1 R_2 R_3] \tag{2.1e}$$
$$\langle G| := \langle G_1 G_2 G_3| := [G_1 G_2 G_3] \tag{2.1f}$$

and their (here real) transposes are denoted by ket-vectors $|\cdots\rangle := \langle\cdots|^T$, the latter are thus *columns.* In such a notation, the bra-ket vectors may be considered as coordinate vectors of the chains. One must be careful about context, i.e the physical meaning of the bra-kets, the chain spaces must be distinguished according to the physical units. Also, not all chains represent the *physical* states.

3. Circuit metrics & scalar product

By definition, the circuit metrical matrix is symmetric, *positively* defined and reads

$$R = \begin{bmatrix} R_1 & 0 & 0 \\ 0 & R_2 & 0 \\ 0 & 0 & R_3 \end{bmatrix}, \quad \det R > 0, \quad R^T = R \quad \text{(symmetry)} \tag{3.1}$$

and its inverse, defined by $GR = 1_{3\times 3} = RG$ is

$$G := G = \frac{1}{R_1 R_2 R_3} \begin{bmatrix} R_2 R_3 & 0 & 0 \\ 0 & R_1 R_3 & 0 \\ 0 & 0 & R_1 R_2 \end{bmatrix} = \begin{bmatrix} G_1 & 0 & 0 \\ 0 & G_2 & 0 \\ 0 & 0 & G_3 \end{bmatrix} = G^T \tag{3.2}$$

Let ρ denote either R or G. Then the *non-euclidean* elliptic (iso)scalar product $\langle\cdot|\cdot\rangle_\rho$ is defined by $\langle\cdot|\cdot\rangle_\rho := \langle\cdot|\rho|\cdot\rangle$. One must be careful with limits (contractions) $\det \rho \to 0$ and remember that every *physical* wire and voltage source has at least its (nontrivial) *positive self-resistance*, that may be included in ρ, so that $\det \rho > 0$.

With respect to the circuit metrics we may define the (iso)*norm* function $|\cdot|_\rho$ by $|x|_\rho := \sqrt{\langle x|x\rangle_\rho}$. Then, the Cauchy-Schwartz (CS) inequality $|\langle x|y\rangle_\rho| \le |x|_\rho |y|_\rho$ is evident whenever the scalar product exists for given vectors, as CS inequality holds for every scalar (inner) product.

4. Boundary operator

Now construct the boundary operator $\partial := (\partial_n)_{n=0,1,2,3}$ of the electric circuit presented on Fig. 2.1 and its matrix representation. In what follows, we identify the chains with their coordinate ket-vectors.

First construct $\partial_0 : C_0 \to C_{-1} := \langle 0\rangle_{\mathbb{R}}$. By definition, the nodes (vertices) are elementary elements of circuits with trivial boundaries, thus

$$\partial_0 v_1 := 0 := \partial_0 v_2 \quad \Longrightarrow \quad \partial_0 = \begin{bmatrix} 0 & 0 \end{bmatrix} := 0_{1\times 2} \tag{4.1}$$

Next define $\partial_1 : C_1 \to C_0$, which acts on the directed contours (edges, branches) by

$$\partial_1 e_1 := v_2 - v_1 := |-1;1\rangle, \quad \partial_1 e_2 := v_2 - v_1 := |-1;1\rangle, \quad \partial_1 e_3 := v_2 - v_1 := |-1;1\rangle \tag{4.2}$$

In coordinate (matrix) representation one has

$$\partial_1 = \begin{bmatrix} -1 & -1 & -1 \\ 1 & 1 & 1 \end{bmatrix} \quad \Longrightarrow \quad \partial_1^T = \begin{bmatrix} -1 & 1 \\ -1 & 1 \\ -1 & 1 \end{bmatrix} := \delta^0 \tag{4.3}$$

Evidently, $\partial_0 \partial_1 = 0$. Note that $\operatorname{rank} \partial_1 = 1$.

Now define $\partial_2 : C_2 \to C_1$, which acts on the closed clockwise directed contours by

$$\partial_2 m_1 := e_1 - e_2 := |1;-1;0\rangle, \quad \partial_2 m_2 := e_2 - e_3 := |0;1;-1\rangle \tag{4.4}$$

from which it follows that

$$\partial_2 := \begin{bmatrix} 1 & 0 \\ -1 & 1 \\ 0 & -1 \end{bmatrix} \quad \Longrightarrow \quad \partial_2^T = \begin{bmatrix} 1 & -1 & 0 \\ 0 & 1 & -1 \end{bmatrix} := \delta^1 \tag{4.5}$$

One again can easily check that $\partial_1 \partial_2 = 0_{2\times 2}$ as well as $\delta^1 \delta^0 = 0$. Note that $\operatorname{rank} \partial_2 = 2$.

We finalize the construction by defining $\partial_3 := 0_{2\times 1}$, which means that $C_3 := \langle 0\rangle_{\mathbb{R}}$.

Remark 4.1. One must be careful when comparing our representation with [2, 3, 4], where the mesh space C_2 is identified with $\operatorname{Im} \partial_2 \subset C_1$.

5. Homology

The boundary operator is defined by its action on the geometrical elements of the circuit, thus not depending on the particular electrical parameters - potentials, voltages, circuit and mesh currents - but only on the topology of the circuit under consideration. One can visualize the boundary operator and its (mathematical) domains and codomains by the following complex:

$$(C_3 :=) \;\; 0 \xrightarrow{(\partial_3 = 0)} C_2 \xrightarrow{\partial_2} C_1 \xrightarrow{\partial_1} C_0 \xrightarrow{(\partial_0 = 0)} 0 \;\; (:= C_{-1}) \tag{5.1}$$

As we have seen, the boundary operator is *nilpotent*,

$$\operatorname{Im} \partial_{n+1} \subseteq \operatorname{Ker} \partial_n \quad \Longleftrightarrow \quad \partial_n \partial_{n+1} = 0, \quad n = 0,1,2 \tag{5.2}$$

which is concisely denoted as $\partial^2 = 0$. A complex $(C, \partial) := (C_n, \partial_n)_{n=0,1,2,3}$ is said to be *exact* at C_n if $\operatorname{Im} \partial_{n+1} = \operatorname{Ker} \partial_n$. To study the *exactness* of this complex, the following three conditions must be inquired:

(i) $0 \stackrel{?}{=} \operatorname{Ker} \partial_2$

(ii) $\operatorname{Im} \partial_2 \stackrel{?}{=} \operatorname{Ker} \partial_1$

(iii) $\operatorname{Im} \partial_1 \stackrel{?}{=} C_0$

For a short exact sequence one can write:

$$C_0 \cong \frac{C_1}{\operatorname{Ker} \partial_1} \cong \frac{C_1}{\operatorname{Im} \partial_2}, \quad \dim C_1 = \dim C_0 + \dim \operatorname{Ker} \partial_1 = \dim C_0 + \dim \operatorname{Im} \partial_2 \tag{5.3}$$

The deviation of a complex from exactness can be described by the *homology* concept. The *homology* of the complex $C := (C_n, \partial_n)_{n=0,1,2,3}$ is the sequence $H(C) := (H_n(C))_{n=0,1,2}$ with homogeneous components $H_n(C)$ called the *homology spaces* that are defined as quotient spaces

$$H_n(C) := \frac{Z_n(C) := \operatorname{Ker} \partial_n}{B_n(C) := \operatorname{Im} \partial_{n+1}}, \quad \dim Z_n = \dim H_n + \dim B_n, \quad n = 0, 1, 2 \tag{5.4}$$

Chains from $Z(C) := \operatorname{Ker} \partial$ are called *cycles* and from $B(C) := \operatorname{Im} \partial$ *boundaries.*

Note that correctness of the homology construction is based on the inclusion (5.2). One can easily see from (5.4) that in homological terms the exactness conditions may be presented as follows:

(i) $0 = \operatorname{Ker} \partial_2 \iff H_2 = 0 \iff \dim H_2 = 0$

(ii) $\operatorname{Im} \partial_2 = \operatorname{Ker} \partial_1 \iff H_1 = 0 \iff \dim H_1 = 0$

(iii) $\operatorname{Im} \partial_1 = C_0 \iff H_0 = 0 \iff \dim H_0 = 0$

6. Homological Kirchhoff Laws

As we can see, not all chains represent the *physical* states. The real electrical configurations are prescribed by the Kirchhoff Laws, which in contemporary terms can be seen [5] as the *discrete* Yang-Mills equations. Hence, it is not surprising that the homological (form of the) Kirchhoff Laws (HKL) read (see e.g [2, 3, 4, 5]

(i) [HKCL] $\partial_1 |i\rangle = 0 \iff |i\rangle \in \operatorname{Ker} \partial_1$

(ii) [HKVL] $R|i\rangle = |\varepsilon\rangle - \delta^0 |\phi\rangle \iff |\varepsilon\rangle - R|i\rangle \in \operatorname{Im} \delta^0$

Thus, in homological terms, the Kirchhoff Laws can compactly be presented by using the boundary (and coboundary) operators of a particular circuit.

It must be noted that the Kirchhoff Laws are the *physical* laws and as other physical laws these can not be fully proved mathematically or other theoretical discussions, but tested only via the physical measurements and observations. This concerns the HKL as well. Here one can observe certain analogy with variational principles of physics and the differential equations of the dynamical systems.

Below we present the homological *branch*, *mesh* and *nodal* analyses, as well as explain how the KCL and KVL follow from the HKL.

The HKCL tells us that the *physical* branch currents are realized only in $Z_1(C) := \operatorname{Ker} \partial_1$ (cycles). To describe the latter, recall the circuit notations (2.1a). We have

$$|\varepsilon\rangle - \delta^0 |\phi\rangle = \begin{bmatrix} \varepsilon_1 \\ \varepsilon_2 \\ \varepsilon_3 \end{bmatrix} - \begin{bmatrix} -1 & 1 \\ -1 & 1 \\ -1 & 1 \end{bmatrix} \begin{bmatrix} \phi_1 \\ \phi_2 \end{bmatrix} = \begin{bmatrix} \varepsilon_1 + \phi_1 - \phi_2 \\ \varepsilon_2 + \phi_1 - \phi_2 \\ \varepsilon_3 + \phi_1 - \phi_2 \end{bmatrix} \tag{6.1}$$

and one can see that

$$|i\rangle := |i_1 i_2 i_3\rangle \tag{6.2a}$$

Journal of Physics: Conference Series **532** (2014) 012022 doi:10.1088/1742-6596/532/1/012022

$$= G\left(|\varepsilon\rangle - \delta^0 |\phi\rangle\right) \tag{6.2b}$$

$$= \begin{bmatrix} G_1 & 0 & 0 \\ 0 & G_2 & 0 \\ 0 & 0 & G_3 \end{bmatrix} \begin{bmatrix} \varepsilon_1 + \phi_1 - \phi_2 \\ \varepsilon_2 + \phi_1 - \phi_2 \\ \varepsilon_3 + \phi_1 - \phi_2 \end{bmatrix} \tag{6.2c}$$

$$= \begin{bmatrix} G_1(\varepsilon_1 + \phi_1 - \phi_2) \\ G_2(\varepsilon_2 + \phi_1 - \phi_2) \\ G_3(\varepsilon_3 + \phi_1 - \phi_2) \end{bmatrix} \tag{6.2d}$$

$$= |\underbrace{G_1(\varepsilon_1 + \phi_1 - \phi_2)}_{i_1}; \underbrace{G_2(\varepsilon_2 + \phi_1 - \phi_2)}_{i_2}; \underbrace{G_3(\varepsilon_3 + \phi_1 - \phi_2)}_{i_3}\rangle \tag{6.2e}$$

which results

$$i_1 = \frac{\varepsilon_1 + \phi_1 - \phi_2}{R_1}, \quad i_2 = \frac{\varepsilon_2 + \phi_1 - \phi_2}{R_2}, \quad i_3 = \frac{\varepsilon_3 + \phi_1 - \phi_2}{R_3} \tag{6.2f}$$

Now, the conventional KVL around the closed loops easily follow:

$$\bar{\delta}\phi := \phi_1 - \phi_2 = -\varepsilon_1 + R_1 i_1 \tag{6.3a}$$

$$= -\varepsilon_2 + R_2 i_2 \tag{6.3b}$$

$$= -\varepsilon_3 + R_3 i_3 \tag{6.3c}$$

while consistency is evident. Hence, the HKCL reads

$$\partial_1 |i_1 i_2 i_3\rangle = 0 \quad \Longleftrightarrow \quad i_1 + i_2 + i_3 = 0 \quad \text{(branch currents plane)}, \quad \dim \operatorname{Ker} \partial_1 = 2 \tag{6.4}$$

that we can rewrite as

$$G_1(\varepsilon_1 + \phi_1 - \phi_2) + G_2(\varepsilon_2 + \phi_1 - \phi_2) + G_3(\varepsilon_3 + \phi_1 - \phi_2) = 0 \tag{6.5}$$

from which it easily follows

$$G_1\varepsilon_1 + G_2\varepsilon_2 + G_3\varepsilon_3 = -(G_1 + G_2 + G_3)(\phi_1 - \phi_2) \quad \text{(voltage plane)} \tag{6.6}$$

Hence, the voltage drop $\bar{\delta}\phi$ between nodes v_1, v_2 is given by

$$-\bar{\delta}\phi = \frac{\langle G|\varepsilon\rangle}{\operatorname{Tr} G} = \frac{\langle G_1 G_2 G_3 | \varepsilon_1 \varepsilon_2 \varepsilon_3\rangle}{\operatorname{Tr} G} \quad \text{(Millman's formula, } n = 3\text{)} \tag{6.7}$$

The latter tells us that for fixed $\bar{\delta}\phi$ the algebraic voltages are not fully arbitrary, because the circuit voltage point $V_\varepsilon := (\varepsilon_1; \varepsilon_2; \varepsilon_3)$ lies on the voltage 2-plane (6.6) as the result of the HKCL and HKVL while the circuit branch current point $I_i := (i_1; i_2; i_3)$ lies on the current 2-plane (6.4). Alternatively, one can consider the Millman's formula (6.7) as a generator of $\bar{\delta}\phi$ as well.

We know from the HKCL that the *physical* branch currents are cycles, i.e $\partial_1 |i\rangle = 0$. Thus it is natural to search the latter as a boundary

$$|i\rangle := \partial_2 |\mu\rangle \quad \in B_1(C) := \operatorname{Im} \partial_2 \quad \text{(converse Poincaré lemma)} \tag{6.8}$$

where $|\mu\rangle$ is called the *mesh current*. Calculate:

$$|i_1 i_2 i_3\rangle = \begin{bmatrix} 1 & 0 \\ -1 & 1 \\ 0 & -1 \end{bmatrix} |\mu_1 \mu_2\rangle \tag{6.9a}$$

$$= |\mu_1; -\mu_1 + \mu_2; -\mu_2\rangle \tag{6.9b}$$

from which we obtain

$$i_1 = \mu_1, \quad i_2 = -\mu_1 + \mu_2, \quad i_3 = -\mu_2 \tag{6.10}$$

and hence the HKCL becomes automatic. The mesh currents may be used to construct bases in Z_1.

7. Homological analysis and modeling

Now we may collect the homological properties of the electric circuit on Fig. 2.1 as follows.

Theorem 7.1 (cf [2]). *The electric circuit of Fig. 2.1 can be represented by the following short exact sequence:*

$$0 \longrightarrow C_2 \xrightarrow{\partial_2} C_1 \xrightarrow{\partial_1} \operatorname{Im}\partial_1 \longrightarrow 0 \tag{7.1}$$

Proof. We compactly collect the basic points of the proof and for convenience use notations of (5.1) as well as the dimensional considerations in (5.4).

(i) Exactness at C_2: $0 = \operatorname{Ker}\partial_2 \overset{?}{\iff} H_2 = 0$.
Note that $\operatorname{Im}\partial_3 = 0$ and $\operatorname{rank}\partial_2 = 2$ (maximal). Hence, $\dim\operatorname{Im}\partial_3 = 0 = \dim\operatorname{Ker}\partial_2$, which is equivalent to $H_2 = 0$.

(ii) $\operatorname{Im}\partial_2 = \operatorname{Ker}\partial_1 \overset{?}{\iff} H_1 = 0$.
Note that $\dim\operatorname{Im}\partial_2 = 2 = \dim\operatorname{Ker}\partial_1$, which is equivalent to $H_1 = 0$.

(iii) $\operatorname{Im}\partial_1 = \operatorname{Ker}\partial_0 \overset{?}{\iff} H_0 = 0$.
Note that $\dim\operatorname{Im}\partial_1 = 1 = \dim\operatorname{Ker}\partial_0$, which is equivalent to $H_0 = 0$. □

Corollary 7.2. *Thus we have*

$$\operatorname{Im}\partial_1 \cong \frac{C_1}{\operatorname{Ker}\partial_1} \cong \frac{C_1}{\operatorname{Im}\partial_2} \tag{7.2}$$

Remark 7.3 (correctness). As usual in the mathematical physics, by the *correctness* of a (modeling) problem one means:

(i) Existence of the solution.
(ii) Uniqueness of the solution.
(iii) Stability of the solution under the infinitesimal deformation of the physical parameters.

Note that conditions 1 and 2 are related to the short exact sequence (7.1) while the explicit form of the solution and its stability is given by the Millman formula (6.7). For more careful study one has to apply the *cohomological* analysis as well, the latter can be realized by the *dual* to the short exact sequence 7.1. Here, we omit the cohomological analysis, one can find more details in [2, 3].

8. Numerical example

As a simple example determine the branch and mesh currents in the circuit shown in Fig. 2.1. Take the electric parameters as (we follow [1], Problem 4.10):

$$\varepsilon_1 = 40V, \quad \varepsilon_2 = 12V, \quad \varepsilon_3 = -24V \tag{8.1a}$$

$$R_1 = 6\Omega, \quad R_2 = 4\Omega, \quad R_3 = 12\Omega \tag{8.1b}$$

First calculate the voltage drop

$$\bar{\delta}\phi := \phi_1 - \phi_2 \tag{8.2a}$$

$$= -\frac{40V \cdot 4\Omega \cdot 12\Omega + 12V \cdot 6\Omega \cdot 12\Omega - 24V \cdot 6\Omega \cdot 4\Omega}{6\Omega \cdot 4\Omega + 4\Omega \cdot 12\Omega + 12\Omega \cdot 6\Omega} \tag{8.2b}$$

$$= -\frac{(1920 + 864 - 576)V\Omega^2}{(24 + 48 + 72)\Omega^2} \tag{8.2c}$$

$$= -\frac{184}{12}V \tag{8.2d}$$

Then calculate the circuit and mesh currents

$$i_1 = \frac{\varepsilon_1 + \bar{\delta}\phi}{R_1} = \frac{12 \cdot 40V - 184V}{6\Omega \cdot 12} = \frac{296V}{72\Omega} = \frac{148}{36}A \qquad = \mu_1 \tag{8.3a}$$

$$i_2 = \frac{\varepsilon_2 + \bar{\delta}\phi}{R_2} = \frac{12 \cdot 12V - 184V}{4\Omega \cdot 12} = -\frac{40V}{48\Omega} = -\frac{40V \cdot \frac{3}{4}}{48\Omega \cdot \frac{3}{4}} = -\frac{30}{36}A \qquad = -\mu_1 + \mu_2 \tag{8.3b}$$

$$i_3 = \frac{\varepsilon_3 + \bar{\delta}\phi}{R_3} = \frac{12 \cdot (-24V) - 184V}{12\Omega \cdot 12} = \frac{-472V}{144\Omega} = -\frac{118}{36}A \qquad = -\mu_2 \tag{8.3c}$$

Finally, check the KCL

$$i_1 + i_2 + i_3 = \frac{148 - 30 - 118}{36}A = 0 \tag{8.4}$$

Computer simulation for this particular circuit can easily be arranged.

Acknowledgments

The research was in part supported by the Estonian Research Council, grant ETF9038. Authors are grateful to participants of MOD II *Moduli, Operads, Dynamics II* [6] (Tallinn, 3-6 June 2014) and, in particular, to A. Siqveland and A. A. Voronov for the thorough discussions and useful remarks.

References

[1] O'Malley J 2011 *Basic Circuit Analysis* 2nd Ed (McGraw-Hill)
[2] Roth J P 1955 *Proc. Nat. Acad. Sci. USA* **41** 518
[3] Bamberg P and Sternberg S 1990 *A Course in Mathematics for Students of Physics* (Cambridge Univ. Press) Vol 2
[4] Frankel T 1997 *The Geometry of Physics* (Cambridge Univ. Press) Appendix B
[5] Zeidler E 2011 *Quantum Field Theory III: Gauge Theory* (Springer-Verlag) sec 22
[6] *AstrAlgo cWeb* 2014 MOD II

3Quantum: Algebra Geometry Information (QQQ Conference 2012) IOP Publishing
Journal of Physics: Conference Series **532** (2014) 012023 doi:10.1088/1742-6596/532/1/012023

Operadic quantization as a tool for discrete geometry

E Paal and J Virkepu

Tallinn University of Technology, Ehitajate tee 5, 19086 Tallinn, Estonia

Abstract. The operadic Lax representations of the harmonic oscillator are used to construct the quantum counterparts of 3d real Lie algebras in the Bianchi classification. The Jacobi operators of these quantum algebras are studied. It is shown how the energy conservation is related to the Jacobi identity and how the quantization leads to an anomaly – the quantum violation of the Jacobi relations. By using the non-vanishing quantum Jacobi operators, the derivative quantum algebra for a triple of 3d real Lie algebras is defined. It is proposed that the derivative algebra is the 3d real Heisenberg algebra. From this it follows that in this model only the discrete values of the spatial coordinates are physically allowed.

1. Introduction and outline of the paper

In Hamiltonian formalism, a mechanical system is described by the canonical variables q^s, p_s and their time evolution is prescribed by the Hamiltonian equations

$$\frac{dq^s}{dt} = \frac{\partial H}{\partial p_s}, \quad \frac{dp_s}{dt} = -\frac{\partial H}{\partial q^s} \tag{1.1}$$

By a Lax representation [1] of a mechanical system one means such a pair (L, A) of matrices (linear operators) L, A that the above Hamiltonian system may be represented as the Lax equation

$$\frac{dL}{dt} = [A, L] := AL - LA \tag{1.2}$$

On the other hand, according to quantum theory the time evolution of a quantum observable $\hat{F}$ is prescribed by the Heisenberg equation

$$\frac{\hbar}{i}\frac{d\hat{F}}{dt} = [\hat{H}, \hat{F}], \quad i = \sqrt{-1} \tag{1.3}$$

Thus, from the algebraic point of view, both the classical and quantum observables may be represented by linear operators, i.e by linear maps $V \to V$ of a linear space V. As a generalization of this one can pose the following question [2]: how can the time evolution of the linear *operations* (multiplications) $V^{\otimes n} \to V$ be described as the physical observables?

The algebraic operations (multiplications) and co-operations (comultiplications) can be seen as an example of the *operadic* variables [3] representing operadic systems. If the latter depend on time one can speak about *operadic dynamics* [2]. The classical operadic dynamics may be introduced by simple and natural analogy with the Hamiltonian dynamics. In particular, the time evolution of the operadic variables may be given by the operadic Lax equation. In

[4, 5, 6], the low-dimensional binary operadic Lax representations for the harmonic oscillator were constructed and in [7] it was shown how the operadic Lax representations are related to the conservation of energy.

In [8], the operadic Lax representations were used to construct the quantum counterparts of the real three-dimensional Lie algebras in the Bianchi classification, see e.g [9, 10], over the harmonic oscillator. In [11], the Jacobi operators of some of these quantum algebras were explicitly calculated.

In this paper, the operadic Lax representations of the harmonic oscillator are used to construct the quantum counterparts of 3d real Lie algebras in the the Bianchi classification. The Jacobi operators of these quantum algebras are studied. It is shown how the energy conservation is related to the Jacobi identity and how the quantization leads to an anomaly – the quantum violation of the Jacobi relations. By using the non-vanishing quantum Jacobi operators, the derivative quantum algebra for a triple of 3d real Lie algebras is defined. It is conjectured that the derivative algebra is the 3d real Heisenberg algebra. From this it follows that in this model only the discrete values of the spatial coordinates are physically allowed.

2. Endomorphism operad and Gerstenhaber brackets

Let K be a unital associative commutative ring, V be a unital K-module, and $\mathcal{E}_V^n := \mathcal{E}nd_V^n := \operatorname{Hom}(V^{\otimes n}, V)$ $(n \in \mathbb{N})$. For an *operation* $f \in \mathcal{E}_V^n$, we refer to n as the *degree* of f and often write (when it does not cause confusion) f instead of $\deg f$. For example, $(-1)^f := (-1)^n$, $\mathcal{E}_V^f := \mathcal{E}_V^n$ and $\circ_f := \circ_n$. Also, it is convenient to use the *reduced* (desuspended) degree $|f| := n-1$. Throughout this paper, we assume that $\otimes := \otimes_K$.

Definition 2.1 (endomorphism operad [3]). For $f \otimes g \in \mathcal{E}_V^f \otimes \mathcal{E}_V^g$ define the (*partial*) *compositions*

$$f \circ_i g := (-1)^{i|g|} f \circ (\mathrm{id}_V^{\otimes i} \otimes g \otimes \mathrm{id}_V^{\otimes(|f|-i)}) \quad \in \mathcal{E}_V^{f+|g|}, \quad 0 \le i \le |f|$$

The sequence $\mathcal{E}_V := (\mathcal{E}_V^n)_{n \in \mathbb{N}}$ equipped with the partial compositions $\circ_i$, is called the *endomorphism operad* of V.

Definition 2.2 (total composition [3]). The *total composition* $\circ : \mathcal{E}_V^f \otimes \mathcal{E}_V^g \to \mathcal{E}_V^{f+|g|}$ is defined by

$$f \circ g := \sum_{i=0}^{|f|} f \circ_i g \quad \in \mathcal{E}_V^{f+|g|}, \quad |\circ| = 0$$

The pair $\operatorname{Com} \mathcal{E}_V := (\mathcal{E}_V, \circ)$ is called the *composition algebra* of $\mathcal{E}_V$.

It turns out that $\operatorname{Com} \mathcal{E}_V$ is not associative but is a graded *Lie-admissible algebra* in a sense given below.

Definition 2.3 (Gerstenhaber brackets [3]). The *Gerstenhaber brackets* $[\cdot,\cdot]_G$ are defined in $\operatorname{Com} \mathcal{E}_V$ as a graded commutator by

$$[f,g]_G := [f,g] := f \circ g - (-1)^{|f||g|} g \circ f = -(-1)^{|f||g|}[g,f], \quad |[\cdot,\cdot]| = 0$$

The *commutator algebra* of $\operatorname{Com} \mathcal{E}_V$ is denoted as $\operatorname{Com}^- \mathcal{E}_V := \{\mathcal{E}_V, [\cdot,\cdot]\}$.

It is easy to see that the Gerstenhaber bracket naturally extend the Lie brackets to operadic realm: if the degrees $|f| = 0 = |g|$ then one receive the standard Lie commutator bracketing as it is used in Lax representation and the Heisenberg equation. The Gerstenhaber extension of

Journal of Physics: Conference Series **532** (2014) 012023 doi:10.1088/1742-6596/532/1/012023

the Lie bracketing is even more impressive due to the observation [3] that $\mathrm{Com}^-\mathcal{E}_V$ is a graded *Lie algebra*. The Jacobi identity reads

$$(-1)^{|f||h|}[f,[g,h]] + (-1)^{|g||f|}[g,[h,f]] + (-1)^{|h||g|}[h,[f,g]] = 0$$

The main principle of the operadic dynamics states that operadic observables form a structure that is a graded analogue of the Poisson algebra, nowadays called the Gerstenhaber algebra. For algebraic modeling of the operadic systems the cohomology spaces of associative algebras are the same natural representation spaces as using the matrix multiplication for modeling the Lie bracketing - the matter of ultimate representation conveniency.

3. Operadic Lax pair and Heisenberg equation

Assume that $K := \mathbb{R}$ or $K := \mathbb{C}$ and operations are differentiable. Dynamics in operadic systems (operadic dynamics) may be introduced by

Definition 3.1 (operadic Lax pair [2]). Allow a classical dynamical system to be described by the Hamiltonian system (1.1). An *operadic Lax pair* is a pair (L, A) of homogeneous operations $L, A \in \mathcal{E}_V$, such that the Hamiltonian system (1.1) may be represented as the *operadic Lax equation*

$$\frac{dL}{dt} = [A, L]_G := A \circ L - (-1)^{|M||A|} L \circ A$$

The pair (L, A) is also called an *operadic Lax representations* of the Hamiltonian system (1.1).

Remark 3.2. Evidently, the degree constraints $|L| = |A| = 0$ give rise to ordinary Lax equation (1.2) [1]. In this paper we assume that $|A| = 0$, which means that d/dt is the time derivative operator with a *scalar* valued time t.

Definition 3.3 (operadic Heisenberg equation [2], cf (1.3)). The *operadic Heisenberg equation* for a quantum (operadic) observable is

$$\frac{\hbar}{i}\frac{d\hat{F}}{dt} = [\hat{H}, \hat{F}]_G, \quad i = \sqrt{-1}$$

In this paper, we do not consider this model.

Example 3.4 (operadic HO). The Hamiltonian of the harmonic oscillator (HO) is

$$H(q,p) = \frac{1}{2}(p^2 + \omega^2 q^2)$$

and its Hamiltonian system reads

$$\frac{dq}{dt} = \frac{\partial H}{\partial p} = p, \qquad \frac{dp}{dt} = -\frac{\partial H}{\partial q} = -\omega^2 q$$

If L is a linear algebraic operation we can use the above Hamilton equations to obtain

$$\frac{dL}{dt} = \frac{\partial L}{\partial q}\frac{dq}{dt} + \frac{\partial L}{\partial p}\frac{dp}{dt} = p\frac{\partial L}{\partial q} - \omega^2 q \frac{\partial L}{\partial p} = [A, L]_G$$

Therefore, we get the following linear partial differential equation for $L(q,p)$:

$$p\frac{\partial L}{\partial q} - \omega^2 q \frac{\partial L}{\partial p} = [A, L]_G$$

By integrating the latter equation one can get collections of operations called [2] the *operadic* (Lax representations of) the harmonic oscillator.

Journal of Physics: Conference Series **532** (2014) 012023 doi:10.1088/1742-6596/532/1/012023

4. Bianchi classification of 3d real Lie algebras

We use the Bianchi classification of 3d real Lie algebras [9], see also [10]. The structure equations

$$[e_j, e_k] = \overset{\circ}{\mu}{}^s_{jk} e_s$$

of the latter can be presented as follows:

$$[e_1, e_2] = -\alpha e_2 + n^3 e_3, \quad [e_2, e_3] = n^1 e_1, \quad [e_3, e_1] = n^2 e_2 + \alpha e_3$$

where the values of the parameters α, n^1, n^2, n^3 and the corresponding structure constants are presented in Table 1.

Table 1. 3d real Lie algebras in Bianchi classification. Here $a > 0$.

Bianchi type	α	$(n^1; n^2; n^3)$	$\overset{\circ}{\mu}{}^1_{12}$	$\overset{\circ}{\mu}{}^2_{12}$	$\overset{\circ}{\mu}{}^3_{12}$	$\overset{\circ}{\mu}{}^1_{23}$	$\overset{\circ}{\mu}{}^2_{23}$	$\overset{\circ}{\mu}{}^3_{23}$	$\overset{\circ}{\mu}{}^1_{31}$	$\overset{\circ}{\mu}{}^2_{31}$	$\overset{\circ}{\mu}{}^3_{31}$
I	0	(0; 0; 0)	0	0	0	0	0	0	0	0	0
II	0	(1; 0; 0)	0	0	0	1	0	0	0	0	0
VII	0	(1; 1; 0)	0	0	0	1	0	0	0	1	0
VI	0	(1; −1; 0)	0	0	0	1	0	0	0	−1	0
IX	0	(1; 1; 1)	0	0	1	1	0	0	0	1	0
VIII	0	(1; 1; −1)	0	0	−1	1	0	0	0	1	0
V	1	(0; 0; 0)	0	−1	0	0	0	0	0	0	1
IV	1	(0; 0; 1)	0	−1	1	0	0	0	0	0	1
VII_a	a	(0; 1; 1)	0	$-a$	1	0	0	0	0	1	a
$\mathrm{III}_{a=1}$	1	(0; 1; −1)	0	−1	−1	0	0	0	0	1	1
$\mathrm{VI}_{a\neq 1}$	a	(0; 1; −1)	0	$-a$	−1	0	0	0	0	1	a

5. Dynamical deformations of 3D real Lie algebras

Definition 5.1 (quasi-canonical coordinates). For the HO, define its *quasi-canonical coordinates* $A_\pm$ by

$$A_+^2 - A_-^2 = 2p, \quad A_+ A_- = \omega q$$

Remark 5.2. Note that these constraints easily imply

$$A_+^2 + A_-^2 = 2\sqrt{2H}$$

Theorem 5.3 (Poisson brackets of quasi-canonical coordinates). *The quasi-canonical coordinates $A_\pm$ satisfy the relations*

$$\{A_+ A_+\} = 0 = \{A_-, A_-\}, \quad \{A_+, A_-\} = \varepsilon := \omega/2\sqrt{2H}$$

Proof. Calculate:

$$\begin{aligned}
2\omega = 2\omega\{p, q\} &= \{A_+^2 - A_-^2, A_+ A_-\} \\
&= \{A_+^2, A_+ A_-\} - \{A_-^2, A_+ A_-\} \\
&= A_+\{A_+^2, A_-\} - \{A_-^2, A_+\} A_- \\
&= A_+\{A_+ A_- +, A_- A_-\} - \{A_- A_-, A_+\} A_- \\
&= 2(A_+^2 + A_-^2)\{A_+, A_-\} \\
&= 4\sqrt{2H}\{A_+, A_-\}
\end{aligned}$$

□

Journal of Physics: Conference Series **532** (2014) 012023 doi:10.1088/1742-6596/532/1/012023

By using the structure constants of the 3d real Lie algebras in the Bianchi classification, one can prescribe [7] the time evolution of the 3d real Lie algebras as given in Table 2.

Table 2. Time evolution of 3d real Lie algebras. Here $p_0 = \sqrt{2E} > 0$.

Dynamical Bianchi type	μ_{12}^1	μ_{12}^2	μ_{12}^3	μ_{23}^1	μ_{23}^2	μ_{23}^3	μ_{31}^1	μ_{31}^2	μ_{31}^3
I^t	0	0	0	0	0	0	0	0	0
II^t	0	0	0	$\frac{p+p_0}{2p_0}$	$\frac{\omega q}{2p_0}$	0	$\frac{\omega q}{2p_0}$	$\frac{p-p_0}{-2p_0}$	0
VII^t	0	0	0	1	0	0	0	1	0
VI^t	0	0	0	$\frac{p}{p_0}$	$\frac{\omega q}{p_0}$	0	$\frac{\omega q}{p_0}$	$-\frac{p}{p_0}$	0
IX^t	0	0	1	1	0	0	0	1	0
VIII^t	0	0	-1	1	0	0	0	1	0
V^t	$\frac{A_-}{\sqrt{2p_0}}$	$\frac{-A_+}{\sqrt{2p_0}}$	0	0	0	$\frac{-A_-}{\sqrt{2p_0}}$	0	0	$\frac{A_+}{\sqrt{2p_0}}$
IV^t	$\frac{A_-}{\sqrt{2p_0}}$	$\frac{-A_+}{\sqrt{2p_0}}$	1	0	0	$\frac{-A_-}{\sqrt{2p_0}}$	0	0	$\frac{A_+}{\sqrt{2p_0}}$
VII_a^t	$\frac{aA_-}{\sqrt{2p_0}}$	$\frac{-aA_+}{\sqrt{2p_0}}$	1	$\frac{p-p_0}{-2p_0}$	$\frac{\omega q}{-2p_0}$	$\frac{-aA_-}{\sqrt{2p_0}}$	$\frac{\omega q}{-2p_0}$	$\frac{p+p_0}{2p_0}$	$\frac{aA_+}{\sqrt{2p_0}}$
$\mathrm{III}_{a=1}^t$	$\frac{A_-}{\sqrt{2p_0}}$	$\frac{-A_+}{\sqrt{2p_0}}$	-1	$\frac{p-p_0}{-2p_0}$	$\frac{\omega q}{-2p_0}$	$\frac{-A_-}{\sqrt{2p_0}}$	$\frac{\omega q}{-2p_0}$	$\frac{p+p_0}{2p_0}$	$\frac{A_+}{\sqrt{2p_0}}$
$\mathrm{VI}_{a\neq1}^t$	$\frac{aA_-}{\sqrt{2p_0}}$	$\frac{-aA_+}{\sqrt{2p_0}}$	-1	$\frac{p-p_0}{-2p_0}$	$\frac{\omega q}{-2p_0}$	$\frac{-aA_-}{\sqrt{2p_0}}$	$\frac{\omega q}{-2p_0}$	$\frac{p+p_0}{2p_0}$	$\frac{aA_+}{\sqrt{2p_0}}$

6. Quantum counterparts of 3d real Lie algebras

Let now the harmonic oscillator be quantized, i.e its canonical coordinates satisfy the CCR

$$[\hat{q}, \hat{q}] = 0 = [\hat{p}, \hat{p}], \quad [\hat{p}, \hat{q}] = \hbar/i$$

Then the classical observables $A_\pm(q,p)$ will be quantized as well and their quantum counterparts are denoted by $\hat{A}_\pm := A_\pm(\hat{q}, \hat{p})$. As the result, the quantum counterparts of the 3d real Lie algebras can be listed as presented in Table 3.

One can easily check that $\mathrm{I}^\hbar$, $\mathrm{II}^\hbar$, $\mathrm{VII}^\hbar$, $\mathrm{VI}^\hbar$, $\mathrm{IX}^\hbar$, $\mathrm{VIII}^\hbar$ are Lie agebras. Thus, in what follows, we will only focus on the algebras $\mathrm{V}^\hbar$, $\mathrm{IV}^\hbar$, $\mathrm{VII}_a^\hbar$, $\mathrm{III}_{a=1}^\hbar$, $\mathrm{VI}_{a\neq1}^\hbar$, and present the latter more compactly in a separate table.

Let β, γ, a, b be real-valued parameters from Table 4 and and let $L^\hbar$ denote an entry from the first column of Table 3. Then the algebras $\mathrm{V}^\hbar$, $\mathrm{IV}^\hbar$, $\mathrm{VII}_a^\hbar$, $\mathrm{III}_{a=1}^\hbar$, $\mathrm{VI}_{a\neq1}^\hbar$ from Table 3 can be presented as Table 5.

Table 4. Values of β, γ, a, b for $L^\hbar$. Here $a > 0$.

$L^\hbar$	β	γ	a	b
$\mathrm{V}^\hbar$	0	0	1	0
$\mathrm{IV}^\hbar$	0	0	1	1
$\mathrm{VII}_a^\hbar$	1	1	a	1
$\mathrm{III}_{a=1}^\hbar$	1	1	1	1
$\mathrm{VI}_{a\neq1}^\hbar$	1	1	$a \neq 1$	-1

Journal of Physics: Conference Series **532** (2014) 012023 doi:10.1088/1742-6596/532/1/012023

Table 3. Quantum counterparts of 3d real Lie algebras over HO. Here $a > 0$.

Quantum Bianchi type	$\hat{\mu}_{12}^1$	$\hat{\mu}_{12}^2$	$\hat{\mu}_{12}^3$	$\hat{\mu}_{23}^1$	$\hat{\mu}_{23}^2$	$\hat{\mu}_{23}^3$	$\hat{\mu}_{31}^1$	$\hat{\mu}_{31}^2$	$\hat{\mu}_{31}^3$
$\mathrm{I}^\hbar$	0	0	0	0	0	0	0	0	0
$\mathrm{II}^\hbar$	0	0	0	$\frac{\hat{p}+p_0}{2p_0}$	$\frac{\omega\hat{q}}{2p_0}$	0	$\frac{\omega\hat{q}}{2p_0}$	$\frac{\hat{p}-p_0}{-2p_0}$	0
$\mathrm{VII}^\hbar$	0	0	0	1	0	0	0	1	0
$\mathrm{VI}^\hbar$	0	0	0	$\frac{\hat{p}}{p_0}$	$\frac{\omega\hat{q}}{p_0}$	0	$\frac{\omega\hat{q}}{p_0}$	$-\frac{\hat{p}}{p_0}$	0
$\mathrm{IX}^\hbar$	0	0	1	1	0	0	0	1	0
$\mathrm{VIII}^\hbar$	0	0	-1	1	0	0	0	1	0
$\mathrm{V}^\hbar$	$\frac{\hat{A}_-}{\sqrt{2}p_0}$	$\frac{-\hat{A}_+}{\sqrt{2}p_0}$	0	0	0	$\frac{-\hat{A}_-}{\sqrt{2}p_0}$	0	0	$\frac{\hat{A}_+}{\sqrt{2}p_0}$
$\mathrm{IV}^\hbar$	$\frac{\hat{A}_-}{\sqrt{2}p_0}$	$\frac{-\hat{A}_+}{\sqrt{2}p_0}$	1	0	0	$\frac{-\hat{A}_-}{\sqrt{2}p_0}$	0	0	$\frac{\hat{A}_+}{\sqrt{2}p_0}$
$\mathrm{VII}_a^\hbar$	$\frac{a\hat{A}_-}{\sqrt{2}p_0}$	$\frac{-a\hat{A}_+}{\sqrt{2}p_0}$	1	$\frac{\hat{p}-p_0}{-2p_0}$	$\frac{\omega\hat{q}}{-2p_0}$	$\frac{-a\hat{A}_-}{\sqrt{2}p_0}$	$\frac{\omega\hat{q}}{-2p_0}$	$\frac{\hat{p}+p_0}{2p_0}$	$\frac{a\hat{A}_+}{\sqrt{2}p_0}$
$\mathrm{III}_{a=1}^\hbar$	$\frac{\hat{A}_-}{\sqrt{2}p_0}$	$\frac{-\hat{A}_+}{\sqrt{2}p_0}$	-1	$\frac{\hat{p}-p_0}{-2p_0}$	$\frac{\omega\hat{q}}{-2p_0}$	$\frac{-\hat{A}_-}{\sqrt{2}p_0}$	$\frac{\omega\hat{q}}{-2p_0}$	$\frac{\hat{p}+p_0}{2p_0}$	$\frac{\hat{A}_+}{\sqrt{2}p_0}$
$\mathrm{VI}_{a\neq 1}^\hbar$	$\frac{a\hat{A}_-}{\sqrt{2}p_0}$	$\frac{-a\hat{A}_+}{\sqrt{2}p_0}$	-1	$\frac{\hat{p}-p_0}{-2p_0}$	$\frac{\omega\hat{q}}{-2p_0}$	$\frac{-a\hat{A}_-}{\sqrt{2}p_0}$	$\frac{\omega\hat{q}}{-2p_0}$	$\frac{\hat{p}+p_0}{2p_0}$	$\frac{a\hat{A}_+}{\sqrt{2}p_0}$

Table 5. Quantum counterparts $L^\hbar$ of 3d real Lie algebras

Quantum Bianchi type	$\hat{\mu}_{12}^1$	$\hat{\mu}_{12}^2$	$\hat{\mu}_{12}^3$	$\hat{\mu}_{23}^1$	$\hat{\mu}_{23}^2$	$\hat{\mu}_{23}^3$	$\hat{\mu}_{31}^1$	$\hat{\mu}_{31}^2$	$\hat{\mu}_{31}^3$
$L^\hbar$	$\frac{a\hat{A}_-}{\sqrt{2}p_0}$	$\frac{-a\hat{A}_+}{\sqrt{2}p_0}$	b	$\frac{-\gamma(\hat{p}-p_0)}{2p_0}$	$\frac{-\beta\omega\hat{q}}{2p_0}$	$\frac{-a\hat{A}_-}{\sqrt{2}p_0}$	$\frac{-\beta\omega\hat{q}}{2p_0}$	$\frac{\gamma(\hat{p}+p_0)}{2p_0}$	$\frac{a\hat{A}_+}{\sqrt{2}p_0}$

Let $\mathcal{S}_{HO}$ denote the state space of the quantum harmonic oscillator and select a triple b_1, b_2, b_3 of real orthonormal state vectors. By using Table 5 we define the structure equations on their linear span $S_3 := \langle b_1; b_2; b_3 \rangle_{\mathbb{R}} \subset \mathbb{R}^3$ by

$$[b_i, b_j]_\hbar := \hat{\mu}_{ij}^s b_s$$

For $x, y \in \mathcal{S}_{HO}$, their quantum multiplication reads

$$[x, y]_\hbar := \hat{\mu}_{jk}^i x^j y^k b_i = \hat{\mu}_{jk}^1 x^j y^k b_1 + \hat{\mu}_{jk}^2 x^j y^k b_2 + \hat{\mu}_{jk}^3 x^j y^k b_3$$

Here on must note that, in general, $[x, y]_\hbar \notin S_3$, thus we have an *open* algebra.

7. Jacobi operators

For $x, y, z \in \mathcal{S}_{HO}$, their quantum Jacobi operator is defined by

$$\hat{J}_\hbar(x; y; z) = [x, [y, z]_\hbar]_\hbar + [y, [z, x]_\hbar]_\hbar + [z, [x, y]_\hbar]_\hbar = \hat{J}_\hbar^1(x; y; z)e_1 + \hat{J}_\hbar^2(x; y; z)e_2 + \hat{J}_\hbar^3(x; y; z)e_3$$

In [8, 11], the quantum Jacobi operators were calculated for all real three-dimensional Lie algebras. In this paper, we need only the result as follows.

Denote

$$(x, y, z) := \begin{vmatrix} x^1 & x^2 & x^3 \\ y^1 & y^2 & y^3 \\ z^1 & z^2 & z^3 \end{vmatrix}, \quad \xi_\pm := \beta\omega\hat{q}\hat{A}_\mp \pm \gamma(\hat{p} \mp p_0)\hat{A}_\pm$$

Then we have

Theorem 7.1 ([8, 11]). *The Jacobi operator components of* $L^\hbar$ *read*

$$\hat{J}^1_\hbar(x; y; z) = -\frac{a(x, y, z)}{\sqrt{2p_0^3}}\hat{\xi}_+, \quad \hat{J}^2_\hbar(x; y; z) = -\frac{a(x, y, z)}{\sqrt{2p_0^3}}\hat{\xi}_-, \quad \hat{J}^3_\hbar(x; y; z) = \frac{a^2(x, y, z)}{p_0}[\hat{A}_+, \hat{A}_-]$$

Remark 7.2. Note that quantum Jacobi operator depends on the (signed) *volume* function $(\cdot, \cdot, \cdot) := \det_{3\times 3}$. Once we expect the existence of the *elementary*, i.e the *quantum* length, the Jacobi operator is inevitable non-vanishing, i.e $\hat{J} \neq 0$. In what follows, we try to share light on this phenomenon.

8. Quasi-canonical quantum conditions

When performing the quantization of the quasi-canonical variables, we shall use the Schrödinger picture, i.e the operators $\hat{q}, \hat{p}, \hat{H}$ and $A_\pm$ do not depend on time. Denote by $[\cdot, \cdot]$ the ordinary Lie commutator bracketing. Following the canonical quantization prescription, the quasi-canonical coordinates satisfy the following symmetric constraints:

$$\hat{A}_+^2 + \hat{A}_-^2 = 2\sqrt{2\hat{H}}, \quad \hat{A}_+^2 - \hat{A}_-^2 = 2\hat{p}, \quad \hat{A}_+\hat{A}_- + \hat{A}_-\hat{A}_+ = 2\omega\hat{q}$$

and the *quasi-canonical* commutation relations (*quasi*-CCR) for $A_\pm$ read as follows:

$$[\hat{A}_+, \hat{A}_+] = 0 = [\hat{A}_-, \hat{A}_-], \quad [\hat{A}_+, \hat{A}_-] = \frac{\hbar}{i}\hat{\varepsilon} := \frac{\hbar}{i}\frac{\omega}{2\sqrt{2\hat{H}}}$$

9. Quantum Noether theorem and anomaly

Theorem 9.1 (quantum Noether theorem). *One has*

$$\begin{aligned} \hat{J}^1_\hbar(x; y; z) &= \frac{a(x, y, z)}{\sqrt{2p_0^3}}\left[\hat{A}_+\left(\sqrt{2E} - \sqrt{2\hat{H}}\right) - \frac{\hbar}{i}\hat{A}_-\frac{\hat{\varepsilon}}{2}\right] \\ \hat{J}^2_\hbar(x; y; z) &= \frac{a(x, y, z)}{\sqrt{2p_0^3}}\left[\hat{A}_-\left(\sqrt{2E} - \sqrt{2\hat{H}}\right) + \frac{\hbar}{i}\hat{A}_+\frac{\hat{\varepsilon}}{2}\right] \\ \hat{J}^3_\hbar(x; y; z) &= \frac{\hbar}{i}\frac{a^2(x, y, z)}{p_0}\hat{\varepsilon} \end{aligned}$$

Proof. Use the quantum relations from Sec. 8. We have

$$\begin{aligned} \hat{\xi}_+ &= \omega\hat{q}\hat{A}_- + (\hat{p} - p_0)\hat{A}_+ \\ &= \frac{1}{2}(\hat{A}_+\hat{A}_- + \hat{A}_-\hat{A}_+)\hat{A}_- + \frac{1}{2}(\hat{A}_+^2 - \hat{A}_-^2)\hat{A}_+ - p_0\hat{A}_+ \\ &= \frac{1}{2}(\hat{A}_+\hat{A}_-^2 + \hat{A}_-\hat{A}_+\hat{A}_- + \hat{A}_+^3 - \hat{A}_-^2\hat{A}_+) - p_0\hat{A}_+ \\ &= \frac{1}{2}\left[\hat{A}_-(\hat{A}_+\hat{A}_- - \hat{A}_-\hat{A}_+) + \hat{A}_+(\hat{A}_-^2 + \hat{A}_+^2)\right] - p_0\hat{A}_+ \end{aligned}$$

$$
\begin{aligned}
&= \frac{1}{2}\hat{A}_-[\hat{A}_+, \hat{A}_-] + \frac{1}{2}\hat{A}_+(\hat{A}_+^2 + \hat{A}_+^2) - p_0\hat{A}_+ \\
&= \frac{\hbar}{i}\hat{A}_-\frac{\hat{\varepsilon}}{2} + \hat{A}_+\sqrt{2\hat{H}} - \sqrt{2E}\hat{A}_+ \\
&= \frac{\hbar}{i}\hat{A}_-\frac{\hat{\varepsilon}}{2} + \hat{A}_+\left(\sqrt{2\hat{H}} - \sqrt{2E}\right)
\end{aligned}
$$

In the same way one can calculate $\hat{\xi}_-$. □

Corollary 9.2 (energy conservation and anomaly). *Let the energy conservation law $\hat{H} = E$ holds. Then one has*

$$
\begin{aligned}
\hat{J}_\hbar^1(x;y;z) &= -\frac{\hbar}{i}\frac{a(x,y,z)}{\sqrt{(2p_0)^3}}\frac{\omega}{2\sqrt{2E}}\hat{A}_- \\
\hat{J}_\hbar^2(x;y;z) &= \frac{\hbar}{i}\frac{a(x,y,z)}{\sqrt{(2p_0)^3}}\frac{\omega}{2\sqrt{2E}}\hat{A}_+ \\
\hat{J}_\hbar^3(x;y;z) &= \frac{\hbar}{i}\frac{a^2(x,y,z)}{p_0}\frac{\omega}{2\sqrt{2E}}
\end{aligned}
$$

These formulae tell us that we have an *anomaly*, i.e the quantum symmetry breaking: only in the classical limit $\hbar \to 0$ the quantum counterparts of the three-dimensional real Lie algebras turn out to be the Lie algebras. In the quantum realm, we are confronted with the *non-Lie* multiplications.

10. Derivative quantum algebra

Theorem 10.1. *The Jacobi operator components generate a 3-dimensional real anti-commutative algebra with sturucture equations*

$$
[\hat{J}_\hbar^1, \hat{J}_\hbar^3] = 0 = [\hat{J}_\hbar^2, \hat{J}_\hbar^3], \quad [\hat{J}_\hbar^1, \hat{J}_\hbar^2] = C\hat{J}_\hbar^3
$$

where

$$
C := -\left(\frac{\hbar\omega}{2E}\right)^2 \frac{(x,y,z)}{32}
$$

Proof. Use Corollary 9.2. □

Definition 10.2 (derivative quantum algebra). The anti-commutative algebra given by the structure equations from Theorem 10.1 is called the *derivative algebra* of $\mathrm{VII}_a^\hbar$, $\mathrm{III}_{a=1}^\hbar$, $\mathrm{VI}_{a\neq 1}^\hbar$.

Corollary 10.3. *Define the new basis in the derivative algebra:*

$$
e_1 := -(x,y,z)\hat{J}_\hbar^3, \quad e_2 := -(x,y,z)\hat{J}_\hbar^1, \quad e_3 := -(x,y,z)\hat{J}_\hbar^2
$$

Then

$$
[e_1, e_3] = 0 = [e_2, e_3], \quad [e_2, e_3] = \beta^2 e_1
$$

where

$$
\beta := \frac{\hbar\omega}{2E}\frac{|(x,y,z)|}{4\sqrt{2}}
$$

Proof. Calculate

$$[e_2, e_3] = (x, y, z)(x, y, z)[\hat{J}^1_\hbar, \hat{J}^2_\hbar] = C(x, y, z)(x, y, z)\hat{J}^3_\hbar = -C(x, y, z)e_1 = \beta^2 e_1$$ □

Theorem 10.4 (operadic quantum condition (OQC)). *The derivative algebra of* $\mathrm{VII}^\hbar_a$, $\mathrm{III}^\hbar_{a=1}$ *and* $\mathrm{VI}^\hbar_{a\neq1}$ *is the 3-dimensional real Heisenberg algebra.*

Proof. By elementary calculus one can see that the Jacobi identity holds. As the only non-vanishing structure constant is $\overset{\circ}{\mu}{}^1_{23}$, one can easily see from the Bianchi Table 1 that $\beta = 1$ perfectly suits. □

Corollary 10.5 (volume quantization and elementary quantum length). *In* $\mathrm{VII}^\hbar_a$, $\mathrm{III}^\hbar_{a=1}$, $\mathrm{VI}^\hbar_{a\neq1}$ *we have* $|(x, y, z)| = 4\sqrt{2}(2n + 1)$, $n = 0, 1, 2, \ldots$. *Thus, the minimal volume is* $V_0 = 2^{5/2}$ *and the elementary length is* $l_0 = 2^{5/6}$.

Proof. Use $\beta = 1$ and the physically allowed energy values $E = E_n = \hbar\omega(n + 1/2)$ of the quantum harmonic oscillator. □

Discussion 10.6 (triality and OQC). Note that the triple of quantum algebras $\mathrm{VII}^\hbar_a$, $\mathrm{III}^\hbar_{a=1}$, $\mathrm{VII}^\hbar_{a\neq1}$, while being non-isomorphic even in the classical limit, still have something common in the *quantum* realm, an invariance property of the triple - the unique derivative quantum algebra which turns out to be the 3d real *Heisenberg algebra*, which is responsible for their quantization. One may thus consider about using the OQC instead of CCR.

Acknowledgments
The research was in part supported by the Estonian Research Council, grants ETF6912 and ETF9038.

References
[1] Lax P D 1968 *Comm. Pure Appl. Math.* **21** 467
[2] Paal E 2007 *J. Gen. Lie Theory Appl.* **1** 57
[3] Gerstenhaber M 1963 *Ann. Math.* **78** 267
[4] Paal E and Virkepu J 2008 *Rep. Math. Phys.* **61** 207
[5] Paal E and Virkepu J 2009 *J. Gen. Lie Theory Appl.* **3** 53
[6] Paal E and Virkepu J 2009 *Noncommutative Structures in Mathematics and Physics* eds S Caenepeel et al (Brussels: K. Vlaam. Acad. Belgie Wet. Kunsten (KVAB)) p 209
[7] Paal E and Virkepu J 2009 *J. Math. Phys.* **50** 053523
[8] Paal E and Virkepu J 2009 *Centr. Eur. J. Phys.* **8** 289
[9] Landau L and Lifshitz E 1973 *Theoretical Physics* Vol 2 *Field Theory* (Moscow: Nauka) (Russian)
[10] Grøn Ø and Gervik S 2007 *Einstein's General Theory of Relativity* (Springer-Verlag)
[11] Paal E and Virkepu J 2011 *Banach Center Publ.* **93** 199

Journal of Physics: Conference Series **532** (2014) 012024 doi:10.1088/1742-6596/532/1/012024

PPN parameters for multiscalar-tensor gravity without a potential

E Randla

Institute of Physics, University of Tartu, Ravila 4c, Tartu 50411, Estonia

Abstract. A generic multiscalar-tensor gravity action functional is proposed in the Jordan conformal frame. After redefining the scalar fields, the parameterised post-Newtonian parameters are calculated, neglecting the potential term. The field equations are shown to be nearly identical with those of scalar-tensor gravity up to the first post-Newtonian approximation, differing only in the definitions of constant terms. This work is a part of a pursuit to chart the landscape of observationally viable scalar-tensor theories.

1. Introduction

The Brans-Dicke theory was developed half a century ago in an attempt to introduce the elusive Mach's principle in a mathematically rigorous relativistic description of gravity [1]. Since then scalar-tensor gravity (STG) theories have been put to use in various other contexts, such as theories of inflation and more recently, to give an account of the accelerating expansion of the universe [2, 3, 4, 5]. Scalar-tensor gravity introduces a new degree of freedom not present in general relativity. When formulating a STG, one has to specify the conformal frame. Common classification distinguishes between the Jordan frame and the Einstein frame; their equivalence in describing nature has remained a subject of academic discussion [2, 3, 6].

A natural generalisation of STG theories is obtained by considering multiple scalar fields. Multiscalar-tensor gravity (MSTG) theories have been first formulated and thoroughly studied in the Einstein conformal frame, including a simplified parameterised post-Newtonian (PPN) analysis revealing the PPN parameters γ and β [7, 8]. From a theoretical point of view, the MSTG action can also be viewed as a special case of the more general multiscalar Horndeski theory action [9, 10]. This paper presents a full PPN analysis of MSTG in the Jordan conformal frame. As a simplification, no potential term is considered. It is commonly argued that a potential would assume the role of the cosmological constant and therefore its contribution to solar system physics would be negligible in comparison with the effects of local matter [11, 12].

The paper is organised as follows. In Sec. 2 the multiscalar-tensor action functional is introduced. The scalar fields are suitably redefined and the equations of motion are presented. In Sec. 3 the parameterised post-Newtonian formalism is introduced and all relevant quantities are assigned orders of magnitude. The field equations are presented up to the first post-Newtonian order in Sec. 4 and solved in Sec. 5, revealing the full set of PPN parameters. The results are summarised in Sec. 6.

2. Action functional and equations of motion

We start with the following general form of the multiscalar-tensor gravity Jordan frame action with N scalar fields Φ^α [13],

$$S = \frac{1}{2\kappa^2}\int_{V_4} d^4x\sqrt{-g}\left(\mathcal{F}R - \mathcal{Z}_{\alpha\beta}g^{\mu\nu}\partial_\mu\Phi^\alpha\partial_\nu\Phi^\beta - 2\kappa^2\mathcal{U}\right) + S_m[g_{\mu\nu},\chi_m] \tag{2.1}$$

We have adopted a system of units where the speed of light and Planck's constant are set to equal one, $c = h = 1$. The action (2.1) is a natural extension to multiple scalar fields of the general Jordan frame STG action [6, 14, 15, 16, 17]. Here indices $\alpha, \beta, \gamma, \ldots = 1, 2, \ldots, N$ label the scalar fields Φ, indices $\mu\nu \ldots = 0123$ belong to spacetime coordinates, while $ij \ldots = 123$ are reserved for spatial coordinates. The arbitrary functions

$$\mathcal{F} = \mathcal{F}(\Phi^1, \Phi^2, \ldots, \Phi^N), \quad \mathcal{Z}_{\alpha\beta} = \mathcal{Z}_{\alpha\beta}(\Phi^1, \Phi^2, \ldots, \Phi^N), \quad \mathcal{U} = \mathcal{U}(\Phi^1, \Phi^2, \ldots, \Phi^N) \tag{2.2}$$

determine a distinct MSTG if specified. Matter is represented by the matter action $S_m[g_{\mu\nu},\chi_m]$, where χ_m denotes matter fields.

For a more straightforward physical interpretation of the theory it is convenient to define a new set of scalar fields $\{\phi^1, \phi^2, \ldots, \phi^{N-1}, \phi^N \equiv \Psi\}$ by setting $\Psi = \mathcal{F}(\Phi^1, \Phi^2, \ldots, \Phi^N)$. Taking into account that

$$\mathcal{Z}_{\alpha\beta}\partial_\rho\Phi^\alpha\partial^\rho\Phi^\beta = \mathcal{Z}_{\alpha\beta}\left(\frac{\partial\Phi^\alpha}{\partial\phi^a}\frac{\partial\Phi^\beta}{\partial\phi^b}\partial_\rho\phi^a\partial^\rho\phi^b + 2\frac{\partial\Phi^\alpha}{\partial\phi^a}\frac{\partial\Phi^\beta}{\partial\Psi}\partial_\rho\phi^a\partial^\rho\Psi + \frac{\partial\Phi^\alpha}{\partial\Psi}\frac{\partial\Phi^\beta}{\partial\Psi}\partial_\rho\Psi\partial^\rho\Psi\right) \tag{2.3}$$

we denote

$$Z_{ab} = \mathcal{Z}_{\alpha\beta}\frac{\partial\Phi^\alpha}{\partial\phi^a}\frac{\partial\Phi^\beta}{\partial\phi^b}, \quad Z_{aN} = \mathcal{Z}_{\alpha\beta}\frac{\partial\Phi^\alpha}{\partial\phi^a}\frac{\partial\Phi^\beta}{\partial\Psi}, \quad Z_{NN} = \mathcal{Z}_{\alpha\beta}\frac{\partial\Phi^\alpha}{\partial\Psi}\frac{\partial\Phi^\beta}{\partial\Psi} \tag{2.4}$$

$$U(\phi^1, \phi^2, \ldots, \phi^{N-1}, \Psi) = \mathcal{U}(\Phi^1, \Phi^2, \ldots, \Phi^N) \tag{2.5}$$

where $a, b, \ldots = 1, 2, \ldots, N-1$ index the scalar fields ϕ. The target space of scalar fields Φ^α can be considered as a N-dimensional space with a metric tensor $\mathcal{Z}_{\alpha\beta}$ and we can use suitable transformations of the coordinates Φ^α for imposing the following $N-1$ conditions:

$$Z_{aN}(\phi^1, \phi^2, \ldots, \phi^{N-1}, \Psi) = 0 \tag{2.6}$$

Denote

$$Z_{NN} = \omega(\phi^1, \phi^2, \ldots, \phi^{N-1}, \Psi)/\Psi \tag{2.7}$$

The action (2.1) now reads

$$S = \frac{1}{2\kappa^2}\int_{V_4} d^4x\sqrt{-g}\left(\Psi R - Z_{ab}\partial_\rho\phi^a\partial^\rho\phi^b - \frac{\omega}{\Psi}\partial_\rho\Psi\partial^\rho\Psi - 2\kappa^2 U\right) + S_m[g_{\mu\nu},\chi_m] \tag{2.8}$$

Note that the functions

$$\omega = \omega(\phi^1, \phi^2, \ldots, \Psi), \quad Z_{ab} = Z_{ab}(\phi^1, \phi^2, \ldots, \Psi), \quad U = U(\phi^1, \phi^2, \ldots, \Psi) \tag{2.9}$$

depend on all scalar fields. The scalar field Ψ acts as a variable part of the gravitational constant, and thus it will be assumed that $\Psi > 0$, while the conditions $2\omega + 3 > 0$ and $Z_{ab} > 0$ are imposed to make the redefinition of the scalar fields in the Einstein conformal frame possible [13]. In this work, the potential term $-2\kappa^2 U$ is neglected in the following. The field equations corresponding to the action (2.8) are

$$\Psi R_{\mu\nu} = \kappa^2\left(T_{\mu\nu} - \frac{\omega+1}{2\omega+3}g_{\mu\nu}T\right) + \nabla_\mu\partial_\nu\Psi + Z_{ab}\partial_\mu\phi^a\partial_\nu\phi^b + \frac{\omega}{\Psi}\partial_\mu\Psi\partial_\nu\Psi + \\ + \frac{g_{\mu\nu}}{4\omega+6}\left[\left(\Psi\frac{\partial Z_{ab}}{\partial\Psi} - Z_{ab}\right)\partial_\rho\phi^a\partial^\rho\phi^b - \frac{\partial\omega}{\partial\Psi}\partial_\rho\Psi\partial^\rho\Psi - 2\frac{\partial\omega}{\partial\phi^a}\partial_\rho\phi^a\partial^\rho\Psi\right] \tag{2.10a}$$

$$(2\omega+3)\Box\Psi = \kappa^2 T + \left(\Psi\frac{\partial Z_{ab}}{\partial\Psi} - Z_{ab}\right)\partial_\rho\phi^a\partial^\rho\phi^b - \frac{\partial\omega}{\partial\Psi}\partial_\rho\Psi\partial^\rho\Psi - 2\frac{\partial\omega}{\partial\phi^a}\partial_\rho\phi^a\partial^\rho\Psi \tag{2.10b}$$

$$Z_{ac}\Box\phi^a = \left(\frac{1}{2}\frac{\partial Z_{ab}}{\partial\phi^c} - \frac{\partial Z_{ac}}{\partial\phi^b}\right)\partial_\rho\phi^a\partial^\rho\phi^b + \frac{1}{2\Psi}\frac{\partial\omega}{\partial\phi^c}\partial_\rho\Psi\partial^\rho\Psi - \frac{\partial Z_{ac}}{\partial\Psi}\partial_\rho\phi^a\partial^\rho\Psi \tag{2.10c}$$

In the rest of the paper Eqs. (2.10a) - (2.10c) will be our principal object of investigation.

3. PPN formalism

The parameterised post-Newtonian formalism is a useful calculation tool to confront metric gravity theories with solar system observations [11, 12]. The gravitational field is assumed to be slowly varying and weak. In a metric description of gravity these assumptions translate to a perturbed empty space-time metric. Matter is described as a perfect fluid and the matter variables giving rise to the metric perturbations via the field equations are assigned orders of smallness in accord with their observed solar system values. Further, the assumption of slow variation of the physical situation is expressed as an additional order of smallness associated with taking time derivatives. The scalar fields also have to be expanded in series around their background values. Substituting the series expansions of all relevant quantities into the field equations (2.10a)–(2.10c), one arrives at the weak field, slow motion field equations of MSTG, which are solved for the metric perturbations. On the other hand, the metric perturbations have been probed by solar system measurements. In order to simplify the comparison between metric gravity theories and the solar system measurements, a generic perturbed metric, the PPN metric, has been postulated [11, 12]. The coefficients of the metric components in this standardised form algebraically determine the PPN parameters, which are restricted by observational data. These statements will be further quantified in the following.

The perfect fluid energy-momentum tensor is given by

$$T^{\mu\nu} = (\rho + \rho\Pi + p)u^\mu u^\nu + pg^{\mu\nu} \tag{3.1}$$

where ρ is the rest energy density, Π is the specific internal energy, p is the pressure and u^μ is the four-velocity of matter. Orders of magnitude are ascribed to these quantities relative to the velocity $v^i = u^i/u^0$ of the source matter, which is taken to be a first order small quantity in units where the speed of light is equal to one. Denoting $\mathcal{O}(n) \propto |\vec{v}|^n$, we assign orders $\mathcal{O}(2)$ to ρ and Π and $\mathcal{O}(4)$ to p. Time derivatives ∂_0 of the metric components and other fields are weighed with a velocity order $\mathcal{O}(1)$. The space-time metric is taken to be a perturbed Minkowski metric

$$g_{\mu\nu} = \eta_{\mu\nu} + h_{\mu\nu} \tag{3.2}$$

The Newtonian potential

$$U = \int \rho'/|\mathbf{x} - \mathbf{x}'|d^3x' \tag{3.3}$$

has the same order of magnitude $\mathcal{O}(2)$ as ρ. The Newtonian description of gravity can be cast in metric form by setting $h^{(2)}_{00} = 2G_N U$ and setting all other perturbations to zero. Here and in the following a superscript in parentheses denotes the order of magnitude of the quantity it is attached to. Thus, in order to describe gravity at the first post-Newtonian level, higher order perturbations have to be considered. The detailed analysis shows that the non-vanishing post-Newtonian order metric components are $h_{ij}^{(2)}$, $h_{0j}^{(3)}$ and $h_{00}^{(4)}$. In standard PPN formalism the so-called geometrised units are chosen such that the coefficient of $2U$ in $h_{00}^{(2)}$, acting as an effective Newtonian constant, is set to equal one. The generic PPN metric is postulated with components

$$g_{00} = -1 + 2U - 2\beta U^2 - 2\xi\Phi_W + (2\gamma + 2 + \alpha_3 + \zeta_1 - 2\xi)\Phi_1 \tag{3.4a}$$

$$+ 2(3\gamma - 2\beta + 1 + \zeta_2 + \xi)\Phi_2 + 2(1 + \zeta_3)\Phi_3 + 2(3\gamma + 3\zeta_4 - 2\xi)\Phi_4 - (\zeta_1 - 2\xi)\mathcal{A} \tag{3.4b}$$

$$g_{0j} = -\frac{1}{2}\left[(4\gamma + 3 + \alpha_1 - \alpha_2 + \zeta_1 - 2\xi)V_j + (1 + \alpha_2 - \zeta_1 + 2\xi)W_j\right] \tag{3.4c}$$

$$g_{ij} = (1 + 2\gamma U)\delta_{ij} \tag{3.4d}$$

where the post-Newtonian potentials have their standard definitions

$$\Phi_W = \int \frac{\rho'\rho''(\mathbf{x}-\mathbf{x}')}{|\mathbf{x}-\mathbf{x}'|^3}\left(\frac{\mathbf{x}'-\mathbf{x}''}{|\mathbf{x}-\mathbf{x}''|} - \frac{\mathbf{x}-\mathbf{x}''}{|\mathbf{x}'-\mathbf{x}''|}\right) d^3x' d^3x'', \quad \mathcal{A} = \int \frac{\rho'[\mathbf{v}'\cdot(\mathbf{x}-\mathbf{x}')]^2}{|\mathbf{x}-\mathbf{x}'|^3} d^3x' \tag{3.5a}$$

$$\Phi_1 = \int \rho' v'^2/|\mathbf{x}-\mathbf{x}'| d^3x', \quad \Phi_2 = \int \rho' U'/|\mathbf{x}-\mathbf{x}'| d^3x' \tag{3.5b}$$

$$\Phi_3 = \int \rho'\Pi'/|\mathbf{x}-\mathbf{x}'| d^3x', \quad \Phi_4 = \int p'/|\mathbf{x}-\mathbf{x}'| d^3x' \tag{3.5c}$$

$$V_i = \int \rho' v_i'/|\mathbf{x}-\mathbf{x}'| d^3x', \quad W_i = \int \rho'[\mathbf{v}'\cdot(\mathbf{x}-\mathbf{x}')](x-x')_i/|\mathbf{x}-\mathbf{x}'|^3 d^3x' \tag{3.5d}$$

while the PPN parameters $\alpha_1, \alpha_2, \alpha_3, \gamma, \beta, \zeta_1, \zeta_2, \zeta_3, \zeta_4$ and ξ remain to be determined by the theory [11, 12].

The perturbed energy-momentum tensor has components

$$T_{00} = \rho\left(1 + \Pi + v^2 - h_{00}^{(2)}\right) + \mathcal{O}(6), \quad T_{0j} = -\rho v_j + \mathcal{O}(5), \quad T_{ij} = \rho v_i v_j + p\delta_{ij} + \mathcal{O}(6) \tag{3.6}$$

The scalar fields also have to be expanded around their constant background values. The order of magnitude of source terms in equations (2.10b) and (2.10c) limits the possible perturbations to

$$\Psi = \Psi_0 + \psi^{(2)} + \mathcal{O}(4), \quad \phi^a = \phi_0^a + \varphi^{a(4)} + \mathcal{O}(6) \tag{3.7}$$

Now we are set to expand the functions ω and Z_{ij} in their Taylor series. We assume the leading terms and constant coefficients to be of order $\mathcal{O}(0)$. For ω the first terms in the Taylor series are

$$\omega(\Psi, \phi^a) = \omega(\Psi_0, \phi_0^a) + \left.\frac{\partial\omega}{\partial\Psi}\right|_{\substack{\Psi=\Psi_0 \\ \phi^a=\phi_0^a}} \psi^{(2)} + \mathcal{O}(4) \equiv \omega_0 + \left.\frac{\partial\omega}{\partial\Psi}\right|_0 \psi^{(2)} + \mathcal{O}(4) \tag{3.8}$$

We could expand Z_{ab} in the same manner, but from Eqs. (2.10a)–(2.10c) it is evident that Z_{ab} or its derivatives appear only as coefficients of quantities that are beyond the order of magnitude necessary for calculations at the first post-Newtonian level. That simplifies the calculations significantly and it will be shown in the next section that the resulting approximate field equations are nearly identical to those of the case with just one scalar field.

4. Approximate field equations

Now all relevant quantities have been expanded in terms of perturbations with known orders of magnitude and Eqs. (2.10a)–(2.10c) can be written up to the first post-Newtonian order. We begin by writing Eq. (2.10b) up to the second order,

$$\nabla^2\psi^{(2)} = -\kappa^2\rho/(2\omega_0+3) \tag{4.1}$$

which is readily solved by

$$\psi^{(2)} = \kappa^2 U/4\pi(2\omega_0+3) \tag{4.2}$$

It is useful to choose the following gauge to fix the coordinates [18]:

$$h_i{}^j{}_{,j} - \frac{1}{2}h_\mu{}^\mu{}_{,i} = \frac{1}{\Psi_0}\psi_{,i} \quad h_0{}^j{}_{,j} - \frac{1}{2}h_j{}^j{}_{,0} = \frac{1}{\Psi_0}\psi_{,0} \tag{4.3}$$

Making use of Eqs. (4.3) the components of the Ricci tensor are calculated for the metric (3.2) from their definition,

$$R_{00} = -\frac{1}{2}\nabla^2 h^{(2)}_{00} - \frac{1}{2}\nabla^2 h^{(4)}_{00} + \frac{1}{\Psi_0}\psi^{(2)}_{,00} + \frac{1}{2\Psi_0}h^{(2)}_{00,j}\psi^{(2)}_{,j} - \frac{1}{2}h^{(2)}_{00,j}h^{(2)}_{00,j} + \frac{1}{2}h^{(2)}_{jk}h^{(2)}_{00,jk} + \mathcal{O}(6) \tag{4.4a}$$

$$R_{0j} = -\frac{1}{2}\nabla^2 h^{(3)}_{0j} - \frac{1}{4}h^{(2)}_{00,0j} + \frac{1}{\Psi_0}\psi^{(2)}_{,0j} + \mathcal{O}(5) \tag{4.4b}$$

$$R_{ij} = -\frac{1}{2}\nabla^2 h^{(2)}_{ij} + \frac{1}{\Psi_0}\psi^{(2)}_{,ij} + \mathcal{O}(4) \tag{4.4c}$$

On the other hand, expanding Eq. (2.10a) results in the following equations for the Ricci tensor components,

$$R_{00} = \frac{\kappa^2}{\Psi_0}\left[\frac{\omega_0+2}{2\omega_0+3}\left(1+\Pi-h^{(2)}_{00}-\psi^{(2)}\right)+v^2+\frac{3\omega_0+3}{2\omega_0+3}\frac{p}{\rho}-\frac{\frac{\partial\omega}{\partial\Psi}\big|_0}{(2\omega_0+3)^2}\psi^{(2)}\right]\rho$$
$$+\frac{1}{2\Psi_0}\left(h^{(2)}_{00,i}\psi^{(2)}_{,i}+2\psi^{(2)}_{,00}+\frac{\frac{\partial\omega}{\partial\Psi}\big|_0}{2\omega_0+3}\psi^{(2)}_{,i}\psi^{(2)}_{,i}\right)+\mathcal{O}(6) \tag{4.5a}$$

$$R_{0j} = \frac{1}{\Psi_0}\left(\psi^{(2)}_{,0j}-\kappa^2 v_j\rho\right)+\mathcal{O}(5) \tag{4.5b}$$

$$R_{ij} = \frac{\kappa^2}{\Psi_0}\left[\frac{\omega_0+1}{2\omega_0+3}\rho\delta_{ij}+\frac{\psi^{(2)}_{,ij}}{\kappa^2}\right]+\mathcal{O}(4) \tag{4.5c}$$

These differ from the well-known case with only one scalar field Ψ by merely the definitions of the constants ω_0 and $\frac{\partial\omega}{\partial\Psi}\big|_0$. Matching the right hand sides of Eqs. (4.4a) and (4.5a), (4.4b) and (4.5b), and (4.4c) and (4.5c) respectively, one arrives at equations for the metric perturbations, which are solved in the next section. Since the equations we are about to solve are identical to the corresponding equations of the single field case up to the definitions of constants, it is reasonable to expect the solutions to be identical with the same exceptions. It is nevertheless instructive to present the calculation, as it forms a basis for doing the calculation for the more general case with a potential.

5. PPN parameters

In this section the approximate field equations are solved for the metric perturbations. The solutions have to be obtained in steps of growing orders of smallness. In the end the metric is compared to the generic PPN metric (3.4d) and the PPN parameters are presented. The interpretation of the parameters is not presented here, since it is thoroughly analysed elsewhere [11, 12]. From Eqs. (4.4a) and (4.5a) it is straightforward to derive the equation for $h^{(2)}_{00}$,

$$\nabla^2 h^{(2)}_{00} = -\frac{2\kappa^2}{\Psi_0}\frac{\omega_0+2}{2\omega_0+3}\rho + \mathcal{O}(4) \tag{5.1}$$

which is solved by

$$h^{(2)}_{00} = \frac{\kappa^2}{2\pi\Psi_0}\frac{\omega_0+2}{2\omega_0+3}U \tag{5.2}$$

revealing the effective gravitational constant

$$G_{\text{eff}} = \frac{\kappa^2}{4\pi\Psi_0}\frac{\omega_0+2}{2\omega_0+3} \tag{5.3}$$

In choosing geometrised units, G_{eff} is set to equal one, resulting in an expression for Ψ_0

$$\Psi_0 = \frac{\kappa^2}{4\pi}\frac{\omega_0+2}{2\omega_0+3} \tag{5.4}$$

In the following the expression (5.4) will be used to eliminate Ψ_0.

From Eqs. (4.4c) and (4.5c) one arrives at an equation for $h^{(2)}_{ij}$,

$$\nabla^2 h^{(2)}_{ij} = -8\pi\frac{\omega_0+1}{\omega_0+2}\rho\delta_{ij} \tag{5.5}$$

which can be readily solved, yielding

$$h^{(2)}_{ij} = \frac{2\omega_0+2}{\omega_0+2}\delta_{ij}U \tag{5.6}$$

Equations (4.4b) and (4.5b) are combined to arrive at

$$\nabla^2 h^{(3)}_{0j} = \frac{2\kappa^2}{\Psi_0}v_j\rho - \frac{1}{2}h^{(2)}_{00,0j} \tag{5.7}$$

solved by

$$h^{(3)}_{0j} = -\frac{7\omega_0+10}{2\omega_0+4}V_j - \frac{1}{2}W_j \tag{5.8}$$

Making further use of the definitions of the post-Newtonian potentials (3.5) and the lower order solutions (4.2), (5.2) and (5.6), Eq. (4.4a) assumes the form

$$R_{00} = -\nabla^2 U - \frac{1}{2}\nabla^2 h^{(4)}_{00} + \frac{1}{\Psi_0}\psi^{(2)}_{,00} + \frac{1}{\omega_0+2}U_{,i}U_{,i} - \nabla^2 U^2 + \frac{4\omega_0+6}{\omega_0+2}\nabla^2\Phi_2 \tag{5.9}$$

while Eq. (4.5a) is rewritten as

$$R_{00} = -\nabla^2 U + \frac{\kappa^2 \left.\frac{\partial\omega}{\partial\Psi}\right|_0}{16\pi(\omega_0+2)(2\omega_0+3)^2}\nabla^2 U^2 + \frac{1}{\Psi_0}\psi^{(2)}_{,00} + \frac{1}{\omega_0+2}U_{,i}U_{,i} - \frac{2\omega_o+3}{\omega_0+2}\nabla^2\Phi_1$$

$$+\left(2+\frac{\kappa^2\,\frac{\partial\omega}{\partial\Psi}\big|_0}{8\pi(\omega_0+2)(2\omega_0+3)^2}+\frac{1}{\omega_0+2}\right)\nabla^2\Phi_2-\nabla^2\Phi_3-\frac{3\omega_0+3}{\omega_0+2}\nabla^2\Phi_4 \tag{5.10}$$

We have made use of the identities

$$U_{,i}U_{,i}=\nabla U\cdot\nabla U=\frac{1}{2}\nabla^2U^2-U\nabla^2U=\frac{1}{2}\nabla^2U^2-\nabla^2\Phi_2 \tag{5.11}$$

Equating (5.9) and (5.10) results in an equation for $h^{(4)}_{00}$, from which the solution can be read as

$$h^{(4)}_{00}=-2(1+\Lambda)U^2+4\left(\frac{2\omega_0+3}{2\omega_0+4}\right)\Phi_1+4\left(\frac{2\omega_0+1}{2\omega_0+4}-\Lambda\right)\Phi_2+2\Phi_3+6\left(\frac{\omega_0+1}{\omega_0+2}\right)\Phi_4 \tag{5.12}$$

where

$$\Lambda=\frac{\kappa^2\,\frac{\partial\omega}{\partial\Psi}\big|_0}{8\pi(2\omega_0+4)(2\omega_0+3)^2} \tag{5.13}$$

Eqs. (5.2), (5.6), (5.8) and (5.12) make up the complete perturbed space-time metric up to the first post-Newtonian order of magnitude. Lastly, the coefficients of the respective PPN potentials in the standard PPN metric (3.4d) and our solution are equated to reveal the full set of PPN parameters

$$\beta=1+\Lambda,\quad \gamma=(\omega_0+1)/(\omega_0+2),\quad \xi=\alpha_1=\alpha_2=\alpha_3=\zeta_1=\zeta_2=\zeta_3=\zeta_4=0 \tag{5.14}$$

It was pointed out earlier that the PPN analysis of a potential-free multiscalar-tensor gravity theory should coincide with the single scalar field case, save for the definitions of the constants ω_0 and $\frac{\partial\omega}{\partial\Psi}\big|_0$, embedded here in the constant Λ. The result (5.14) confirms that claim.

6. Concluding remarks

The parameterised post-Newtonian parameters were calculated for a general multiscalar-tensor theory of gravity with no potential term. Since, after suitably redefining the scalar fields, the structure of the field equations restricts the possible orders of magnitude of perturbations of all but one scalar field to beyond what is relevant to PPN calculations, the equations that determine the PPN parameters are nearly identical to those of the case with a single scalar field, differing only by the definitions of constants ω_0 and $\frac{\partial\omega}{\partial\Psi}\big|_0$, which here encode the background values of all scalar fields. Solving the perturbed field equations up to the first post-Newtonian order confirms the close analogy of MSTG with the one field case. The presented calculation forms a foundation for generalising the presented results to the case with an arbitrary potential, extending the PPN analysis of the single field case with a potential [19] to multiple scalar fields.

Acknowledgements

This work was supported by the Estonian Science Foundation Grant No. 8837, by the Estonian Ministry for Education and Science Institutional Research Support Project IUT02-27 and by the European Union through the European Regional Development Fund (Centre of Excellence TK114).

References

[1] Brans C and Dicke R H 1961 *Phys. Rev.* **124** 925
[2] Faraoni V 2004 *Cosmology in Scalar-Tensor Gravity* Fund. Theor. Phys. vol 139 (Kluwer Acad. Publ.)
[3] Faraoni V and Capozziello S 2010 *Beyond Einstein gravity: A Survey of gravitational theories for cosmology and astrophysics* Fund. Theor. Phys. vol 170 (Dordrecht: Springer)

[4] Tsujikawa S 2010 *Lecture Notes in Phys.* **800** 99
[5] Clifton T, Ferreira P G, Padilla A and Skordis C 2012 *Phys. Rept.* **513** 1
[6] Flanagan E E 2004 *Classical Quantum Gravity* **21** 3817
[7] Damour T and Esposito-Farese G 1992 *Classical Quantum Gravity* **9** 2093
[8] Damour T and Nordtvedt K 1993 *Phys. Rev.* D **48** 3436.
[9] Padilla A and Sivanesan V 2013 *J. High Energy Phys.* **1304** 032
[10] Kobayashi T, Tanahashi N and Yamaguchi M 2013 *Phys. Rev.* D **88** 083504
[11] Will C M 1993 *Theory and Experiment in Gravitational Physics* (Cambridge Univ. Press)
[12] Will C M 2006 *Liv. Rev. Rel.* **9** 3
[13] Kuusk P, Järv L and Randla E 2014 *Algebra, Geometry and Mathematical Physics* Springer Proceedings in Mathematics & Statistics eds A Makhlouf, E Paal, S D Silvestrov and A Stolin (Springer-Verlag) p 661
[14] Bergmann P G 1968 *Int. J. Theor. Phys.* **1** 25
[15] Nordtvedt K 1970 *Astrophys. J.* **161** 1059
[16] Wagoner R 1970 *Phys. Rev.* D **1** 3209
[17] Esposito-Farese G and Polarski D 2001 *Phys. Rev.* D **63** 063504
[18] Nutku Y 1969 *Astrophys. J.* **155** 999
[19] Hohmann M, Järv L, Kuusk P and Randla E 2013 *Phys. Rev.* D **88** 084054

Journal of Physics: Conference Series **532** (2014) 012025 doi:10.1088/1742-6596/532/1/012025

Examples of cosmological rips in the scalar-tensor cosmology

M Saal and A Frantskjavitšius

Füüsika Instituut, Tartu Ülikool, Ravila 14c, 50411 Tartu, Estonia

Abstract. Plenitude of models incorporating a dynamical dark energy equation of state are proposed and the scalar-tensor gravity is certainly one possible candidate. In this paper we examine whether a cosmological rip scenarios are possible in the scalar-tensor gravity by parametrizing the equation of state of dark energy $\mathrm{w}_{de}(z)$ in an appropriate way. The parameters of the model are constrained by the observational data of type Ia supernovas collected to *Union 2.1* compilation.

1. Introduction

According to the contemporary observational data the expansion of the universe has been accelerating for about five billion years. Adding a cosmological constant Λ into the dynamical equations (derived in the framework of general relativity (GR)) of the cosmological model accounts for the accelerating expansion. This is equivalent to an additional matter component with an equation of state (EoS) parameter $\mathrm{w}_\Lambda = p_\Lambda/\rho_\Lambda$, $\mathrm{w}_\Lambda \equiv -1$. However, recent analyses point out the possibility that dark energy is dynamical, in which case the EoS parameter is not constant, for a review see [1]. Moreover, observationally it cannot be excluded [2] that the EoS parameter of dark energy may cross the so-called phantom divide line (PDL) $\mathrm{w}_{de} = -1$ [3, 4]. Phenomenological models where the EoS parameter of dark energy is assumed to be oscillatory [5, 6, 7] are also analysed and they are not completely contradicted by recent data.

If the energy density of dark energy ρ is growing and the EoS parameter w_{de} is crossing the PDL the consequences on the future dynamics of the universe are significant. In the most dramatic case, called a big rip [8], the energy density of dark energy will grow uncontrollably and will quickly become infinite. The acceleration of the universe is fast enough to break apart all bounded structures before the evolution ends in a spacetime singularity. According to a more moderate scenario, known as a little rip [9], the energy density approaches infinity as $t \to \infty$. Recently, a pseudo-rip [10] and a quasi-rip scenarios [11] were proposed. Neither of them contain spacetime singularities, for a classification of future singularities, see [12, 13], but the energy density of dark energy still becomes sufficiently high to rip apart certain bound systems while leaving the other systems bounded. The key difference is that in the case of the pseudo-rip systems that were once disintegrated remain unbounded, whereas in the case of the quasi-rip once disintegrated systems may recover and form bounded systems again [11]. The latter scenario can be realized only in the case where the energy density of dark energy is not monotonically increasing but is increasing and decreasing in turns or even oscillating steadily. A similar behaviour occurs also in the quintom scenario where the EoS parameter is able to evolve across the cosmological constant boundary, for a review see [14].

The unknown nature of dark energy encourages us to look for extensions of the GR. Among the others, scalar-tensor gravity (STG) [15] where a tensor and a scalar degrees of freedom are non-minimally coupled is a substantial candidate. In STG the gravitational coupling depends on scalar field Ψ, and free functions $\omega(\Psi)$ and $V(\Psi)$ provide additional freedom to incorporate a number of competitive theories of gravity. Crossing the PDL in the framework of STG is possible [16, 17, 18] and the question about the future dynamics of dark energy is relevant. In this paper we examine whether a big rip, a little rip, a pseudo-rip and a quasi-rip scenarios are possible in the general STG by parametrizing the EoS parameter $\mathrm{w}_{de}(z)$ of dark energy in an appropriate way. For comparable approaches see also [19, 20, 21].

The outline of the paper is as follows. We start in Sec. 2 with the equations for the STG cosmology and define the energy density and the pressure of dark energy so that the usual conservation equation is satisfied. In Sec. 3 we list some parametrizations to deal with the unknown EoS parameter $\mathrm{w}_{de}(z)$ of dark energy. In Sec. 4 we use the observational data of type Ia supernova (SN Ia) from the *Union 2.1* compilation [22] the parameters of the EoS w_{de}. We propose a modified parametrization of the effective gravitational constant G_{eff} and calculate the best fit parameters for different parametrizations of w_{de} (see Table 1), but using only the *Un on 2.1* compilation. In Sec. 5 we present a general condition for cosmological rips to happen and give some examples of rip scenarios which are possible in the framework of STG and are consistent with the constraints found in section 4. Finally, Sec. 6 provides a brief summary.

2. General scalar-tensor cosmology

In what follows we consider a general STG in the Brans-Dicke-Bergmann-Wagoner (BDBW) parametrization in the Jordan frame representation. The action of the STG is given by

$$S = \frac{1}{2\kappa^2}\int d^4x\sqrt{-g}\left[\Psi R(g) - \frac{\omega(\Psi)}{\Psi}\nabla^\rho\Psi\,\nabla_\rho\Psi - 2\kappa^2 V(\Psi)\right] + S_m[g_{\mu\nu},\chi_m] \tag{2.1}$$

where $\omega(\Psi)$ is the coupling function, $V(\Psi) > 0$ is the scalar field self-interaction, κ^2 is the constant part of the effective gravitational constant κ^2/Ψ and $S_m[g_{\mu\nu},\chi_m]$ is the matter contribution term. Since we require that the gravitational constant remains positive it follows that $\Psi > 0$.

Assuming that the geometry of the universe is described by the flat ($k = 0$) Friedmann-Lemaître-Robertson-Walker (FLRW) metric and the ordinary matter is modelled by a perfect barotropic fluid we can write the field equations derived from the action (2.1) as follows

$$H^2 = -H\frac{\dot\Psi}{\Psi} + \frac{1}{6}\frac{\dot\Psi^2}{\Psi^2}\omega(\Psi) + \frac{\kappa^2}{\Psi}\frac{\rho_m}{3} + \frac{\kappa^2}{\Psi}\frac{V(\Psi)}{3} \tag{2.2a}$$

$$2\dot H + 3H^2 = -2H\frac{\dot\Psi}{\Psi} - \frac{1}{2}\frac{\dot\Psi^2}{\Psi^2}\omega(\Psi) - \frac{\ddot\Psi}{\Psi} - \frac{\kappa^2}{\Psi}\mathrm{w}\rho_m + \frac{\kappa^2}{\Psi}V(\Psi) \tag{2.2b}$$

$$\ddot\Psi + 3H\dot\Psi = -\frac{1}{2\omega(\Psi)+3}\left[\frac{d\omega(\Psi)}{d\Psi}\dot\Psi^2 - \kappa^2(1-3\mathrm{w})\rho_m - 2\kappa^2\left(2V(\Psi) - \Psi\frac{dV(\Psi)}{d\Psi}\right)\right] \tag{2.2c}$$

Since in the Jordan frame representation the matter part of the action does not contain a scalar field Ψ the usual conservation law holds

$$\dot\rho_m + 3H\left(\rho_m + p_m\right) = 0 \tag{2.3}$$

It turns out that defining the energy density of dark energy ρ_{de} and the pressure of dark energy p_{de} are as follows (for a discussion, see [23, 17]):

$$\rho_{de} = \frac{3}{\kappa^2}\left(-H\dot\Psi + \frac{1}{6}\frac{\dot\Psi^2}{\Psi}\omega - H^2(\Psi - \Psi_0)\right) + V \tag{2.4a}$$

$$p_{de} = \frac{3}{\kappa^2}\left(\frac{2}{3}H\dot{\Psi} + \frac{1}{6}\frac{\dot{\Psi}^2}{\Psi}\omega + \frac{1}{3}\ddot{\Psi} + \left(\frac{2}{3}\dot{H} + H^2\right)(\Psi - \Psi_0)\right) - V \tag{2.4b}$$

The constraint equation (2.2a) and the dynamical equation (2.2b) can be written in a similar way as in the case of the standard Friedmann model

$$H^2 = \frac{\kappa^2}{3\Psi_0}(\rho_m + \rho_{de}) \tag{2.5a}$$

$$2\dot{H} + 3H^2 = -\frac{\kappa^2}{\Psi_0}(p_m + p_{de}) \tag{2.5b}$$

Here Ψ_0 corresponds to the value of the scalar field Ψ at the present moment t_0. Moreover, definitions (2.4a) and (2.4b) provide that the usual conservation law holds also for the dark energy

$$\dot{\rho}_{de} + 3H(\rho_{de} + p_{de}) = 0 \tag{2.6}$$

Note that equation (2.6) is nothing but the scalar field equation (2.2c). Using the relation between the scale factor a and the readshift z

$$da/a = -dz/(1+z) \tag{2.7}$$

we can write the solution of equation (2.6) as an integral

$$\rho_{de} = \rho_{de,0}\exp\left[\int_0^z 3(1 + \mathrm{w}_{de}(\tilde{z}))/(1+\tilde{z})\, d\tilde{z}\right] \tag{2.8}$$

where the EoS parameter of dynamical dark energy $\mathrm{w}_{de}(z) = p_{de}/\rho_{de}$ is an unknown function in general.

3. Different parametrizations

In this section we will list some parametrizations for the dark energy EoS parameter $\mathrm{w}_{de}(z)$. An appropriate parametrization allows us to complete the integration (2.8) and represent the Hubble parameter as $H(z; p_i)$ where parameters p_i should be determined from observations. All parametrizations employed below contain two free parameters $(\mathrm{w}_0, \mathrm{w}_1)$ and can be represented in a general form

$$\mathrm{w}_{de}(z) = \mathrm{w}_0 + \mathrm{w}_1 f(z) \tag{3.1}$$

The first free parameter w_0 corresponds to the EoS parameter at redshift $z = 0$ and the second free parameter w_1 describes the dynamics of EoS parameter. By construction the functional form of $f(z)$ describes the asymptotic behaviour of parametrization.

3.1. The Che all er-Polansk -L nder CPL parametr zat on

The well known parametrization proposed by Chevallier and Polanski [24] and by Linder [25] is represented as follows

$$\mathrm{w}_{de}(z) = \mathrm{w}_0 + \mathrm{w}_1(1-a) = \mathrm{w}_0 + \mathrm{w}_1 z/(1+z) \tag{3.2}$$

The parametrization (3.2) works very well at high redshifts and has the limiting value at the limit $z \to \infty$. However, the EoS described by the parametrization (3.2) blows up quickly if we approach to $z \to -1$ and is singular at this limit. Therefore, the CPL parametrization can be used in the case if negative redshifts are small enough. Integrating equation (2.8) we get for the energy density of dark energy

$$\rho_{de} = \rho_{de,0}\,(1+z)^{3(1+\mathrm{w}_0+\mathrm{w}_1)}\,\exp\left[-3\mathrm{w}_1 z/(1+z)\right] \tag{3.3}$$

The dimensionless density parameters are defined as

$$\Omega_m = \frac{\kappa^2\,\rho_m}{3H^2\Psi_0} \tag{3.4a}$$

$$\Omega_{de} = \frac{\kappa^2\,\rho_{de}}{3H^2\Psi_0} \tag{3.4b}$$

and the constraint equation (2.5a) can be written as (in the case of flat universe)

$$\Omega_m + \Omega_{de} = 1 \tag{3.5}$$

Now the Hubble parameter $H(z;p_i)$ can be represented as

$$H^2(z;\Omega_{m,0},\mathrm{w}_0,\mathrm{w}_1) = H_0^2\left[\Omega_{m,0}(1+z)^3 + (1-\Omega_{m,0})(1+z)^{3(1+\mathrm{w}_0+\mathrm{w}_1)}\exp\left[-\frac{3\mathrm{w}_1 z}{1+z}\right]\right] \tag{3.6}$$

3.2. The Ma-Zhang MZ parametr zat on: type I and type II
Ma and Zhang proposed [26] two types of divergence-free parametrizations

$$\mathrm{w}_{de}(z) = \mathrm{w}_0 + \mathrm{w}_1\left[\frac{\ln(2+z)}{1+z} - \ln 2\right] \qquad \text{(MZ type I)} \tag{3.7a}$$

$$\mathrm{w}_{de}(z) = \mathrm{w}_0 + \mathrm{w}_1\left[\frac{\sin(1+z)}{1+z} - \sin(1)\right] \qquad \text{(MZ type II)} \tag{3.7b}$$

Note that the parametrization of type II allows the oscillating behaviour of EoS parameter. Both parametrizations (3.7a) and (3.7b) are bounded at high redshifts as well as negative redshifts. MZ type I approaches to the finite limiting value $\mathrm{w}(z) \to \mathrm{w}(-1) = \mathrm{w}_0 + \mathrm{w}_1(1-\ln 2)$ and MZ type II approaches to the finite limiting value $\mathrm{w}(z) \to \mathrm{w}(-1) = \mathrm{w}_0 + \mathrm{w}_1(1-\sin(1))$ as $z \to -1$.

3.2.1. MZ type I From equation (2.8) we get for the energy density

$$\rho_{de} = \rho_{de,0}\left\{2^{6\mathrm{w}_1}\,(1+z)^{3(1+\mathrm{w}_0+\mathrm{w}_1(1-\ln 2))}\,\exp\left[-3\mathrm{w}_1\left(\frac{z+2}{z+1}\right)\ln(z+2)\right]\right\} \tag{3.8}$$

and using dimensionless density parameters (3.4a) we can write $H(z;p_i)$ as

$$H^2(z;\Omega_{m,0},\mathrm{w}_0,\mathrm{w}_1) = H_0^2\left[\Omega_{m,0}(1+z)^3\right] + H_0^2\,(1-\Omega_{m,0})2^{6\mathrm{w}_1}\,(1+z)^{3(1+\mathrm{w}_0+\mathrm{w}_1(1-\ln 2))}\times$$
$$\times\exp\left[-3\mathrm{w}_1\left(\frac{z+2}{z+1}\right)\ln(z+2)\right] \tag{3.9}$$

3.2.2. MZ type II Similarly to the previous case we get relations for the energy density and the Hubble parameter

$$\rho_{de} = \rho_{de,0}\left[(1+z)^{3(1+\mathrm{w}_0-\mathrm{w}_1\sin(1))}\right]\times$$

$$\times\exp\left\{3\mathrm{w}_1\left[(\mathrm{Ci}(1+\mathrm{z})-\mathrm{Ci}(1))-\left(\frac{\sin(z+1)}{(z+1)}-\sin(1)\right)\right]\right\} \tag{3.10a}$$

$$H^2(z;\Omega_{m,0},\mathrm{w}_0,\mathrm{w}_1) = H_0^2\left[\Omega_{m,0}(1+z)^3\right] + H_0^2\left(1-\Omega_{m,0}\right)\left[(1+z)^{3(1+\mathrm{w}_0-\mathrm{w}_1\sin(1))}\right]\times$$

$$\times\exp\left\{3\mathrm{w}_1\left[(\mathrm{Ci}(1+\mathrm{z})-\mathrm{Ci}(1))-\left(\frac{\sin(z+1)}{(z+1)}-\sin(1)\right)\right]\right\} \tag{3.10b}$$

where the cosine integral $\mathrm{Ci}(z)$ is defined by

$$\mathrm{Ci(z)} = -\int_z^\infty \frac{\cos z}{z}\,dz \tag{3.11}$$

3.3. The L -Zhang LZ parametr zat on

In order to take into account possible oscillations of the dark energy EoS parameter between the redshift range 0 to 2 Li and Zhang have proposed [27] to use the parametrization

$$\mathrm{w}_{de}(z) = \mathrm{w}_0 - \mathrm{w}_1\sin\left[A\ln(1+z)\right] \tag{3.12}$$

Here an additional parameter A determines the frequency of oscillations and was chosen to be $A = \frac{3}{2}\pi$ in [27]. The parametrization (3.12) is bounded at the limit $z \to -1$ but the frequency of oscillations becomes infinite (the amplitude remains constant). Different parametrizations for the oscillating EoS parameter of dark energy is also considered by Kurek *et al* [28].

As previously the expressions for the energy density and for the Hubble parameter are presented as follows:

$$\rho_{de} = \rho_{de,0}\left\{(1+z)^{3(1+\mathrm{w}_0)}\exp\left[3\frac{\mathrm{w}_1}{A}\left[\cos[A\ln(1+z)]-1\right]\right]\right\} \tag{3.13a}$$

$$H^2(z;\Omega_{m,0},\mathrm{w}_0,\mathrm{w}_1) = H_0^2\left[\Omega_{m,0}(1+z)^3\right] +$$

$$+ H_0^2\left(1-\Omega_{m,0}\right)\left\{(1+z)^{3(1+\mathrm{w}_0)}\exp\left[3\frac{\mathrm{w}_1}{A}\left[\cos[A\ln(1+z)]-1\right]\right]\right\} \tag{3.13b}$$

The constant EoS parameter ($\mathrm{w}_{de} = -\mathrm{w}_0$) can be considered as a special case of parametrizations (3.2), (3.7a), (3.7b) and (3.12) if $\mathrm{w}_1 = 0$, i.e the dark energy is not dynamical.

4. Constraints from observations

At the present paper we will use the observations of SN Ia to constrain the free parameters of models presented above. Specifically, we apply the Bayesian analysis to *Union 2.1* compilation which contains data sets for 580 supernova. Since the likelihood function $\mathcal{L}$ is defined to be proportional to $\sim e^{-\chi^2/2}$ it follows that we are looking for the values of parameters which are minimizing the value of χ^2. For *Union 2.1* compilation χ^2_{SN} is given as follows

$$\chi^2_{SN} = \sum_{i=1}^{580}\left[\mu_i^{the}(z;p_i) - \mu_i^{obs}(z)\right]^2/\sigma_i^2(z) \tag{4.1}$$

Here μ_i^{obs} is the observed distance modulus and $\mu_i^{the}(z;p_i)$ is the theoretical distance modulus, p_i stands for the parameters of the model and σ_i is the 1σ uncertainty of the ith data point.

Distance modulus μ is defined as the difference between an apparent magnitude m and an absolute magnitude M and can be expressed by means of the luminosity distance D_L [29, 30]

$$\mu^{the}(z;p_i) \equiv 5\log_{10}[D_L(z)] + \mu_0 \tag{4.2}$$

where the nuisance parameter $\mu_0 = -5\log_{10} h + 42.38$ ($h = 100/H_0\ \mathrm{km\,s^{-1}\,Mpc^{-1}}$) is independent of observed data points. The luminosity distance $D_L(z)$ in the case of the flat universe is given by

$$D_L(z) = (1+z)/H_0 \cdot \int_0^z d\tilde{z}/H(\tilde{z}) \tag{4.3}$$

Since we consider the dynamics of the universe in the framework of STG it is necessary to take into account an additional effect. Namely, the effective gravitational constant G_{eff} depends on the scalar field (i.e cosmological time) and is different from the value $\frac{\kappa^2}{\Psi}$ due to contribution from the interaction of the scalar field and a test particle. The effective gravitational constant G_{eff} measured (on scales where the scalar field Ψ is effectively massless) between two test masses in the Cavendish-type experiment is given in the BDBW parametrization by [31, 32]

$$8\pi G_{eff} = \frac{\kappa^2}{\Psi}\frac{2\omega(\Psi)+4}{2\omega(\Psi)+3} \tag{4.4}$$

Using the observational data of type Ia supernovas as a probe to fix the free parameters of the model p_i we should take into account that the Chandrasekhar mass depends on G_{eff} given by (4.4) and not on G_N. This changes the peak luminosity of supernovas [?, 33, 34, 35] see also [Preprint astro-ph/9907222]) and the expression of the distance modulus (4.2) should be modified by an additional term

$$\mu^{the}(z;p_i) = 5\log_{10}[D_L(z)] + \mu_0 + \frac{15}{4}\log_{10}\frac{G_{eff}(z)}{G_{eff,0}} \tag{4.5}$$

where $G_{eff,0}$ is the effective gravitational constant at the redshift $z = 0$. As in the case of $\mathrm{w}_{de}(z)$ the functional form of $G_{eff}(z)$ is unknown in general and using some appropriate parametrization is essential. From different considerations it is known that the effective gravitational constant should be nearly constant starting from the early universe until today [36]. The weak field observations give also a strong constraint for the present variation of the effective gravitational constant $|\dot{G}_{eff}/G_{eff}| \leq 10^{-12}$ [37].

Nesseris and Perivolaropoulos have proposed [38, 39] to use the parametrization for G_{eff} as

$$G_{eff}(z) = G_{eff,0}\left[1+\alpha\,(1-a)^2\right] = G_{eff,0}\left[1+\alpha\, z^2/(1+z)^2\right] \tag{4.6}$$

where α is the parameter of order $\alpha \sim 0.2$ [39]. The parametrization (4.6) has a singularity at the limit $z = -1$ and grows fast if we approach to this limit. Inspired by the parametrization of MZ type I (3.7a) for the $\mathrm{w}_{de}(z)$ we will propose a divergence-free ansatz for $G_{eff}(z)$

$$G_{eff}(z) = G_{eff,0}\left[1 + G_1\left(\frac{\ln(2+z)}{1+z} - \ln 2\right)\right] \tag{4.7}$$

which has the asymptotic values as follows:

$$G_{eff}(z) = \begin{cases} G_{eff,0} & \text{if } z = 0 \\ G_{eff,0}[1 - G_1 \ln 2] & \text{if } z \to \infty \\ G_{eff,0}[1 + G_1(1-\ln 2)] & \text{if } z \to -1 \end{cases} \tag{4.8}$$

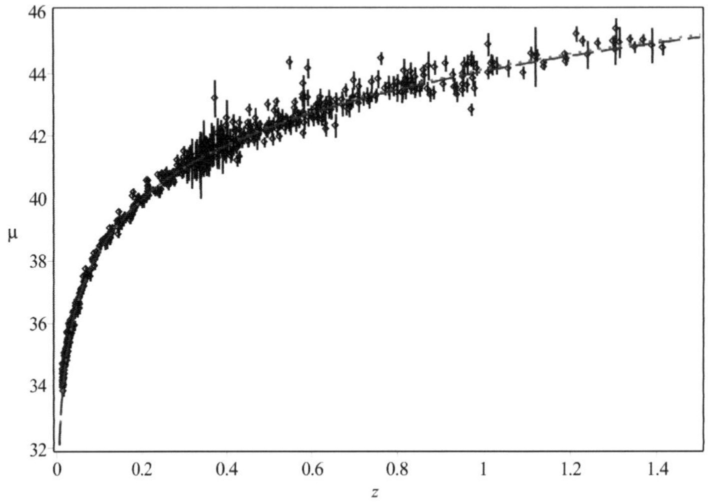

Figure 1. Hubble diagram for the ΛCDM (red dotted line) with $\Omega_{m,0} = 0.277$ and for the STG with Li-Zhang parametrization $w_0 = -1.10$, $w_1 = -0.25$, $\Omega_{m,0} = 0.289$ (blue line).

Note that negative G_1, i.e decreasing of the effective gravitational constant supports the accelerated expansion of the universe.

Next, adopting the idea of the Li-Zang parametrization (3.12) we propose the following ansatz for G_{eff}

$$G_{eff}(z) = G_{eff,0}\left[1 - \tilde{G}_1 \sin\left(\tilde{A}\ln(1+z)\right)\right] \tag{4.9}$$

As in the case of EoS parameter, at the limit $z \to -1$ the frequency of oscillations becomes infinite but the amplitude remains constant. Here we take $\tilde{A} = \frac{2\pi}{ln2}$, i.e the frequency of oscillation of G_{eff} is smaller than for the EoS parameter.

Next we use the data from SN Ia observations collected to *Union 2.1* compilation to restrict the parameter space of selected models. The Hubble diagram for the *Union 2.1* sample is presented in Figure 1. The best fit values of parameters for the parametrizations described above are presented in Table 1 and 1σ and 2σ confidence level contours in the (w_0, w_1) parameter space are presented in Figure 2.

Table 1. Best fit values of parameters for different parametrizations

Parametrization \ Parameter	$\Omega_{m,0}$	w_0	w_1	$\tilde{G}_1$	χ^2_{SN}/dof
CPL	0.280	−1.008	−0.016	0.001	0.9761
MZ I	0.279	−1.007	0.011	0.001	0, 9761
MZ II	0.281	−1.012	0.015	0.001	0.9761
LZ	0.289	−0.970	0.158	0.003	0.9760

Journal of Physics: Conference Series **532** (2014) 012025 doi:10.1088/1742-6596/532/1/012025

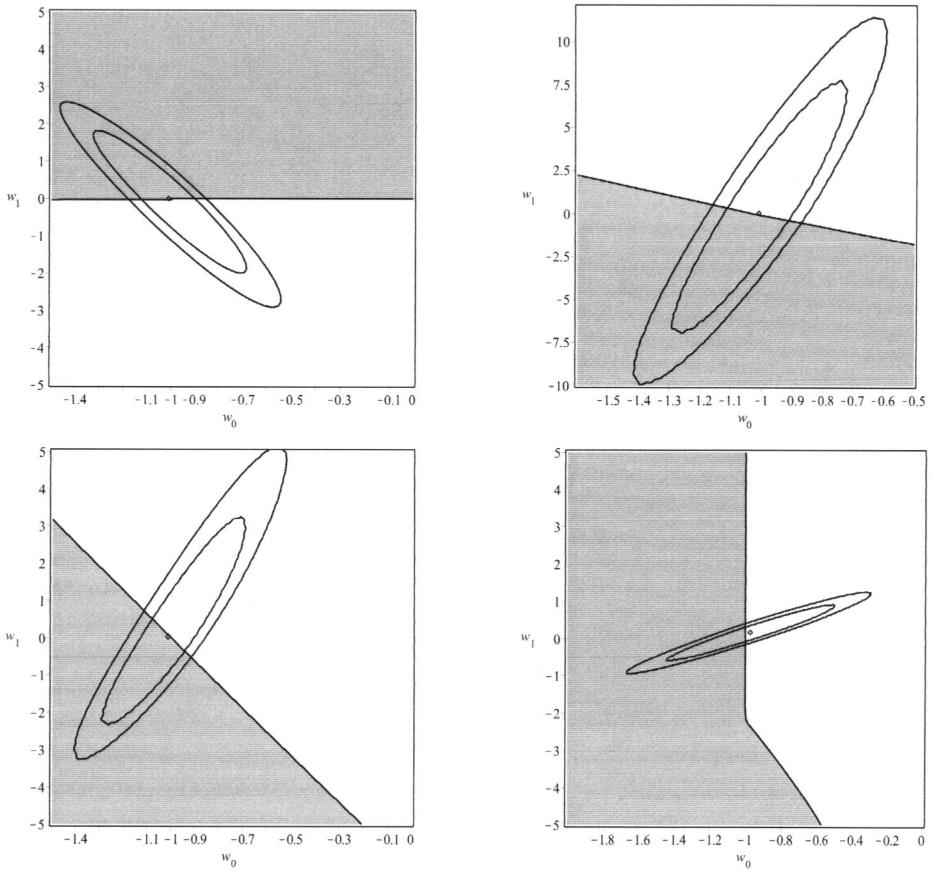

Figure 2. Likelihood contours at 1σ and 2σ confidence level in the $(\mathrm{w}_0, \mathrm{w}_1)$ plane for different parametrizations: CPL (up left), MZ type I (up right), MZ type II (down left) and LZ (down right). The point denotes the best fit value of the corresponding parametrization. Shaded area in the parameter space corresponds to the limit where the little rip scenario realizes.

5. Classification of cosmological rips and rip scenarios

As mentioned in the introduction there is at least four different scenarios for which the evolution of the universe results in the dissolution (eternally or temporarily) of bound systems: a big rip, a little rip, a pseudo-rip and a quasi-rip. It was argued by Frampton et al. [10] that if the dark energy density is either constant or monotonically increasing, then there are three possible rip scenarios. Using the time asymptotic of the Hubble parameter, rip conditions can be classified as

$$\begin{array}{lll} \text{Big rip:} & H(t) \to \infty, & t \to t_{rip} < \infty \\ \text{Little rip:} & H(t) \to \infty, & t \to \infty \\ \text{Pseudo-rip:} & H(t) \to H_{max}, & t \to \infty \end{array}$$

Here H_{max}, t_{rip} are finite constants. Monotonic evolution of the dark energy is certainly not the only possibility (for a review, see [14]). As pointed out by Wei et al. [11] most interesting is the case where energy density of dark energy is monotonically increasing, crossing the PDL and then monotonically decreasing in the second stage. This being so, it is possible that some structures dissolve in the first stage and then form again a bounded structure in the second stage. The scenario is known as a quasi-rip [11].

The situation where the energy density of dark energy behaves non-monotonically seem to be possible in the framework of STG. Even in the so-called limit of GR the STG contains oscillating solutions as shown in [40, 41, 42, 43] and hence the oscillating EoS parameter of dark energy.

In order to find a boundary condition for massive structures to be dissolved Nesseris and Perivolaropoulos [44] proposed to use in the Newtonian limit a hybrid metric which interpolates between Schwarzschild and FLRW metrics

$$ds^2 = -\left(1 - \frac{2G_N M}{a(t)\,b}\right) dt^2 + a(t)^2 \left[db^2 + b^2(d\theta^2 + sin^2\theta d\phi^2)\right] \tag{5.1}$$

Here M is the characteristic mass of the bounded system, b is the comoving radial coordinate and G_N is the Newton gravitational constant. Using the definition $r = a(t) \cdot b$ and inserting the metric (5.1) into the geodesic equation it reads

$$\ddot{r} = \frac{\ddot{a}}{a} r + \frac{L^2}{r^3} - \frac{G_N M}{r^2} \equiv -\frac{dV_{eff}}{dr} \tag{5.2}$$

where L is the constant angular momentum per unit mass. Using Kepler's third law the square of constant angular momentum can be written as $L^2 = G_N M r_0$. The equation for the radius of an initially stable ($\ddot{r} = 0$, $\dot{r} = 0$) circular bound system reads (effective potential V_{eff} has a minimum)

$$\frac{\ddot{a}}{a} r^4 - G_N M r + G_N M r_0 = 0 \tag{5.3}$$

The fourth order equation (5.3) has a solution only if the following condition is fulfilled [44]

$$\frac{\ddot{a}}{a} \equiv \dot{H}(t) + H(t)^2 \leq \frac{27}{256} \frac{G_N M}{r_0^3} \tag{5.4}$$

where r_0 is the characteristic scale of corresponding system. The condition (5.4) can be considered as a criterion for systems to stay bounded. If the expansion rate exceeds this limit the minimum of effective potential will vanish and systems start to break apart. It is worth to note, that the criterion (5.4) contains due to construction a constant gravitational coupling G_N which is not accurate in the framework of STG. At present we do not consider this question as a serious problem and assume that condition (5.4) is sufficiently accurate in STG and even if we are looking for the redshift values approaching $z \to -1$.

Finally, taking into account the relation between the cosmic time t and the redshift z

$$d/dt = -H(1+z)\, d/dz \tag{5.5}$$

we can write the condition (5.4) in a dimensionless form as follows

$$F(z; p_i) \equiv \dot{h} + h^2 = -(1+z)h(z)\,h'(z) + h^2(z) \leq \frac{27}{256} \frac{G_N M}{r_0^3} H_0^{-2} \tag{5.6}$$

where $' = d/dz$, $h(z) = H(z)/H_0$ and p_i refers to parameters of the model. The derivative of $F(z; p_i)$ with respect to redshift z determines the redshift value z_{ext} when the expansion rate is extremal. However, since the interesting dynamics takes place at the close limit $z \to -1$ it is more convenient to use the variable $N = \ln a$. Upon using the relation between the cosmic time t and the number of e-folds N

$$d/dt = H\, d/dN \tag{5.7}$$

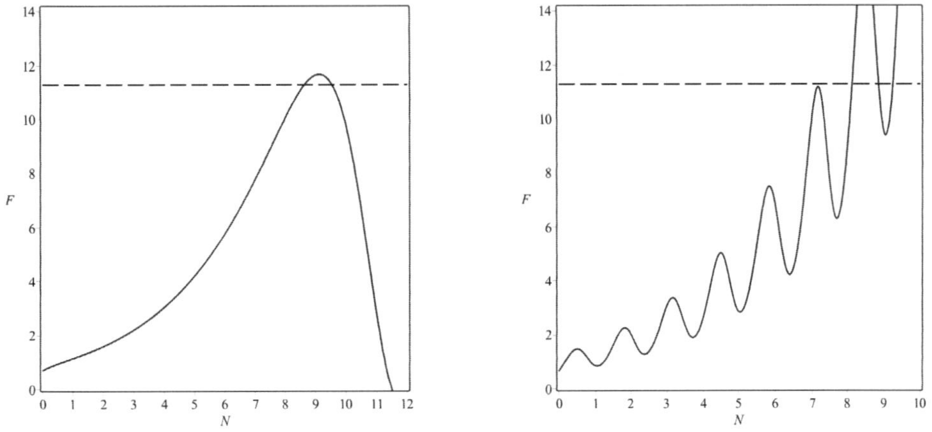

Figure 3. The quasi-rip realized by CPL (left) with parameter values $h = 0.6989$, $\mathrm{w}_0 = -1.108$, $\mathrm{w}_1 = -8 \cdot 10^{-6}$, $\Omega_{m,0} = 0.280$ and little rip realized by LZ (right) with parameter values $h = 0.6989$, $\mathrm{w}_0 = -1.10$, $\mathrm{w}_1 = -0.25$, $\Omega_{m,0} = 0.289$. Parameters $(\mathrm{w}_0, \mathrm{w}_1)$ are choosen to be inside 1σ likelihood contour (see Figure 2) and $\Omega_{m,0}$ is choosen as the best-fit value. The dashed line corresponds to the disintegration limit of the Coma cluster.

we can represent the rip condition (5.4) by

$$F(N, p_i) \equiv h(N)\, h'(N) + h^2(N) \leq \frac{27}{256} \frac{G_N M}{r_0^3} H_0^{-2} \tag{5.8}$$

where $' = d/dN$. In Figure 3 we present two examples to illustrate the dynamics allowed by parametrizations CPL (3.2) and LZ (3.12).

6. Summary

We considered different parametrizations of the EoS parameter $\mathrm{w}_{de}(z)$ of dark energy and the oscillating parametrization of the effective gravitational constant G_{eff} in the framework of general STG. Parameters of the model are fitted with the observational data of type Ia supernovas collected to *Union 2.1* compilation. We also notice the following.

- Not surprisingly the data from SN Ia observation do not constrain the parameter space sufficiently to make significant conclusion about the future dynamics of the universe.
- As expected the results depend strongly on the parametrizations applied. The 1σ and 2σ likelihood contours for different parametrizations allow various dynamics, including the quasi-rip scenario. Moreover, it was shown [Frantskjavitšius A 2011 Cosmological rips in the general relativity limit (Tartu: BSc Theses)] that quasi-rip scenario is possible (without fitting the parameters of the model) in the framework of potential dominated STG cosmology at the limit of GR [41]. However, the best fit values of the parameters $(\mathrm{w}_0, \mathrm{w}_1)$ are in a range where the little rip scenario can be realized.

Acknowledgments

This work was supported by the Estonian Science Foundation Grant No. 8837, by the Estonian Ministry for Education and Science Institutional Research Support Project IUT02-27 and by the European Union through the European Regional Development Fund (Centre of Excellence TK114).

References

[1] Copeland E J, Sami M and Tsujikawa S 2006 *Int. J. Mod. Phys.* D **15** 1753
[2] Cai R G, Su Q and Zhang H B 2010 *JCAP* **1004** 012
[3] Caldwell R R 2002 *Phys. Lett.* B **545** 23
[4] Carroll S M, Hoffman M and Trodden M 2003 *Phys. Rev.* D **68** 023509
[5] Linder E V 2006 *Astropart. Phys.* **25** 167
[6] Kurek A, Hrycyna O and Szydlowski M 2008 *Phys. Lett.* B **659** 14
[7] Lazkoz R, Salzano V and Sendra I 2010 *Phys. Lett.* B **694** 198
[8] Caldwell R R, Kamionkowski M and Weinberg N N 2003 *Phys. Rev. Lett.* **91** 071301
[9] Frampton P H, Ludwick K J and Scherrer R J 2011 *Phys. Rev.* D **84** 063003
[10] Frampton P H, Ludwick K J and Scherrer R J 2012 *Phys. Rev.* D **85** 083001
[11] Wei H, Wang L F and Guo X J 2012 *Phys. Rev.* D **86** 083003
[12] Stefancic H 2005 *Phys. Rev.* D **71** 084024
[13] Nojiri S I, Odintsov S D and Tsujikawa S 2005 *Phys. Rev.* D **71** 063004
[14] Cai Y F, Saridakis E N, Setare M R and Xia J Q 2010 *Phys. Rept.* **493** 1
[15] Faraoni V 2004 *Cosmology in Scalar-Tensor Gravity* (Kluwer)
[16] Elizalde E, Nojiri S I and Odintsov S D 2004 *Phys. Rev.* D **70** 043539
[17] Gannouji R, Polarski D, Ranquet A and Starobinsky A A 2006 *JCAP* **0609** 016
[18] Järv L, Kuusk P and Saal M 2010 *Proc. Est. Acad. Sci. Phys. Math.* **59** 306
[19] Nojiri S I and Odintsov S D 2005 *Phys. Rev.* D **72** 023003
[20] Nojiri S, Odintsov S D and Saez-Gomez D 2011 *AIP Conf. Proc.* **1458** 207
[21] Frampton P H, Ludwick K J, Nojiri S I, Odintsov S D and Scherrer R J 2012 *Phys. Lett.* B **708** 204
[22] Suzuki N 2012 *et al Astrophys. J.* **746** 85
[23] Torres D F 2002 *Phys. Rev.* D **66** 043522
[24] Chevallier M and Polarski D 2001 *Int. J. Mod. Phys.* D **10** 213
[25] Linder E V 2003 *Phys. Rev. Lett.* **90** 091301
[26] Ma J Z and Zhang X 2011 *Phys. Lett.* B **699** 233
[27] Li H and Zhang X 2011 *Phys. Lett.* B **703** 119
[28] Kurek A, Hrycyna O and Szydlowski M 2010 Phys. Lett. B **690** 337
[29] Perivolaropoulos L 2005 *Phys. Rev.* D **71** 063503
[30] Nesseris S and Perivolaropoulos L 2005 *Phys. Rev.* D **72** 123519
[31] Barrow J D and Parsons P 1997 *Phys. Rev.* D **55** 1906
[32] Will C M 2005 *Living Rev. Rel.* **93**
[33] Gaztanaga E, Garcia-Berro E, Isern J, Bravo E and Dominguez I 2002 *Phys. Rev.* D **65** 023506
[34] Riazuelo A and Uzan J P 2002 *Phys. Rev.* D **66** 023525
[35] Gannouji R and Polarski D 2008 *JCAP* **0805** 018
[36] Uzan J P 2003 *Rev. Mod. Phys.* **75** 403
[37] Bertotti B, Iess L and Tortora P 2003 *Nature* **425** 374
[38] Nesseris S and Perivolaropoulos L 2007 *JCAP* **0701** 018
[39] Nesseris S and Perivolaropoulos L 2007 *Phys. Rev.* D **75** 023517
[40] Järv L, Kuusk P and Saal M 2010 *Phys. Rev.* D **81** 104007
[41] Järv L, Kuusk P and Saal M 2010 *Phys. Lett.* B **694** 1
[42] Järv L, Kuusk P and Saal M 2012 *Phys. Rev.* D **85** 064013
[43] Saal M, Järv L and Kuusk P 2012 *J. Phys.: Conf. Ser.* **384** 012029
[44] Nesseris S and Perivolaropoulos L 2004 *Phys. Rev.* D **70** 123529

On quantum groups and Lie bialgebras related to $sl(n)$

A Stolin and I Pop

Department of Mathematical Sciences, University of Gothenburg, SE-405 30 Gothenburg, Sweden

Abstract. Given an arbitrary field $\mathbb{F}$ of characteristic 0, we study Lie bialgebra structures on $sl(n,\mathbb{F})$, based on the description of the corresponding classical double. For any Lie bialgebra structure δ, the classical double $D(sl(n,\mathbb{F}),\delta)$ is isomorphic to $sl(n,\mathbb{F})\otimes_{\mathbb{F}} A$, where A is either $\mathbb{F}[\varepsilon]$, with $\varepsilon^2=0$, or $\mathbb{F}\oplus\mathbb{F}$ or a quadratic field extension of $\mathbb{F}$. In the first case, the classification leads to quasi-Frobenius Lie subalgebras of $sl(n,\mathbb{F})$. In the second and third cases, a Belavin-Drinfeld cohomology can be introduced which enables one to classify Lie bialgebras on $sl(n,\mathbb{F})$, up to gauge equivalence. The Belavin-Drinfeld untwisted and twisted cohomology sets associated to an r-matrix are computed.

2010 Mathematics Subject Classification: 17B37, 17B62.
Keywords: Quantum group, Lie bialgebra, classical double, r-matrix, admissible triple.

1. Introduction

Following [3], we recall that a quantized universal enveloping algebra (or a quantum group) over a field k of characteristic zero is a topologically free topological Hopf algebra H over the formal power series ring $k[[\hbar]]$ such that $H/\hbar H$ is isomorphic to the universal enveloping algebra of a Lie algebra $\mathfrak{g}$ over k.

The quasi-classical limit of a quantum group is a Lie bialgebra. A Lie bialgebra is a Lie algebra $\mathfrak{g}$ together with a cobracket δ which is compatible with the Lie bracket. Given a quantum group H, with comultiplication Δ, the quasi-classical limit of H is the Lie bialgebra $\mathfrak{g}$ of primitive elements of $H/\hbar H$ and the cobracket is the restriction of the map $(\Delta-\Delta^{21})/\hbar(\mathrm{mod}\hbar)$ to $\mathfrak{g}$.

The operation of taking the semiclassical limit is a functor $SC: QUE \to LBA$ between categories of quantum groups and Lie bialgebras over k. The existence of universal quantization functors was proved by Etingof and Kazhdan [4, 5]. They used Drinfeld's theory of associators to construct quantization functors for any field k of characteristic zero. More precisely, let $(\mathfrak{g},\delta)$ be a Lie bialgebra over k. Then one can associate a Lie bialgebra $\mathfrak{g}_\hbar$ over $k[[\hbar]]$ defined as $(\mathfrak{g}\otimes_k k[[\hbar]],\hbar\delta)$. According to Theorem 2.1 of [5] there exists an equivalence $\widehat{Q}$ between the category $LBA_0(k[[\hbar]])$ of topologically free over $k[[\hbar]]$ Lie bialgebras with $\delta\equiv 0\,(\mathrm{mod}\,\hbar)$ and the category $HA_0(k[[\hbar]])$ of topologically free Hopf algebras cocommutative modulo $\hbar$. Moreover, for any $(\mathfrak{g},\delta)$ over k, one has the following: $\widehat{Q}(\mathfrak{g}_\hbar)=U_\hbar(\mathfrak{g})$.

Due to this equivalence, the classification of quantum groups whose quasi-classical limit is $\mathfrak{g}$ is equivalent to the classification of Lie bialgebra structures on $\mathfrak{g}\otimes_{\mathbb{C}}\mathbb{C}[[\hbar]]$. Since any cobracket over $\mathbb{C}[[\hbar]]$ can be extended to one over $\mathbb{C}((\hbar))$ and conversely, any cobracket over $\mathbb{C}((\hbar))$, multiplied by an appropriate power of $\hbar$, can be restricted to a cobracket over $\mathbb{C}[[\hbar]]$, this in turn reduces to the problem of finding Lie bialgebras on $\mathfrak{g}\otimes_{\mathbb{C}}\mathbb{C}((\hbar))$. Denote, for the sake of simplicity, $\mathbb{K}:=\mathbb{C}((\hbar))$ and $\mathfrak{g}(\mathbb{K}):=\mathfrak{g}\otimes_{\mathbb{C}}\mathbb{K}$.

As a first step towards classification, following ideas of [6], we proved in [8] that for any Lie bialgebra structure on $\mathfrak{g}(\mathbb{K})$, the associated classical double is of the form $\mathfrak{g}(\mathbb{K})\otimes_{\mathbb{K}} A$, where A

is one of the following associative algebras: $\mathbb{K}[\varepsilon]$, where $\varepsilon^2 = 0$, $\mathbb{K} \oplus \mathbb{K}$ or $\mathbb{K}[j]$, where $j^2 = \hbar$.

As it was shown in [8], the classification of Lie bialgebras with classical double $\mathfrak{g}(\mathbb{K}[\varepsilon])$ leads to the classification of quasi-Frobenius Lie algebras over $\mathbb{K}$, which is a complicated and still open problem.

Unlike this case, the classification of Lie bialgebras with classical double $\mathfrak{g}(\mathbb{K}) \oplus \mathfrak{g}(\mathbb{K})$ can be achieved by cohomological and combinatorial methods. In [8], we introduced a Belavin-Drinfeld cohomology theory which proved to be useful for the study of Lie bialgebra structures. To any non-skewsymmetric r-matrix r_{BD} from the Belavin-Drinfeld list [1], we associated a cohomology set $H^1_{BD}(\mathfrak{g}, r_{BD})$. We proved the existence of a one-to-one correspondence between any Belavin-Drinfeld cohomology and gauge equivalence classes of Lie bialgebra structures on $\mathfrak{g}(\mathbb{K})$. In case $\mathfrak{g} = sl(n)$, we showed that for any non-skewsymmetric r-matrix r_{BD}, the cohomology set $H^1_{BD}(sl(n), r_{BD})$ has only one class, which is represented by the identity.

Regarding the classification of Lie bialgebras whose classical double is isomorphic to $\mathfrak{g}(\mathbb{K}[j])$, with $j^2 = \hbar$, a cohomology theory can be introduced too. Our result states that there exists a one-to-one correspondence between Belavin-Drinfeld twisted cohomology and gauge equivalence classes of Lie bialgebra structures on $\mathfrak{g}(\mathbb{K})$ whose classical double is isomorphic to $\mathfrak{g}(\mathbb{K}[j])$. In [8], we proved that the twisted cohomology corresponding to the Drinfeld-Jimbo r-matrix has only one class, represented by a certain matrix J (not the identity). A deeper investigation was done in the subsequent article [9] where twisted cohomologies for $sl(n)$ associated to generalized Cremmer-Gervais r-matrices were studied.

The aim of the present article is the study of Lie bialgebra structures on $sl(n, \mathbb{F})$, for an arbitrary field $\mathbb{F}$ of characteristic zero. Again the idea is to use the description of the classical double. We will show that for any Lie bialgebra structure δ, the classical double $D(sl(n, \mathbb{F}), \delta)$ is isomorphic to $sl(n, \mathbb{F}) \otimes_{\mathbb{F}} A$, where A is either $\mathbb{F}[\varepsilon]$, with $\varepsilon^2 = 0$, or $\mathbb{F} \oplus \mathbb{F}$ or a quadratic extension of $\mathbb{F}$. In the first case, the classification leads to quasi-Frobenius Lie subalgebras of $sl(n, \mathbb{F})$. In the second and third cases, we will introduce a Belavin-Drinfeld cohomology which enables one to classify Lie bialgebras on $sl(n, \mathbb{F})$, up to gauge equivalence. In the particular case $\mathbb{F} = \mathbb{C}((\hbar))$ we recover the classification of quantum groups whose classical limit is $sl(n, \mathbb{C})$ obtained in [8,9].

2. Description of the classical double

From the general theory of Lie bialgebras it is known that for each Lie bialgebra structure δ on a fixed Lie algebra L one can construct the corresponding classical double $D(L, \delta)$. As a vector space, $D(L, \delta) = L \oplus L^*$. Moreover, since the cobracket of L induces a Lie bracket on L^*, there exists a Lie algebra structure on $L \oplus L^*$, induced by the bracket and cobracket of L, and such that the canonical symmetric nondegenerate bilinear form Q on this space is invariant.

Let $\mathbb{F}$ be an arbitrary field of zero characteristic. Let us assume that δ is a Lie bialgebra structure on $sl(n, \mathbb{F})$. Then one can construct the corresponding classical double $D(sl(n, \mathbb{F}), \delta)$.

Similarly to Lemma 2.1 from [6], one can prove that $D(sl(n, \mathbb{F}), \delta)$ is a direct sum of regular adjoint $sl(n)$-modules. Combining this result with Prop. 2.2 from [2], one obtains the following

Theorem 2.1. *There exists one associative, unital, commutative algebra A of dimension 2 over $\mathbb{F}$, such that $D(sl(n, \mathbb{F}), \delta) \cong sl(n, \mathbb{F}) \otimes_{\mathbb{F}} A$.*

The symmetric invariant nondegenerate bilinear form Q on $sl(n, \mathbb{F}) \otimes_{\mathbb{F}} A$ is given in the following way. For arbitrary elements $f_1, f_2 \in sl(n, \mathbb{F})$ and $a, b \in A$ we have

$$Q(f_1 \otimes a, f_2 \otimes b) = K(f_1, f_2) \cdot t(ab)$$

where K denotes the Killing form on $sl(n, \mathbb{F})$ and $t : A \longrightarrow \mathbb{F}$ is a trace function.

Let us now investigate the algebra A. Since A is unital and of dimension 2 over $\mathbb{F}$, one can choose a basis $\{e, 1\}$, where 1 denotes the unit. Moreover, there exist p and q in $\mathbb{F}$ such that $e^2 + pe + q = 0$. Let $\Delta = p^2 - 4q \in \mathbb{F}$. We distinguish the following cases:

(i) Assume $\Delta = 0$. Let $\varepsilon := (e+p)/2$. Then $\varepsilon^2 = 0$ and $A = \mathbb{F}\varepsilon \oplus \mathbb{F} = \mathbb{F}[\varepsilon]$.

(ii) Assume Δ is the square of a nonzero element of $\mathbb{F}$. In this case, one can choose $e' \in \mathbb{F}^*$ such that $e'^2 = \Delta$. Then $A = \mathbb{F} \oplus e'\mathbb{F} = \mathbb{F} \oplus \mathbb{F}$.

(iii) Assume Δ is not a square of an element of $\mathbb{F}$. Then $A = \mathbb{F} + e'\mathbb{F}$, where $e' = (e+p)/2$ and $e'^2 = \Delta/4 \in \mathbb{F}$. Thus A is a quadratic field extension of $\mathbb{F}$.

Summing up the above observations, we get

Theorem 2.2. *Let δ be an arbitrary Lie bialgebra structure on $sl(n,\mathbb{F})$. Then $D(sl(n,\mathbb{F}),\delta) \cong sl(n,\mathbb{F}) \otimes_{\mathbb{F}} A$, where $A = \mathbb{F}[\varepsilon]$ and $\varepsilon^2 = 0$, $A = \mathbb{F} \oplus \mathbb{F}$ or A is a quadratic field extension of $\mathbb{F}$.*

The classification of Lie bialgebras with classical double $sl(n,\mathbb{F}[\varepsilon])$ leads to the classification of quasi-Frobenius Lie algebras over $\mathbb{F}$. More precisely, due to the correspondence between Lie bialgebras and Manin triples (see [3]), the following result holds:

Proposition 2.3. *There exists a one-to-one correspondence between Lie bialgebra structures on $sl(n,\mathbb{F})$ whose corresponding double is isomorphic to $sl(n,\mathbb{F}[\varepsilon])$ and Lagrangian subalgebras W of $sl(n,\mathbb{F}[\varepsilon])$ complementary to $sl(n,\mathbb{F})$.*

Similarly to Theorem 3.2 from [7], one can prove

Proposition 2.4. *Any Lagrangian subalgebra W of $sl(n,\mathbb{F}[\varepsilon])$ complementary to $sl(n,\mathbb{F})$ is uniquely defined by a subalgebra L of $sl(n,\mathbb{F})$ together with a nondegenerate 2-cocycle B on L.*

We recall that a Lie algebra is called quasi-Frobenius if there exists a nondegenerate 2-cocycle on it. The complete classification of quasi-Frobenius Lie subalgebras of $sl(n,\mathbb{F})$ is not generally known for large n.

3. Belavin-Drinfeld untwisted cohomologies

Unlike the previous case, the classification of Lie bialgebras with classical double $sl(n,\mathbb{F}) \oplus sl(n,\mathbb{F})$ can be achieved by cohomological and combinatorial methods.

Lemma 3.1. *Any Lie bialgebra structure δ on $sl(n,\mathbb{F})$ for which the associated classical double is isomorphic to $sl(n,\mathbb{F}) \oplus sl(n,\mathbb{F})$ is a coboundary $\delta = dr$ given by an r-matrix satisfying $r + r^{21} = f\Omega$, where $f \in \mathbb{F}^*$ and $\mathrm{CYB}(r) = 0$.*

We may suppose that $f = 1$. Naturally we want to classify all such r up to $GL(n,\mathbb{F})$-equivalence. Let $\overline{\mathbb{F}}$ denote the algebraic closure of $\mathbb{F}$. Any Lie bialgebra structure δ over $\mathbb{F}$ can be extended to a Lie bialgebra structure $\overline{\delta}$ over $\overline{\mathbb{F}}$.

According to [1], the Lie bialgebra structures on a simple Lie algebra $\mathfrak{g}$ over an algebraically closed field are coboundaries given by non-skewsymmetric r-matrices. Suppose we have fixed a Cartan subalgebra $\mathfrak{h}$ and the corresponding root system. Any r-matrix depends on a discrete and a continuous parameter. The discrete parameter is an admissible triple $(\Gamma_1, \Gamma_2, \tau)$, i.e. an isometry $\tau : \Gamma_1 \longrightarrow \Gamma_2$ where $\Gamma_1, \Gamma_2 \subset \Gamma$ such that for any $\alpha \in \Gamma_1$ there exists $k \in \mathbb{N}$ satisfying $\tau^k(\alpha) \notin \Gamma_1$. The continuous parameter is a tensor $r_0 \in \mathfrak{h} \otimes \mathfrak{h}$ satisfying

$$r_0 + r_0^{21} = \Omega_0, \quad (\tau(\alpha) \otimes 1 + 1 \otimes \alpha)(r_0) = 0, \quad \forall \alpha \in \Gamma_1$$

Here Ω_0 denotes the Cartan part of the quadratic Casimir element Ω. Then the associated r-matrix is given by the following formula

$$r = r_0 + \sum_{\alpha > 0} e_\alpha \otimes e_{-\alpha} + \sum_{\alpha \in (Span\Gamma_1)^+} \sum_{k \in \mathbb{N}} e_\alpha \wedge e_{-\tau^k(\alpha)}$$

Now, let us assume that δ is a Lie bialgebra structure on $sl(n,\mathbb{F})$. Then its extension $\overline{\delta}$ has a corresponding r-matrix. Up to $GL(n,\overline{\mathbb{F}})$-equivalence, we have the Belavin-Drinfeld classification. We may therefore assume that our r-matrix is of the form $r_X = (\mathrm{Ad}_X \otimes \mathrm{Ad}_X)(r)$, where $X \in GL(n,\overline{\mathbb{F}})$ and r satisfies the system $r + r^{21} = \Omega$ and $\mathrm{CYB}(r) = 0$.

Let $\sigma \in Gal(\overline{\mathbb{F}}/\mathbb{F})$. Since

$$\delta(a) = [r_X, a \otimes 1 + 1 \otimes a]$$

for any $a \in sl(n,\mathbb{F})$ we have

$$(\sigma \otimes \sigma)(\delta(a)) = [\sigma(r_X), a \otimes 1 + 1 \otimes a]$$

and $(\sigma \otimes \sigma)(\delta(a)) = \delta(a)$. Consequently, $\sigma(r_X) = r_X + \lambda\Omega$, for some $\lambda \in \overline{\mathbb{F}}$. Let us show that $\lambda = 0$. Really,

$$\Omega = \sigma(\Omega) = \sigma(r_X) + \sigma(r_X^{21}) = r_X + r_X^{21} + 2\lambda\Omega$$

Thus $\lambda = 0$ and $\sigma(r_X) = r_X$. Consequently,

$$(\mathrm{Ad}_{X^{-1}\sigma(X)} \otimes \mathrm{Ad}_{X^{-1}\sigma(X)})(\sigma(r)) = r$$

We recall the following

Definition 3.2. Let r be an r-matrix. The *centralizer* $C(r)$ of r is the set of all $X \in GL(n,\overline{\mathbb{F}})$ satisfying $(\mathrm{Ad}_X \otimes \mathrm{Ad}_X)(r) = r$.

Using the same argumets as in the proof of Theorem 4.3 [8], it follows that $\sigma(r) = r$ and $X^{-1}\sigma(X) \in C(r)$, for any $\sigma \in Gal(\overline{\mathbb{F}}/\mathbb{F})$.

Definition 3.3. Let r be a non-skewsymmetric r-matrix from the Belavin-Drinfeld list and $C(r)$ its centralizer. We say that $X \in GL(n,\overline{\mathbb{F}})$ is a *Belavin-Drinfeld cocycle* associated to r if $X^{-1}\sigma(X) \in C(r)$, for any $\sigma \in Gal(\overline{\mathbb{F}}/\mathbb{F})$.

The set of Belavin-Drinfeld cocycles associated to r will be denoted by $Z_{BD}(sl(n,\mathbb{F}),r)$. Note that this set contains the identity.

Definition 3.4. Two cocycles X_1 and X_2 in $Z_{BD}(sl(n,\mathbb{F}),r)$ are called *equivalent* if there exists $Q \in GL(n,\mathbb{F})$ and $C \in C(r)$ such that $X_1 = QX_2C$.

Definition 3.5. Let $H^1_{BD}(sl(n,\mathbb{F}),r)$ denote the set of equivalence classes of cocycles from $Z_{BD}(sl(n,\mathbb{F}),r)$. We call this set the *Belavin-Drinfeld cohomology* associated to the r-matrix r. The Belavin-Drinfeld cohomology is said to be *trivial* if all cocycles are equivalent to the identity, and *non-trivial* otherwise.

Combining the above definitions with the preceding discussion, we obtain

Proposition 3.6. *For any non-skewsymmetric r-matrix r, there exists a one-to-one correspondence between $H^1_{BD}(sl(n,\mathbb{F}),r)$ and gauge equivalence classes of Lie bialgebra structures on $sl(n,\mathbb{F})$ with classical double isomorphic to $sl(n,\mathbb{F}) \oplus sl(n,\mathbb{F})$ and $\overline{\mathbb{F}}$-isomorphic to $\delta = dr$.*

The Belavin-Drinfeld cohomology set can be computed as in [8] and the following result holds.

Theorem 3.7. *For any non-skewsymmetric r-matrix r, $H^1_{BD}(sl(n,\mathbb{F}),r)$ is trivial. Any Lie bialgebra structure on $sl(n,\mathbb{F})$ with classical double $sl(n,\mathbb{F}) \oplus sl(n,\mathbb{F})$ is of the form $\delta = dr$, where r is an r-matrix which is, up to a multiple from $\mathbb{F}^*$, $GL(n,\mathbb{F})$-equivalent to a non-skewsymmertric r-matrix from the Belavin-Drinfeld list.*

4. Belavin-Drinfeld twisted cohomologies

We focus on the study of Lie bialgebra structures on $sl(n,\mathbb{F})$ whose classical double is isomorphic to $sl(n,\mathbb{F})\otimes_{\mathbb{F}} A$, where A is a quadratic extension of $\mathbb{F}$. We may suppose that $A=\mathbb{F}(\sqrt{d})$, where d is not a square in $\mathbb{F}$. We will show that Lie bialgebras of this type can also be classified by means of certain cohomology sets.

Twisted cohomologies associated to r-matrices for $sl(n,\mathbb{F})$ can be defined as in [8], where we studied the particular case $\mathbb{F}=\mathbb{C}((\hbar))$. First, similarly to Prop. 5.3 of [8], one can prove the following

Proposition 4.1. *Any Lie bialgebra structure on $sl(n,\mathbb{F})$ with classical double isomorphic to $sl(n,\mathbb{F}[\sqrt{d}])$ is given by an r-matrix r' which satisfies $CYB(r')=0$ and $r'+r'_{21}=\sqrt{d}\Omega$.*

Over $\overline{\mathbb{F}}$, all r-matrices are gauge equivalent to the ones from Belavin-Drinfeld list. It follows that there exists a non-skewsymmetric r-matrix r and $X\in GL(n,\overline{\mathbb{F}})$ such that $r'=\sqrt{d}(\mathrm{Ad}_X\otimes\mathrm{Ad}_X)(r)$.

The field $\mathbb{F}[\sqrt{d}]$ is endowed with a conjugation $\overline{a+b\sqrt{d}}=a-b\sqrt{d}$. Denote by σ_2 its lift to $Gal(\overline{\mathbb{F}}/\mathbb{F})$. If $X\in GL(n,\mathbb{F}[\sqrt{d}])$, then $\sigma_2(X)=\overline{X}$. Now let us consider the action of σ_2 on r'. We have $\sigma_2(r')=r'+\lambda\Omega$, for some $\lambda\in\overline{\mathbb{F}}$. Let us show that $\lambda=-\sqrt{d}$. Indeed, since $r'+r'_{21}=\sqrt{d}\Omega$, we also have $\sigma_2(r')+\sigma_2(r'_{21})=-\sqrt{d}\Omega$. Combining these relations with $\sigma_2(r')=r'+\lambda\Omega$, we get $\lambda=-\sqrt{d}$ and therefore $\sigma_2(r')=r'-\sqrt{d}\Omega=-r'_{21}$.

Recall now that $r'=\sqrt{d}(\mathrm{Ad}_X\otimes\mathrm{Ad}_X)(r)$. Then condition $\sigma_2(r')=-r'_{21}$ implies

$$(\mathrm{Ad}_{X^{-1}\sigma_2(X)}\otimes\mathrm{Ad}_{X^{-1}\sigma_2(X)})(\sigma_2(r))=r^{21}$$

For any $\sigma\in Gal(\overline{\mathbb{F}}/\mathbb{F}[\sqrt{d}])$, $\sigma(r')=r'$, which in turn implies

$$(\mathrm{Ad}_{X^{-1}\sigma(X)}\otimes\mathrm{Ad}_{X^{-1}\sigma(X)})(\sigma(r))=r$$

Now, using the same type of arguments as in the proof of Theorem 4.3 [8], one can deduce that $\sigma(r)=r$ and therefore the following result holds.

Proposition 4.2. *Any Lie bialgebra structure on $sl(n,\mathbb{F})$ with classical double isomorphic to $sl(n,\mathbb{F}[\sqrt{d}])$ is given by $r'=\sqrt{d}(\mathrm{Ad}_X\otimes\mathrm{Ad}_X)(r)$, where r is, up to a multiple from $\mathbb{F}^*$, a non-skewsymmetric r-matrix from the Belavin-Drinfeld list and $X\in GL(n,\overline{\mathbb{F}})$ satisfies $(\mathrm{Ad}_{X^{-1}\sigma_2(X)}\otimes\mathrm{Ad}_{X^{-1}\sigma_2(X)})(r)=r^{21}$ and, for any $\sigma\in Gal(\overline{\mathbb{F}}/\mathbb{F}[\sqrt{d}])$, $(\mathrm{Ad}_{X^{-1}\sigma(X)}\otimes\mathrm{Ad}_{X^{-1}\sigma(X)})(r)=r$.*

Definition 4.3. Let r be a non-skewsymmetric r-matrix from the Belavin-Drinfeld list. We say that $X\in GL(n,\overline{\mathbb{F}})$ is a *Belavin-Drinfeld twisted cocycle* associated to r if

$$(\mathrm{Ad}_{X^{-1}\sigma_2(X)}\otimes\mathrm{Ad}_{X^{-1}\sigma_2(X)})(r)=r^{21}$$

and for any $\sigma\in Gal(\overline{\mathbb{F}}/\mathbb{F}[\sqrt{d}])$,

$$(\mathrm{Ad}_{X^{-1}\sigma(X)}\otimes\mathrm{Ad}_{X^{-1}\sigma(X)})(r)=r$$

The set of Belavin-Drinfeld twisted cocycle associated to r will be denoted by $\overline{Z}_{BD}(sl(n,\mathbb{F}),r)$. Let us analyse for which admissible triples this set is non-empty.

Let $S\in GL(n,\mathbb{F})$ be the matrix with 1 on the second diagonal and 0 elsewhere. Let us denote by s the automorphism of the Dynkin diagram given by $s(\alpha_i)=\alpha_{n-i}$ for all $i\leq n-1$.

Proposition 4.4. *Let r be a non-skewsymmetric r-matrix associated to an admissible triple (Γ_1,Γ_2,τ). If $\overline{Z}_{BD}(sl(n,\mathbb{F}),r)\neq\emptyset$, then $s(\Gamma_1)=\Gamma_2$ and $s\tau=\tau^{-1}s$.*

Definition 4.5. Let X_1 and X_2 be two Belavin-Drinfeld twisted cocycles associated to r. We say that they are *equivalent* if there exist $Q \in GL(n, \mathbb{F})$ and $C \in C(r)$ such that $X_2 = QX_1C$.

The set of equivalence classes of twisted cocycles corresponding to a non-skewsymmetric r-matrix r will be denoted by $\overline{H}^1_{BD}(sl(n,\mathbb{F}), r)$.

Remark 4.6. If two twisted cocycles X_1 and X_2 are equivalent, then the corresponding r-matrices $\sqrt{d}(\mathrm{Ad}_{X_1} \otimes \mathrm{Ad}_{X_1})(r)$ and $\sqrt{d}(\mathrm{Ad}_{X_2} \otimes \mathrm{Ad}_{X_2})(r)$ are gauge equivalent via Q.

Remark 4.7. In fact, by obvious reasons it is better to denote $\overline{H}^1_{BD}(sl(n,\mathbb{F}), r)$ by $\overline{H}^1_{BD}(sl(n,\mathbb{F}), r, d)$. However, we fix d and the notation $\overline{H}^1_{BD}(sl(n,\mathbb{F}), r)$ is not misleading.

Proposition 4.8. *There exists a one-to-one correspondence between the twisted cohomology set $\overline{H}^1_{BD}(sl(n,\mathbb{F}), r)$ and gauge equivalence classes of Lie bialgebra structures on $sl(n,\mathbb{F})$ with classical double isomorphic to $sl(n,\mathbb{F}[\sqrt{d}])$ and $\overline{\mathbb{F}}$-isomorphic to $\delta = dr$.*

Let r_{DJ} be the Drinfeld-Jimbo r-matrix. Having fixed a Cartan subalgebra $\mathfrak{h}$ of $\mathfrak{g}$ and the associated root system, we choose a system of generators e_α, $e_{-\alpha}$, h_α where $\alpha > 0$ such that $K(e_\alpha, e_{-\alpha}) = 1$. Denote by Ω_0 the Cartan part of Ω. Then

$$r_{DJ} = \sum_{\alpha>0} e_\alpha \otimes e_{-\alpha} + \frac{1}{2}\Omega_0$$

The twisted cohomology corresponding to r_{DJ} can be studied in the same manner as was done in [8] (see Prop. 7.15). Let $J \in GL(n, \mathbb{F}[\sqrt{d}])$ denote the matrix with entries $a_{kk} = 1$ for $k \le m$, $a_{kk} = -\sqrt{d}$ for $k \ge m+1$, $a_{k,n+1-k} = 1$ for $k \le m$, $a_{k,n+1-k} = \sqrt{d}$ for $k \ge m+1$, where $m = [(n+1)/2]$.

Theorem 4.9. *The Belavin-Drinfeld twisted cohomology $\overline{H}^1_{BD}(sl(n), r_{DJ})$ is non-empty and consists of one element, the class of J.*

Proof. Let X be a twisted cocycle associated to r_{DJ}. Then X is equivalent to a twisted cocycle $P \in GL(n, \mathbb{F}[\sqrt{d}])$, associated to r_{DJ}. We may therefore assume from the beginning that $X \in GL(n, \mathbb{F}[\sqrt{d}])$ and it remains to prove that all such cocycles are equivalent. The proof will be done by induction.

For $n = 2$, consider

$$J = \begin{pmatrix} 1 & 1 \\ \sqrt{d} & -\sqrt{d} \end{pmatrix}$$

and let $X = (a_{ij}) \in GL(2, \mathbb{F}[\sqrt{d}])$ satisfy $\overline{X} = XSD$ with

$$D = \mathrm{diag}(d_1, d_2) \quad \in GL(2, \mathbb{F}[\sqrt{d}])$$

The identity is equivalent to the following system:

$$\overline{a_{11}} = a_{12}d_1, \quad \overline{a_{12}} = a_{11}d_2, \quad \overline{a_{21}} = a_{22}d_1, \quad \overline{a_{22}} = a_{21}d_2$$

Assume that $a_{21}a_{22} \ne 0$. Let $a_{11}/a_{21} = a' + b'\sqrt{d}$. Then $a_{12}/a_{22} = a' - b'\sqrt{d}$. One can immediately check that $X = QJD'$, where

$$Q = \begin{pmatrix} a' & b' \\ 1 & 0 \end{pmatrix} \in GL(2, \mathbb{F}), \quad D' = \mathrm{diag}(a_{21}, a_{22}) \in \mathrm{diag}(2, \mathbb{F}[\sqrt{d}])$$

Journal of Physics: Conference Series **532** (2014) 012026 doi:10.1088/1742-6596/532/1/012026

For $n = 3$, set

$$J = \begin{pmatrix} 1 & 0 & 1 \\ 0 & 1 & 0 \\ \sqrt{d} & 0 & -\sqrt{d} \end{pmatrix}$$

and let $X = (a_{ij}) \in GL(3, \mathbb{F}[\sqrt{d}])$ satisfy $\overline{X} = XSD$ with $D = \mathrm{diag}(d_1, d_2, d_3) \in GL(3, \mathbb{K}[\sqrt{d}])$. The identity is equivalent to the following system:

$$\begin{aligned} &\overline{a_{11}} = d_1 a_{13}, \quad \overline{a_{21}} = d_1 a_{23}, \quad \overline{a_{31}} = d_1 a_{33}, \\ &\overline{a_{12}} = d_2 a_{12}, \quad \overline{a_{22}} = d_2 a_{22}, \quad \overline{a_{32}} = d_2 a_{32}, \\ &\overline{a_{13}} = d_3 a_{11}, \quad \overline{a_{23}} = d_3 a_{21}, \quad \overline{a_{33}} = d_3 a_{31} \end{aligned}$$

Assume that $a_{21}a_{22}a_{23} \neq 0$. Let

$$a_{11}/a_{21} = b_{11} + b_{13}\sqrt{d}, \quad a_{31}/a_{21} = b_{31} + b_{33}\sqrt{d}$$

Then

$$a_{13}/a_{23} = b_{11} - b_{13}\sqrt{d}, \quad a_{33}/a_{23} = b_{31} - b_{33}\sqrt{d}$$

On the other hand, let $b_{12} := a_{12}/a_{22}$ and $b_{32} := a_{32}/a_{22}$. Note that $b_{12} \in \mathbb{F}$, $b_{32} \in \mathbb{F}$. One can immediately check that $X = QJD'$, where

$$Q = \begin{pmatrix} b_{11} & b_{12} & b_{13} \\ 1 & 1 & 0 \\ b_{31} & b_{32} & b_{33} \end{pmatrix} \in GL(3, \mathbb{F}), \quad D' = \mathrm{diag}(a_{21}, a_{22}, a_{23}) \in \mathrm{diag}(3, \mathbb{F}[\sqrt{d}])$$

Assume $n > 3$. Denote the constructed above $J \in GL(n, \mathbb{F}[\sqrt{d}])$ by J_n. We are going to prove that if $X \in GL(n, \mathbb{F}[\sqrt{d}])$ satisfies $\overline{X} = XSD$, then using elementary row operations with entries from $\mathbb{F}$ and multiplying columns by proper elements from $\mathbb{F}[\sqrt{d}]$ we can bring X to J_n.

We will need the following operations on a matrix $M = \{m_{pq}\} \in \mathrm{Mat}(n)$

1. $u_n(M) = \{m_{pq},\ p, q = 2, 3, \ldots, n-1\} \in \mathrm{Mat}(n-2)$;
2. $g_n(M) = \{m_{pq}\} \in \mathrm{Mat}(n+2)$, where m_{pq} are already defined for $p, q = 1, 2, \ldots n$, $m_{00} = m_{n+1,n+1} = 1$ and the rest $m_{0,a} = m_{a,0} = m_{n+1,a} = m_{a,n+1} = 0$.

It is clear that $u_n(X)$ satisfies the twisted cocycle condition. However, its determinant might vanish. To avoid this complication, we note that columns $2, 3, \ldots, n-1$ of X are linearly independent. Applying elementary row operations (in fact, they are permutations) we obtain a new cocycle X_1, which is equivalent to X and such that $u_n(X_1)$ is a cocycle in $GL(n-2, \mathbb{F}[\sqrt{d}])$. Then, by induction, there exist $Q_{n-2} \in GL(n-2, \mathbb{F})$ and a diagonal matrix D_{n-2} such that

$$Q_{n-2} \cdot u_n(X_1) \cdot D_{n-2} = J_{n-2}$$

Consider

$$X_n = g_{n-2}(Q_{n-2}) \cdot X_1 \cdot g_{n-2}(D_{n-2})$$

Clearly, X_n is a twisted cocycle equivalent to X and $u_n(X_n) = J_{n-2}$.

Applying elementary row operations with entries from $\mathbb{F}$ and multiplying by a proper diagonal matrix we can obtain a new cocycle $Y_n = (y_{pg})$ equivalent to X with the following properties:

1. $u_n(Y_n) = J_{n-2}$;
2. $y_{12} = y_{13} = \ldots = y_{1,n-1} = 0$ and $y_{n2} = y_{n3} = \ldots = y_{n,n-1} = 0$;
3. $y_{11} = y_{1n} = 1$, here we use the fact that if $y_{pq} = 0$, then $y_{p,n+1-q} = 0$.

It follows from the cocycle condition $\overline{Y_n} = Y_n \cdot S \cdot \mathrm{diag}(h_1, \ldots, h_n)$ that $h_1 = h_n = 1$ and hence, $y_{n1} = \overline{y_{nn}}$.

Now, we can use the first row to achieve $y_{n1} = -y_{nn} = \sqrt{d}$ and after that, we use the first and the last rows to "kill" $\{y_{k1},\ k = 2, \ldots, n-1\}$. Then the set $\{y_{kn},\ k = 2, \ldots, n-1\}$ will be "killed" automatically. We have obtained J_n from X and thus, have proved that X is equivalent to J_n. □

Now investigate twisted cohomologies associated to arbitrary non-skewsymmetric r-matrices. The following two results will prove to be useful for our study.

Lemma 4.10. *Assume $X \in \overline{Z}_{BD}(sl(n), r)$. Then there exists a twisted cocycle $Y \in GL(n, \mathbb{F}[\sqrt{d}])$, associated to r, and equivalent to X.*

Proof. We have $X \in GL(n, \overline{\mathbb{F}})$ and for any

$$\sigma \in Gal(\overline{\mathbb{F}}/\mathbb{F}[\sqrt{d}]), \quad X^{-1}\sigma(X) \in C(r)$$

On the other hand, the Belavin-Drinfeld cohomology for $sl(n)$ associated to r is trivial. This implies that X is equivalent to the identity, where in the equivalence relation we consider $\mathbb{F}[\sqrt{d}]$ instead of $\mathbb{F}$. So there exists $Y \in GL(n, \mathbb{F}[\sqrt{d}])$ and $C \in C(r)$ such that $X = YC$. On the other hand, $Y \in \overline{Z}_{BD}(sl(n), r)$ since

$$(\mathrm{Ad}_{X^{-1}\sigma_2(X)} \otimes \mathrm{Ad}_{X^{-1}\sigma_2(X)})(r) = r^{21} \quad \Longrightarrow \quad (\mathrm{Ad}_{Y^{-1}\sigma_2(Y)} \otimes \mathrm{Ad}_{Y^{-1}\sigma_2(Y)})(r) = r^{21}$$ □

Proposition 4.11. *Let r be a non-skewsymmetric r-matrix associated to an admissible triple $(\Gamma_1, \Gamma_2, \tau)$ satisfying $s(\Gamma_1) = \Gamma_2$ and $s\tau = \tau^{-1}s$. If $X \in \overline{Z}_{BD}(sl(n, \mathbb{F}), r)$, then there exist $R \in GL(n, \mathbb{F})$ and $D \in \mathrm{diag}(n, \overline{\mathbb{F}})$ such that $X = RJD$.*

Proof. According to Lemma 4.10, $X = YC$, where $Y \in GL(n, \mathbb{F}[\sqrt{d}])$ and $C \in C(r)$. Since

$$(\mathrm{Ad}_{Y^{-1}\sigma_2(Y)} \otimes \mathrm{Ad}_{Y^{-1}\sigma_2(Y)})(r) = r^{21}, \quad (\mathrm{Ad}_S \otimes \mathrm{Ad}_S)(r) = r^{21}$$

it follows that $S^{-1}Y^{-1}\sigma_2(Y) \in C(r)$. On the other hand, by Lemma 4.11 from [8], $C(r) \subset \mathrm{diag}(n, \overline{\mathbb{F}})$. We get $S^{-1}Y^{-1}\sigma_2(Y) \in \mathrm{diag}(n, \overline{\mathbb{F}})$. Now Theorem 4.9 implies that $Y = RJD_0$, where $R \in GL(n, \mathbb{F})$ and $D_0 \in \mathrm{diag}(n, \overline{\mathbb{F}})$. Consequently, $X = RJD_0C = RJD$ with $D = D_0C \in \mathrm{diag}(n, \overline{\mathbb{F}})$. □

Let T denote the automorphism of $\mathrm{diag}(n, \overline{\mathbb{F}})$ defined by $T(D) = SD^{-1}S\overline{D}$.

Lemma 4.12. *Let r be a non-skewsymmetric r-matrix with centralizer $C(r)$. Let $X = RJD$, with $R \in GL(n, \mathbb{F})$ and $D \in \mathrm{diag}(n, \mathbb{F}[\sqrt{d}])$. Then $X \in \overline{Z}_{BD}(sl(n, \mathbb{F}), r)$ if and ony if $D \in T^{-1}(C(r))$.*

Proof. Let us first note that $X \in \overline{Z}_{BD}(sl(n, \mathbb{F}), r)$ if and only if for any $\sigma \in Gal(\overline{\mathbb{F}}/\mathbb{F}[\sqrt{d}])$, $X^{-1}\sigma(X) \in C(r)$ and $SX^{-1}\overline{X} \in C(r)$. We have $X = RJD$ which implies

$$\overline{X} = \overline{RJD} = RJS\overline{D} = RJDD^{-1}S\overline{D} = XD^{-1}S\overline{D} = XST(D)$$

We immediately get that $SX^{-1}\overline{X} \in C(r)$ if and only if $T(D) \in C(r)$. □

Lemma 4.13. *Let $X_1 = R_1JD_1$ and $X_2 = R_2JD_2$ be two Belavin-Drinfeld twisted cocycles associated to r. Then X_1 and X_2 are equivalent if and only if $D_2D_1^{-1} \in C(r) \cdot Ker(T)$.*

Proof. Assume the two cocycles are equivalent. There exist $Q \in GL(n,\mathbb{F})$ and $C \in C(r)$ such that $X_2 = QX_1C$. Then

$$Q = R_2 J D_2 C^{-1} D_1^{-1} J^{-1} R_1^{-1}$$

Since $Q = \overline{Q}$ and $\overline{J} = JS$, we get

$$D_2 C^{-1} D_1^{-1} = S\overline{D_2 C^{-1} D_1^{-1}}S$$

Thus $D_2C^{-1}D_1^{-1} \in Ker(T)$. On the other hand, $C \in C(r) \subset \mathrm{diag}(n,\overline{\mathbb{F}})$, so $D_2C^{-1}D_1^{-1} = D_2D_1^{-1}C^{-1}$. We have obtained that $D_2D_1^{-1} \in C(r)\cdot Ker(T)$. Conversely, if this condition is satisfied, then write $D_2D_1^{-1} = D_0C$, where $C \in C(r)$ and $D_0 \in Ker(T)$. Denote $Q := R_2JD_0J^{-1}R_1^{-1}$. Then, by construction, $Q = \overline{Q}$ and $X_2 = QX_1C$. □

By lemmas 4.12 and 4.13, we get

Proposition 4.14. *Let r be a non-skewsymmetric r-matrix associated to an admissible triple (Γ_1,Γ_2,τ) satisfying $s(\Gamma_1) = \Gamma_2$ and $s\tau = \tau^{-1}s$. Then*

$$\overline{H}^1_{BD}(sl(n,\mathbb{F}),r) = \frac{T^{-1}(C(r))}{C(r)\cdot Ker(T)}$$

At this point, one needs the explicit description of the centralizer and its preimage under T.

Lemma 4.15. *Let r be a non-skewsymmetric r-matrix associated to an admissible triple (Γ_1,Γ_2,τ). Then the following hold:*

(a) $C(r)$ consists of all diagonal matrices $D = \mathrm{diag}(d_1,\ldots,d_n)$ such that $d_i = s_is_{i+1}\ldots s_n$, where $s_i \in \overline{\mathbb{F}}$ satisfy the condition: $s_i = s_j$ if $\alpha_i \in \Gamma_1$ and $\tau(\alpha_i) = \alpha_j$.

(b) $T^{-1}(C(r))$ consists of all diagonal matrices $D = \mathrm{diag}(d_1,\ldots,d_n)$ such that $d_i = s_is_{i+1}\ldots s_n$, where $s_i \in \overline{\mathbb{F}}$ satisfy the condition: $\overline{s}_is_{n-i} = \overline{s}_js_{n-j}$ if $\alpha_i \in \Gamma_1$ and $\tau(\alpha_i) = \alpha_j$.

Proof. Part (a) can be proved in the same way as Lemma 5.5 from [8] and (b) follows immediately from (a). □

Let us make the following remark. Any admissible triple (Γ_1,Γ_2,τ) can be viewed as a union of strings

$$\alpha_{i_1} \xrightarrow{\tau} \alpha_{i_2} \xrightarrow{\tau} \ldots \xrightarrow{\tau} \alpha_{i_k}, \quad \tau(\alpha_{i_k}) \notin \Gamma_1$$

The above lemma implies that elements of $C(r)$ have the property that $s_{i_1} = s_{i_2} = \ldots = s_{i_k}$, i.e. s_i is constant on each string. In turn, elements of $T^{-1}(C(r))$ satisfy

$$\overline{s}_{i_1}s_{n-i_1} = \overline{s}_{i_2}s_{n-i_2} = \ldots = \overline{s}_{i_k}s_{n-i_k}$$

i.e. $\overline{s}_is_{n-i}$ is constant on each string.

Theorem 4.16. *Suppose r is a non-skewsymmetric r-matrix with admissible triple (Γ_1,Γ_2,τ) satisfying $s\tau = \tau^{-1}s$. Let $str(\Gamma_1,\Gamma_2,\tau)$ denote the number of symmetric strings not containing the middlepoint. Then*

$$\overline{H}^1_{BD}(sl(n,\mathbb{F}),r) = \left(\frac{\mathbb{F}^*}{N_{\mathbb{F}(\sqrt{d})/\mathbb{F}}(\mathbb{F}(\sqrt{d}))^*}\right)^{str(\Gamma_1,\Gamma_2,\tau)}$$

Proof. Let $\varphi : (\overline{\mathbb{F}}^*)^n \to \mathrm{diag}(n, \overline{\mathbb{F}})$ be the map

$$\varphi(s_1, \ldots, s_{n-1}, s_n) = \mathrm{diag}(s_1 \ldots s_n, s_2 \ldots s_n, \ldots, s_{n-1}s_n, s_n)$$

Consider $\widetilde{T} = \varphi^{-1}T\varphi$. Since $Ker(T) = \varphi Ker(\widetilde{T})$, we have

$$\frac{T^{-1}(C(r))}{Ker(T) \cdot C(r)} \cong \frac{\widetilde{T}^{-1}\varphi^{-1}(C(r))}{Ker(\widetilde{T}) \cdot \varphi^{-1}(C(r))}$$

We make the following remarks:

(i) $(s_1, \ldots, s_n) \in Ker(\widetilde{T})$ if and only if $\overline{s}_i s_{n-i} = 1$ for all $i \leq n-1$ and $\overline{s}_n = s_1 \ldots s_n$.
(ii) $(s_1, \ldots, s_n) \in \varphi^{-1}(C(r))$ is equivalent to s_i is constant on each string of the given triple.
(iii) $(s_1, \ldots, s_n) \in \widetilde{T}^{-1}\varphi^{-1}(C(r))$ implies that $\overline{s}_i s_{n-i}$ is constant on each string.

Step 1.
Suppose that the admissible triple is the disjoint union of two symmetric strings

$$\alpha_{i_1} \xrightarrow{\tau} \alpha_{i_2} \xrightarrow{\tau} \ldots \xrightarrow{\tau} \alpha_{i_k}, \quad \alpha_{n-i_k} \xrightarrow{\tau} \alpha_{n-i_{k-1}} \xrightarrow{\tau} \ldots \xrightarrow{\tau} \alpha_{n-i_1}$$

Here we recall that τ has the property that $\tau(\alpha_{n-j}) = \alpha_{n-i}$ if $\tau(\alpha_i) = \alpha_j$.

Let $(s_1, \ldots, s_n) \in \widetilde{T}^{-1}\varphi^{-1}(C(r))$. Then

$$\overline{s}_{i_1} s_{n-i_1} = \ldots = \overline{s}_{i_k} s_{n-i_k} =: t, \quad \overline{s}_{n-i_1} s_{i_1} = \ldots = \overline{s}_{n-i_k} s_{i_k} = \overline{t}$$

One can check that $(s_1, \ldots, s_n) \in Ker(\widetilde{T}) \cdot \varphi^{-1}(C(r))$. Indeed, let us assume first that $n = 2m+1$. Then $(s_1, \ldots, s_n)$ is the product of the following elements:

$$(s_1, \ldots, s_m, (\overline{s}_m)^{-1}, \ldots, (\overline{s}_1)^{-1}, \overline{s_1 \ldots s_m}), \quad (1, \ldots, 1, s_{m+1}\overline{s}_m, \ldots, s_{n-1}\overline{s}_1, s_n(\overline{s_1 \ldots s_m})^{-1}$$

The first factor belongs to $Ker(\widetilde{T})$ and the second is in $\varphi^{-1}(C(r))$ since the $n - i_1, \ldots, n - i_k$ coordinates have the constant value t.

Suppose that $n = 2m$. Consider

$$\left(s_1, \ldots, s_{m-1}, r_m, (\overline{s}_{m+1})^{-1}, \ldots, (\overline{s}_1)^{-1}, \overline{s}_n\right), \quad (1, \ldots, 1, s_m/r_m, s_{m+1}\overline{s}_{m-1}, \ldots, s_{n-1}\overline{s}_1)$$

where

$$r_m = \frac{\overline{s_1 \ldots s_{m-1} s_n}}{s_1 \ldots s_{m-1}\overline{s}_n}$$

The first factor is in $Ker(\widetilde{T})$ since $r_m \overline{r}_m = 1$ and the second is in $\varphi^{-1}(C(r))$ since neither $n - i_1, \ldots, n - i_k$ can be m, and the corresponding coordinates all equal t.

Step 2.
Let us assume that the admissible triple includes a symmetric string

$$\alpha_{i_1} \xrightarrow{\tau} \alpha_{i_2} \xrightarrow{\tau} \ldots \xrightarrow{\tau} \alpha_{i_k} \xrightarrow{\tau} \ldots \xrightarrow{\tau} \alpha_{n-i_k} \xrightarrow{\tau} \alpha_{n-i_{k-1}} \xrightarrow{\tau} \ldots \xrightarrow{\tau} \alpha_{n-i_1}$$

not containing the middlepoint. Let $(s_1, \ldots, s_n) \in \widetilde{T}^{-1}\varphi^{-1}(C(r))$. Then

$$\overline{s}_{i_1} s_{n-i_1} = \ldots = \overline{s}_{i_k} s_{n-i_k} = \overline{s}_{n-i_1} s_{i_1} = \ldots = \overline{s}_{n-i_k} s_{i_k} = t$$

We note that $t \in \mathbb{F}$ since $t = \overline{t}$.

Case 1. Assume there exists $q \in \mathbb{F}(\sqrt{d})$ such that $t = q\overline{q}$. Then $(s_1, \ldots, s_n) \in Ker(\widetilde{T}) \cdot \varphi^{-1}(C(r))$. Indeed, one can make the same construction as in Step 1, except for the positions $i_1, \ldots, i_k$, $n - i_1, \ldots, n - i_k$ where we consider instead the decomposition

$$(\ldots, s_{i_l}, \ldots, s_{n-i_l}, \ldots) = (\ldots, s_{i_l}/q, \ldots, s_{n-i_l}/q, \ldots) \cdot (\ldots, q, \ldots, q, \ldots)$$

Case 2. Assume for any $q \in \mathbb{F}(\sqrt{d})$, $t \neq q\overline{q}$. Then it follows that $(s_1, \ldots, s_n) \notin Ker(\widetilde{T}) \cdot \varphi^{-1}(C(r))$. Indeed, let us assume the contrary, i.e. we may write $s_i = p_i r_i$, where $\overline{p}_i p_{n-i} = 1$ for all $i \leq n-1$, $\overline{p}_n = p_1 \ldots p_n$ and

$$r_{i_1} = \ldots = r_{i_k} = r_{n-i_1} = \ldots = r_{n-i_k}$$

It follows that

$$t = \overline{s}_{i_1} s_{n-i_1} = \overline{r}_{i_1} r_{n-i_1} = \overline{r}_{i_1} r_{i_1}$$

which is a contradiction.

Step 3.
Let us suppose that the admissible triple includes a symmetric string

$$\alpha_{i_1} \xrightarrow{\tau} \alpha_{i_2} \xrightarrow{\tau} \ldots \xrightarrow{\tau} \alpha_{i_k} \xrightarrow{\tau} \ldots \xrightarrow{\tau} \alpha_{n-i_k} \xrightarrow{\tau} \alpha_{n-i_{k-1}} \xrightarrow{\tau} \ldots \xrightarrow{\tau} \alpha_{n-i_1}$$

containing the middlepoint. In this case

$$\overline{s}_{i_1} s_{n-i_1} = \ldots = \overline{s}_{i_k} s_{n-i_k} = \overline{s}_{n-i_1} s_{i_1} = \ldots = \overline{s}_{n-i_k} s_{i_k} = t$$

Moreover, $t = s_m \overline{s}_m$, where s_m is the coordinate corresponding to the middlepoint α_m. Then again $(s_1, \ldots, s_n) \in Ker(\widetilde{T}) \cdot \varphi^{-1}(C(r))$ since we may proceed as in Step 2, case 1 by taking $q = s_m$. □

Example 4.17. For $\mathbb{F} = \mathbb{R}$ and $d = -1$, it follows that given an r-matrix r with admissible triple $(\Gamma_1, \Gamma_2, \tau)$ we have

$$\overline{H}^1_{BD}(sl(n, \mathbb{R}), r) = (\mathbb{Z}_2)^{str(\Gamma_1, \Gamma_2, \tau)}$$

Example 4.18. Let us consider $\mathbb{F} = \mathbb{C}((\hbar))$ and $d = \hbar$. Then $N(\mathbb{F}(\sqrt{d})) = \mathbb{F}$ and Theorem 4.16 implies that $\overline{H}^1_{BD}(sl(n, \mathbb{C}((\hbar))), r)$ is trivial (consists of one element) for any r-matrix r satisfying the condition of Proposition 4.4 and empty otherwise. We have thus generalized our previous results [9], where we proved that twisted cohomologies for $sl(n)$ associated to generalized Cremmer-Gervais r-matrices are trivial.

This result completes classification of quantum groups which have $sl(n, \mathbb{C})$ as the classical limit. Summarizing, we have the following picture:

1. According to [4, 5], there exists an equivalence between the category $HA_0(\mathbb{C}[[\hbar]])$ of topologically free Hopf algebras cocommutative modulo $\hbar$ and the category $LBA_0(\mathbb{C}[[\hbar]])$ of topologically free over $\mathbb{C}[[\hbar]]$ Lie bialgebras with $\delta \equiv 0 \,(\mathrm{mod}\, \hbar)$.
2. To describe the category $LBA_0(\mathbb{C}[[\hbar]])$, it is sufficient (multiplying by a proper power of $\hbar^N$) to classify Lie bialgebra structures on the Lie algebra $\mathfrak{g} \otimes_{\mathbb{C}} \mathbb{C}((\hbar))$.

3. Following [6], only three classical Drinfeld doubles are possible, namely

$$D(\mathfrak{g} \otimes_{\mathbb{C}} \mathbb{C}((\hbar))) = \mathfrak{g} \otimes_{\mathbb{C}} A_k, \quad k = 1, 2, 3$$

Here

$$A_1 = \mathbb{K}[\varepsilon], \quad \varepsilon^2 = 0, \quad A_2 = \mathbb{K} \oplus \mathbb{K}, \quad A_3 = \mathbb{K}(\sqrt{\hbar}) \quad \text{with} \quad \mathbb{K} = \mathbb{C}((\hbar))$$

4. Lie bialgebra structures related to the case A_1 are in a one-to-one correspondence with quasi-Frobenius subalgebras of $\mathfrak{g} \otimes_{\mathbb{C}} \mathbb{C}((\hbar))$.
5. Now we turn to the case $D(\mathfrak{g} \otimes_{\mathbb{C}} \mathbb{C}((\hbar))) = \mathfrak{g} \otimes_{\mathbb{C}} A_2$ with $\mathfrak{g} = sl(n)$. Up to multiplication by $\hbar^N$ and conjugation by an element of $GL(n, \mathbb{K})$, the related Lie bialgebra structures are defined by the Belavin-Drinfeld data (see [1] and Section 2, the main ingredient is the triple $\tau : \Gamma_1 \to \Gamma_2$) and an additional data called a Belavin-Drinfeld cohomology. In the case $\mathfrak{g} = sl(n)$, the cohomology consists of *one element* independently of the Belavin-Drinfeld data. As a representative of this cohomology class one can choose the *identity matrix*.
6. Finally, in the case A_3 and $\mathfrak{g} = sl(n)$ the description is as follows. Up to multiplication by $\hbar^N$ and conjugation by an element of $GL(n, \mathbb{K})$, the related Lie bialgebra structures are defined by the Belavin-Drinfeld data and an additional data called a *twisted* Belavin-Drinfeld cohomology. In this case the twisted cohomology consists of *one element* if $\tau : \Gamma_1 \to \Gamma_2$ satisfies the condition of Proposition 4.4 and is *empty* otherwise (no Lie bialgebra structures of the type A_3 if τ does not satisfy the condition of Proposition 4.4). If the cohomology class is non-empty, it can be represented by the matrix J introduced before Theorem 4.9.

Appendix A.

Throughout the paper we use the following convenient notations for the arXiv references:

[8] Stolin A and Pop I 2013 arXiv:1303.4046

[9] Stolin A and Pop I 2013 arXiv:1309:7133

References

[1] Belavin A and Drinfeld V 1984 *Math. Phys. Rev.* **4** 93
[2] Benkart G and Zelmanov E 1996 *Invent. Math.* **126** 1
[3] Drinfeld V G 1997 Quantum groups *Proc. ICM, Berkeley 1996* (Providence, RI: AMS) **1** 798
[4] Etingof P and Kazhdan D 1996 *Sel. Math. (NS)* **2** 1
[5] Etingof P and Kazhdan D 1998 *Sel. Math. (NS)* **4** 213
[6] Montaner F, Stolin A and Zelmanov E 2010 *Sel. Math. (NS)* **16** 935
[7] Stolin A 1999 *Comm. Algebra* **27** 4289
[8] Bibliographic description is given in Appendix A
[9] Bibliographic description is given in Appendix A

Fréchet-algebraic deformation quantizations

S Waldmann

Institute for Mathematics, Julius Maximilian University Würzburg
Emil-Fischer-Straße 31, 97074 Würzburg, Germany

Abstract. In this review I present some recent results on the convergence properties of formal star products. Based on a general construction of a Fréchet topology for an algebra with countable vector space basis I discuss several examples from deformation quantization: the Wick star product on the flat phase space $\mathbb{R}^{2n}$ gives a first example of a Fréchet algebraic framework for the canonical commutation relations. More interesting, the star product on the Poincaré disk can be treated along the same lines, leading to a non-trivial example of a convergent star product on a curved Kähler manifold.

1. Introduction

In this review we recall some recent progress in the understanding of the passage from formal deformation quantization [1] to a more analytic framework for star products as obtained in [2].

Let us recall the basic definitions of formal deformation quantization from [1], see also [3] for a pedagogical introduction. We consider a Poisson manifold (M, π) where $\pi \in \Gamma^\infty(\Lambda^2 TM)$ is the Poisson bivector field. This means that $\{f, g\} = \pi(\mathrm{d}f, \mathrm{d}g)$ defines a Poisson bracket for $f, g \in \mathscr{C}^\infty(M)$. We will use complex-valued functions throughout this paper. The Jacobi identity for this bracket is equivalent to the condition $[\![\pi, \pi]\!] = 0$ using the Schouten bracket. Then a *star product* $\star$ for (M, π) is a $\mathbb{C}[[\lambda]]$-bilinear associative multiplication for $\mathscr{C}^\infty(M)[[\lambda]]$ written as

$$f \star g = \sum_{r=0}^{\infty} \lambda^r C_r(f, g) \tag{1.1}$$

such that $C_0(f, g) = fg$ is the pointwise product,

$$C_1(f, g) - C_1(g, f) = \mathrm{i}\{f, g\} \tag{1.2}$$

gives the Poisson bracket in the first order commutator, the C_r are bidifferential operators, and $1 \star f = f = f \star 1$ for all $f \in \mathscr{C}^\infty(M)[[\lambda]]$.

The classification and existence of such star products is by now well-established: starting with the early works of deWilde and Lecomte [4] and Fedosov [5] in the symplectic case, Kontsevich proved the existence and classification in the general Poisson case in [arXiv:q-alg/9709040] that later appeared as [6]. In addition to these existence and classification results one has many explicit examples, a good understanding of the quantization of symmetries and phase space reduction in this context, and a well-working representation theory for the resulting algebras, see e.g [7] for a review.

However, from a physical point of view the situation is still unsatisfactory: the deformation parameter λ plays the role of Planck's constant $\hbar$ which one would like to substitute by a positive real number. Since one can easily show that the series (1.1) typically diverges for all $\lambda \neq 0$, one faces a serious convergence problem.

There are several different approaches to overcome this problem. Most notably is the idea to replace the deformation aspect as above by the notion of a continuous field of topological algebras. Here mainly continuous fields of C^*-algebras have been considered, see e.g Rieffel's definition of strict deformation quantization [8]. In this context one knows various examples and some general constructions based on actions of $\mathbb{R}^d$ and some more general groups [9–13] but a general picture seems to be missing.

In this review we present a different approach, namely to investigate the convergence of the series (1.1) more directly within a Fréchet-algebraic framework. The reason for this choice is that for $M = \mathbb{R}^{2n}$ with the usual Weyl-Moyal star product the polynomials form a subalgebra where convergence is trivially fulfilled as the series simply terminate. To extend the star product beyond polynomials one would like to establish a topology for which it is continuous. Then the completion with respect to this topology should yield a hopefully large and interesting class of functions. However, since the canonical commutation relations will be realised in this larger algebra, no Banach topology will be possible.

In this work we do not get any general existence or classification results as this may even be too much to ask for. Instead, we focus on some non-trivial example, the star product on the Poincaré disk. Topologically, the disk is still trivial but from the Kähler geometry point of view, we have curvature compared to the case of the flat $\mathbb{R}^{2n}$.

Though we only have examples so far, we can at least split the question about convergence of star products in two separate problems. The first will be a general construction of a locally convex topology for an algebra having a countable basis. This construction will not work in general but requires some rather technical conditions on the multiplication to be satisfied. Nevertheless, it is conceptually rather easy and clear. The second step will consist in finding good examples where the above construction actually works. Surprisingly, we have many interesting examples even from beyond deformation quantization which stress that the general construction can be used to produce examples of locally convex algebras in various areas of mathematics.

The paper is organized as follows: in the next section we present the general construction of a Fréchet topology for a complex algebra with countable vector space basis in two versions. We discuss several simple examples where the necessary technical conditions for the constructions are fulfilled. In Section 3 we pass to the first non-trivial example from deformation quantization: the Wick star product on $\mathbb{R}^{2n}$. Already here we get a quite non-trivial completion and a Fréchet algebraic framework for the canonical commutation relations. Section 4 contains a brief review of the explicit construction of a star product of Wick type on the Poincaré disk. In Section 5 we show that the general construction can be applied to the star product on the disk. In fact, the whole phase space reduction of the star product can be formulated within our Fréchet algebraic context. The resulting Fréchet algebra on the disk enjoys many nice features some of which we explain.

2. The general construction

Consider a complex algebra $\mathcal{A}$ with a vector space basis $\{e_\alpha\}_{\alpha \in I}$ such that the index set I is *countable*. Then the product $\star$ of $\mathcal{A}$ is completely determined by its values on the basis vectors

$$e_\alpha \star e_\beta = \sum\nolimits_\gamma C^\gamma_{\alpha\beta} e_\gamma \tag{2.1}$$

Here we introduced the structure constants $C^\gamma_{\alpha\beta}$. For given α and β we have only finitely many γ for which $C^\gamma_{\alpha\beta} \neq 0$. A general element $a \in \mathcal{A}$ can be written as $a = \sum_\alpha a_\alpha e_\alpha$ with only finitely many a_α being different from zero, thereby defining the evaluation functionals $e^\alpha \colon a \mapsto a_\alpha$.

The aim is now to find a locally convex topology on $\mathcal{A}$ such that $\star$ becomes continuous *and* all the evaluation functionals e^α become continuous, too.

The first step is therefore to consider the seminorm $\|a\|_{0,0,\gamma} = |a_\gamma|$. which has to be among the continuous seminorms of the searched-for topology if we want e^γ to be continuous. While this makes e^γ continuous, the product $\star$ has no reason to be continuous with respect to all the $\|\cdot\|_{0,0,\gamma}$. In order to see the possible failure of continuity we estimate

$$\|a \star b\|_{0,0,\gamma} = \left|\sum\nolimits_{\alpha,\beta} C^\gamma_{\alpha\beta} a_\alpha b_\beta\right| \leq \sqrt{\sum\nolimits_{\alpha,\beta} |a_\alpha|^2 |C^\gamma_{\alpha\beta}|}\sqrt{\sum\nolimits_{\alpha,\beta} |b_\beta|^2 |C^\gamma_{\alpha\beta}|} \tag{2.2}$$

Thus we get the following simple candidate for additional seminorms in order to control the continuity of $\star$. We define

$$h_{1,0,\gamma})(a) = \sum\nolimits_\alpha |a_\alpha|^2 C^\gamma_{\alpha,\cdot} \quad \text{with} \quad C^\gamma_{\alpha,\cdot} = \sum\nolimits_\beta |C^\gamma_{\alpha\beta}| \tag{2.3}$$

and

$$h_{1,1,\gamma})(b) = \sum\nolimits_\beta |b_\beta|^2 C^\gamma_{\cdot,\beta} \quad \text{with} \quad C^\gamma_{\cdot,\beta} = \sum\nolimits_\alpha |C^\gamma_{\alpha\beta}| \tag{2.4}$$

Then $\|a\|_{1,\ell,\gamma} = \sqrt{h_{1,\ell,\gamma}(a)}$ for $\ell = 0, 1$ will be seminorms estimating the continuity requirement for $\star$ with respect to $\|\cdot\|_{0,0,\gamma}$. But then we have to iterate this procedure to get continuity estimates for $\star$ with respect to $\|\cdot\|_{1,\ell,\gamma}$ and so on. Thus we define recursively for $m = 0, 1, \ldots$, $\ell = 0, \ldots 2^m - 1$, and $\gamma \in I$

$$h_{m+1,2\ell,\gamma}(a) = \sum\nolimits_\alpha h_{m,\ell,\alpha}(a)^2 C^\gamma_{\alpha,\cdot} \quad h_{m+1,2\ell+1,\gamma}(a) = \sum\nolimits_\beta h_{m,\ell,\beta}(a)^2 C^\gamma_{\cdot,\beta} \tag{2.5}$$

for every $a \in \mathcal{A}$ and set $\|a\|_{m,\ell,\gamma} = \sqrt[2^m]{h_{m,\ell,\gamma}(a)}$. By construction, it is now clear that we have the continuity estimate

$$\|a \star b\|_{m,\ell,\gamma} \leq \|a\|_{m+1,\ell,\gamma} \|b\|_{m+1,2^m+\ell,\gamma} \tag{2.6}$$

However, it may very well happen that all these quantities diverge to $+\infty$. In particular, the numbers $C^\gamma_{\alpha,\cdot}$ and $C^\gamma_{\cdot,\beta}$ need not to be finite. In this situation our construction *fails*. In any case, we consider those elements of $\mathcal{A}$ where we have convergence of all the above series and set

$$\mathcal{A}_{\text{nice}} = \left\{a \in \mathcal{A} \;\middle|\; \|a\|_{m,\ell,\gamma} < \infty \text{ for all } m, \ell, \gamma\right\} \tag{2.7}$$

It is then easy to see that $\mathcal{A}_{\text{nice}}$ is a subalgebra of $\mathcal{A}$ on which the product is continuous with respect to the locally convex topology induced by all the seminorms $\|\cdot\|_{m,\ell,\gamma}$. Nevertheless, it might happen that $\mathcal{A}_{\text{nice}} = \{0\}$ is trivial.

Theorem 2.1 *Suppose we have $\mathcal{A} = \mathcal{A}_{\text{nice}}$.*

i.) The completion $\widehat{\mathcal{A}}$ of $\mathcal{A}$ becomes a Fréchet algebra.

ii.) The evaluation functionals $e^\gamma : \mathcal{A} \longrightarrow \mathbb{C}$ are continuous and extend continuously to $\widehat{\mathcal{A}}$.

iii.) The vectors $\{e_\alpha\}_{\alpha \in I}$ form an unconditional Schauder basis of $\widehat{\mathcal{A}}$, i.e $a = \sum_\gamma e^\gamma(a) e_\gamma$ converges unconditionally.

Let us comment on this construction: the first problem is of course that we have to choose a basis and the construction *depends* on this choice in a rather obscure way. In many examples we will see that there are good reasons for such a choice, so this will be only a minor drawback. More severe is that it might happen that $\mathcal{A}_{\text{nice}} = \{0\}$ is trivial. In this case, it will turn out that sometimes a rescaling of the given basis yields better convergence properties and can make the construction work. It turns out that by an appropriate rescaling one even can control properties of the resulting topology.

The remarkable feature of the construction is that it does not require an *associative* algebra at all: it would be very interesting to apply this to some of the standard candidates of infinite-dimensional Lie algebras with countable vector space basis like e.g the Witt algebra or the Virasoro algebra. We also note that there are alternative constructions of locally convex topologies for algebras having a countable vector space basis, see [14, Prop. 2.1].

While the above construction already yields several interesting examples, we can still refine it. The idea is that on $\mathcal{A}$ there might be a collection of linear functionals Ω of special interest. We are now interested in constructing a finer topology such that the functionals $\omega \in \Omega$ also become continuous. Of course, we can just add the seminorms $\|a\|_\omega = |\omega(a)|$ to achieve the continuity of ω and then start the recursion again, in order to guarantee the continuity of the product. However, this will not be necessary as the seminorms

$$\|a\|_{m,\ell,\omega} = \sqrt[2^m]{\sum\nolimits_\gamma h_{m,\ell,\gamma}(a)|\omega(e_\gamma)|} \tag{2.8}$$

will do the job. Again, we have the problem that these new seminorms might all diverge to $+\infty$. In this case the construction fails again. Thus we consider

$$\mathcal{A}_{\Omega\text{-nice}} = \left\{a \in \mathcal{A}_{\text{nice}} \;\middle|\; \|a\|_{m,\ell,\omega} < \infty \text{ for all } m, \ell, \omega\right\} \tag{2.9}$$

which turns out to be a subalgebra of $\mathcal{A}_{\text{nice}}$. Under the assumption $\mathcal{A}_{\Omega\text{-nice}} = \mathcal{A}$ we get the same conclusions as in Theorem 2.1 also in this case.

We conclude this section with some examples where the above general construction has been applied successfully:

Example 2.2 (polynomials I) Consider $\mathcal{A} = \mathbb{C}[z]$ with the standard basis $\{e_n = z^n\}_{n\in\mathbb{N}_0}$. In this case the construction can be show to work. The resulting completion is the algebra of formal power series $\mathbb{C}[[z]]$ with its usual Fréchet algebra structure. Now we can no longer interpret the elements of the completion as functions on $\mathbb{C}$. Thus we pass to the second version of the construction and take $\Omega = \{\delta_p\}$ for a point $p \in \mathbb{C}$ different from 0. Denote by $R = |p| > 0$ its absolute value. For $R \geq 1$ the second version of the construction fails but for $R < 1$ we get a non-trivial completion which is a subalgebra of the Banach algebra of holomorphic functions on the closed disk $\mathrm{B}_R(0)^{\mathrm{cl}} \subseteq \mathbb{C}$, endowed with a Fréchet topology strictly finer than the Banach space topology of uniform convergence.

Example 2.3 (polynomials II) Again, we consider $\mathcal{A} = \mathbb{C}[z]$ but now with the basis $e_n = \frac{z^n}{n!}$. The first version of the construction still works the same way and yields the same completion $\widehat{\mathcal{A}} = \mathbb{C}[[z]]$. However, the second version changes drastically. The resulting Fréchet algebra $\widehat{\mathcal{A}}$ does not depend on $R > 0$ at all and it is explicitly given by those series $a = \sum_n \tilde{a}_n z^n/n!$ with

$$\|a\|_\epsilon = \sup_n \frac{|\tilde{a}_n|}{n!^\epsilon} < \infty \tag{2.10}$$

for all $\epsilon > 0$. With other words, the Taylor coefficients of a have *sub-factorial* growth. In fact, the (quite complicated) seminorms for the second version of the construction can be shown to lead to the same topology as the one described by the seminorms (2.10). This example can be extended to the case of Laurent polynomials $\mathbb{C}[z^{-1}, z]$ where one gets a similar completion: the Laurent coefficients are required to have sub-factorial growth, too. This is again a strongly nuclear Köthe space with the monomials being an absolute Schauder basis.

Example 2.4 (matrices) The next example starts with the noncommutative non-unital algebra $\mathrm{M}_\infty(\mathbb{C})$ of matrices with only finitely many non-zero entries. Here we have an obvious basis to start with: the elementary matrices E_{ij} where we have a 1 at the (i, j)-th position and zeros elsewhere. Taking them as a first try, the construction fails for this basis. A rescaling to $\hat{E}_{ij} = \frac{1}{\sqrt{i!j!}} E_{ij}$ makes the construction work. The completion is a Fréchet algebra containing matrices with entries having sub-factorial growth. Moreover, the trace functional is continuous and hence the completion can be seen as a particular subalgebra of the trace class operators. The alternative we investigated starts with the basis $\tilde{E}_{ij} = \frac{1}{ij} E_{ij}$ instead. Also in this case the construction works and we get a larger completion: the completion contains at least those infinite matrices A whose coefficients $\tilde{A}_{ij}$ with respect to $\tilde{E}_{ij}$ are bounded. The trace functional is not continuous and the completion does contain operators which are not of trace class.

Example 2.5 (group algebras) A last class of examples is given by the group algebra $\mathbb{C}[G]$ of a finitely generated group. A first try suggests to take the group elements $g \in G$ as the basis, but then the construction fails. A rescaling can be done by means of the length $L(g)$ of $g \in G$ with respect to a fixed choice of generators of the group. Then the basis $e_g = 1/(L(g)!)^\epsilon$ makes the construction work. Here $\epsilon > 0$ is a parameter of the construction. The resulting completion gives a Fréchet algebra inside $\ell^1(G)$. Many properties can be determined explicitly. A further study of this group algebra is work in progress [Cahen M, Gutt S and Waldmann S].

As a conclusion one can say that the procedure described in Theorem 2.1 and in its alternative version gives a rather general framework to produce interesting Fréchet algebras, even though for the particular example there is typically still some work to be done in order to identify the precise topology.

3. The Wick star product on $\mathbb{C}^n$

We come now to a first non-trivial example from deformation quantization: on $\mathbb{C}^n$ one considers the Wick star product

$$f \star_{\mathrm{Wick}} g = \sum_{I,J=0}^{\infty} \frac{(2\hbar)^{|I|}}{I!} g^{IJ} \frac{\partial^{|I|} f}{\partial z^I} \frac{\partial^{|J|} g}{\partial \overline{z}^J} \tag{3.1}$$

where $f, g \in \mathcal{A}_{\mathrm{Wick}} = \mathbb{C}[z^1, \ldots, z^n, \overline{z}^1, \ldots, \overline{z}^n]$ are polynomials, $I, J \in \mathbb{N}_0^n$ are multiindices, and $g = \mathrm{diag}(+1, \ldots, +1)$ is the canonical Kähler structure on $\mathbb{C}^n$. Clearly, the Wick star product converges for polynomials as the series is actually a finite sum. We take now the following basis

$$e_{I,J} = \frac{z^I \overline{z}^J}{(2\hbar)^{|I|+|J|} I! J!} \tag{3.2}$$

to start our construction. Here $I, J \in \mathbb{N}_0^n$ are as before and $a \in \mathcal{A}_{\mathrm{Wick}}$ will be written as $a = \sum_{I,J} a_{IJ} e_{IJ}$. The choice for this normalization of the basis is motivated on one hand by physics as the $(2\hbar)^{|I|+|J|}$ in the denominator will ensure that the basis is "dimensionless", and on the other hand by the Taylor formula. It is now a quite involved computation of all the relevant structure constants to show that the first version of the construction actually works. The resulting completion can be described in a rather simple way: first we introduce the following seminorms measuring again a sub-factorial growth of the coefficients a_{IJ}, i.e

$$\|a\|_\epsilon = \sup_{I,J} \frac{|a_{IJ}|}{(|I| + |J|)!^\epsilon} \tag{3.3}$$

Then the completion can be viewed as the subset of all formal series $a = \sum_{IJ} a_{IJ} e_{IJ}$ with finite seminorms $\|a\|_\epsilon < \infty$. In fact, we have the following theorem:

Theorem 3.1 *The completion $\widehat{\mathcal{A}}_{\mathrm{Wick}}$ of the Wick star product algebra $\mathcal{A}_{\mathrm{Wick}} = (\mathbb{C}[z^1, \ldots, \overline{z}^n], \star_{\mathrm{Wick}})$ with respect to the construction of Theorem 2.1 is a strongly nuclear Fréchet algebra with absolute Schauder basis $\{e_{IJ}\}_{I,J \in \mathbb{N}_0^n}$ consisting of real-analytic functions on $\mathbb{R}^{2n}$ with Taylor coefficients a_{IJ} having sub-factorial growth. An equivalent system of seminorms is given by $\{\|\cdot\|_\epsilon\}_{\epsilon>0}$.*

Remark 3.2 The (completed) Wick star product algebra has been studied in detail in [2, 15] resulting in many additional properties some of which we list here:

i.) Having an absolute Schauder basis, the completion is a Köthe sequence space. This can also be seen directly, as the isomorphism to a Köthe sequence space is given by mapping a to the sequence of Taylor coefficients $\{a_{IJ}\}_{I,J \in \mathbb{N}_0^n}$. This gives also the proof of the strong nuclearity thanks to the Grothendieck-Pietsch criterion.

ii.) The Fréchet topology is independent of $\hbar$. Moreover, the Wick star product $f \star_{\mathrm{Wick}} g$ of two functions in the completion $\widehat{\mathcal{A}}_{\mathrm{Wick}}$ depends holomorphically on $\hbar$. In fact, the series expression (3.1) converges absolutely in the Fréchet topology. This shows that we have a holomorphic (even an entire) deformation in the sense of [16].

iii.) The $\star_{\mathrm{Wick}}$-exponential of linear polynomials converges absolutely. This is non-trivial since the completion has *no* general entire functional calculus, it is *not* locally multiplicatively convex. In fact, due to the presence of elements obeying the canonical commutation relations this can not be the case.

iv.) The additive group $(\mathbb{R}^{2n}, +)$ acts on the completion by continuous algebra automorphism in a smooth way. In fact, the smooth topology with respect to this action coincides with the original topology. Moreover, the action is inner by means of the $\star_{\mathrm{Wick}}$-exponentials of linear polynomials.

v.) For $\hbar > 0$ the δ-functionals are continuous positive linear functionals and the corresponding GNS construction with respect to δ_0 is the usual Bargmann-Fock representation in normal ordering of the z's and $\overline{z}$'s.

vi.) Since all evaluation functionals at points in $\mathbb{R}^{2n}$ are continuous, we can view the elements of the completion as functions. They are real-analytic and have a holomorphic/anti-holomorphic extension to $\mathbb{C}^n \times \mathbb{C}^n$. The Fréchet topology of $\widehat{\mathcal{A}}_{\mathrm{Wick}}$ is then finer as the Fréchet topology of locally uniform convergence of the extensions on $\mathbb{C}^n \times \mathbb{C}^n$.

As a last remark we would like to mention that the construction of the topology by means of Theorem 2.1 of course looks very ad-hoc. However, *a posteriori*, we showed in [17] that there is a much more conceptual approach to the Wick algebra, based on an arbitrary continuous bilinear form on a locally convex space: this enlarges the construction to infinite dimensions and proves some very nice functorial properties.

4. The star product on the Poincaré disk

We come now to the main example for a convergent star product beyond the flat situation. We base the construction on the results from [18, 19] where a star product on $\mathbb{CP}^n$ and on the

Poincaré disk $\mathbb{D}_n$ was constructed by means of a phase space reduction. We have to recall some parts of this construction in order to introduce some notation.

Consider $\mathbb{C}^{n+1}$ with the pseudo Kähler metric defined by $g = \mathrm{diag}(-1, +1, \ldots, +1)$ and define the Kähler potential

$$y(z) = -g(z, \overline{z}) = |z^0|^2 - |z^1|^2 - \cdots - |z^n|^2 \tag{4.1}$$

Then one considers the open cone

$$C^+_{n+1} = \{z \in \mathbb{C}^{n+1} \mid y(z) > 0\} \tag{4.2}$$

We use y as a momentum map for the canonical U(1)-action on $\mathbb{C}^{n+1}$ to construct the Marsden-Weinstein quotient for the momentum value 1. Thus let $C = \{z \in C^+_{n+1} \mid y(z) = 1\}$ be the constraint surface and let $M_{\mathrm{red}} = C/\mathrm{U}(1)$ be the reduced phase space. It turns out that this is the Poincaré disk $M_{\mathrm{red}} = \mathbb{D}_n$ with the symplectic structure being the one from its canonical Kähler structure of constant negative holomorphic curvature. Alternatively, we can view $\mathbb{D}_n$ as an open subset of $\mathbb{CP}^n$ by $\mathbb{D}_n = \{[z] \in \mathbb{CP}^n \mid z \in C^+_{n+1}\}$ where $[z]$ denotes the complex line through the point $z \in \mathbb{C}^{n+1} \setminus \{0\}$. This gives the correct complex structure on $\mathbb{D}_n$ but of course not the correct Kähler structure. Note that in this particular situation we also have a direct map going from the big phase space C^+_{n+1} to the reduced one, i.e the holomorphic quotient map $\pi\colon C^+_{n+1} \longrightarrow \mathbb{D}_n$ by mapping z to $[z]$.

Using the metric g on C^+_{n+1} this gives us a Wick star product $\star_{\mathrm{Wick}}$ by the very same formula (3.1), now of course involving some additional signs due to the changed g. This Wick star product inherits the correct $\mathrm{SU}(1, n)$ symmetry from the pseudo Kähler structure of C^+_{n+1}. The basic idea is to modify the Wick star product on C^+_{n+1} in a suitable way such that it passes to the quotient $\mathbb{D}_n$.

To this end, one first observes that the U(1)-invariant functions form a subalgebra with respect to the Wick star product. Next, one calls a function $R \in \mathscr{C}^\infty(C^+_{n+1})$ *radial* if it depends only on the "radius" y, i.e if there is a smooth function $\varrho \in \mathscr{C}^\infty(\mathbb{R}^+)$ with $R = \varrho \circ y$. Clearly, a radial function is U(1)-invariant. Moreover, a function $F \in \mathscr{C}^\infty(C^+_{n+1})$ is called *homogeneous* if it is of the form $F = \pi^* u$ with $u \in \mathscr{C}^\infty(\mathbb{D}_n)$. Finally, we introduce the global vector field

$$\frac{\partial}{\partial y} = \frac{1}{2y} \sum_{k=0}^{n} \left(z^k \frac{\partial}{\partial z^k} + \overline{z}^k \frac{\partial}{\partial \overline{z}^k} \right) \tag{4.3}$$

which on radial functions indeed acts as the partial derivative by y. A homogeneous function F can be characterized infinitesimally by $\frac{\partial}{\partial y} F = 0 = X_y F$, where X_y is the Hamiltonian vector field of y.

A simple computation shows that for a radial function R and a U(1)-invariant function F one has

$$R \star_{\mathrm{Wick}} F = F \star_{\mathrm{Wick}} R = \sum_{r=0}^{\infty} \frac{(2\hbar)^r}{r!} y^r \frac{\partial^r R}{\partial y^r} \frac{\partial^r F}{\partial y^r} \tag{4.4}$$

This shows that inside the U(1)-invariant functions the radial ones are central. However, the product is not the pointwise product. Hence a restriction to the constraint surface C is not possible directly: the pointwise ideal generated by $y - 1$ is not the $\star_{\mathrm{Wick}}$-ideal. Thus the Wick star product does not restrict directly to $\mathbb{D}_n$. Now this flaw can be cured by passing to an equivalent star product $\tilde{\star}$ such that for this new star product the U(1)-invariant functions are still a subalgebra and the $\tilde{\star}$-product with a radial function becomes the pointwise product. There is an explicit formula for the equivalence transformation and the resulting formula for $\tilde{\star}$ is

$$F \tilde{\star} G = \sum_{r=0}^{\infty} \frac{1}{r!} \left(\frac{2\hbar}{y}\right)^r \prod_{k=1}^{r} \left(1 - k\frac{2\hbar}{y}\right)^{-1} y^r g^{i_1 j_1} \cdots g^{i_r j_r} \frac{\partial^r F}{\partial z^{i_1} \cdots \partial z^{i_r}} \frac{\partial^r G}{\partial \overline{z}^{j_1} \cdots \partial \overline{z}^{j_r}} \tag{4.5}$$

for $F, G \in \mathscr{C}^\infty(C^+_{n+1})^{\mathrm{U}(1)}[[\hbar]]$ being U(1)-invariant. Now the radial functions behave like scalars and hence the $\tilde{\star}$-ideal generated by $y - 1$ coincides with the pointwise ideal, i.e the vanishing ideal of C. Hence we can induce a star product $\star_{\mathbb{D}_n}$ on the disk by the identification

$$\mathscr{C}^\infty(\mathbb{D}) \cong \mathscr{C}^\infty(C^+_{n+1})^{\mathrm{U}(1)} / (y-1)\mathscr{C}^\infty(C^+_{n+1})^{\mathrm{U}(1)} \tag{4.6}$$

This gives the very explicit formula for $u, v \in \mathscr{C}^\infty(\mathbb{D}_n)$: we take $F = \pi^* u$ and $G = \pi^* v$, compute their star product upstairs according to (4.5) and set $y = 1$ afterwards. Then the resulting function on the constraint surface C is the pull-back of $u \star_{\mathbb{D}_n} v$.

We use now this star product $\star_{\mathbb{D}_n}$ on the disk to implement our construction of a Fréchet topology according to Theorem 2.1.

5. The basis and the algebra on the disk

The first step is of course to find a subalgebra of $\mathscr{C}^\infty(C^+_{n+1})^{\mathrm{U}(1)}$ with a countable vector space basis such that the product $\tilde{\star}$ is well-defined for some $\hbar \neq 0$. Then we can implement our construction, still on the big phase space, and pass to the quotient by the same identification as in (4.6) after having established the topological context.

We start with the U(1)-invariant monomials

$$\mathsf{e}_{P,Q,\alpha} = \left(z^0\right)^{\alpha - |P|} \left(\overline{z}^0\right)^{\alpha - |Q|} z^P \overline{z}^Q \tag{5.1}$$

where $P, Q \in \mathbb{N}_0^n$ are multiindices for the last n coordinates and $\alpha \geq |P|, |Q|$. We call such (P, Q, α) an *index triple.* Then we consider the functions

$$\mathsf{f}_{P,Q,\alpha} = \frac{1}{P!(\alpha - |P|)!Q!(\alpha - |Q|)!} \left(\frac{y}{2\hbar}\right)_\alpha \frac{\mathsf{e}_{P,Q,\alpha}}{y^\alpha} \tag{5.2}$$

where $(x)_\alpha = x(x+1)\cdots(x+\alpha-1)$ is the Pochhammer symbol. Here and in the following $2\hbar$ will be a complex number from the set

$$2\hbar \in \mathbb{C} \setminus \{0, -1, -1/2, -1/3, \ldots\} \tag{5.3}$$

which guarantees that the Pochhammer symbol $\left(\frac{y}{2\hbar}\right)_\alpha$ will be different from zero on the constraint surface for all $\alpha \in \mathbb{N}_0$.

Proposition 5.1 *The functions $\{\mathsf{f}_{P,Q,\alpha}\}_{P,Q,\alpha}$ form a basis of a $\tilde{\star}$-subalgebra $\mathcal{A}_\hbar(C^n_{n+1})$ where the star product is convergent. The structure constants can be computed explicitly and one obtains $\mathcal{A}_\hbar(C^n_{n+1}) = (\mathcal{A}_\hbar(C^n_{n+1}))_{\mathrm{nice}}$.*

While the computation of the structure constants and the relevant estimates to guarantee $\mathcal{A}_\hbar(C^n_{n+1}) = (\mathcal{A}_\hbar(C^n_{n+1}))_{\mathrm{nice}}$ according to Theorem 2.1 are quite involved, the above first version of our construction is not yet sufficient. The reason is that the evaluation functionals at points $w \in C^+_{n+1}$ are *not* continuous. In fact, one can show that the topology given by Theorem 2.1 on the subalgebra spanned by the basis $\{\mathsf{f}_{P,Q,\alpha}\}_{P,Q,\alpha}$ is just the Cartesian product topology with respect to the index triples. Thus it is much too coarse to interpret the elements of the completion as functions on C^+_{n+1}.

Journal of Physics: Conference Series **532** (2014) 012027 doi:10.1088/1742-6596/532/1/012027

The second version of our general construction provides a way out of this: we enforce the continuity of all the δ-functionals at points in C^{+}_{n+1} by considering the set $\Omega = \{\delta_w \mid w \in C^{+}_{n+1}\}$. It is now again a nontrivial estimation to see that the procedure works and yields a reasonable Fréchet topology. Indeed, the result is again very simple. Consider the seminorms

$$\|a\|_{\epsilon} = \sup_{P,Q,\alpha} |a_{P,Q,\alpha}|/\alpha!^{\epsilon} \tag{5.4}$$

where the supremum is taken over all index triples and the $a_{P,Q,\alpha}$ are the coefficients of a with respect to the basis $\{\mathsf{f}_{P,Q,\alpha}\}_{P,Q,\alpha}$. Then one has the following result:

Theorem 5.2 *For Ω as above we have $\mathcal{A}_{\hbar}(C^{n}_{n+1}) = (\mathcal{A}_{\hbar}(C^{n}_{n+1}))_{\Omega\text{-nice}}$ and the induced topology can equivalently be described by the seminorms $\{\|\cdot\|_{\epsilon}\}_{\epsilon>0}$. The completion with respect to this topology yields a Fréchet algebra such that every evaluation functional δ_w for $w \in C^{+}_{n+1}$ is continuous. The completion is a strongly nuclear Köthe space with $\{\mathsf{f}_{P,Q,\alpha}\}_{P,Q,\alpha}$ being an absolute Schauder basis, consisting of a subalgebra of real-analytic functions on C^{+}_{n+1}.*

The last step is now to pass to the disk: as before in the formal setting we can simply quotient by the $\tilde{\star}$-ideal generated by the function $y-1$ which coincides with the vanishing ideal of the constraint surface C inside $\mathcal{A}_{\hbar}(C^{n}_{n+1})$. This will only be possible for the allowed values of $\hbar$ according to (5.3). Since the δ-functionals are continuous the ideal is closed and hence we obtain a Fréchet algebra $\mathcal{A}_{\hbar}(\mathbb{D}_n)$ as quotient. While this works by abstract arguments we also have a more explicit description of the elements of $\mathcal{A}_{\hbar}(\mathbb{D}_n)$ as functions on the disk $\mathbb{D}_n$.

In fact, we have again a vector space basis becoming a absolute Schauder basis after completion given by the functions

$$\mathsf{f}_{P,Q}(v) = [\mathsf{f}_{P,Q,\alpha}](v) = \frac{1}{P!(\alpha-|P|)!Q!(\alpha-|Q|)!}\left(\frac{1}{2\hbar}\right)_{\alpha}\frac{v\overline{v}^{Q}}{(1-|v|^2)^{\alpha}} \tag{5.5}$$

where $P, Q \in \mathbb{N}_0^n$ are multiindices, $\alpha = \max(|P|,|Q|)$, and $v = \frac{z}{z^0} \in \mathbb{D}_n$ is a point in the disk corresponding to the complex line $[z]$. These functions form a linearly independent set and we write $[a] = \sum_{P,Q} a_{P,Q}\mathsf{f}_{P,Q}$ for an element in the span of them, where $a \in \mathcal{A}_{\hbar}(C^{n}_{n+1})$ is a representative upstairs. Then we define the seminorms

$$\|[a]\|_{\epsilon} = \sup_{P,Q}\frac{|a_{P,Q}|}{\max(|P|,|Q|)!^{\epsilon}} \tag{5.6}$$

again controlling a sub-factorial growth of the "Taylor coefficients" $a_{P,Q}$. The final result is then the following:

Theorem 5.3 *The algebra $\mathcal{A}_{\hbar}(\mathbb{D}_n)$ consists of those real-analytic functions $[a]$ on the disk with $\|[a]\|_{\epsilon} < \infty$ for all $\epsilon > 0$. It is a strongly nuclear Köthe space with absolute Schauder basis $\{\mathsf{f}_{P,Q}\}_{P,Q}$ and all evaluation functionals at $v \in \mathbb{D}_n$ are continuous.*

Remark 5.4 We list some further properties of the algebra on the disk:

i.) The class of functions we obtain is a subalgebra of those real-analytic functions on $\mathbb{D}_n$ which allow for a holomorphic/anti-holomorphic extension to $\mathbb{D}_n \times \mathbb{D}_n$ and not just to a small open neighbourhood of the diagonal. Then the topology of $\mathcal{A}_{\hbar}(\mathbb{D}_n)$ is (strictly) finer than the Fréchet topology of locally uniform convergence on $\mathbb{D}_n \times \mathbb{D}_n$ of the extensions.

ii.) The symmetry group $\mathrm{SU}(1,n)$ of the Kähler structure of $\mathbb{D}_n$ acts via pull-backs also on the algebra $\mathcal{A}_{\hbar}(\mathbb{D}_n)$ by continuous automorphisms. Moreover, the action is smooth and the smooth topology coincides with the original topology of $\mathcal{A}_{\hbar}(\mathbb{D}_n)$. The classical momentum map for the corresponding Lie algebra action of $\mathfrak{su}(1, n)$ provides also a *quantum momentum map*, i.e the $\star_{\mathbb{D}_n}$-commutators with the components of the classical momentum map generate the $\mathfrak{su}(1, n)$-action.

iii.) If $\hbar$ is real then the complex conjugation provides a continuous *-involution for the algebra $\mathcal{A}_\hbar(\mathbb{D}_n)$. The $\mathrm{SU}(1, n)$-symmetry acts by *-automorphisms in this case.

iv.) Let $\hbar$ be in the set (5.3) of allowed values. Then the Fréchet space $\mathcal{A}_\hbar(\mathbb{D}_n)$ does *not* depend on $\hbar$. This is not true upstairs: here even the vector spaces $\mathcal{A}_\hbar(C^n_{n+1})$ viewed as subspaces of $\mathscr{C}^\infty(C^+_{n+1})$ do depend on $\hbar$. Since the topology of $\mathcal{A}_\hbar(\mathbb{D}_n)$ is independent of $\hbar$, the following statement makes sense: for every two functions $a, b \in \mathcal{A}_\hbar(\mathbb{D}_n)$ the map $\hbar \mapsto a \star_{\mathbb{D}_n} b$ is holomorphic for the allowed values of $\hbar$. Thus we have a holomorphic deformation of the functions of the disk in the sense of [16]. Note however, that the classical case $\hbar = 0$ is *not* in the domain where we have a holomorphic dependence. This shows that the singularities for the classical limit $\hbar \longrightarrow 0$ are much more subtle than in the flat case discussed in Remark 3.2.

v.) For $\hbar > 0$ the evaluation functionals δ_v at points $v \in \mathbb{D}_n$ are continuous positive functionals. Thanks to the transitive action of $\mathrm{SU}(1, n)$ it will be sufficient to consider one of them, say δ_0. The Gel'fand ideal can be computed explicitly. It has a complementary closed subspace

$$\mathfrak{D}_\hbar = \left\{ \psi = \sum\nolimits_Q \psi_Q \mathsf{f}_{0,Q} \;\middle|\; (\psi_Q)_{Q \in \mathbb{N}_0} \text{ has sub-factorial growth} \right\} \subseteq \mathcal{A}_\hbar(\mathbb{D}_n) \tag{5.7}$$

in $\mathcal{A}_\hbar(\mathbb{D}_n)$ which can be identified with the GNS pre-Hilbert space. The inner product as well as the GNS representation can be computed explicitly. It is an ongoing project to identify the corresponding Hilbert space with more familiar Hilbert spaces from coherent states quantizations like in [20].

Acknowledgments
It is a pleasure to thank the organizers of the *QQQ* conference *3Quantum: Algebra Geometry Information* (Tallinn, July 2012) for their kind invitation and the stimulating conference.

References

[1] Bayen F, Flato M, Frønsdal C, Lichnerowicz A and Sternheimer D 1978 *Ann. Phys.* **111** 61
[2] Beiser S and Waldmann S 2014 *J. Reine Angew. Math.* **688** 147
[3] Waldmann S 2007 *Poisson-Geometrie und Deformationsquantisierung. Eine Einführung* (Springer-Verlag)
[4] DeWilde M and Lecomte P B A 1983 *Lett. Math. Phys.* **7** 487
[5] Fedosov B V 1986 *Sov. Phys. Dokl.* **31** 877
[6] Kontsevich M 2003 *Lett. Math. Phys.* **66** 157
[7] Waldmann S 2005 *Rev. Math. Phys.* **17** 15
[8] Rieffel M A 1989 *Comm. Math. Phys.* **122** 531
[9] Rieffel M A 1993 *Mem. Amer. Math. Soc.* **106** 93 pp
[10] Natsume T, Nest R and Peter I 2003 *Lett. Math. Phys.* **66** 73
[11] Bieliavsky P, Detournay S and Spindel P 2009 *Comm. Math. Phys.* **289** 529
[12] Bieliavsky P, Bonneau P and Maeda Y 2007 *Pacific J. Math.* **230** 41
[13] Bieliavsky P 2002 *J. Symplectic Geometry* **1** 269
[14] Cuntz J 2005 *K-Theory* **35** 93
[15] Beiser S, Römer H and Waldmann S 2007 *Comm. Math. Phys.* **272** 25
[16] Pflaum M J and Schottenloher M 1998 *J. Geom. Phys.* **28** 31
[17] Waldmann S 2014 *J. Geom. Phys.* **81** 10
[18] Bordemann M, Brischle M, Emmrich C and Waldmann S 1996 *Lett. Math. Phys.* **36** 357
[19] Bordemann M, Brischle M, Emmrich C and Waldmann S 1996 *J. Math. Phys.* **37** 6311
[20] Cahen M, Gutt S and Rawnsley J 1994 *Lett. Math. Phys.* **30** 291

3Quantum: Algebra Geometry Information (QQQ Conference 2012) IOP Publishing
Journal of Physics: Conference Series **532** (2014) 012028 doi:10.1088/1742-6596/532/1/012028

Permutations of cubical arrays

G P Wene

Department of Mathematics, University of Texas at San Antonio
6900 North Loop 1604 West, San Antonio, Texas 78249-0697, U.S.A.

Abstract. The structure constants of an algebra determine a cube called the cubical array associated with the algebra. The permuted indices of the cubical array associated with a finite semifield generate new division algebras. We do not not require that the algebra be finite and ask "Is it possible to choose a basis for the algebra such any permutation of the indices of the structure constants leaves the algebra unchanged?" What are the associated algebras? Author shows that the property "weakly quadratic" is invariant under all permutations of the indices of the corresponding cubical array and presents two algebras for which the cubical array is invariant under all permutations of the indices.

1. Introduction

The structure constants Γ_{ijk} of an n-dimensional algebra $\mathfrak{A}$ over a field F with basis $\{e_0, e_1, e_2, \ldots, e_{n-1}\}$ can be defined by

$$e_i e_j = \sum_{k=0}^{n-1} \Gamma_{ijk} e_k$$

These structure constants Γ_{ijk} determine a $n \times n \times n$ cube A called the cubical array (Knuth [1]) associated with the algebra $\mathfrak{A}$. Let σ be a permutation of the set of indices, $\{i, j, k\}$. Then A^σ will represent the 3-cube A with subscripts permuted by σ; i. e., the $k\sigma$th subscript of A^σ is the kth subscript of A. If $A = (A_{ijk})$, then if $B = A^{(123)} = (B_{ijk})$ we have $B_{ijk} = A_{kij}$.

The algebra corresponding to $A^{(12)}$ is called the opposite algebra and is denoted by $\mathfrak{A}^{op}$. Pairing each algebra with its opposite gives

Table 1. The algebra and opposite.

$\mathfrak{A}$	$\mathfrak{A}^{op}$
A	$A^{(12)}$
$A^{(123)}$	$A^{(23)} = A^{(123)(12)}$
$A^{(132)}$	$A^{(13)} = A^{(132)(12)}$

If the algebra $\mathfrak{A}$ is associative, then so is the algebra $\mathfrak{A}^{op}$.

The group of permutations on a three element set is usually denoted by S_3 and is generated by the permutations (12) and (13). The cubical array was introduced to study finite geometries;

if π is a finite projective plane coordinatized by the division ring S and A is the cubical array associated with S, then the six permutations of the indices of A determine a series of at most six non-isotopic planes. The permutations (12) and (13) have the geometrical interpretations: the dual of the of the semifield is generated by the action of (12) on the associated cubical array and the transpose is generated by (13). The same series of permutation can be applied to the cubical array corresponding to any arbitrary division ring and will result in division rings.

Kantor [2] proves that the transposition and dualization of a commutative semifield is a sympletic semifield and, conversely, the dualization and transposition of a sympletic semifield is a commutative semifield. Ball and Brown [3] constructed potentially six non-isotopic semifields from a Cohen-Ganley pair of functions (f, g). Ball, Ebert and Lavrauw [4] use the cubical array to compute the twelve isotopy classes of semifields associated with a semifield of rank 2 over its left nucleus. Ball and Lavrauw [5] examine the series of division rings arising from semifields two dimensional over a weak nucleus and Kantor [6] notes that the planes determined by the Hughes Kleinfeld semifields [7] have exactly one "image" under the action of S_3.

We seek algebraic characterizations of the sequence of algebras obtained by cubical permutations.

There are two approaches: one addresses properties of the algebra; the second specifies the non-zero elements of the array.

Lemma 1.1. *Let the finite dimensional algebra $\mathfrak{A}$ be the algebra direct sum of the algebras $\mathfrak{A}_1$ and $\mathfrak{A}_2$. The cube A^σ determines a direct sum for all $\sigma \in S_3$.*

Proof. Let $\mathfrak{B}_1$ and $\mathfrak{B}_2$ be basies for $\mathfrak{A}_1$ and $\mathfrak{A}_2$ respectively. Construct a basis for $\mathfrak{A}_1 \oplus \mathfrak{A}_2$ as $\{e_i, e_j : e_i \in \mathfrak{B}_1, e_j \in \mathfrak{B}_2\}$. Then $\Gamma_{ijk} = 0$ when i, j, k are not all contained in the same $\mathfrak{B}_i$, $i = 1$ or 2. □

We wish to find a property that is invariant under the action of S_3 shared by a large class of algebras. Such a property is a generalization of the quadratic algebras.

2. Weakly quadratic algebras

A *quadratic algebra* $\mathfrak{A}$ over a field F of characteristic not 2, is a unital algebra in which every $x \in \mathfrak{A}$ satisfies the condition $x^2 \in span\,\{1, x\}$. A generalization of the quadratic algebras are the *weakly quadratic algebra.*

A finite dimensional algebra $\mathfrak{A}$ with a basis $\{e_0, e_1, \ldots, e_{n-1}\}$, such that e_0 is idempotent, is called *weakly quadratic* (relative to that basis) if for any $x \in \mathfrak{A}$ we have $x^2 \in span\{e_0, x\}$ and

(i) $\Gamma_{0ik} = a\delta_{ik}\,,\quad i \neq 1$ for some $a \in F$
(ii) $\Gamma_{i0k} = b\delta_{ik}\,,\quad i \neq 1$ for some $b \in F$
(iii) $\Gamma_{iik} = c\delta_{k0}\,,\quad c \in F$
(iv) $\Gamma_{ijk} = -\Gamma_{jik} = \Gamma_{jki}\,,$ where no two of i, j, k are the same and none is equal to 0

Example 2.1. Table 3 gives the multiplication constants for a weakly quadratic algebra.

Table 2. A weakly quadratic algebra.

	e	I	J	K
e	e	aI	aJ	aK
I	bI	ce	K	$-J$
J	bJ	$-K$	ce	I
K	bK	J	$-I$	ce

If $a = b = 1$ and $c = -1$ the table defines the real quaternion division ring.

Theorem 2.2. *If the algebra $\mathfrak{A}$ is weakly quadratic then so is $\mathfrak{A}^\sigma$ for all $\sigma \in S_3$.*

Proof. Clearly if $\mathfrak{A}$ is weakly quadratic then so is $\mathfrak{A}^{(12)}$. The structure constants Φ_{ijk} of $\mathfrak{A}^{(13)}$ are given in table 3. □

Table 3. Structure constants for A and $A^{(13)}$.

Γ	Φ
$\Gamma_{0jk} = a\delta_{jk},\ j \neq 1$ for $a \in F$	$\Phi_{kj0} = a\delta_{jk},\ j \neq 1$ for $A \in F$
$\Gamma_{i0k} = b\delta_{ik},\ i \neq 1$ for $b \in F$	$\Phi_{k0i} = b\delta_{ik},\ i \neq 1$ for $b \in F$
$\Gamma_{iik} = c\delta_{ik},\ c \in F$	$\Phi_{kii} = c\delta_{ik},\ c \in F$
$\Gamma_{ijk} = -\Gamma_{jik}$	$\Phi_{kji} = -\Phi_{kij}$
$\Gamma_{jik} = -\Gamma_{jki}$	$\Phi_{kij} = -\Phi_{ikj}$
$\Gamma_{iij} = \delta_{0j}$	$\Phi_{iij} = \delta_{0j}$

Example 2.3. The Jordan algebra $\mathfrak{J} = F1 \oplus \mathfrak{B}$ of the symmetric bilinear form q on $\mathfrak{B}$ is a Type D Jordan algebra (Jacobson [8]). Albert [9] showed any Type D algebra has a basis $\{e_0, e_1, \ldots, e_{n-1}\}$ such that

$$e_0 e_i = e_i, \quad e_i e_j = \alpha_i \delta_{ij} e_0$$

If $\alpha_i = 1$ non-zero structure constants are

$$\Gamma_{iik} = \Gamma_{kii} = \Gamma_{iki} = \delta_{k0}$$

We note that the associated cubical array is invariant under all $\sigma \in S_3$.

Definition 2.4 (Osborn [10]). If $\mathfrak{A}$ is an F algebra and η an element of F not equal to $\frac{1}{2}$. Let $\mathfrak{A}^*$ denote the vector space $\mathfrak{A}$ under a new multiplication $*$ defined in terms of the old multiplication by

$$a * b = \eta ab + (1 - \eta) ba$$

The algebra $\mathfrak{A}^*$ is said to be strongly quasi-equivalent to $\mathfrak{A}$.

Theorem 2.5. *If the algebra $\mathfrak{A}$ is weakly quadratic then so is $\mathfrak{A}^*$.*

Proof. We must show that if the multiplication constants of the weakly quadratic algebra $\mathfrak{A}$ satisfy the conditions 1 through 4 then the multiplication constants Φ_{ijk} of $\mathfrak{A}^*$ satisfy analogous conditions

$$e_0 * e_i = \eta e_0 e_i + (1 - \eta) e_i e_0 = [\eta a + (1 - \eta) b] e_i, \quad \Phi_{0ii} = [\eta a + (1 - \eta) b]$$

Similarly,

$$\Phi_{i0i} = [\eta b + (1 - \eta) a]$$

Table 4. A quasi-equivalent algebra.

	e	I	J	K
e	e	$(6a-5b)\,I$	$(6a-5b)\,J$	$(6a-5b)\,K$
I	$(6b-5a)\,I$	ce	$11K$	$-11J$
J	$(6b-5a)\,J$	$-11K$	ce	$11I$
K	$(6b-5a)\,K$	$11J$	$-11I$	ce

If $i \neq 0$, then $e_i * e_i = e_i e_i = ce_0$ and $\Phi_{iik} = c\delta_{0k}$. If $i = 0$, then $\Phi_{00k} = \delta_{0k}$ and

$$e_i * e_j = \eta\Gamma_{ijk}e_k + (1-\eta)\Gamma_{jik}e_k = \eta\Gamma_{ijk}e_k - (1-\eta)\Gamma_{ijk}e_k = (2\eta-1)\Gamma_{ijk}e_k, \quad \Phi_{ijk} = (2\eta-1)\Gamma_{ijk}$$

Similarly

$$\Phi_{jik} = (2\eta-1)\Gamma_{jik}$$

From the above relations we have

$$\Phi_{ijk} = (2\eta-1)\Gamma_{ijk} = -(2\eta-1)\Gamma_{ikj} = -\Phi_{ijk}$$

Note that it is sufficient to prove that $\mathfrak{A}^\sigma$ is quadratic for $\sigma = (12)$ and $\sigma = (13)$, since these two permutations generate S_3. □

Example 2.6. Let $\mathfrak{A}$ be the algebra of Example 3. Let $\eta = 6$ The multiplication in $\mathfrak{A}^*$ is given in table 4.

3. Algebras invariant under S_3

Knuth [1] was interested in the cubical arrays associated with finite semifields, that is, finite division rings with unit element.

Oehmke [11] showed that any finite semifield S has a cyclic basis $\{1, x, x^2, \dots, x^n\}$ for some $x \in S$. Using this basis and $n \geq 2$, the structure constants become $\Gamma_{0ik} = \Gamma_{i0k} = \delta_{ik}$ and $\Gamma_{11k} = \delta_{k2}$. If the cubical array is to be invariant under the action of S_3, we must have $\Gamma_{110} = \delta_{11}$ and $\Gamma_{110} = \delta_{02}$. Hence if the dimension is greater than or equal to 2, then cubical array is not invariant under S_3. If the dimension is one or two, the semifields is a finite field.

We examine three patterns of non-zero entries in the cubical array. The first has non-zero elements given by

$$\Gamma_{iii} = \alpha_i \neq 0, \quad i = 1, \dots, n$$

If the algebra $\mathfrak{A}$ is a finite sum of copies of the field F, then $\mathfrak{A}$ has a basis $\{e_i : i = 0, 1, .., n-1\}$ of pairwise orthogonal idempotents.

We assume that the algebra $\mathfrak{A}$ has a basis $\{e_0, e_1, \dots, e_n, e_{n+1}, \dots, e_m\}$.

Theorem 3.1. *$\mathfrak{A}$ is an n-dimensional algebra whose set of nonzero Γ_{ijk}s invariant under S_3 is the set $\{\Gamma_{iii} = \alpha_i \neq 0 : i = 0, 1, \dots, n-1\}$ if and only if the algebra $\mathfrak{A}$ is a direct sum of fields.*

Proof. $\Gamma_{iii} = \alpha_i \neq 0$ says that for each element e_i of the basis, $e_i^2 = \alpha_i e_i \neq 0$. The set $\{\widehat{b_i} = e_i/\alpha_i : i = 1, \dots, n\}$ of pairwise orthogonal idempotent will also be a basis of $\mathfrak{A}$ over F. Relative to this new basis $\Gamma_{ijk} = 0$ if $i \neq j$ for all k, and in general, $\hat{b}_i\hat{b}_j = \delta_{ij}$. Then $\mathfrak{A} = \oplus_{i=0}^{n-1} Fb_i$. □

These algebras will all be associative, commutative, and have an identity element

$$e = \hat{b}_0 + \hat{b}_2 + \cdots + \hat{b}_{n-1}$$

Corollary 3.2. *If $\mathfrak{A}$ is an m-dimensional algebra whose set of nonzero Γ_{ijk}s invariant under S_3 is the set $\{\Gamma_{iii} = \alpha_i \neq 0 : i = 0, \ldots, n \leq m\}$ then $\mathfrak{A}$ is a direct sum of a finite direct sum of copies of the field F and a zero ring.*

Our next collection of algebras will not be associative. Since there are usually many basises of a vector space, we choose one that is particularly useful.

Let $\mathfrak{A}$ denote an unital algebra $\mathfrak{A}$ with basis $\{e_0, e_1, e_2, \ldots, e_{n-1}\}$ over the field F. Relative to this basis the cubical array has elements $\Gamma_{0ii} = \Gamma_{i0i} = 1$ for all $i = 0, \ldots, n$. This time the non-zero elements are generated by the set

$$\Gamma_{ii0} = 1, \quad i = 0, \ldots, n-1$$

with the axes permuted by S_3. The set of triples

$$\{(0, i, i), (i, 0, i), (i, i, 0) : i = 0, \cdots, n-1\}$$

is invariant under the action of any $\sigma \in S_n$.

This condition and invariance under S_3 implies that the algebra is commutative, unital and, if $n \geq 2$, the algebra will not be associative since $(e_1, e_1, e_2) = e_2$. The multiplication for the algebra is given by

$$e_0 e_i = e_i = e_i e_0, \quad e_i e_j = \delta_{ij}, \quad i, j = 0, \ldots, n-1$$

In summary, we have

Theorem 3.3. *The set of nonzero entries is $\Gamma_{iij}, \Gamma_{jii}, \Gamma_{iji}$, all equal to 1, if and only if the algebra is a Type D Jordan algebra with multiplication as given above.*

By Jacobson [12], these algebras will be semisimple.

4. Conclusion

To summarize the results about the cubical arrays associated with finite dimensional algebra under actions of S_3:

(i) A finite division ring remains a finite division (not necessarily the same division ring).
(ii) A direct sum of algebras remains a direct sum (again, not necessarily the same direct sum).
(iii) A subclass of weakly quadratic algebras remains weakly quadratic.
(iv) The planes determined by the Hughes Kleinfeld semifields have exactly one "image" under the action of S_3.
(v) A finite direct sum of fields is invariant under the action of S_3.
(vi) A Type D Jordan algebra is invariant if and only if the multiplication satisfies

$$e_0 e_i = e_i e_0 = e_i, \quad 1 \leq i \leq n \quad \text{and} \quad e_i e_j = \delta_{ij} e_0$$

(vii) A quadratic algebra is invariant if and only if it is a Type D Jordan algebra with the multiplication described in 5.

There are additional patterns that can be analyzed, i.e, the pattern generated by $\{(i, i, i), (i, i, 0)\}$. A slight modification of the above arguments show that the algebras need not be finite dimensional.

References

[1] Knuth D E 1965 *J. Algebra* **2** 182
[2] Kantor W M 2003 *J. Algebra* **270** 96
[3] Ball S and Brown M R 2004 *Adv. Math* **189** 68
[4] Ball S, Ebert G and Lavrauw M 2007 *J. Algebra* **311** 117
[5] Ball S and Lavrauw M 2007 *Des. Codes Cryptogr.* **44** 63
[6] Kantor W M 2006 Finite semifields *Finite Geometries, Groups and Computation* eds A Hulpke *et al* (de Gruyter) p 103
[7] Hughes D R and Kleinfeld E 1960 *Amer. J. Math.* **82** 389
[8] Jacobson N 1968 *Structure and Representations of Jordan Algebras* (Providence, RI: AMS)
[9] Albert A A 1946 *Trans. Amer. Math. Soc.* **59** 524
[10] Osborn J M *Adv. Math.* **8** 163
[11] Oehmke R H 1980 *Proc. Amer. Math. Soc.* **79** 174
[12] Jacobson F D and Jacobson N 1949 *Trans. Amer. Math. Soc.* **65** 141

Geometrodynamics and Lorentz symmetry

D K Wise

Department Mathematik, FAU Erlangen-Nürnberg, Cauerstr. 11, 91058 Erlangen, Germany

Abstract. We study the dynamics of gauge theory and general relativity using fields of local observers, thus maintaining local Lorentz symmetry despite a space/time splitting of fields. We start with Yang-Mills theory, where observer fields are defined as normalized future-timelike vector fields. We then define observers without a fixed geometry, and find these play two related roles in general relativity: splitting fields into spatial and temporal parts, and 'breaking' gauge symmetry, effectively reducing the spacetime $\mathrm{SO}(n,1)$ connection to an observer-dependent spatial $\mathrm{SO}(n)$ connection. In both gauge theory and gravity, the observer field reduces the action to canonical form, without using gauge fixing. In the 4d gravity case, the result is a manifestly Lorentz covariant counterpart of the Ashtekar-Barbero formulation. We also explain how this leads geometrically to a picture of general relativity in terms of 'observer space' rather than spacetime - a setting where both spacetime symmetry and the dynamical description are simultaneously available.

1. Introduction: geometrodynamics and observers

Geometrodynamics is the picture of general relativity as *evolving spatial geometry*. Since Einstein's equations are not directly about space but about spacetime, the geometrodynamic viewpoint traditionally involves picking an arbitrary 'time' coordinate t on spacetime and decomposing equations into a part that describes the field configurations at fixed t and a part describing time evolution. Two well-studied approaches of this kind are the ADM [2] and Ashtekar-Barbero [3] formulations, which focus respectively on the dynamics of a spatial metric and the dynamics of a spatial $\mathrm{SO}(3)$ connection.

Unfortunately, the significant advantages of geometrodynamics, including the Hamiltonian description of gravity, usually come at the expense of manifest local Lorentz symmetry. When local reference frames are bound to the time slicing, and fields put in 'time gauge', local Lorentz transformation of frames no longer act in any obvious way.

This is further complicated in the Ashtekar-Barbero picture, where it is not obvious how local $\mathrm{SO}(3,1)$ transformations should act on the Lie algebra part of the $\mathrm{SO}(3)$ connection. While Barbero's Hamiltonian formulation derives from a Lorentz-covariant action [9], the usual derivation explicitly breaks Lorentz symmetry using a time gauge. This has resulted in significant controversy, especially in loop quantum gravity, which is founded on the non-covariant Ashtekar-Barbero approach. Restoring Lorentz covariance in the quantum theory has been a complicated and sensitive issue, see e.g [1, 4, 5, 11], suggesting the need for a simple geometric way of maintaining Lorentz covariance from the outset.

Here we present a recent alternative picture of geometrodynamics, relying not on a global time coordinate, but on a *local field of observers*. This field plays two complementary roles: it splits spacetime into spatial and temporal directions, but also gives an observer-dependent breaking of $\mathrm{SO}(3,1)$ symmetry to $\mathrm{SO}(3)$. Despite this 'broken' symmetry, full Lorentz gauge symmetry is still manifest, since local Lorentz transformations act not only on the physical fields, but also on the observer fields.

In general relativity and related geometric gauge theories, the appearance of symmetry breaking often hints at geometrical foundations based on *Cartan geometry* [14]. We show that the symmetry breaking introduced by the observer field gives a 'spatial Cartan geometry' on spacetime, which nonetheless respects spacetime Lorentz symmetry. When the observer field is normal to a hypersurface, this induces a Cartan geometry in the usual sense on the hypersurface, leading to a picture of gravity as evolving Cartan geometries on space, or '*Cartan geometrodynamics*'.

The full geometric picture involves symmetry breaking at two levels: first reducing Poincaré symmetry to Lorentz symmetry to give spacetime Cartan geometry [13], and then further from Lorentz symmetry to rotational symmetry [6] giving Cartan geometrodynamics. We will see these geometric pieces fit nicely together in the Cartan geometry of *observer space* [8], the space of all possible observers in spacetime. Observer space provides a setting for general relativity with the advantages of both covariant and canonical approaches.

2. Observer fields in Yang-Mills theory

To help set the stage for general relativity, and build up most of the tools we will need, we first consider a context in which the spacetime geometry is a fixed background structure.

2.1. Observers

Let M be an $(n+1)$-dimensional manifold with Lorentzian metric g, equipped with an orientation and a time orientation. An *observer* in M is a unit future-directed timelike tangent vector, and the space of all observers, the *observer space* of M, is the unit future tangent bundle $O \subset TM$ [8]. An *observer field* u is just a section of O–a unit future-timelike vector field.

An observer field u has a corresponding *co-observer field* $\hat{u} := -g(u, \cdot)$, a 1-form whose kernel defines 'spatial directions' at each point in M. An n-dimensional submanifold of M with tangent spaces in $\ker \hat{u}$ will be called a *spatial hypersurface*; these exist only where the Frobenius condition

$$\hat{u} \wedge d\hat{u} = 0 \tag{2.1}$$

holds, in which case we can locally write $\hat{u} = N\,dt$ for some functions t and N corresponding to the time coordinate and lapse function in foliation-based approaches. [6]

Whether or not an observer field induces a spatial foliation, it gives a canonical way to split any physical field on M into 'temporal' and 'spatial' parts. In particular, for the complex of differential forms $\Omega^\bullet(M)$, we define the *spatial forms* $\Omega^\bullet_\perp$ to be the kernel of interior multiplication ι_u, and *temporal forms* $\Omega^\bullet_\parallel$ to be the kernel of wedging with $\hat{u}$. This puts an additional grading on $\Omega^\bullet(M)$:

$$\Omega^\bullet(M) = \Omega^\bullet_\perp(M) \oplus \Omega^\bullet_\parallel(M)$$

where $\Omega_\perp$ has grade 0 and $\Omega_\parallel$ has grade 1. The corresponding projections into spatial and temporal parts are

$$X^\perp := \iota_u(\hat{u} \wedge X), \quad X^\parallel := \hat{u} \wedge \iota_u X$$

The spatial projection is a homomorphism; the temporal projection is a derivation.

Likewise, observers have preferred spatial and temporal derivatives. The natural time derivative for the observer field u is the Lie derivative $\pounds_u = \iota_u d + d\iota_u$, and the Lie derivative of a spatial form is spatial, since $\pounds_u$ commutes with ι_u. The *spatial differential*

$$d^\perp X = dX - \hat{u} \wedge \pounds_u X$$

is always a graded derivation, but squares to zero precisely when the Frobenius condition (2.1) holds; in this case it is indeed the usual spatial differential on each leaf of the corresponding foliation.

We also define a *spatial Hodge star* $*_\perp\colon \Omega^p_\perp \to \Omega^{n-p}_\perp$, related to the usual spacetime Hodge star operator $*\colon \Omega^p \to \Omega^{n+1-p}$ by

$$*_\perp X := -*(\hat{u}\wedge X), \quad \hat{u}\wedge *_\perp X = (-1)^p *X \quad \text{for } X\in\Omega^p_\perp$$

This works because the spacetime Hodge star automatically interchanges spatial and temporal forms for any observer. Just as $X\wedge *Y = \langle X, Y\rangle \mathrm{vol}$, where vol is the volume form on M, we have $X \wedge *_\perp Y = \langle X, Y\rangle \iota_u \mathrm{vol}$ whenever X and Y are both spatial.

2.2. Yang-Mills

We can apply all of this, for example, to Yang-Mills theory. Introducing an observer field, the action $S[A] = -\int \mathrm{tr}(F\wedge *F)$ now falls almost effortlessly into a familiar-looking form:

$$S[A] = \int \hat{u}\wedge \mathrm{tr}\left(E\wedge *_\perp E - B\wedge *_\perp B\right)$$

Here $E := -\iota_u F$ and $B := F^\perp$ are the nonabelian electric and magnetic fields, so that $F = B - \hat{u}\wedge E$. For Minkowski spacetime with $u = \partial_t$ and $\hat{u} = dt$, we thus get the usual action $\int dt(E^2 - B^2)$.

Further analysis of Yang-Mills theory using observer fields yields familiar formulas, but with some unexpected new terms involving derivatives of $\hat{u}$. For example, writing $A = a - \hat{u}\phi$, where $a := A^\perp$ is the nonabelian magnetic vector potential and $\phi := -\iota_u A$ is the nonabelian electric potential, we find

$$B = d^\perp a + a\wedge a - (d^\perp \hat{u})\phi, \quad E = -d^\perp_a \phi - \pounds_u a + (\pounds_u \hat{u})\phi$$

which are familiar except for the $d^\perp\hat{u}$ and $\pounds_u\hat{u}$ terms. In fact, $d^\perp\hat{u}$ vanishes whenever the Frobenius integrability condition holds, and $\pounds_u\hat{u}$ vanishes whenever u is a Killing vector field.

Writing E and B in terms of their potentials, action again takes familiar form, this time revealing the Hamiltonian analysis:

$$S = -\int \hat{u}\wedge \mathrm{tr}\left(E\wedge *_\perp \pounds_u a + \ldots\right)$$

We have written just the first term, from which we can see that E is the momentum of the vector potential a, as usual. The omitted terms give the usual constraints, though again with new terms proportional to derivatives of $\hat{u}$.

The observer-based decomposition of the Yang-Mills equations $d_A F = 0$ and $d_A{*}F = 0$ is also straightforward:

Maxwell-like	Nonabelian self-sourcing	Terms that vanish if Frobenious condition (2.1) holds	Terms that vanish if u is a Killing vector field	
$d^\perp B$	$+[a, B]$	$-d^\perp\hat{u}\wedge E$		$=0$
$d^\perp E + \pounds_u B$	$+[a, E] - [\phi, B]$		$-\pounds_u\hat{u}\wedge E$	$=0$
$d^\perp *_\perp E$	$+[a, *_\perp E]$	$+d^\perp\hat{u}\wedge *_\perp B$		$=0$
$d^\perp *_\perp B - \pounds_u *_\perp E$	$+[a, *_\perp B] + [\phi, *_\perp E]$		$-\pounds_u\hat{u}\wedge *_\perp B$	$=0$

The real reason for the terms involving $d^\perp\hat{u}$ and $\pounds_u\hat{u}$ is that the observer-based decomposition has local Lorentz symmetry much less evident in the foliation-based approach.

2.3. Local Lorentz symmetry

If spacetime is Lorentzian, then physics should be invariant under local Lorentz transformations relating local observers to each other. An *observer transformation* [6] is simply a gauge transformation $\lambda\colon O \to O$, acting on sections by composition: $u \mapsto \lambda \circ u$. In a local trivialization, this amounts to left multiplication by an $\mathrm{SO}(n,1)$-valued function. Any two observer fields are related by some observer transformation, and obviously all of the equations we have written are invariant under observer transformations.

Foliation invariance is a special case of invariance under observer transformations. An observer field satisfying the Frobenius condition (2.1) is equivalent to a spacelike foliation of M. Thus, any two spacelike foliations are related by some observer transformation. However, there is no subgroup of the group of gauge transformations that preserves the property (2.1) of being a foliation, since this depends on the initial section u. Thus the transformations going from one *foliation* to another form not a group but a groupoid–a sub-groupoid of the transformation groupoid.

3. Observers in general relativity

Two issues arise when we attempt a similar decomposition of general relativity. First, we face a dilemma of priority: we want the dynamics of gravity as seen by an observer, but the very definition of an observer requires the metric, which is determined by the dynamics. Second, unlike in Yang-Mills theory, where the gauge group is unaffected by the decomposition, here we start with an $\mathrm{SO}_o(n,1)$ connection on spacetime, but each observer should see a spatial $\mathrm{SO}(n)$ connection. These issues are related, and we deal with each in turn.

3.1. Observers without a background geometry

With no fixed geometry on M, an $(n+1)$-dimensional manifold, we will create a 'fake' geometry for observers to live in. Pick a vector bundle $\mathcal{T}$ over M, isomorphic to the tangent bundle TM, but so far not in any specified way. We call $\mathcal{T}$ the *fake tangent bundle* and equip it with all of the structure we demanded of TM in Sec. 2, namely a fixed 'metric' (a smoothly varying Lorentzian inner product on fibers) as well as an orientation and time orientation.

A *fake observer* on M is a unit future-directed timelike element of $\mathcal{T}$, and the space of all fake observers is the *fake observer space* $\mathcal{O} \subset \mathcal{T}$ [8]. A *fake observer field* y is just a section of $\mathcal{O}$. Of course, y lets us do the same things with $\mathcal{T}$ that u lets us do with TM, and $\hat{u}$ with T^*M.

In particular, y splits $\mathcal{T}$ into a part $\mathcal{T}_\perp$ orthogonal to y and a part $\mathcal{T}_\parallel$ parallel to y. More generally, it splits the exterior bundle $\Lambda^\bullet\mathcal{T}$ into a fiberwise direct sum

$$\Lambda^\bullet\mathcal{T} = \Lambda^\bullet_\perp\mathcal{T} \oplus_M \Lambda^\bullet_\parallel\mathcal{T}$$

Here $\Lambda^\bullet_\perp\mathcal{T} = \Lambda^\bullet\mathcal{T}_\perp$ is generated by wedge products of vectors orthogonal to y, while $\Lambda^\bullet_\parallel\mathcal{T}$ is the kernel of the exterior product with y. Since fibers of $\mathcal{T}$ are oriented Lorentzian inner product spaces, there is a Hodge star on $\Lambda^\bullet\mathcal{T}$, which we write as '$\star$' to distinguish it from the Hodge star '$*$' on differential forms. Just as $*$ turns spatial forms into temporal forms, we also have

$$\star(\Lambda^p_\perp\mathcal{T}) = \Lambda^{n+1-p}_\parallel\mathcal{T} \tag{3.1}$$

Fake observers become 'real' once we introduce a *coframe field*–a vector bundle isomorphism $e\colon TM \xrightarrow{\sim} \mathcal{T}$.A coframe field in general relativity determines the spacetime metric by pullback, but here it also conspires with the fake observer field y to give an observer field $u = e^*y$. This observer field splits fields into spatial and temporal parts, just as in Sec. 2.1.

Most notably, the coframe field, itself a $\mathcal{T}$-valued 1-form, splits into spatial and temporal parts:

$$e = E + \hat{u}y \tag{3.2}$$

A crucial observation is that no further splitting using $\mathcal{T} = \mathcal{T}_\perp \oplus \mathcal{T}_\parallel$ is needed: the spatial 1-form E already takes values in $\mathcal{T}_\perp \subset \mathcal{T}$, since y by definition, takes values in $\mathcal{T}_\parallel$. On any spatial hypersurface S, E restricts to an isomorphism $E\colon TS \to \mathcal{T}_\perp|_S$, so E is justly called the *spatial coframe field.*

3.2. Breaking Lorentz symmetry

Since the fake observer field y determines the spatial sub-bundle $\mathcal{T}_\perp \subset \mathcal{T}$, it also 'breaks' local Lorentz symmetry to rotational symmetry. It is easiest to see how all physical fields are affected if we write this symmetry breaking in the language of principal bundles.

The *fake frame bundle* $\mathcal{F}$ is the principal $\mathrm{SO}_o(n,1)$ bundle of orthonormal frames in $\mathcal{T}$ respecting the orientation and time orientation, where $\mathrm{SO}_o(n,1)$ is the connected Lorentz group. A fake observer field y is a section of the fake observer space $\mathcal{O}$, but this just the associated bundle $\mathcal{O} \cong \mathcal{F} \times_{\mathrm{SO}_o(n,1)} H^n$, where $H^n \cong \mathrm{SO}_o(n,1)/\mathrm{SO}(n)$ is hyperbolic space. Such a section is the same as a reduction of $\mathcal{F}$ to a principal $\mathrm{SO}(n)$ bundle $\mathcal{F}_y \to M$, given by pullback:

$$\begin{array}{ccc} \mathcal{F}_y & \longrightarrow & \mathcal{F} \\ \downarrow & & \downarrow \\ M & \xrightarrow{\;y\;} & \mathcal{O} \end{array}$$

Thanks to this reduction, any physical field living in a representation of $\mathrm{SO}_o(n,1)$ now splits into representations of $\mathrm{SO}(n)$. In particular, the fundamental and adjoint representations each split into two irreducible representations of $\mathrm{SO}(n)$:

$$\mathbb{R}^{n,1} \cong \mathbb{R}^n \oplus \mathbb{R}, \quad \mathfrak{so}(n,1) \cong \mathfrak{so}(n) \oplus \mathbb{R}^n \tag{3.3}$$

The two representations labeled $\mathbb{R}^n$ are canonically isomorphic. The group $\mathrm{SO}(n)$ is the stabilizer of a unique 'observer' $\xi \in \mathbb{R}^{n,1}$, a unit timelike vector with positive 'time' component. The $\mathbb{R}^n$ in the first line of (3.3) is the orthogonal complement of ξ; the map $\mathbb{R}^n \to \mathfrak{so}(n,1)$ given by $v \mapsto \eta(\xi,\cdot)v - \eta(v,\cdot)\xi$ is an intertwiner of $\mathrm{SO}(n)$ representations and its image is the $\mathbb{R}^n$ in the second line of (3.3).

3.3. Lorentz equivariant spatial connections

We can now decompose a connection ω on $\mathcal{F}$ in two ways, both into spatial and temporal parts and via the reduction to $\mathrm{SO}(n)$ symmetry. Doing the spatiotemporal splitting first, write $\omega = \Omega + \hat{u}\Xi$,where $\Omega = \omega^\perp$ and $\Xi = \iota_u\omega$. Then, since Ω can be thought of as an $\mathfrak{so}(n,1)$-valued 1-form on $\mathcal{F}$, we can pull it back to $\mathcal{F}_y$, where it splits into two irreducible $\mathrm{SO}(n)$ subrepresentations of $\mathfrak{so}(n,1)$:

$$\Omega = \underbrace{\boldsymbol{\Omega}}_{\mathfrak{so}(n)\text{ part}} + \underbrace{K}_{\mathbb{R}^n\text{ part}} \tag{3.4}$$

corresponding to infinitesimal rotations and boosts, from the perspective of the observer. We call $\boldsymbol{\Omega}$ the *spatial* $\mathbf{SO}(\boldsymbol{n})$ *connection.* It is a spatial form, $\iota_u\boldsymbol{\Omega} = 0$, and remains so under local Lorentz transformations acting both on $\boldsymbol{\Omega}$ and u. It restricts to a connection on any spatial hypersurface.

By splitting the torsion $d_\omega e$ into spatial and temporal parts, and further into representations of $\mathrm{SO}(n)$, one can show that $d^\perp_{\boldsymbol{\Omega}}E$ vanishes whenever $d_\omega e$ does. In particular, when the Frobenius condition holds, $\boldsymbol{\Omega}$ restricts to the unique torsion free connection on any spatial hypersurface.

3.4. Palatini

Using all of this, let us now introduce a fake observer field in the Palatini action for general relativity. The action in $n+1$ dimensions can be written, up to a constant factor, as

$$S[e,\omega] = \int \mathrm{tr}\left(\star(\underbrace{e\wedge\cdots\wedge e}_{n-1})\wedge R\right) \tag{3.5}$$

which resembles Yang-Mills theory, since both $\star(e\wedge\cdots\wedge e)$ and R live in $\mathfrak{so}(n,1)$. More precisely, $\star(e\wedge\cdots\wedge e)$ is a $\Lambda^2\mathcal{T}$-valued $(n-1)$-form, but $\Lambda^2\mathcal{T}\cong\mathrm{Ad}(\mathcal{F})$, thanks to the isomorphism $\Lambda^2\mathbb{R}^{n,1}\cong\mathfrak{so}(n,1)$ between bivectors and infinitesimal pseudorotations. Thus both forms take values in $\mathrm{Ad}(\mathcal{F})$ and the trace is exactly as in Yang-Mills theory.

Obviously nothing changes about the theory if we introduce a fake observer field y, a section of $\mathcal{F}\times_{\mathrm{SO}_o(n,1)}H^n$, but leave the formula for the action unchanged, defining $S[e,\omega,y]:=S[e,\omega]$.

Using the observer field $u=e^*y$, we can split fields into spatial and temporal parts. This breaks the action into a number of terms, of which we write just the first:

$$\mathrm{tr}\left(\star(\underbrace{e\wedge\cdots\wedge e}_{n-1})\wedge R\right) = \hat{u}\wedge\mathrm{tr}\left(\star(\underbrace{E\wedge\cdots\wedge E}_{n-1})\wedge\pounds_u\Omega+\cdots\right)$$

The remaining terms are straightforward to work out, but unimportant for present purposes.

This simplifies when we also split fields according to representations of $\mathrm{SO}(n)$, because $\pounds_u\Omega$ lives in $\mathfrak{so}(3,1)=\mathfrak{so}(3)\oplus\mathbb{R}^n$, whereas $\star(E\wedge\cdots\wedge E)$ lives just in $\mathbb{R}^n$. More precisely, since E is $\mathcal{T}_\perp$-valued, $(E\wedge\cdots\wedge E)$ is $\Lambda^{n-1}\mathcal{T}_\perp$-valued and, by (3.1), $\star(E\wedge\cdots\wedge E)$ is $\Lambda^2_\parallel\mathcal{T}$-valued. But under the isomorphism $\Lambda^2\mathcal{T}\cong\mathcal{F}_y\times_{\mathrm{SO}(n)}\mathfrak{so}(n,1)$, $\Lambda^2_\parallel\mathcal{T}$ is identified with $\mathcal{F}_y\times_{\mathrm{SO}(n)}\mathbb{R}^n$. Thus, in the first term of the action, we can replace Ω with its $\mathbb{R}^n$ part:

$$S[e,\omega] = \int \hat{u}\wedge\mathrm{tr}\left(\star(\underbrace{E\wedge\cdots\wedge E}_{n-1})\wedge\pounds_u K+\cdots\right) \tag{3.6}$$

We have arrived at a hybrid of canonical and covariant formulations. Simply introducing the field y, the action falls into the form $\int dt\,(p\dot{q}+\cdots)$, from which we can read off the canonical momenta and the constraints. In particular the $\mathfrak{so}(n)$ part $\boldsymbol{\Omega}$ of the spatial connection (3.4) is nondynamical, and only the $\mathbb{R}^n$ part K has nonvanishing momentum. This agrees with standard Hamiltonian analysis of the Palatini action (see e.g [12, Sec. 4.2]) where K is identified as the extrinsic curvature of a spatial hypersurface. On the other hand, since Lorentz transformations act on all fields including y, we maintain Lorentz covariance.

3.5. Ashtekar-Barbero

In 3+1 dimensions, a modification [9] of the action (3.5) is possible, since $\star$ maps $\Lambda^2\mathcal{T}\cong\mathrm{Ad}(\mathcal{F})$ to itself:

$$S[e,\omega] = \int \mathrm{tr}\left((\star+\tfrac{1}{\gamma})(e\wedge e)\wedge R\right).$$

Splitting the fields as before into spatial and temporal parts and by SO(3) representations, the analog of (3.6) is

$$S = \int \hat{u}\wedge\mathrm{tr}\left(\star(E\wedge E)\wedge\pounds_u(K-\tfrac{1}{\gamma}\star\boldsymbol{\Omega})+\cdots\right) \tag{3.7}$$

Whereas in (3.6) $\star(E\wedge E)$ was the momentum of K, it is now the momentum of $-\frac{1}{\gamma}\star\mathsf{A}$, where

$$\mathsf{A}=\boldsymbol{\Omega}+\gamma\star K \tag{3.8}$$

is a spatial SO(3) connection, the analog in our framework of the *Ashtekar-Barbero connection.* It is a Lorentz-covariant spatial connection in the same sense as $\mathbf{\Omega}$, but differs from $\mathbf{\Omega}$ in its torsion and curvature. In particular, since $d_{\mathsf{A}} = d_{\mathbf{\Omega}} - \gamma[\star K, \cdot\,]$, the torsion $d_{\mathsf{A}}E$ is propotional to $\star K$, which in the case of a foliation is just the extrinsic curvature of the observer field (cf. [10]).

The Hamiltonian analysis following from (3.7) is similar to the standard analysis using gauge fixing [9], but maintaining Lorentz symmetry using observer fields also avoids the second class constraints that are inevitable in gauge-fixed approaches. Details can be found in our previous work [6]. Here, we continue instead with the geometric interpretation.

4. Cartan geometrodynamics

4.1. Evolving spatial Cartan geometries

Cartan geometry is ubiquitous in gauge-theoretic formulations of gravity [13]. Its appearance in spacetime physics is signaled by a connection with values in some group G, but with gauge symmetry only under a closed subgroup H, and where the homogeneous space G/H has the same type of geometry as spacetime, usually a Lorentzian manifold of the same dimension. In general, a *Cartan geometry* modeled on G/H is a principal H bundle $Q \to M$ equipped with an H-equivariant $\mathfrak{g}$-valued 1-form that is a linear isomorphism at each point of Q and restricts to the Maurer-Cartan form on vertical vectors.

A key example in gravitational physics comes from treating the connection ω and coframe field e in (3.5) as the $\mathfrak{so}(n,1)$ and $\mathbb{R}^{n,1}$ parts of a unified $\mathfrak{iso}(n,1)$-valued connection. This connection can be viewed as 1-form on the principal $\mathrm{SO}(n,1)$ bundle $\mathcal{F}$, and this gives a Cartan geometry modeled on Minkowski spacetime $\mathrm{ISO}(n,1)/\mathrm{SO}(n,1)$.

The 'spatial geometry' studied here nicely parallels this example. Just as $\omega + e$ gives a Cartan geometry modeled on Minkowski spacetime, $\mathbf{\Omega} + E$ gives a 'spatial' Cartan geometry modeled on Euclidean space. More precisely:

Theorem 4.1. *Let* $(\mathcal{F} \to M, \omega + e)$ *be a Cartan geometry with model* $\mathrm{ISO}_o(n,1)/\mathrm{SO}_o(n,1)$*; let* y *be a fake observer field with corresponding observer field* $u = e^*y$*. If* S *is any integral hypersurface of* $\ker \hat{u}$ *then* $(\mathcal{F}_y|_S \to S, \mathbf{\Omega} + E)$ *is a Cartan geometry with model* $\mathrm{ISO}(n)/\mathrm{SO}(n)$*, where* $\mathbf{\Omega}$ *is the spatial* $\mathrm{SO}(n)$ *connection defined in* (3.4) *and* E *is the spatial coframe* (3.2)*. In the case* $n{=}3$*, the same is also true if we replace* $\mathbf{\Omega}$ *by the Ashtekar-Barbero connection* A (3.8).

In particular, if $S = S_t$ is the level set of a time function t on spacetime, then each spatial slice S_t becomes a spatial geometry. There is a canonical correspondence between metrics and torsion-free Cartan geometries modeled on Euclidean space.

4.2. Symmetry breaking and observer space

It is often useful to think about Cartan geometry in physics in terms of spontaneous symmetry breaking. The reason is that a Cartan connection with model G/H can be viewed as an ordinary Ehresmann connection on a principal G bundle, pulled back along a reduction to an H bundle.

In our case, although we have used the observer field to break symmetry, the relation to spatial Cartan geometry is not immediate. In particular, the theorem in the previous section refers to spatial Cartan geometry modeled on Euclidean space $\mathrm{ISO}(n)/\mathrm{SO}(n)$, but $\mathcal{F}_y$ is instead a reduction of the $\mathrm{SO}(n,1)$ bundle $\mathcal{F}$; we have no obvious $\mathrm{ISO}(n)$ bundle in sight. Of course, we can extend $\mathcal{F}_y$ to an $\mathrm{ISO}(n)$ bundle, and extend the Cartan connection $\mathbf{\Omega} + E$ to an Ehresmann connection on the extension, but this seems less than natural. It would be nice to see the spatial Cartan geometry coming directly from $\mathrm{ISO}(n)$ symmetry breaking.

It helps to think more carefully about observers' role in symmetry breaking. In Minkowski spacetime, there are two equivalent ways to specify an observer. One can first pick a point in spacetime, reducing $\mathrm{ISO}(n,1)$ symmetry to $\mathrm{SO}(n,1)$, and then a unit timelike vector there, reducing further to $\mathrm{SO}(n)$. Equivalently, one can pick first a spacelike hyperplane, or surface of simultaneity, reducing symmetry to $\mathrm{ISO}(n)$, and then pick the location of the observer within that hyperplane, reducing further to $\mathrm{SO}(n)$. The features and their corresponding stabilizers are:

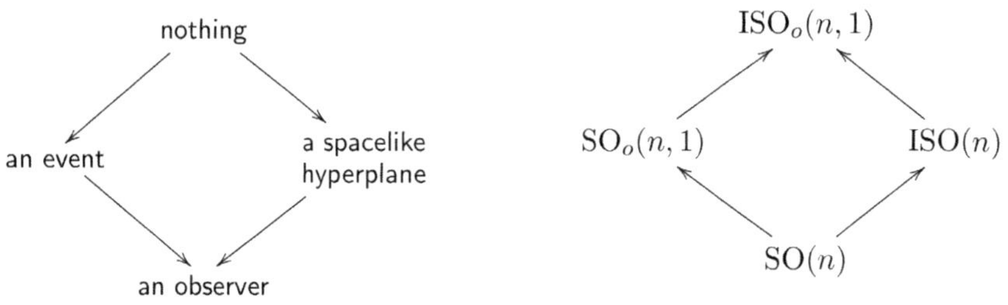

The end result is the same, but the intermediate steps in the two branches are quite different.

Moreover, breaking symmetry all the way from $\mathrm{ISO}_o(n,1)$ down to $\mathrm{SO}(n)$, the Poincaré Lie algebra $\mathfrak{iso}(n,1)$ splits into four irreducible $\mathrm{SO}(n)$ representations. This can be viewed either in terms of spacetime or the space of spacelike hyperplanes, and the relationship between these two perspectives is best explained with the diagram:

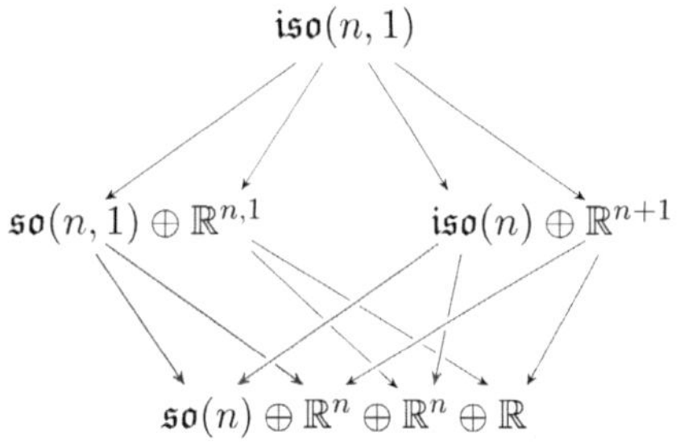

The adjacent pairs of arrows are the two projections of a binary direct sum of representations; in the bottom row, each direct summand is the intersection of the two representations pointing to it.

Let us write the decomposition of just the spatial part of the spacetime Cartan connection:

$$\begin{array}{rccccccc} & \mathfrak{so}(n) & \oplus & \mathbb{R}^n & \oplus & \mathbb{R}^n & \oplus & \mathbb{R} \\ (\Omega+e)^\perp = & \boldsymbol{\Omega} & + & K & + & E & + & 0 \end{array}$$

From the spacetime perspective, it seems rather ad hoc to join $\boldsymbol{\Omega}$, the spatial $\mathfrak{so}(n)$ part of a Lorentz connection, with E, the spatial part of a coframe, into an $\mathrm{ISO}(n)$ connection. But on the other side of the picture, based on the space of spatial hyperplanes, these pieces seem to come naturally from broken $\mathrm{ISO}(n)$ symmetry. In particular, Cartan geometrodynamics is clearer when we consider both levels of symmetry breaking at once.

What is the *geometric* meaning of breaking symmetry all the way from $\mathrm{ISO}_o(n,1)$ to $\mathrm{SO}(n)$? Evidently, this should lead to Cartan geometry modeled on the homogeneous space

$$O(\mathbb{R}^{n,1}) \cong \mathrm{ISO}_o(n,1)/\mathrm{SO}(n)$$

which is the $(2n+1)$-dimensional *observer space* of Minkowski spacetime. We call a Cartan geometry modeled on $O(\mathbb{R}^{n,1})$ an *observer space geometry* [8]. While observer space has played a mostly superficial role in this paper, as the bundle whose sections are observer fields, we now see that the full geometric picture of covariant canonical gravity gives observer space a life of its own.

In fact, the spacetime Cartan geometry $(\mathcal{F}, \omega + e)$ on M induces an observer space geometry on $\mathcal{O}$. Different descriptions of the same physical situation in terms of Cartan geometrodynamics are given by pulling this Cartan geometry back along sections of the bundle $\mathcal{O} \to M$. In fact, we almost saw this this before, when we noted how y breaks Lorentz symmetry:

$$\begin{array}{ccc} \mathcal{F}_y & \longrightarrow & \mathcal{F} \\ \downarrow & & \downarrow \\ M & \xrightarrow{\;y\;} & \mathcal{O} \end{array}$$

The bundle on the right is the principal $\mathrm{SO}(n)$ bundle over observer space, on which the canonical observer space geometry is built. Each 'canonical' description of a solution of general relativity corresponds to one such pullback square. However, it is the observer space geometry itself that is truly 'canonical' in the mathematical sense of being constructed without arbitrary choices [7, 8].

Acknowledgments
This paper is based largely on work with Steffen Gielen [6, 7, 8], to whom I am also thankful for comments and suggestions.

References
[1] Alexandrov S and Livine E R 2003 *Phys. Rev.* D **67** 044009
[2] Arnowitt R L, Deser S and Misner C W 1962 *Gravitation: an Introduction to Current Research* ed L Witten (John Wiley & Sons)
[3] Barbero G J F 1995 *Phys. Rev.* D **51** 5507
[4] Cianfrani F and Montani G 2009 *Phys. Rev. Lett.* **102** 091301
[5] Geiller M, Lachièze-Rey M, Noui K and Sardelli F 2011 *SIGMA Symmetry Integrability Geom. Methods Appl.* **7** 083
[6] Gielen S and Wise D K 2012 *Phys. Rev.* D **85** 104013
[7] Gielen S and Wise D K 2012 *Gen. Relativity Gravitation* **44** 3103
[8] Gielen S and Wise D K 2013 *J. Math. Phys.* **54** 052501
[9] Holst S 1996 *Phys. Rev.* D **53** 5966
[10] Montesinos M 1999 *J. Math. Phys.* **40** 4084
[11] Rovelli C and Speziale S 2011 *Phys. Rev.* D **83** 104029
[12] Thiemann T 2008 *Modern Canonical Quantum General Relativity* (Cambridge Univ. Press)
[13] Wise D K 2009 *SIGMA Symmetry Integrability Geom. Methods Appl.* **5** 080; 2010 *Classical Quantum Gravity* **27** 155010
[14] Wise D K 2012 *J. Phys.: Conf. Ser.* **360** 012017

Lie-algebraic symmetries of generalized Davey-Stewartson equations

A Zuevsky
Max-Planck Institut für Mathematik, Vivatsgasse 7, 53111 Bonn, Germany

Abstract. We identify the full Lie-algebraic structure of the generalized Davey-Stewartson (GDS) system of equations with symmetries of a specific of continual Lie algebras. In particular, we show that they are related to two copies of the Poisson bracket continual Lie algebra.

1. Generalized Davey-Stewartson equations

A system of nonlinear partial differential equations in 2+1 dimensions as a model of wave propagation in a bulk medium of an elastic material with couple stresses has recently been derived in [5], namely

$$\begin{aligned} &i\psi_t + \delta\psi_{xx} + \psi_{yy} = \chi|\psi|^2\psi + \gamma(w_x + \phi_y)\psi \\ &w_{xx} + n\phi_{xy} + m_2 w_{yy} = (|\psi|^2)_x \\ &n w_{xy} + \lambda\phi_{xx} + m_1\phi_{yy} = (|\psi|^2)_y \end{aligned} \tag{1.1}$$

with the condition $(\lambda - 1)(m_1 - m_2) = n^2$. The subscripts in (1.1) denote corresponding partial derivatives. Here $\psi(t,x,y)$ is a complex function, $w(t,x,y)$ and $\phi(t,x,y)$ are real functions, and δ, n, m_1,m_2, λ, χ, γ are real constants. The authors of [5] demonstrated that when the parameters are related by $n = 1-\lambda = m_1 - m_2$, then (1.1) can be reduced to the standard Davey-Stewartson (DS) equations (in general not integrable) by a non-invertible point transformation of dependent variables. Therefore, they called (1.1) the generalized Davey-Stewartson (GDS) equations. In [5] some travelling type solutions of (1.1) in terms of elementary and elliptic functions were obtained. Global existence results were studied in [6]. In another recent work [7], under some constraints on the physical parameters, the so-called hyperbolic-elliptic-elliptic case of the system (1.1) (in [7] the system is classified into different types according to the signs of parameters $(\delta, m_1, m_2, \lambda)$) was shown to admit singular solutions that blow up in a finite time.

In [2] authors work with a differential form of the equations (1.1). Differentiating the last two equations of (1.1) by x and y respectively, performing the substitutions $w_x \to w$, $\phi_y \to \phi$, and rewriting the corresponding system in a real form by separating $\psi = u + iv$ into real and imaginary parts, one obtains a system of four real partial differential equations

$$\begin{aligned} &u_t + \delta v_{xx} + v_{yy} = \chi v(u^2 + v^2) + \gamma v(w + \phi) \\ &-v_t + \delta u_{xx} + u_{yy} = \chi u(u^2 + v^2) + \gamma u(w + \phi) \\ &w_{xx} + n\phi_{xx} + m_2 w_{yy} = 2(u_x^2 + u u_{xx} + v_x^2 + v v_{xx}) \\ &n w_{yy} + \lambda\phi_{xx} + m_1\phi_{yy} = 2(u_y^2 + u u_{yy} + v_y^2 + v v_{yy}) \end{aligned} \tag{1.2}$$

In the sequel, following [2], we call (1.2) the GDS (generalized Davey-Stewartson) equations.

In [2] the Lie symmetry algebra of the generalized Davey-Stewartson (GDS) equations (1.2) was computed. In particular, it was shown that for special choice of parameters it is a centerless Kac-Moody-Virasoro algebra. It was also shown that under certain conditions imposed on parameters in the system is infinite-dimensional and isomorphic to that of the standard integrable Davey-Stewartson equations which is known to have a very specific Kac-Moody-Virasoro loop algebra structure. The main result of the paper [2] was to show that, when some conditions on physical parameters δ, n, m_1, m_2, λ are imposed, the Lie algebra of the symmetry group of the GDS system has a Kac-Moody-Virasoro (KMV) loop structure (same as for the usual integrable DS equations [8]). In particular, they show in that this algebra is isomorphic to that of DS equations:

$$i\psi_t + \delta_1\psi_{xx} + \psi_{yy} = \delta_2|\psi|^2\psi + w\psi, \quad \varepsilon_1 w_{xx} + w_{yy} = \varepsilon_2(|\psi|^2)_{yy} \tag{1.3}$$

with $\delta_1 = \pm 1$, $\delta_2 = \pm 1$. The Lie algebra of the symmetry group of the integrable DS system is referred to as the DS algebra.

The purpose of this note is some further study of GDS equations from group theoretical point of view. We show that the system (1.2) possesses a conitnual Lie algebra symmetry. In particular, the systems of generators of corresponding Lie algebras is a special case of a continual Lie algebra [18]. Finally, let us also mention that the GDS is related to the generalized Alber's equation discissed in [1, 22].

2. Continual Lie algebra symmetries of GDS system

In appendix we recall the definition of a continual Lie algebra. Two basic examples relevant to the Lie algebra of symmetries describing the GDS equations are the Witt and Poisson bracket continual Lie algebras [18] are as follows.

2.1. *Witt algebra*

The Witt algebra [9] is the centerless Virasoro algebra. The commutation relation on the single generator $X(\phi)$, for two ϕ, $\psi \in \mathcal{E}$, are

$$[X(\phi), X(\psi)] = X(\phi\partial\psi - \psi\partial\phi) \equiv X\left(\left[\phi,^{\partial}\psi\right]\right), \quad K(\phi,\psi) = \phi\partial\psi - \psi\partial\phi \tag{2.1}$$

where ∂ denotes the differentiation with respect to a real parameter with obvious notation after the last equality in (2.1). The only condition that the mapping $K(\phi,\psi)$ satisfies is (A.2) with $k = m = n = 0$.

2.2. *Poisson bracket algebra*

As the second example we consider in this section is a continual Lie algebra defined by the mappings

$$K_{0,0}(\phi,\psi) = 0, \quad K_{0,\pm 1}(\phi,\psi) = \mp i\partial\phi\cdot\psi, \quad K_{1,-1}(\phi,\psi) = -i\partial(\phi\cdot\psi) \tag{2.2}$$

and $K_{n,m}(\phi,\psi) = i(n\ \partial\psi\cdot\phi - m\ \partial\phi\cdot\psi)$, n, $m \in \mathbb{Z}$. In [16] it was proved that this continual Lie algebra is isomorphic to the Poisson bracket algebra under an appropriate substitution of variables.

2.3. Generators of the GDS continual Lie algebra

In [2] the commutation relations of the symmetry algebra of the system of equations of (1.2) were obtain. They have the following form:

$$[T(f_1), T(f_2)] = T(f_1 f_2' - f_1' f_2) \tag{2.3a}$$

$$[T(f), X(g)] = X(fg' - \frac{1}{2} f' g) \tag{2.3b}$$

$$[T(f), Y(h)] = Y(fh' - \frac{1}{2} f' h) \tag{2.3c}$$

$$[T(f), W(m)] = W(fm') \tag{2.3d}$$

$$[X(g_1), X(g_2)] = -\frac{1}{2\delta} W(g_1 g_2' - g_1' g_2) \tag{2.3e}$$

$$[Y(h_1), Y(h_2)] = -\frac{1}{2} W(h_1 h_2' - h_1' h_2) \tag{2.3f}$$

$$[X(g), Y(h)] = [X(g), W(m)] = [Y(h), W(m)] = [W(m_1), W(m_2)] = 0 \tag{2.3g}$$

Using the generators $T(f)$, $X(g)$, $Y(h)$, $W(m)$, let us take

$$X_0 = T(f), \quad X_1(g) = X(g), \quad X_{-1} = Y(h), \quad X_{-2} = W(m), \quad X_2 = \frac{1}{\delta} W(m) \tag{2.4}$$

Then it is easy to see that (2.4) form a a continual Lie algebra [16] with a non-abelian Cartan subalgebra. Then the corresponding kernels in the commutation relations (A.1) have the form

$$K_{0,0}(f_1, f_2) = f_1 f_2' - f_1' f_2 \tag{2.5a}$$

$$K_{0,1}(f, g) = fg' - \frac{1}{2} f' g \tag{2.5b}$$

$$K_{0,-1}(f, h) = fh' - \frac{1}{2} f' h \tag{2.5c}$$

$$K_{0,2}(f, m) = fm' \tag{2.5d}$$

$$K_{0,-2}(f, m) = \frac{1}{\delta} fm' \tag{2.5e}$$

$$K_{1,1}(g_1, g_2) = -\frac{1}{2\delta}(g_1 g_2' - g_1' g_2) \tag{2.5f}$$

$$K_{-1,-1}(h_1, h_2) = -\frac{1}{2} W(h_1 h_2' - h_1' h_2) \tag{2.5g}$$

$$K_{1,-1}(g, h) = K_{1,\pm 2}(g, h) = K_{-1,\pm 2}(g, h) = K_{2,\pm 2}(g, h) = 0 \tag{2.5h}$$

By substitution one checks that the mappings (2.5e) indeed satisfy to the consequences (A.2) of Jacobi identity. The Cartan part of this algebra is isomorphic to the Witt algebra (2.1).

A partial differential operator representation in variables (x, y, t) for the Lie algebra are given by (2.5e) was obtained in [2]:

$$\begin{aligned} X_0(f) = {} & f(t)\partial_t + \frac{1}{2} f'(t)\left(x\partial_x + y\partial_y - u\partial_u - v\partial_v - 2w\partial_w - 2\phi\partial_\phi\right) \\ & - \frac{(x^2 + \delta y^2)}{8\delta}\left[f''(t)(v\partial_u - u\partial_v) + \frac{f'''(t)}{2\gamma}(\partial_w + \partial_\phi)\right] \end{aligned} \tag{2.6a}$$

$$X(g) = g(t)\partial_x - \frac{x}{2\delta}\left[g'(t)(v\partial_u - u\partial_v) + \frac{g''(t)}{2\gamma}(\partial_w + \partial_\phi)\right] \tag{2.6b}$$

$$Y(h) = h(t)\partial_y - \frac{y}{2}\left[h'(t)(v\partial_u - u\partial_v) + \frac{h''(t)}{2\gamma}(\partial_w + \partial_\phi)\right] \tag{2.6c}$$

$$W(m) = m(t)(v\partial_u - u\partial_v) + \frac{m'(t)}{2\gamma}(\partial_w + \partial_\phi) \tag{2.6d}$$

The functions $g(t)$, $h(t)$, and $m(t)$ are arbitrary functions of class $C^\infty(I)$, $I \subseteq \mathbb{R}$.

In [2] the standard infinitesimal procedure [14] to find the symmetry algebra L and hence the symmetry group G of (1.2) was applied. The article [2] is mainly focused on the case when $f(t)$ is allowed to be arbitrary. The symmetry algebra realized by (2.6a) is then infinite-dimensional and has the structure of a Kac-Moody-Virasoro algebra [2]. This is called in [2] the GDS symmetry algebra and the corresponding system.

On the other hand, it is easy to see that the algebra defined by the mappings (2.5e) represents $m = 2$ generalized local part of two copies of the Poisson bracket continual Lie algebra (see appendix for the definitions). Let us identify

$$x_1(f) = iX_1(f), \quad x_{-1}(f) = iX_{-1}(f), \quad x_2(g) = \frac{1}{2\delta}W(g), \quad x_{-2}(g) = -W(g) \tag{2.7}$$

where x_i, $|i| \leq 2$, denote generators of the Poisson bracket algebra (2.2). Then we put

$$T(f) = X_0(f) = x_0(f) + \frac{1}{2}\tilde{x}_0(f) \tag{2.8}$$

where the generator $\tilde{x}_0$ belongs to the second copy of the Poisson bracket continual Lie algebra defined by the Poisson bracket continual algebra mappings with inverted grading indices, i.e., $\widetilde{K}_{i,j} = K_{j,i}$. Then in commutation relations (A.1) one has, for example,

$$[X_0(f), X_1(g)] = \left[x_0(f) + \frac{1}{2}\tilde{x}_0(f),\ x_1(g)\right] = x_1\left(K_{0,1}(f,g) + \frac{1}{2}K_{1,0}(f,g)\right) \tag{2.9a}$$

$$= x_1\left(K_{0,1}(f,g) - \frac{1}{2}K_{0,1}(g,f)\right) = x_1\left(fg' - \frac{1}{2}gf'\right) \tag{2.9b}$$

and similarly for generators $x_{-1} = Y$, $x_2 = -\frac{1}{2\delta}W$ and $x_{-2} = W$, i.e., one can check that the relations (A.2) are also fulfilled.

Thus we can see that the full algebra describing the GDS system of equations is given by two copies of the Poisson bracket continual Lie algebra with "dual" mappings.

3. Conclusions

In this paper we have derived the continual Lie symmetries of the generalized Davey-Stewartson equations. Provided with the continual Lie algebraic formulation of the generalized DS system, we are able, in principle, to formulate the general algebraic procedure of solving such a system according to [12]. This will be a topic of a separate paper.

Appendix A. Continual Lie algebras

Let $\mathcal{E}$ be a vector space. A continual Lie algebra [11, 15, 16, 17] is generated by the generalized local part $\mathcal{G}^{m_0} = \oplus_{|n|\leq m_0}\mathcal{G}_n$, $\mathcal{G}_n = \{X_n(\phi), \phi \in \mathcal{E}\}$, $n \in \mathbb{Z}$, satisfying the defining relations for all $\phi, \psi \in \mathcal{E}$, and $|n|, |m|, |n+m| \leq m_0$,

$$[X_n(\phi), X_m(\psi)] = X_{n+m}(K_{n,m}(\phi,\psi)) \tag{A.1}$$

where $K_{n,m} : \mathcal{E} \times \mathcal{E} \to \mathcal{E}$, $n,m \in \mathbb{Z}$, are bilinear mappings. In analogy with classical case of discrete root space Lie algebra, we call $\mathcal{E}$ *the root space*. Ordinary Jacobi identity applied to elements $X_i(\phi)$ imply the following conditions on $K_{n,m}$:

$$K_{k,m+n}(\phi, K_{m,n}(\psi,\chi)) + K_{m,n+k}(\psi, K_{n,k}(\chi,\phi)) + K_{n,k+m}(\chi, K_{k,m}(\phi,\psi)) = 0 \tag{A.2}$$

$$K_{n,m}(\phi,\psi) = -K_{m,n}(\psi,\phi) \tag{A.3}$$

for all $\phi, \psi, \chi \in \mathcal{E}$, and $|l| \leq m_0$, where l denotes an index (or a sum of indeces) in (A.2). Then an infinite dimensional algebra $\mathcal{G}(\mathcal{E};K) = \mathcal{G}'(\mathcal{E};K)/J$ is called a continual contragredient Lie algebra where $\mathcal{G}'(\mathcal{E};K)$ is a Lie algebra freely generated by the minimal (in accordance with m_0) generalized local part $\mathcal{G}^{m_0}$, and J is the largest homogeneous ideal with trivial intersection with $\mathcal{G}_0$ (consideration of the quotient is equivalent to imposing the Serre relations in an ordinary Lie algebra case) [16, 17]. When $|m_0| \leq 1$, the commutation relations (A.1) have the form

$$[X_0(\phi), X_0(\psi)] = X_0(K_{0,0}(\phi,\psi)), \quad [X_0(\phi), X_{\pm1}(\psi)] = X_{\pm1}(K_{0,\pm1}(\phi,\psi)) \tag{A.4}$$
$$[X_1(\phi), X_{-1}(\psi)] = X_0(K_{1,-1}(\phi,\psi)) \tag{A.5}$$

for all ϕ, and $\psi \in \mathcal{E}$. Then conditions (A.2-A.3) reduce to

$$K_{0,0}(\phi,\psi) = -K_{0,0}(\psi,\phi) \tag{A.6a}$$
$$K_{\pm1}(K_{0,0}(\phi,\psi),\chi) = K_{\pm1}(\phi, K_{\pm1}(\psi,\chi)) - K_{\pm1}(\psi, K_{\pm}(\phi,\chi)) \tag{A.6b}$$
$$K_{0,0}(\psi, K_0(\phi,\chi)) = K_{1,-1}(K_1(\psi,\phi),\chi) + K_{1,-1}(\phi, K_{-1}(\psi,\chi)) \tag{A.6c}$$
$$K_{0,0}(\phi, K_{0,0}(\psi,\chi)) + K_{0,0}(\psi, K_{0,0}(\chi,\phi)) + K_{0,0}(\chi, K_{0,0}(\phi,\psi)) = 0 \tag{A.6d}$$

References

[1] Alber I E 1978 *Proc. R. Soc. Lond.* A **363** 525
[2] Aykanat Ö and Güngör F 2006 *J. Math. Phys.* **47** 1
[3] Babaoglu C and Erbay S 2001 *Chaos Solitons Fractals* **12** 381
[4] Güngör F 2006 vol 2 *SIGMA Symmetry Integrability Geom. Methods Appl.* **2** 014
[5] Babaoglu C and Erbay S 2004 *Int. J. Nonlin. Mech.* **39** 942
[6] Babaoglu C *et al* 2004 *J. Phys.* A **37** 11531
[7] Eden A *et al* 2006 *Nonlinear Anal.* A **64** 979
[8] Champagne B and Winternitz P 1988 *J. Math. Phys.* **29** 1
[9] Devchand Ch and Saveliev M V 1991 *Phys. Lett. B* **258** 364
[10] Kac V G 1991 *Infinite-Dimensional Lie Algebras* 3rd edn (Cambridge Univ. Press)
[11] Kashaev R et al 1992 *Ideas and Methods in Mathematical Analysis, Stochastics, and Applications* Oslo 1988 (Cambridge Univ. Press) p 295
[12] Leznov A N and Saveliev M V 1992 *Group-Theoretical Methods for Integration of Non-Linear Dynamical Systems* Progress in Physics Series **15** (Birkhauser-Verlag)
[13] Bayen F et all 1978 *Ann. Phys.* **110-111** 61
[14] Olver P J 1991 *Applications of Lie Groups to Differential Equations* (Springer-Verlag)
[15] Saveliev M V 1989 *Comm. Math. Phys.* **121** 283
[16] Saveliev M V and Vershik A M 1989 *Comm. Math. Phys.* **126** 367
[17] Saveliev M V and Vershik A M 1990 *Phys. Lett.* A **143** 121
[18] Zuevsky A 2009 *SIGMA: Symmetry, Integrability, Geom. Methods Appl.* **5** 105
[19] Zuevsky A 2005 *Czech J. Phys.* **55** 1545
[20] Zuevsky A 2004 *J. Phys.* A **37** 537
[21] Zuevsky A 2006 *Ann. Henri Poincaré* **7** 1569
[22] Zuevsky A 2008 *Rep. Math. Phys.* **61** 189

Institute of Physics
Dirac House, Temple Back
Bristol BS1 6BE UK

Curran Associates Inc
proceedings.com